Chemical Production Scheduling

Understand common scheduling as well as other advanced operational problems with this valuable reference from a recognized leader in the field. Beginning with basic principles and an overview of linear and mixed-integer programming, this unified treatment introduces the fundamental ideas underpinning most modeling approaches, and will allow you to easily develop your own models. With more than 150 figures, the basic concepts and ideas behind the development of different approaches are clearly illustrated. The book addresses a wide range of problems arising in diverse industrial sectors, from oil and gas to fine chemicals, and from commodity chemicals to food manufacturing. A perfect resource for engineering and computer science students, researchers working in the area, and industrial practitioners.

Christos T. Maravelias is the Anderson Family Professor in Energy and the Environment and Professor of Chemical and Biological Engineering at Princeton University.

Cambridge Series in Chemical Engineering

SERIES EDITOR
Arvind Varma, *Purdue University*

EDITORIAL BOARD
Juan de Pablo, *University of Chicago*
Michael Doherty, *University of California-Santa Barbara*
Ignacio Grossmann, *Carnegie Mellon University*
Jim Yang Lee, *National University of Singapore*
Antonios Mikos, *Rice University*

BOOKS IN THE SERIES

Baldea and Daoutidis, *Dynamics and Nonlinear Control of Integrated Process Systems*
Chamberlin, *Radioactive Aerosols*
Chau, *Process Control: A First Course with Matlab*
Cussler, *Diffusion: Mass Transfer in Fluid Systems, Third Edition*
Cussler and Moggridge, *Chemical Product Design, Second Edition*
De Pablo and Schieber, *Molecular Engineering Thermodynamics*
Deen, *Introduction to Chemical Engineering Fluid Mechanics*
Denn, *Chemical Engineering: An Introduction*
Denn, *Polymer Melt Processing: Foundations in Fluid Mechanics and Heat Transfer*
Dorfman and Daoutidis, *Numerical Methods with Chemical Engineering Applications*
Duncan and Reimer, *Chemical Engineering Design and Analysis: An Introduction 2E*
Fan, *Chemical Looping Partial Oxidation Gasification, Reforming, and Chemical Syntheses*
Fan and Zhu, *Principles of Gas-Solid Flows*
Fox, *Computational Models for Turbulent Reacting Flows*
Franses, *Thermodynamics with Chemical Engineering Applications*
Grossmann, *Advanced Optimization for Process Systems Engineering*
Leal, *Advanced Transport Phenomena: Fluid Mechanics and Convective Transport Processes*
Lim and Shin, *Fed-Batch Cultures: Principles and Applications of Semi-Batch Bioreactors*
Litster, *Design and Processing of Particulate Products*
Maravelias, *Chemical Production Scheduling*
Marchisio and Fox, *Computational Models for Polydisperse Particulate and Multiphase Systems*
Mewis and Wagner, *Colloidal Suspension Rheology*
Morbidelli, Gavriilidis, and Varma, *Catalyst Design: Optimal Distribution of Catalyst in Pellets, Reactors, and Membranes*
Nicoud, *Chromatographic Processes*
Noble and Terry, *Principles of Chemical Separations with Environmental Applications*
Orbey and Sandler, *Modeling Vapor-Liquid Equilibria: Cubic Equations of State and Their Mixing Rules*
Petyluk, *Distillation Theory and Its Applications to Optimal Design of Separation Units*
Pfister, Nicoud, and Morbidelli, *Continuous Biopharmaceutical Processes: Chromatography, Bioconjugation, and Protein Stability*
Ramkrishna and Song, *Cybernetic Modeling for Bioreaction Engineering*
Rao and Nott, *An Introduction to Granular Flow*
Russell, Robinson, and Wagner, *Mass and Heat Transfer: Analysis of Mass Contactors and Heat Exchangers*
Schobert, *Chemistry of Fossil Fuels and Biofuels*
Shell, *Thermodynamics and Statistical Mechanics*

Sirkar, *Separation of Molecules, Macromolecules and Particles: Principles, Phenomena and Processes*
Slattery, *Advanced Transport Phenomena*
Varma, Morbidelli, and Wu, *Parametric Sensitivity in Chemical Systems*
Vassiliadis et al., *Optimization for Chemical and Biochemical Engineering*
Weatherley, *Intensification of Liquid–Liquid Processes*
Wolf, Bielser, and Morbidelli, *Perfusion Cell Culture Processes for Biopharmaceuticals*
Zhu, Fan, and Yu, *Dynamics of Multiphase Flows*

Chemical Production Scheduling

Mixed-Integer Programming Models and Methods

CHRISTOS T. MARAVELIAS
Princeton University

CAMBRIDGE
UNIVERSITY PRESS

CAMBRIDGE
UNIVERSITY PRESS

University Printing House, Cambridge CB2 8BS, United Kingdom

One Liberty Plaza, 20th Floor, New York, NY 10006, USA

477 Williamstown Road, Port Melbourne, VIC 3207, Australia

314–321, 3rd Floor, Plot 3, Splendor Forum, Jasola District Centre, New Delhi – 110025, India

79 Anson Road, #06–04/06, Singapore 079906

Cambridge University Press is part of the University of Cambridge.

It furthers the University's mission by disseminating knowledge in the pursuit of education, learning, and research at the highest international levels of excellence.

www.cambridge.org
Information on this title: www.cambridge.org/9781107154759
DOI: 10.1017/9781316650998

First published 2021

A catalogue record for this publication is available from the British Library.

Library of Congress Cataloging-in-Publication Data
Names: Maravelias, Christos T., 1973– author.
Title: Chemical production scheduling / Christos T. Maravelias, Princeton University.
Description: First edition. | Cambridge ; New York, NY : Cambridge University Press, [2020] | Series: Cambridge series in chemical engineering | Includes bibliographical references and index.
Identifiers: LCCN 2020037674 (print) | LCCN 2020037675 (ebook) | ISBN 9781107154759 (hardback) | ISBN 9781316650998 (epub)
Subjects: LCSH: Chemical plants–Management. | Chemical engineering. | Production scheduling.
Classification: LCC TP155.6 .M37 2020 (print) | LCC TP155.6 (ebook) | DDC 660–dc23
LC record available at https://lccn.loc.gov/2020037674
LC ebook record available at https://lccn.loc.gov/2020037675

ISBN 978-1-107-15475-9 Hardback

To the memory of my father.

Contents

Preface

Background and Motivation

Scheduling is a decision-making process that concerns the allocation of limited resources to competing tasks over time with the goal of optimizing one or more objectives. Scheduling appears in a wide range of sectors, from services to sports, and from education to manufacturing. In the process industries, it arises, for example, in the oil, pharmaceuticals, specialty chemicals, and food and nutraceuticals sectors. Importantly, there is already significant industrial evidence suggesting that the use of advanced optimization methods for scheduling can lead to multimillion-dollar annual savings.

Chemical production scheduling is a relatively new field. The first papers discussing systematic methods appeared in the late 1970s, while the field became one of the major areas of process systems engineering (PSE) only in the late 1990s. Today it is one of the most active research areas in PSE with multiple sessions in national and international chemical engineering conferences dedicated to it and its closely related process operations areas. Its role is only expected to increase as chemical companies move toward product customization and diversification. Despite the volume of papers in the field, however, there is no book discussing the subject as a whole or a book that can be used for a senior-/graduate-level course.

Accordingly, the book is written with two goals in mind. First, it presents a general framework for chemical production scheduling by (1) unifying the notation that has been used by different communities; (2) presenting a classification of the various types of problems and models that have been proposed to address them; and (3) introducing some general principles. Second, it presents the major, modeling and computational, advances in the field over the last 30 years. The book focuses on representative methods and results, but each chapter ends with a discussion of the relevant literature.

Audience

The book is aimed at (1) researchers working in the area of chemical production scheduling or, more broadly, process systems engineering; (2) graduate students interested in the topic; and (3) industrial practitioners. Chemical engineering, industrial engineering, and computer science students are most likely to use this book. The reader is expected to have basic linear algebra knowledge (the equivalent of an undergraduate class in any engineering discipline). Readers with a bachelor of science (BS) degree in engineering or natural sciences will be able to follow the book. The book can also be potentially used in two courses: (1) as one of the main resources for a senior/graduate course on process systems engineering or process optimization/operations, and (2) as the main text for a graduate elective course on chemical production scheduling.

Organization

In broad terms, the book is divided into four parts:

(I) "Background" (Chapters 1 and 2): Chapter 1 presents an introduction to chemical production scheduling, while Chapter 2 presents some background on mixed-integer linear programming.

(II) "Basic Methods" (Chapters 3 through 7): Basic concepts and models for the most encountered classes of problems.

(III) "Advanced Methods" (Chapters 8 through 11): Concepts and models for more complex classes of problems.

(IV) "Special Topics" (Chapters 12 through 15): Advanced solution methods (Chapters 12 and 13), real-time scheduling (Chapter 14), and integration of production planning and scheduling (Chapter 15).

Parts I and II can be used for a senior-/graduate-level semester-long (fifteen-week) course on process operations/optimization. Parts I through III and, potentially, a selection of topics from Part IV can be used for a graduate-level semester-long course on chemical production scheduling.

Approach

One interesting characteristic of scheduling in general, and chemical production scheduling in particular, is that problems arise in many different types of facilities (what we will later define as *production environments*) and can be subject to a wide range of different processing features and constraints, resulting in many different classes of problems. In addition, since the optimization of these systems is challenging, researchers have proposed very different models to address these problems, where a model is typically applicable to a narrow set of problems. Consequently, to present a unified treatment of chemical production scheduling, we had to overcome two major challenges: (1) identify the key concepts, underpinning all problem classes and models, and unifying themes, across all methods; and (2) introduce the reader to different problems and models while keeping the presentation succinct. To address these challenges, the book adopts five basic principles.

First, the presentation is based on, essentially, a road map introduced in Chapter 1. The road map has two components: a classification of scheduling problems (discussed in Section 1.3) and a classification of models (discussed in Section 1.5).

Second, the complexity of covered problems, and corresponding concepts and methods, increases gradually: from single-unit problems, introduced in Chapter 3; to problems in network environments, discussed in Chapter 7; to some *advanced* problems, in Chapters 8 through 11; and then to special topics, in Chapters 12 through 15.

Third, new problem features, and the corresponding concepts, are gradually introduced; for example, batching decisions are introduced in Chapter 3, general resource constraints are introduced in Chapter 4, storage considerations are introduced in Chapter 5, and so on.

Fourth, the book starts with a broad coverage of alternative modeling approaches, so the reader is exposed to most of them, but gradually focuses on fewer approaches, so that more classes of problems can be covered. For example, five ways to model sequencing/timing are discussed in Chapter 3, but only one such approach is discussed in Chapter 8.

Finally, figures are used strategically to explain complex concepts, so that the reader is not distracted by the details pertaining to these concepts. The reader can continue reading and, if interested, return to study the corresponding figures, often containing multiple panels, separately from the text. In that respect, some figures are designed to serve as standalone illustrative examples.

Each chapter includes a "Notes and Further Reading" section where the reader can find additional background information, high-level discussion of extensions of the methods presented in each chapter, and references to related sources. Also, each chapter, except for Chapters 1 and 8, ends with an "Exercises" section, where effort has been made to keep the necessary data to a minimum while covering a wide range of methods discussed in the corresponding chapter. Additional exercises will become available online (see "Online Resources"). Finally, footnotes are used extensively for terminology and notation clarifications; cross-references to related material covered in different chapters; and disambiguation. They are also used to pose questions that are designed to facilitate the understanding of the material.

In terms of notation, each letter is used consistently throughout the book to denote the same parameter or variable, with few exceptions, noted. Also, starting in Chapter 3, we use lowercase Latin characters for indices, uppercase Latin bold letters for sets, uppercase Latin characters for variables, Greek letters for parameters, and regular uppercase Latin letters for set elements.

Online Resources

Additional exercises and some updated auxiliary material (e.g., list of software tools for the development of scheduling methods, all images of the book) are available at cambridge.org/9781107154759. Additional resources can be made available, upon request, to course instructors.

Acknowledgments

I am sincerely grateful to many people who helped me, in many different ways, to write this book.

First, I would like to thank all the PhD students and other group members I was fortunate and privileged to work with on the topic of chemical production scheduling and chemical process operations in general. Specifically, I would like to thank Charles Sung, Arul Sundaramoorthy, Kaushik Subramanian, Andres Merchan, Sara Velez, Yachao Dong, Dhruv Gupta, Michael Risbeck, Ho Jae Lee, Yifu Chen, Venkatachalam Avadiappan, Yaqing Wu, and Giorgos Koponos. The methods and examples they developed as well as, importantly, the lessons they taught me are present in the pages of this book. In addition, I would like to thank two other group members, Dr. Shamik Mishra and Dr. Boeun Kim, for carefully going through the manuscript and helping me to proofread it and generate the index.

Second, I would like to express my deepest gratitude to my colleagues in the Department of Chemical and Biological Engineering at the University of Wisconsin–Madison for creating a highly collegial and intellectually stimulating academic environment; and for continuing to nurture a tradition in which scholarly activities, and book-writing in particular, are highly valued. More specifically, I would like to thank Jim Rawlings for his continuous encouragement, his valuable insights into scheduling and control, and our discussions about book writing; and Manos Mavrikakis for his advice. I would also like to acknowledge financial support from the Olaf Hougen Program and the Paul Elfers Professorship.

Third, I would like to thank a number of colleagues: Ignacio Grossmann, not only for our discussions on production scheduling but also for serving as a role model and giving me advice throughout my academic career; Gabriela Henning, for being willing to challenge the status quo and for sharing with me her outside-the-box thoughts on scheduling; John Wassick and Jeff Kelly for "educating" me with their insights into the industrial application of optimization-based methods for scheduling; and Pedro Castro and Lazaros Papageorgiou for our fruitful discussions on scheduling.

Last but not least, I would like to thank ta *koritsakia*, Emi and Bella, for bringing so much joy every day; and my wife, Vanesa, for her support during this journey – without her love and patience, this book would have been impossible to write.

Part I

Background

1 Introduction

This chapter provides an overview of scheduling and its role within a manufacturing organization, with special focus on the process industries. Section 1.1 introduces some preliminary concepts, including a definition of scheduling, some simple problems, the role of scheduling in a manufacturing supply chain, and a general problem statement for chemical production scheduling. The different chemical production environments, which play a key role in the definition of the different classes of problems, are introduced in Section 1.2. Section 1.3 presents the different classes of problem, and Section 1.4 outlines the different approaches to scheduling. Finally, an outline of the book is presented in Section 1.5.

1.1 Preliminaries

1.1.1 Scheduling: Applications and Definition

Scheduling finds applications in a wide variety of sectors: educational institutions, government organizations, manufacturing, sports, transportation, etc. In the manufacturing sector, scheduling has been practiced for more than a century, though, surprisingly, the first systematic methods appeared in the 1950s.[1] Since then, the field has seen the development of theoretical results regarding the complexity of different scheduling problems and a variety of modeling and solution methods. The aforementioned developments coupled with advances in computing power have enabled the use of optimization-based methods for a number of scheduling problems.

Scheduling is a *decision-making process that concerns the allocation of limited resources to competing tasks over time with the goal of optimizing one or more objectives*. A typical example would be the allocation of (limited) classrooms to a set of (competing) classes with the goal of meeting instruction needs and instructor preferences. In terms of manufacturing, the first methods were developed for discrete manufacturing. Consequently, most books focus on this type of problem, where scheduling and sequencing are viewed as two different types of decision. In this book, we will use the term *scheduling* to describe a more broad decision-making process where,

[1] Interestingly, the widely used Gantt chart was introduced by Henry Gantt in the 1910s (e.g., *Organizing for Work*, pp. 45, 89, and 95), but as a visualization tool. Gantt did not present a method to find schedules.

A

Job	τ_i	ρ_i	ε_i
A	3	0	5
B	4	2	10
C	2	4	8

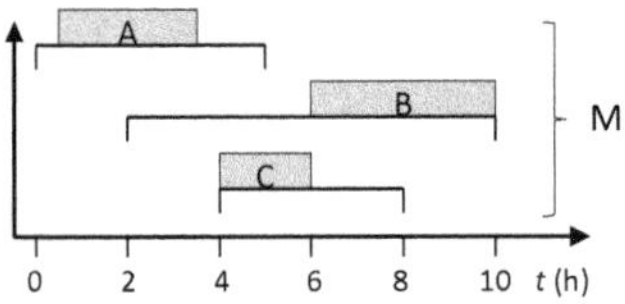

B

Job	$\tau_{i,j}$ M1/M2	ρ_i	ε_i
A	3/3	0	5
B	4/4	2	10
C	2/2	4	8
D	2/4	2	8
E	3/3	4	9

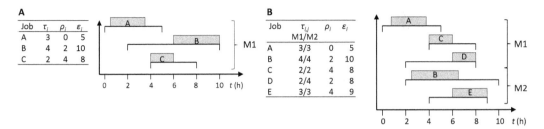

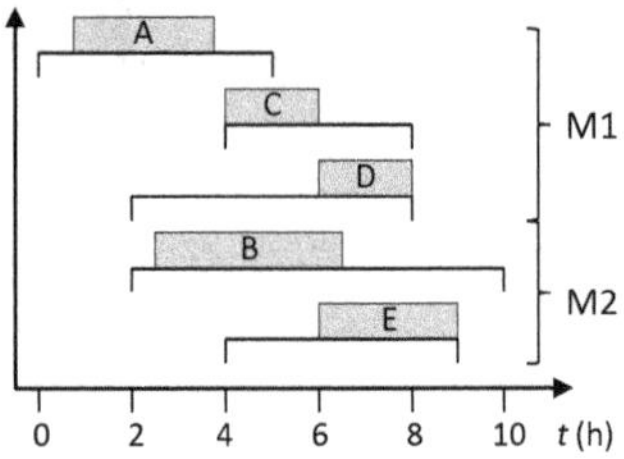

Figure 1.1 Single-machine (A) and parallel-machine (B) problems. Processing times, τ_{ij}, and release/due times, ρ_i/ε_i, are given in the tables. The ordinate of the charts represent time; width of rectangles, representing jobs, is proportional to processing times; and the placement of the rectangles shows the timing of jobs.

among others, we decide (1) what and how many tasks need to be carried out to meet given constraints; (2) what resources will be allocated to the selected tasks; and (3) when the selected tasks will be carried out. Also, as we will discuss in the next section, the necessity to model the amounts of (mostly fluid) materials involved in chemical production is a key feature that distinguishes discrete from chemical manufacturing scheduling.

1.1.2 Some Simple Problems

In this subsection, we discuss simple problems in order to introduce the three basic decisions that need to be made when solving a scheduling problem: (1) The decision on the number of jobs to be scheduled, (2) the assignment of jobs to machines, and (3) the sequencing of jobs assigned to a particular machine. For now, we will use the terms *jobs* and *machines* from discrete manufacturing.

The simplest scheduling problem is the single-machine problem, where N jobs $i \in \mathbf{I} = \{1, 2, \ldots N\}$ subject to release, ρ_i, and due, ε_i, times[2] and processing times, τ_i, have to be scheduled on one machine. Note that, as stated, this is a feasibility problem: Find a schedule that satisfies all constraints. An instance of this problem and a solution to it are shown in Figure 1.1A. Potential objective functions for this problem are (1) minimization of makespan and (2) minimization of lateness (if due times are not hard).

A due time is meant to be a *preferred* delivery time; it can be violated at a cost. If a due time is hard, that is, cannot be violated, then the term *deadline* is used instead. Note that this problem involves only sequencing and timing decisions: the sequence in which jobs are processed and the (exact) time they start. Note that there are infinite timing solutions for a given sequence. In the example shown in Figure 1.1A, the sequencing and timing are A (starting at $t = 1$) $\rightarrow$ C (starting at $t = 4$) $\rightarrow$ B (starting at $t = 6$).

[2] In this book, we will use the terms release/due *time*, instead of the more commonly used release/due *date*. This is because the horizons in the instances we will use to illustrate the basic concepts are short, typically less than 24 hours. Also, we will assume that the orders are due at a granularity finer than one day.

An extension of this problem is the *parallel machines* problem, where N jobs $i \in \mathbf{I} = \{1, 2, \ldots N\}$ subject to release, ρ_i, and due, ε_i, times have to be scheduled on M machines $j \in \mathbf{J} = \{1, 2, \ldots M\}$. The processing times of job i on machine j is τ_{ij}. This problem can also be a feasibility problem or have a time-related objective function. A solution to an instance of this problem is shown in Figure 1.1B. In addition, if the processing costs, γ_{ij}, of job i on the various machines are different, then a minimization of processing cost can be an alternative objective function. Note that there are two decision types to be made in this case: (1) assignment of jobs to machines and (2) sequencing and timing of jobs on each machine. In the instance shown in Figure 1.1B, jobs A, C, and D are assigned to machine M1, and jobs B and E are assigned to machine M2. However, as in the *single-machine* problem, the tasks (jobs) to be scheduled are known prior to optimization.

The natural extension of the previous problem is when the conversion of (external) orders to (internal to the problem) jobs is part of the optimization (scheduling). Consider the case where N orders $i \in \mathbf{I} = \{1, 2, \ldots N\}$ with release/due times, ρ_i/ε_i, and amounts due, φ_i, have to be met. Each order has to be converted into jobs to be processed on exactly one of M parallel machines $j \in \mathbf{J} = \{1, 2, \ldots, M\}$. Each order is for a specific product, so processing times can be expressed in terms of orders, that is, the processing times of (all jobs for) order i on machine j is τ_{ij}. In addition, each machine has capacity β_j, that is, all jobs executed on machine j will have size β_j. Unlike the previous two problems, the number of jobs is unknown – it is essentially an optimization decision.

1.1.3 Scheduling in the Supply Chain

A manufacturing supply chain (SC) is a network of facilities and distribution options that performs the following functions: procurement of raw materials, transformation of raw material into intermediate and finished products, and distribution of finished products to customers. The goal of SC optimization is to maintain high customer satisfaction levels at the minimum total cost. To achieve this goal, SC managers have to optimize material, monetary, and information flows, and in order to do so have to consider a number of *planning functions*. The *supply chain planning matrix* in Figure 1.2 shows the main planning functions, which, in general, can be categorized in terms of (1) the timescale of the planning horizon and the frequency at which the associated decisions are made and (2) the business functions.

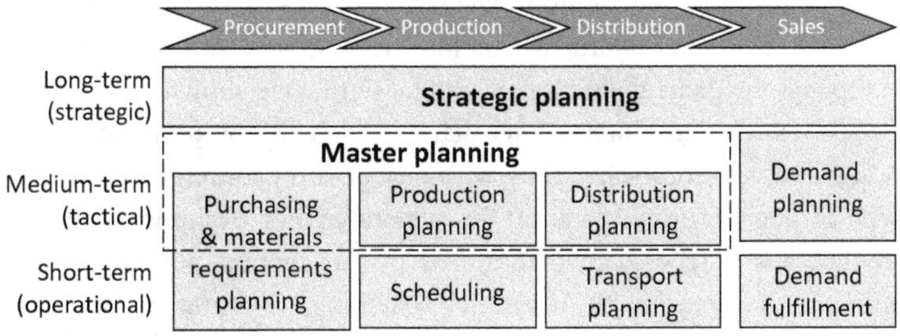

Figure 1.2 Supply chain matrix.

Long-term (strategic) planning decisions concern mostly the structure (design) of the supply chain, for example, location of production sites and warehouses, and facility capacity. The planning horizon is multiple years. The goal of medium-term (tactical) planning is to meet customer demand over a medium-term horizon (3–12 months) at the minimum cost given the current SC structure. Medium-term decisions include, for example, the calculation of monthly production targets and the allocation of these targets to different production sites. Finally, short-term (operational) planning concerns *day-to-day* decisions, for example, the assignment of production tasks to specific equipment units and the timing of the execution of these tasks. The planning horizon can be anywhere between a few days and a few months.

In terms of business functions, planning can be divided into procurement, production, distribution, and sales, though the importance and complexity of the associated problems vary greatly by sector. For instance, in industrial gases, the distribution of liquid gases using trailers under a vendor-managed inventory policy is significantly more complex than the associated production scheduling problem (of air separation units), whereas in specialty products the production of multiple low-volume, high-value products in multi-product batch facilities is more complex than the associated distribution problem.

It is important to note that the planning functions are not independent from each other. They interact because decisions made by one function become inputs to another, for example, production targets, determined at the master planning level, become inputs to the scheduling function. In fact, in some ways, the goal of SC optimization is to effectively integrate organizational units so as to coordinate material, information, and financial flows. However, solving the integrated problem is intractable and business units have different decision-makers and/or different objectives. Thus, decisions have to be made based on the solution of smaller, *local* planning functions.

Chemical production scheduling deals with the short-term planning for the production business function, in the chemicals sector, broadly defined. As it will be discussed in the next subsection, the interactions of the scheduling function with the other planning functions have important implications.

1.1.4 Interactions with Other Planning Functions

In addition to the planning functions shown in Figure 1.2, scheduling also interacts with shop-floor management, or in the case of high-volume (typically continuous) manufacturing, with real-time optimization (RTO) and process control. While the specifics of these interactions are not unique across industrial sectors, not even across plants of the same company, the general structure is similar to the one shown in Figure 1.3. There are four ways in which these interactions shape the scheduling problem at hand.

First, the interaction of scheduling with production planning in particular determines what type of decisions are made by the scheduler. For example, if production targets (orders), but not batches, are determined by production planning, then scheduling encompasses all three major decisions (batching, assignment, sequencing/timing); otherwise, if production targets are converted into batches at the production planning level, then scheduling involves only assignment and sequencing decisions.

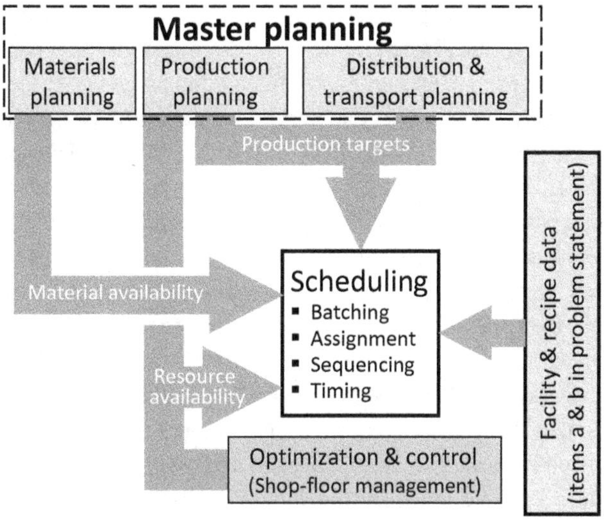

Figure 1.3 Interactions of scheduling with other planning functions.

Second, market considerations (including demand planning) combined with capacity constraints determine the production goal. For example, in a capacity-constrained environment, the production goal, and therefore the objective function of an optimization-based approach, would be the maximization of throughput or profit; otherwise, if excess capacity is available, then the objective function will most likely be the minimization of total cost. As we will see in Section 1.3, the objective function is one of the features determining the class of a problem.

Third, master and demand planning determine the production pattern. The production of high-volume products with relatively constant demand often leads to periodic production (production wheel) that requires periodic scheduling. If the demand is time varying but there are good quality forecasts available, then short-term scheduling is required in order to maintain a *target inventory position* (TIP) for a *make-to-stock* (*push*) policy. Yet, when the demand is time varying and good forecasts are not available (typically for low-volume specialty products), then a *make-to-order* (*pull*) inventory policy is typically adopted, which means that short-term scheduling should be used frequently to react to new orders.

Fourth, input parameters to scheduling (e.g., order data) are determined by other functions. For example, production planning determines production targets; procurement planning determines the availability of feedstocks (or the release date of orders); and process control determines the availability of resources. Conversely, scheduling decisions become inputs to other functions. The flow of information from other planning functions to scheduling is shown in Figure 1.3.

In general, the integration of planning functions and/or organizational units improves the competitiveness of the SC as a whole through the coordination of material, information, and financial flows. The integration with planning has the potential to lead to better solutions, especially when the overall economics are governed by planning decisions, but the ability to meet specific targets depends on scheduling decisions. The integration with control is critical when inputs to scheduling depend on control actions, for example, when changeover times and costs in continuous processes depend on the transition between steady states. The integration with control is not critical when production tasks have processing times (and costs) that do not depend on control decisions, for example, batch processes with fixed processing times and temperature/pH profiles. The integration with design, though the latter is not a planning function, can also lead to better designs because the capacity of a facility often depends on the mix of products it produces, which means that better decisions regarding the capacity of the units can be made when the schedule for the nominal production of the facility is considered. The integration of scheduling with process control and medium-term planning will be discussed in Chapters 14 and 15, respectively. Finally, we note that there are two challenges that may prevent such integration: (1) organization challenges (e.g., multiple decision-makers with conflicting objectives) and (2) challenges pertaining to the formulation and solution of optimization models (e.g., the models cannot be generated due to lack of data or are computationally intractable).

1.1.5 Scheduling in the Process Industries

In the process industries, scheduling problems arise in multiple sectors, such as (1) the petroleum industry (e.g., transportation and unloading of crude oil from ships and pipelines, crude oil distillation column charging schedule); (2) pharmaceutical, nutraceutical, and food manufacturing (e.g., scheduling of small-scale multiproduct batch processes); (3) specialty chemical manufacturing (e.g., batch scheduling); (4) mining; (5) metals processing; and (6) energy generation, etc.

The importance of scheduling has increased due to the recent trend toward product customization and diversification. To produce multiple low-volume products, chemical companies employ multiproduct facilities with often complex process networks and multiple shared resources. To utilize these capital-intensive assets efficiently and at the same time respond to demand fluctuations, manufacturing facilities have to be utilized close to their capacity, which leads to a challenging scheduling problem: Production targets close to system limits have to be met at the minimum cost subject to constraints resulting from the complex nature of these facilities. Another important driver, in high-volume production environments, is cost reduction and maximization of production efficiency.

To address these problems, researchers in the area of process systems engineering (PSE) have developed various approaches. Most of the early papers, in the late 1970s and early 1980s, considered the scheduling of batch processes in what we will later define as *sequential* production environments, that is, problems where batches have to be processed in consecutive stages (with no intermediate mixing or splitting). These problems are in some ways similar to the problems arising in discrete manufacturing. The scope of chemical production scheduling was expanded in the early 1990s to include facilities of arbitrary structure with no material handling (i.e., mixing and splitting) restrictions, additional resource constraints, and various processing restrictions. Since then, numerous modeling and solution methods have been proposed to address an ever-widening set of problems and applications. As a result, chemical production scheduling has become an important research subarea of process operations.

1.1.6 General Problem Statement

The term *scheduling* has been traditionally used to describe only the allocation of tasks to resources over time, that is, it considers only the timing of tasks. In this book, we consider a broader problem that includes the determination of the number and type of tasks to be performed, the assignment of tasks to resources, and the sequencing and/or timing of tasks. Accordingly, the general problem statement is as follows.

Given are the following:

(a) Production facility data (e.g., processing unit, storage vessel capacities)
(b) Production recipes (e.g., processing times/rates, resource requirements)
(c) Production costs (e.g., materials and utilities costs)
(d) Material and resource availability and compatibility with tasks
(e) Production targets or orders with due dates

Note that inputs to scheduling include parameters that can be considered *fixed* (e.g., facility-related or product-related data) as well as parameters defined by other planning functions and thus can be time varying (e.g., material and resource availability, production targets).

The goal is to find the *optimal* schedule that meets production targets while satisfying all resource constraints. The optimization objective is typically the minimization of total cost (raw materials + production + inventory + utilities), though other time-related objective functions (e.g., makespan and lateness minimization) as well as profit or throughput maximization can also be used (see Section 1.3.3 for a detailed discussion of objective functions).

As it will be discussed in the next section, there are different classes of scheduling problems with different sets of optimization decisions. However, in the most general case, the following are the major decisions (see Figure 1.4A for illustration):

(a) Selection and sizing of tasks (batches or lots) to be carried out (batching/lot-sizing)
(b) Assignment of tasks to processing units
(c) Sequencing and/or timing of tasks on each unit

In most problems, these decisions fully define a schedule, that is, material inventory and resource utilization profiles can be determined when we know the number, type, size, and timing of tasks.

Figure 1.4A shows the decisions made to meet demand for four products, A (800 kg), B (500 kg), C (300 kg), and D (250 kg), in a facility with two parallel units, U1 and U2, with capacity equal to 300 and 250 kg, respectively. The batching decisions for product A lead to two batches (A1 and A2) of 300 kg each and one batch (A3) of 200 kg. In terms

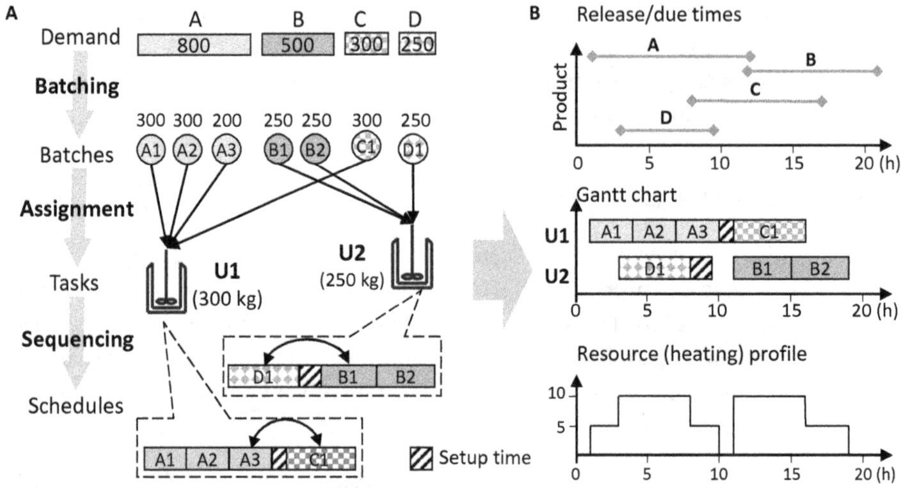

Figure 1.4 Scheduling decisions (A) and solution representation (B).

of batch-unit assignments, they are all assigned to unit U1. The demand for product C is met by one batch (C1) of size 300 kg also assigned to unit U1. The sequencing in U1 then is A1 → A2 → A3 → C1 with a changeover between A3 and C1.

A schedule is graphically represented via a Gantt chart showing the utilization of equipment over time and other auxiliary graphs. Figure 1.4B shows the windows within which batches should be executed (top panel); the Gantt charts for the two processing units (middle panel); and a utility (heating) consumption profile, assuming that all batches require five units of heating during their execution (bottom panel).

As an operational problem, scheduling is solved in a recursive manner. A schedule is generated based on the information available at the time; the early part of the schedule is implemented; and as additional information becomes available (e.g., disturbances, new orders), reoptimization is performed to dynamically adjust. This iterative procedure, in this book termed *real-time* (or *online*) *scheduling,* is discussed in Chapter 14.

1.2 Chemical Production Environments

In this section, we first present a classification of discrete manufacturing scheduling problems (Section 1.2.1) and then discuss some insights that allow us to define the major production environments for chemical manufacturing (Section 1.2.2). Finally, we discuss the main characteristics of sequential (Section 1.2.3), network (Section 1.2.4), and hybrid (Section 1.2.5) environments.

1.2.1 Discrete Manufacturing Machine Environments

The entities to be scheduled are *jobs*, $i \in \mathbf{I}$, and the shared resources are *machines*, $j \in \mathbf{J}$. A job may have to be allocated to a single *step* (*operation*) or it may require a number of steps, $k \in \mathbf{K}$. Typically, a job is not divided into multiple jobs nor is merged with other jobs. A problem is described by a triplet $\alpha/\beta/\gamma$, where α denotes the machine environment, β describes details of processing characteristics and constraints, and γ gives the objective function. Field β may include multiple entries.

Regarding field α, the following are the most common *machine environments* (see Figure 1.5):

- ***Single machine*** (***1***): All jobs are to be processed on a single machine; it is a special case of all other machine environments.
- ***Flow-shop*** (***F***): There are m machines and each job has to be processed on all m machines; all jobs follow exactly the same routing. Equivalently, each job requires m operations (stages), and each operation can be performed on only one machine.
- ***Job-shop*** (***J***): There are m machines; each job has a predetermined routing, $\mathbf{J}_i$, to follow, where $\mathbf{J}_i$ is an ordered subset of machines, which may also include the same machine more than once (*recirculation*).
- ***Open-shop*** (***O***): There are m machines, and each job has to be processed on a subset, $\mathbf{J}_i$; there are no restrictions in the routing of each job; i.e., $\mathbf{J}_i$ is not ordered.

Interestingly, the four aforementioned environments represent different types of job routing, from specific to general. Going from ***1*** (which has no routing) to ***F***, we add multiple steps (and therefore routing), but all routings are (i) predetermined and (ii) identical among all jobs. Going from ***F*** to ***J***, we keep predetermined routings, but the restriction of identical routings is relaxed. Finally, in ***O*** both restrictions are relaxed. Thus, **1** is a special case of ***F*** (with $k = 1$), which is a special case of ***J***, which is a special case of ***O***. This generalization is shown as horizontal differentiation in Figure 1.5.

Another attribute of a machine environment is the number of machines capable of performing an operation. In the simplest case, each operation can be performed by only one machine. Environments ***1***, ***F***, ***J***, and ***O*** fall in this category. If there are multiple machines available for each operation, then we have the following environments:

- ***Machines in parallel*** (***P***): Each job i has to be processed on one of m parallel machines.
- ***Flexible flow-shop*** (***FF***): Each job has to visit a set of steps, $k \in \mathbf{K}$, following a predetermined and common among all jobs routing, where each step consists of a set of parallel machines $j \in \mathbf{J}_k$. The routing is in terms of steps, not machines.
- ***Flexible job-shop*** (***FJ***): This is a generalization of the ***P*** and ***J*** environments. Each job has to go through a number of steps, where each step can be performed by multiple machines in parallel. Each job has to visit a different subset of steps in a job-specific routing. A machine may belong to multiple steps and/or different steps for each job. As in ***J***, the route of a job may include the same step more than once (recirculation).
- ***Flexible open-shop*** (***FO***): This is the generalization of the ***P*** and ***O*** environments.

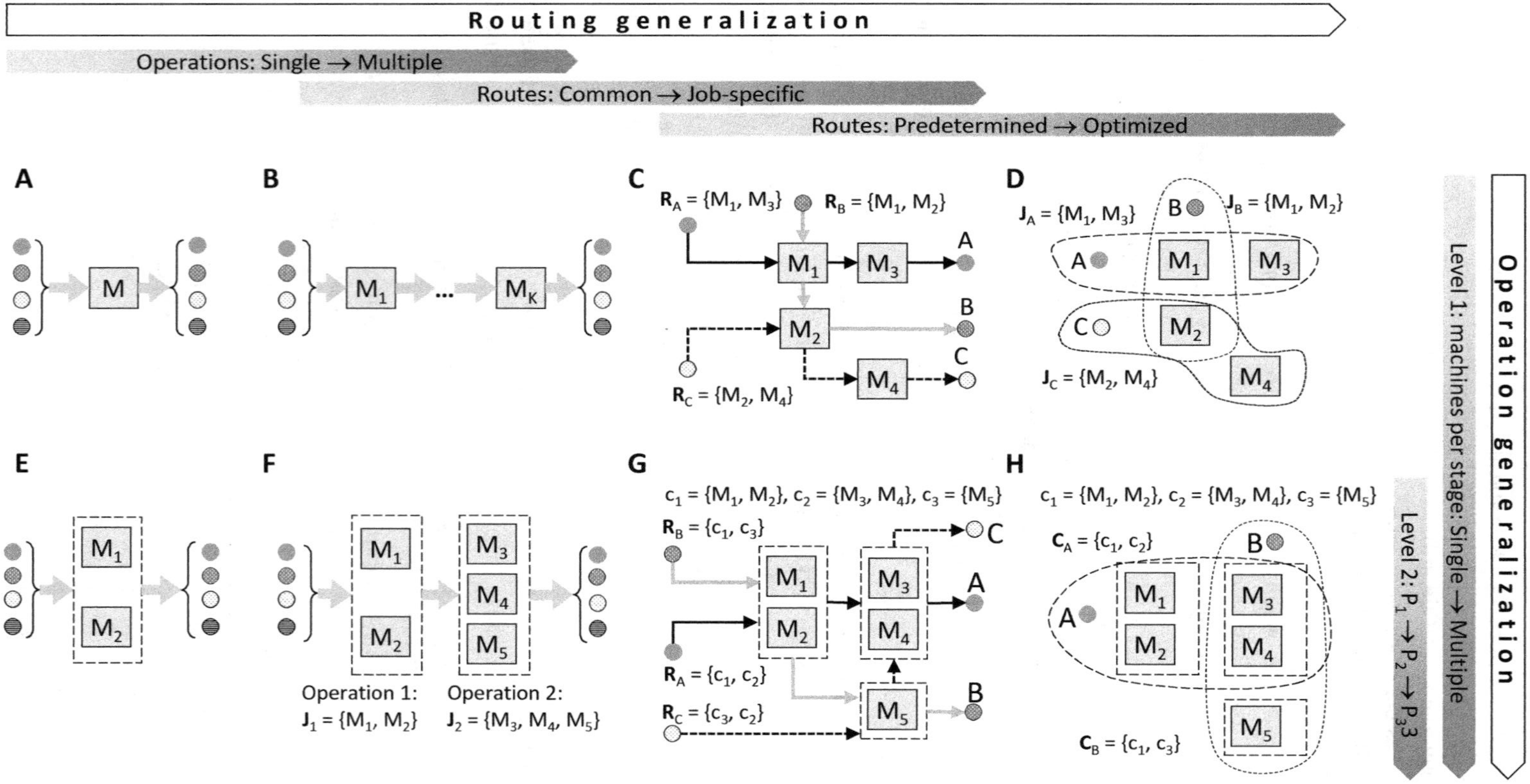

Figure 1.5 Classification of discrete manufacturing scheduling problems. (A) Single machine. (B) Flow-shop. (C) Job-shop. (D) Open-shop. (E) Machines in parallel. (F) Flexible flow-shop. (G) Flexible job-shop. (H) Flexible open-shop.

Environments ***P***, ***FF***, ***FJ***, and ***FO*** are the *flexible* counterparts of environments ***1***, ***F***, ***J***, and ***O***, respectively, which means that ***1*** is a special case of ***P***, ***F*** is a special case of ***FF***, and so on. Note that the flexible environments are generalizations of two environments; for example, ***FF*** is a generalization of the ***P*** (routing generalization) and ***F*** (operation generalization) environments. Also, in flexible environments there is one additional optimization decision: the assignment of jobs to machines.

Finally, for a given operation, there are three types of machines in parallel:

- Parallel identical machines ($\boldsymbol{P_1}$): the processing time of each job i is the same, τ_i, in all machines.
- Different parallel machines with specified *speeds* ($\boldsymbol{P_2}$): if v_j is the speed of machine j, then the processing time of job i in machine j is τ_i/v_j; that is, the speed (effectiveness) of machines is relatively the same for all jobs.
- Unrelated machines ($\boldsymbol{P_3}$): the processing time of job i in machine j is τ_{ij}.

Type $\boldsymbol{P_3}$ is a generalization of $\boldsymbol{P_2}$, which is a generalization of $\boldsymbol{P_1}$. Note that in multistep environments (e.g., ***FF***, ***FJ***) the parallel machines of each step may be of different type (e.g., step 1 in a ***FF*** can be of type $\boldsymbol{P_1}$ while step 2 is of type $\boldsymbol{P_3}$). The generalization of environments in terms of the machines comprising a step is shown as vertical differentiation in Figure 1.5.

Another aspect is the suitability of machines to perform one or more operations (of specific jobs). Since this feature is common in chemical manufacturing, we will discuss the details later. For now, we note that in ***F*** and ***FF*** it is usually assumed that each machine can perform a single operation. In ***F***, this means that there are as many machines as steps; and in ***FF***, it means that $\mathbf{J}_k \cap \mathbf{J}_{k'} = \emptyset$ for $k \neq k'$, and $\cup_{k \in \mathbf{K}} \mathbf{J}_k = \mathbf{J}$; that is, $\mathbf{J}_k$, $k \in \mathbf{K}$ is a partition of $\mathbf{J}$. The same assumption is often made in ***J***, but not all jobs have to go through all machines. In ***FJ*** , the simplest case arises when machines belong to *work centers*, $c \in \mathbf{C}$, and the routings of the jobs are in terms of work centers; that is, the routing, $\mathbf{R}_i$, of job i is an ordered set of centers (see Figure 1.5G). However, in ***FJ***, more complex situations may arise because jobs have different routings.

1.2.2 Critical Insights

An assumption for the basic classification presented in the previous subsection is that the tasks to be scheduled correspond to specific jobs. For example, in a flexible flowshop with three jobs, $\mathbf{I} = \{A, B, C\}$, and two operations, $\mathbf{K} = \{1, 2\}$, each one of which can be carried out by two machines, $\mathbf{J}_1 = \{M1, M2\}$ and $\mathbf{J}_2 = \{M3, M4\}$, there are six tasks to be scheduled: A-1 (job A in operation 1), A-2, B-1, B-2, C-1, and C-2. While this type of discrete-manufacturing-like processing is found in chemical manufacturing (e.g., in batch pharmaceutical manufacturing), more complex environments with, for example, tasks with multiple inputs and outputs and task splitting or merging are also common. Thus, to represent chemical production scheduling problems we will have to expand the environments introduced in the previous subsection. Environments where the identity of a job has to be maintained through a series of operations will be referred to as *sequential* production environments (discussed in Section 1.2.3). Environments where this structure is not present, and therefore the notion of a job going through multiple operations is not defined, will be referred to as *network* environments (discussed in Section 1.2.4). Before we discuss the different environments in detail, we illustrate a key point using Examples 1.1 to 1.4. Specifically, we show that it is the material handing restrictions rather than the configuration of the process that determine the production environment.

With respect to the traditional notation discussed in Section 1.2.1, we note the following:

(a) We will use the term *unit*, instead of *machine*, to describe a resource that can perform at most one task at a time (unary resource); units can be further grouped into processing units and storage units (vessels).

(b) In our discussion of sequential environments, we will use the term *stage*, instead of *operation*, to denote a distinct processing step in a facility (e.g., fermentation, centrifugation, etc).

(c) When discussing sequential environments, we will use the term *order* or *batch*, instead of the term *job*, to denote the entity that has to be processed in different stages.

(d) We will use the term *task* to denote a specific transformation; for example, in sequential environments, the processing of an order in a stage is a task.

(e) In a given solution, a task might be executed multiple times; we will refer to each execution of a task as batch of a task.[3]

Example 1.1 A *Linear* Network Environment Consider the process network shown in Figure 1.6A, where raw material A is converted via task T1, carried out in unit U1, into intermediate INT, which is then converted via task T2, executed on unit U2, to final

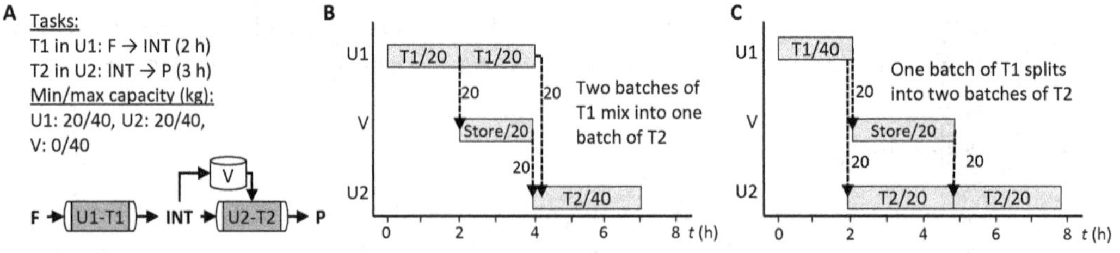

Figure 1.6 Example illustrating that facility configuration (physical layout of equipment) does not determine the production environment (batch mixing and splitting are allowed shown process). (A) Process schematic and data. (B) Schedule with batch mixing. (C) Schedule with batch splitting. In (B) and (C), tasks and batchsizes given inside blocks.

[3] The careful reader might have noticed that the term *batch* is used in two ways, to denote (1) the entity that is processed through different stages in sequential environments, and (2) the specific execution of a task. This is one of the few times we introduce some notation ambiguity in this book. However, no disambiguation will be necessary because the context in which the term will be used to denote the two different concepts will be different. The former will be used when sequential environments are discussed, while the latter will be discussed mostly in the context of network environments.

product P. Intermediate INT can be stored in vessel V, with capacity 40 kg, while the batchsizes of tasks T1 and T2 should be between 20 and 40 kg. The environment appears to be a two-stage sequential environment (flow-shop-like) since all tasks have a single input and output material; each task can be carried out by a single unit; and, in addition, each unit can perform only one task. Nevertheless, if there are no restrictions in the handling of intermediate INT, then two batches of T1 can be merged before being consumed by a single batch of T2 (see Figure 1.6B) or a single batch of T1 can be divided into two batches (see Figure 1.6C). Thus, this process is not a sequential process.

Example 1.2 A *Branched* Sequential Environment Consider the process shown in Figure 1.7A, where a raw material, F, is converted, via task T0, into intermediate, INT, which can then be converted into three different products (P1, P2, P3). All tasks are carried out in dedicated units, their batchsize is 20 kg, and there is no storage for intermediate INT. Given the capacities and lack of intermediate storage, each batch of T0 is converted to one batch of final product; it cannot be (divided to be) converted to multiple products nor can it be mixed with another batch of T0. Thus, this is a *sequential* production environment. Specifically, if we were to use the notation from the discrete manufacturing problem classification (Section 1.2.1), this is a flexible flow-shop with one machine for the first operation and three machines for the second.

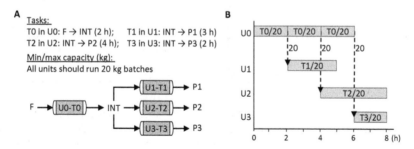

Figure 1.7 Example illustrating that material handling restrictions determine, among others, the type of production environment. (A) Process network and data. (B) Schedule with no batch mixing/splitting (tasks and batchsizes given inside blocks).

These two examples show that it is not the configuration of the process that determines the production environment. Rather, it is the manner in which batches can be handled. In Example 1.1, we have what appears to be a sequential configuration, but the environment is not sequential, whereas in Example 1.2 we have an *arboretum* configuration, but the processing is sequential. In other words, it is the material handling restrictions that lead to sequential processing.

Another key insight, illustrated through Examples 1.3 and 1.4, is that the type of processing is not necessarily determined by the number of materials used as inputs or outputs. Also, we may have materials that are produced in a single batch, but can then be consumed by multiple batches, which means that mixing (blending) is not allowed but splitting is (Example 1.3) and vice versa (Example 1.4). Also, a single facility may have different restrictions, which means that there are subsystems with different type of processing. This is, in fact, quite common in practice. Thus, it is often more accurate to refer to types of processing or types of processes rather than facilities of a single type of environment.

Example 1.3 Complex Material Handling 1 Consider the process shown in Figure 1.8, where two final products, P1 and P2, are produced by mixing intermediate INT, produced by task T1, and additive A, produced by task T0. A key restriction in this process is that one batch of INT has to be consumed by a single batch of T2 (to produce P1) or a single batch of T3 (to produce P2); it cannot be mixed with other batches of INT nor split. This implies that we have sequential type of processing, although a task consumes two materials. However, a batch of additive A, which is produced in the same unit as intermediate INT, can be stored and used in multiple batches of T2 and/or T3. This implies that additive A cannot be treated as a material in a sequential environment. Figure 1.8C shows a Gantt chart in which the no mix/split transfer of INT is denoted with a dashed line, while the splitting and subsequent multiple transfers of A are denoted by the dotted line.

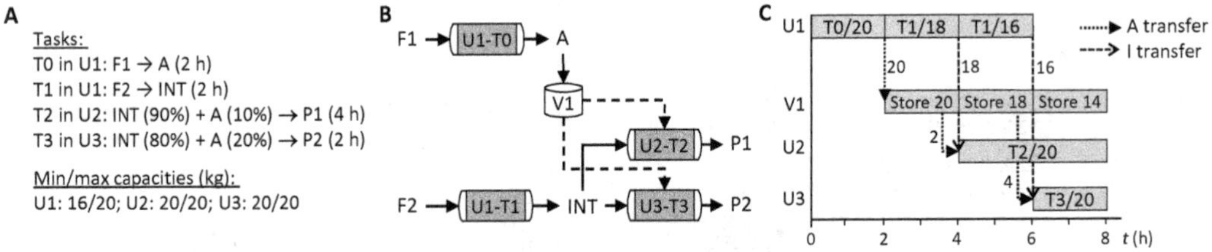

Figure 1.8 Example illustrating that different material handling restrictions, and thus different types of processing, can be present in the same facility: sequential processing for intermediate INT; no batch mixing, but batch splitting allowed for additive A. (A) Processing data. (B) Process schematic. (C) Gantt chart with material transfers (tasks and batchsizes given inside blocks).

Example 1.4 Complex Material Handling 2 Consider the process shown in Figure 1.9A, where task T1 produces a pure intermediate INT, to be converted into product P1, and a mixture U, which consists of unreacted feed F and waste W. The intermediate INT produced by one batch of T1 has to be consumed by exactly one batch of T2; no mixing nor splitting is allowed. Thus, we have sequential processing, although task T1 produces two materials. On the other hand, U from multiple batches of T1 can

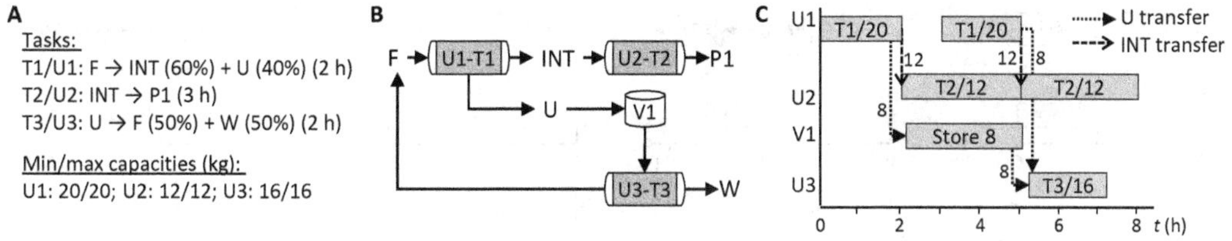

Figure 1.9 Example illustrating that different material restrictions, and thus different types of processing, can be present in the same facility. Sequential processing for intermediate INT; batch mixing allowed for secondary stream, U, of task T1. (A) Processing data. (B) Process network. (C) Gantt chart with material transfers (tasks and batchsizes given inside blocks).

be stored and when enough material is accumulated sent to separation T3, which yields pure F and waste stream W. In this case, a batch of T1 yields two output materials with different material restrictions.

Another key difference between discrete manufacturing and chemical processing, and specifically continuous processing, is the manner in which jobs (tasks) affect the inventory of input and output materials. In discrete manufacturing, as well as chemical batch processing, a task is typically assumed to consume all its input materials instantaneously at the beginning of its execution and produce all of its output materials at its end (see Figure 1.10A). This means that, in discrete manufacturing (and batch processing), if a job has to undergo multiple operations, then its processing in a subsequent operation cannot start unless the previous operation is completed. In continuous processing, however, output materials are produced during the execution of the task (see Figure 1.10B), which means that a downstream task can start before the upstream task is completed (Figure 1.10C). The scheduling of continuous processes will be discussed in Chapter 9.

Finally, in chemical manufacturing the proportions in which input materials are consumed or output materials are produced are not necessarily fixed parameters. For example, depending on the quality of crude oils fed to a distillation unit, the various distillation splits are produced at different ratios (e.g., the percent yield of light gasoil from a light crude is higher than the percent yield from a heavier crude). The scheduling of operations with variable conversion coefficients is a multiperiod blending problem, which is often referred to as *blend scheduling* or simply *blending*. We will discuss blending in Chapter 11. The discussion in the remainder of the present chapter will be based on batch processes, though the major ideas are also applicable to continuous processes, as well as operations involving blending.

1.2.3 Sequential Environments

A process is sequential if it consists of tasks that produce/consume a single material, and batch mixing and splitting are not allowed for both the input and output materials. In other words, removing the flexibility to mix, split, and recycle fluids results in

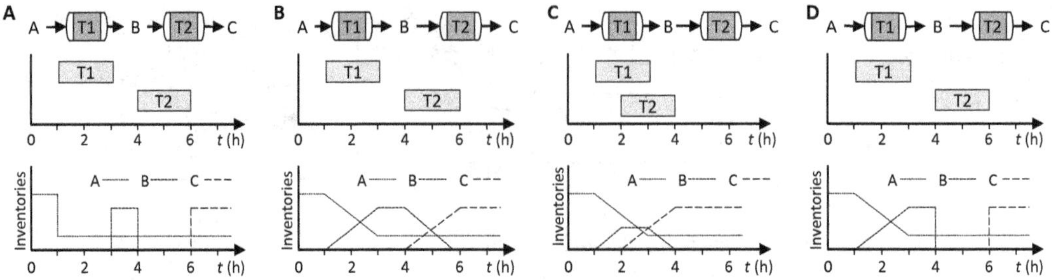

Figure 1.10 Inventory profiles of materials consumed and produced by different types of chemical processes. Profiles drawn assuming instantaneous material transfer to/from units, and zero residence time for continuous processes. (A) T1 and T2 are batch. (B) T1 and T2 are continuous. (C) T1 and T2 are continuous. (D) T1 is continuous, T2 is batch.

environments that, conceptually, are similar to the ones presented in Section 1.2.1. A process can also be modeled as sequential if it includes tasks that consume materials that come from multiple batches (batch mixing) or produces materials that can be used in multiple downstream batches (batch splitting), but these materials do not have to be accounted for; for example, they are always available if they are inputs or can always be disposed/stored if they are outputs. Essentially, (the representation of) a process is sequential if it can be modeled in terms of *batches* (jobs) and *stages* (operations).

With respect to the machine environments in Figure 1.5, ***1*** and ***P*** can be treated as special cases of ***F*** and ***J***, respectively, and ***O*** and ***FO*** are not relevant because chemical transformations and separations have to be carried out in a given sequence. Hence, we will focus our discussion on two multiproduct environments that are similar to the flexible flow-shop (***FF***) and flexible job-shop (***FJ***) environments. The first is the so-called *multistage* environment, where each batch has to be processed in a number of stages, each stage consists of one or more parallel processing units, and a unit belongs to a single stage. All batches go through the stages in the same sequence, unless a batch is not processed in given stages. In the absence of other processing features (e.g., storage and utility constraints), this is equivalent to ***FF***. The second is the *multipurpose* environment, where batches cannot be mixed nor split and have to go again through multiple stages, but the sequence of stages is batch specific. Also, a processing unit may belong to different stages depending on the batch, and/or a unit may belong to multiple stages. In the absence of other constraints, this is equivalent to ***FJ***.

Before we discuss these two environments in detail, we note that there is a discrepancy between the general problem statement (Section 1.1.6) and the manner in which these environments are represented. Specifically, the general problem is expressed in terms of facility (e.g., unit capacities) and product (e.g., recipes) data, raw material and resource availability, and product demand, not in terms of *batches*. For now, we will assume that the batching problem is solved independently prior to scheduling, so the batches to be scheduled to meet the demand of all products are fixed and inputs to the scheduling problem.[4] We will relax this assumption in Chapter 4. If the batching problem is solved a priori, then the scheduling problem can be expressed in terms of batches rather than products. For example, instead of using the routing of products, we directly assume that batches have routings; and similarly, processing times, which would normally be a function of product and batchsize, can be assumed fixed and known.

Building upon the ideas presented in Section 1.2.1, we next discuss the basic elements for the representation of general multistage and multipurpose environments. In addition to routing restrictions (horizontal differentiation in Figure 1.5) and different levels of operation complexity (vertical differentiation), we consider the suitability of a unit to perform a task of a given batch as a third dimension. Figure 1.11 shows a classification of the different subtypes of sequential environments as well as the new sets necessary to define these problems.

We start with the multistage problem. The simplest case arises when all batches go through all the stages and all units are compatible with all batches (see Figure 1.11A). In this case, we only need to define the set of stages, $k \in \mathbf{K}$, and the units in each stage, $j \in \mathbf{J}_k$. The first generalization results when a batch, $i \in \mathbf{I}$, does not have to be processed in one or more stages (see Figure 1.11B) (note that this problem can also be viewed as a multipurpose problem). In this case, we have to define batch-specific routings, $\mathbf{R}_i \subseteq \mathbf{K}$. Note that, in the absence of pre-emption,[5] this problem is not equivalent to the one with $\mathbf{R}_i = \mathbf{K}$ and some processing times being equal to zero. (Why would this be so?) The second generalization results if batch-unit compatibility is considered, that is, only a subset of units in a stage are suitable to process a batch (Figure 1.11C). To represent this, we introduce subsets $\mathbf{J}_{ik}$, which include the units in stage k that are suitable for batch i.

In multipurpose facilities, we have batch-dependent routings, $\mathbf{R}_i$. The simplest case arises when units belong to the same work center (Figure 1.11D). If work centers are viewed as independent stages, then this case can be modeled by defining routings $\mathbf{R}_i$ and stage-specific unit subsets $\mathbf{J}_k$. Note that Figure 1.5G shows the same case with routings defined in terms of centers. In the first generalization, we have routings with reentries; that is, units belong to multiple stages for the same batch. If the reentering batch undergoes the same type of operation, then we simply add the same stage more than once in the routing; that is, $\mathbf{R}_i$ is an ordered multiset (Figure 1.5E). If the reentering batch undergoes a different task, then a new stage is added in the routing $\mathbf{R}_i$, and processing times are represented in terms of stages. For example, consider an environment with two units, U1 and U2, and batch P that has to undergo

[4] The majority of approaches to multistage and multipurpose problems in the process engineering literature have considered the problem with given batches. The introduction of storage and/or utility considerations preserves the batch/stage-based representation but introduces additional features and constraints. However, if batching decisions are considered at the scheduling level, then the problem cannot be viewed as a flexible flow-shop/job-shop because the number of jobs is not known.

[5] Preemption is the act of temporarily interrupting a task being carried out on a unit with the intention to resume it later. Throughout the book, we will assume that we have no preemption, which is a reasonable assumption in chemical production.

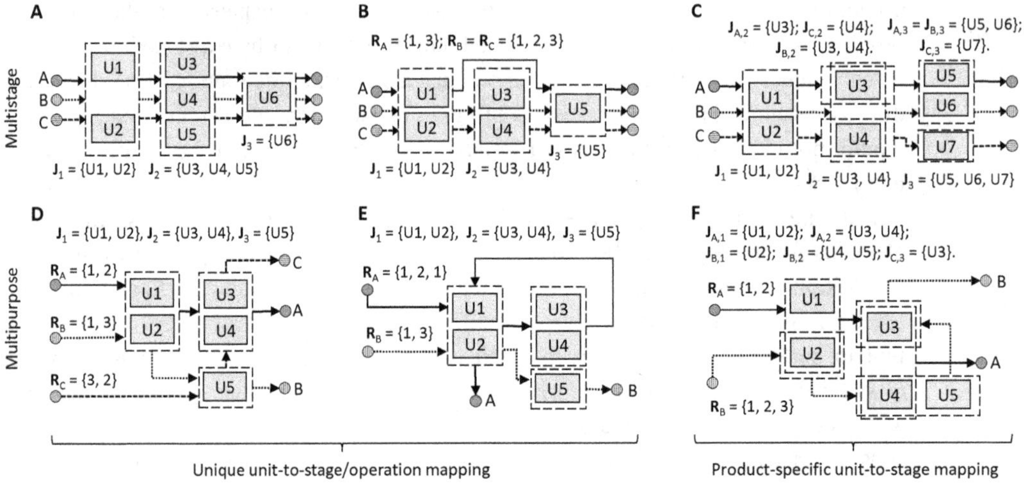

Figure 1.11 Illustration of main types of sequential production environments.

three tasks: O1 in U1 $\rightarrow$ O2 in U2 $\rightarrow$ O3 in U1. This problem can be modeled using routing $\mathbf{R}_P = \{1,2,3\}$ with $\mathbf{J}_{P,1} = \{U1\}$, $\mathbf{J}_{P,2} = \{U2\}$, and $\mathbf{J}_{P,3} = \{U1\}$. Finally, in the most general case units belong to different stages depending on which batch they process (Figure 1.5F). In this case, we use subsets $\mathbf{J}_{ik}$, where stage numbers depend on batches.

Since all sequential environments can be viewed as special cases of the multipurpose environment with reentries and unit suitability constraints that depend on batch and stage (Figure 1.11F), the structure of all such facilities can be stated in terms of batch routings, $\mathbf{R}_i$, and batch/stage-unit suitability information, $\mathbf{J}_{ik}$, which leads to the following general problem representation/statement:

Given are a set of batches $i \in \mathbf{I}$ and a set of units $j \in \mathbf{J}$. In addition:

(a) Each batch has to be processed in stages, possibly including reentries, following a batch-specific sequence given by routing $\mathbf{R}_i$.

(b) The set of units suitable of processing batch i in stage k is denoted by $\mathbf{J}_{ik}$.

For the *simple* multistage environment, we have $\mathbf{R}_i = \mathbf{K}$ for all i, and $\mathbf{J}_{ik} = \mathbf{J}_k$ for all i and k, where $\mathbf{J}_k$ is the subset of units belonging to stake k.

1.2.4 Network Environments

Network processing arises when there are no restrictions in the way input and output materials are handled. As illustrated in Example 1.1, it is not the configuration of the facility but rather the manner in which batches and materials are handled that *define* the type of processing. In the absence of other processing restrictions (e.g., storage limitations, unit connectivity, transfer equipment, etc.), network processing implies that materials can flow freely from storage vessels (suitable for their storage) to processing units where they are consumed, and from processing units where they are produced to other processing units or storage vessels.

Since there is no requirement to maintain batch identity, the notion of a batch going through different stages is irrelevant and processing stages are not necessary to define network environments. Also, traditionally, batching decisions are considered simultaneously with assignment and sequencing/timing decisions. Furthermore, tasks are typically defined in terms of input and output materials; that is, a task is identified by the materials it consumes and produces and the corresponding conversion coefficients rather than a product and a stage (as in sequential processing). Thus, tasks can be defined using a matrix where tasks correspond to columns, materials to rows, and the matrix elements are the corresponding conversion coefficients.

Two widely used representations of network processes are the state-task network (STN) and the resource-task network (RTN), which *rely* on the consumption/production of materials, modeled either as states in STN or resources in RTN, by tasks. The two aforementioned approaches can handle a number of additional characteristics and constraints (e.g., storage and utility constraints), though these are not specific to network processing. In this book, we will use both representations, formally introduced in Chapter 7, to model a range of features.

1.2.5 General Environments

There exist processes where batch identity has to be tracked and preserved for some materials (e.g., intermediate to be further processed downstream), while other materials (additives, promoters, byproducts, etc.) are not subject to special handling constraints. We will refer to such environments as *general* or *hybrid* environments. An environment is general when, for example, a task consumes multiple materials or produces multiple materials and some of them have mixing/splitting restrictions, or a proper subset of materials produced or consumed by a task has mixing or splitting restrictions. Further, in general, as well as in most sequential environments, the size of the batch can change as it moves along the various stages.

We have already seen two examples of general production environments in Examples 1.3 and 1.4. The process in Example 1.3 is representative of problems where a task consumes multiple materials (i.e., we have material mixing), but mixing of batches of the same material is not allowed for at least one input material. This type of processing is common in the process industries; for example, in emulsifier production, an additive is added to a batch prior to spray tower processing. Note the distinction here between mixing of batches (of the same material) and mixing of different materials. The former refers to the case where the output of two batches of the same task is mixed, which if allowed leads to network processing even if a task produces a single material and this material is consumed by a single task (see Figure 1.6B). The latter refers to the case where a task requires two inputs that therefore have to be mixed. Both types of mixing lead to network processing.

The process in Example 1.4 represents facilities where the main output material of a task should be consumed, without being mixed or split, by a downstream task, and a byproduct has to also be handled (but without any special restrictions). For example, centrifugation of a batch coming from fermentation results in a liquid concentrate that has to be processed (without mixing or splitting) further, and a byproduct aqueous stream that has to be treated. Also, in fermentation processes it is common to remove yeast before the end of the batch, which is going to be used in multiple subsequent fermentations. Here, note the difference between batch splitting and a batch producing multiple materials. The former arises when the output of the same batch is consumed in multiple subsequent batches, which leads to network processing even if only one material is produced (see Figure 1.6C). The latter arises when a task has multiple output materials, which can be combined with no mixing/splitting restrictions (Example 1.4).

Finally, we note that general environments can in some cases be treated as sequential ones. For instance, the process in Figure 1.8 can be modeled as a sequential process if the additive is always available (e.g., produced in a unit with excess capacity) and thus does not need to be accounted for. Similarly, the process in Figure 1.9 can be simplified by ignoring the byproduct if it does not require any further treatment or the resources for this treatment are always available. While similar simplifications can be applied to many problems, in the general case we need methods that preserve batch integrity while monitoring inventory levels.

1.3 Classes of Problems

Traditionally, discrete manufacturing scheduling problems have been classified in terms of a triplet $\alpha/\beta/\gamma$: (1) α denotes the *machine* environment; (2) β denotes the processing restrictions and characteristics; and (3) γ denotes the objective function. We will follow the same convention to define the classes of chemical production scheduling problems, but the first entry will denote the *production*, rather than *machine*, environment.

1.3.1 Production Environments (α)

The major types of environments are sequential-multistage (***Sms***), sequential-multipurpose (***Smp***), network (***N***), and general (***G***). Sequential production environments can be further divided into the ones shown in Figure 1.11:

(a) ***Sms-1***: multistage with unique batch routing ($\mathbf{R}_i = \mathbf{R} = \{1, 2, \ldots, |\mathbf{K}|\}$, for all i), and unique partition of units to stages as well as unit-stage compatibility ($\mathbf{J}_{ik} = \mathbf{J}_k$ for all i and k) – see Figure 1.11A.

(b) ***Sms-2***: multistage with batch-specific routing, and unique partition of units to stages and unit-stage compatibility ($\mathbf{J}_{ik} = \mathbf{J}_k$ for all i and $k \in \mathbf{R}_i$) – see Figure 1.11B.

(c) ***Sms-3***: multistage with batch-specific routing and unit-stage compatibility that is batch and stage specific – see Figure 1.11C.

(d) ***Smp1***: multipurpose with unique partition of units to stages and unit-stage compatibility ($\mathbf{J}_{ik} = \mathbf{J}_k$ for all i and $k \in \mathbf{R}_i$) and no reentries – see Figure 1.11D.

(e) ***Smp2*:** multipurpose with unique partition of units to stages and unit-stage compatibility and reentries – see Figure 1.11E.

(f) ***Smp3*:** multipurpose with unit-stage compatibility that is batch and stage specific, and reentries – see Figure 1.11F.

A facility that has different types of sequential subsystems can be treated as a sequential facility of the broadest type. For example, a facility that has both ***Sms-1*** and ***Sms-3*** subsystems can be treated as an ***Sms-3*** environment, since ***Sms-1*** is a special case of ***Sms-3***.

A facility that has a sequential subsystem (of any subtype) and a network subsystem is a general environment. Alternatively, a combination of entries can be used for α to better describe the general environment. For instance, a facility that includes an upstream ***Sms-1*** subsystem followed by a network subsystem can be denoted by ***G*** or ***Sms-1, N***.

1.3.2 Processing Restrictions and Features (β)

Many restrictions and features are similar to those found in discrete manufacturing (e.g., setups and release/due times), while others are specific to chemical production (e.g., storage constraints and material transfers). The most important or commonly found features are the following.

General Resource Constraints. Thus far we have considered processing unit availability as the only resource constraint. However, chemical process facilities almost always involve a range of other shared resources, such as auxiliary units (e.g., storage vessels), utilities (e.g., steam, water, cleaning-in-place) and labor. They can be discrete (e.g., labor) or continuous (e.g., steam). If binding, then resource constraints should be accounted for. Resource constraints will be introduced in Chapter 4.

***Setups*.** Setups are preparatory activities carried out before a processing activity starts. They may involve processing unit retooling, cleaning, and sterilization. They can be sequence independent or sequence dependent, the latter also referred to as *changeovers*. They typically require downtime, during which the unit cannot be used, and often incur a cost. Setups are often defined in terms of product families; that is, they are necessary only between products belonging in different families. Furthermore, setups may require shared resources such as cleaning-in-place (CIP), which means that they are resource constrained. The modeling of setups and changeovers is introduced in Chapter 3, while some additional features and approaches are discussed in Chapter 8.

***Material Storage*.** Storage constraints is one of the main differences between discrete and chemical manufacturing. There are two types of storage constraints: (1) constraints on the duration a material can be stored (in a processing unit or storage vessel), and (2) storage capacity constraints (number and size of storage vessels). The combination of these two aspects is often referred to as storage policy (see detailed discussion in Chapters 5 and 8).

***Material Transfer*.** In many facilities, the transfer of material is highly constrained because it takes time and requires shared resources. In these cases, transfer activities should be modeled and considered simultaneously with processing tasks. Also, resource-constrained material transfer activities often result in limited connectivity among units. They may also require that material charging and withdrawing operations should be modeled as separate activities. Material transfer is discussed in Chapter 8.

Finally, we note that multiple processing characteristics and restrictions can be present simultaneously; that is, β may include multiple entries.

1.3.3 Objective Functions (γ)

Unlike discrete manufacturing, where the objective to be minimized is often a function of job completion times, in chemical production scheduling monetary objective functions are often employed, especially when batching and assignment decisions are considered, and thus (1) the total production is an optimization decision and (2) the same demand can be satisfied by different schedules with different costs.

To illustrate the different time-related objective functions, we consider environment ***Sms-1*** with $|\mathbf{K}| = 1$; that is, we have to process a set of batches, $i \in \mathbf{I}$, with release (ρ_i) and due (ε_i) dates, in parallel units (see Figure 1.12). If C_i is the completion time of batch i, then we define the following variables:

(a) Lateness of batch, i: $L_i = C_i - \delta_i$
(b) Tardiness of batch, i: $T_i = \max\{C_i - \delta_i, 0\} = \max\{L_i, 0\}$
(c) Earliness of batch, i: $E_i = \max\{\delta_i - C_i, 0\}$

Based on these variables, we can define various objective functions. First, we have functions that depend on the maximum, across batches, variable value; for example, for maximum lateness we have the following:

$$L_{MAX} = \max_i \{L_i\}. \tag{1.1}$$

Second, we have the total (weighted) functions; for example, for tardiness we have the following:

$$T_{TOT} = \sum_i \omega_i T_i, \tag{1.2}$$

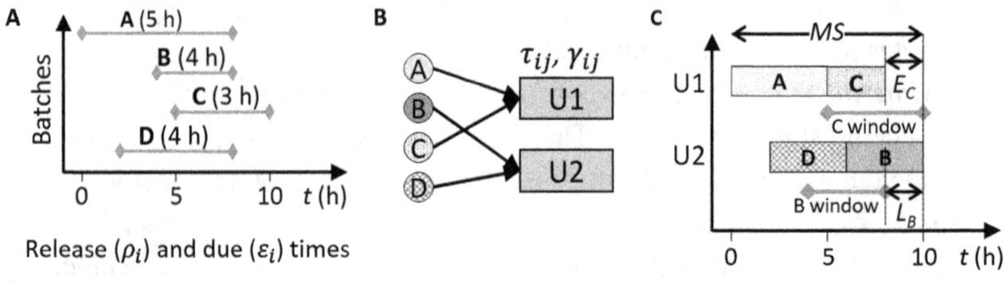

Figure 1.12 Illustration of time-related objective functions. (A) Batch processing windows. (B) Batch-unit assignments (τ_{ij}: processing times; γ_{ij}: processing costs). (C) Gantt charts and metrics; *MS*: makespan, E_C: earliness of batch C, L_B: and lateness of batch B.

where ω_i is a weight used to assign importance. If $\omega_i = 1$ for all i, then we have the total tardiness. Third, we have functions that depend on completion times, but not due dates; for example, for makespan we have the following:

$$MS = \max_i \{C_i\}, \tag{1.3}$$

and for total (weighted) completion time we have the following:

$$C_{TOT} = \sum_i \omega_i C_i. \tag{1.4}$$

In addition, if completion time is not a concern, as long as orders and due times are satisfied, and there are multiple schedules that meet the given orders, then the objective can be cost minimization. For instance, in the single-stage problem, if there are unit-specific batch processing costs, γ_{ij}, then we seek to find the assignment of batches to units that would minimize the total processing cost. Other cost components often considered include (1) inventory cost; (2) total changeover cost; and (3) utility costs, especially when utility prices are time varying (e.g., daily electricity price fluctuation).

Finally, when the capacity of the facility cannot meet total demand, then, in order to optimize resource utilization, we may want to maximize the (weighted) total production, potentially subject to some minimum demand constraints, or the total profit. Combinations of objective functions with weights can also be used.

1.3.4 Problem Classification

Based on the three entries discussed in the previous subsections, we define a problem as shown in Figure 1.13. We note that field α has a single entry, unless we replace G with a combination of environments; field β has multiple entries; and field γ typically has one entry but can have multiple if combinations of objective functions are considered. We note that while this classification is more natural for batch processes, it is also applicable to continuous processing, using the notion of a *run* (i.e., the specific execution of a continuous task) instead of a batch.

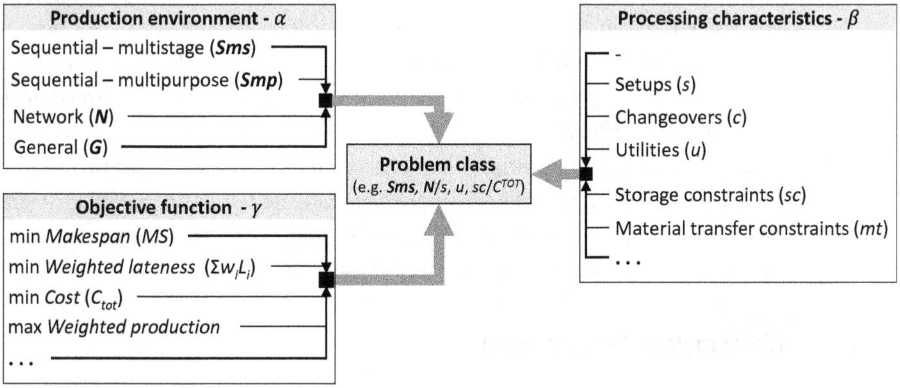

Figure 1.13 Definition of chemical production scheduling problems.

1.4 Approaches to Scheduling

A number of different modeling and solution paradigms have been employed to address scheduling problems.

1.4.1 Problem-Specific Algorithms

Problem-specific algorithms include (i) simple priority rules; (ii) rigorous algorithms guaranteed to obtain the optimal solution if run for sufficient time (e.g., the Carlier–Pinson algorithm for job-shop problems); and (iii) heuristic algorithms (e.g., the shifting bottleneck algorithm for job-shop problems). They are computationally effective but limited to problems with relatively simple structure and not many complicating constraints. Thus, they cannot be readily applied to problems found in most chemical facilities and have received almost no attention in the PSE literature.

1.4.2 Metaheuristics

Metaheuristics are general-purpose algorithms (e.g., genetic algorithms, simulated annealing, particle swarm optimization) applied to scheduling problems. They are effective for large-scale problems where simpler heuristics give poor results while rigorous methods are too expensive. A disadvantage is the inability or difficulty to account for constraints, which may result in the evaluation of many infeasible points and therefore solutions of low quality and long solution times. They have received limited attention in the PSE literature.

1.4.3 Integrated Modeling/Solution Approaches

Integrated approaches include modeling approaches and the associated search algorithms. One such example is constraint programming (CP) constructs and the corresponding propagation algorithms. Another example is timed-automata (TA) representations coupled with reachability analysis algorithms. They are broad methods – they can be used to model a wide range of constraints. For example, global constraints, and finite-domain, interval-valued, and set-valued variables in CP allow us to model problems that are difficult to formulate with other methods. CP methods are very effective at finding feasible solutions fast and at solving highly constrained problems. However, the effectiveness of these methods often deteriorates significantly when instances become larger or problem characteristics change. Methods based on TA have been explored for a few classes of problems, while CP has received also some attention but primarily in conjunction with mixed-integer programming methods toward the development of hybrid methods (see Section 1.4.5).

1.4.4 Mathematical Programming

The problem is represented using algebraic equations (the mathematical programming model or mathematical program) leading to an optimization model. Since discrete decisions are involved, mixed-integer programming (MIP) models are necessary.

A wide range of decomposition solution techniques can be developed, where subproblems are also mathematical programs. Mathematical programming, and MIP in particular, allows us to model different types of processing constraints and thus represent a wide range of chemical production scheduling problems. A potential disadvantage is that MIP solvers are not as easily customizable as CP propagation algorithms, so they cannot be easily fine-tuned to address specific problems. On the other hand, MIP is the method of choice for a variety of problems, from finance to military applications, which means that substantial efforts over the last 30 years have gone into the development of effective MIP solvers, resulting in major computational enhancements (six to eight orders of magnitude, over the last 30 years). Furthermore, a number of high-level modeling languages are available, making the formulation, testing, and deployment of MIP scheduling models easy. Math programming is the paradigm that has received the most attention in PSE.

1.4.5 Hybrid Methods

Hybrid methods exploit the complementary strengths of different modeling and/or solution methods. The most common approach is to decompose the original problem into subproblems that are modeled and solved using different approaches, such as decomposition into an assignment MIP subproblem and a sequencing CP subproblem. They are effective for many classes of problems, especially in sequential environments, where the structure of the problem enables its decomposition into subproblems that can be solved effectively using algorithms with complementary strengths. One disadvantage is that they require advanced domain knowledge and they cannot be readily modified to address different classes of problems.

1.5 Scheduling MIP Model Classification

Before we present a model classification, we discuss three closely related concepts: *problem*, *problem representation*, and *model* (or modeling approach). A scheduler is faced with a *problem*. Problems can be expressed in different ways, depending on the production environment, the interaction of scheduling with the other planning functions, and so on. To be addressed, the problem must be converted to a standard form, understood by the people using the solution. This is what we term *problem representation*. Though it depends on the problem and the method that will be used to address it, the problem representation is, in principle, independent of those two. Next, for a given representation, different models can be adopted; for example, problems in sequential environments, represented using products, stages, and units, can be addressed using different models. In that respect, scheduling MIP models[6] can be classified in terms of three attributes: (1) major scheduling decisions, (2) modeling elements, and (3) modeling of time.

[6] Strictly speaking, there is a subtle difference between *model* and *formulation*. Specifically, multiple formulations can be developed for the same model. However, the two terms have been used interchangeably in the literature, and, in this book, we use primarily the term *model*.

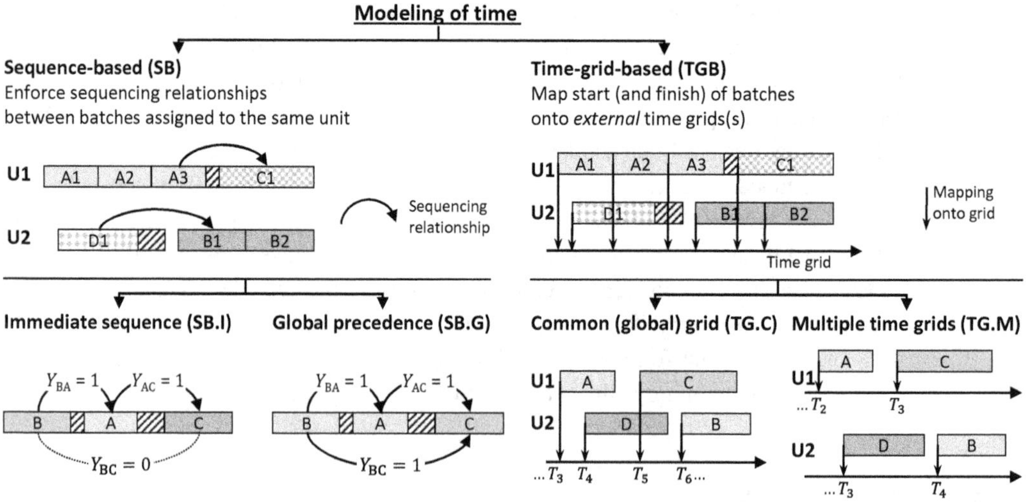

Figure 1.14 Illustration of first two levels in modeling of time; in sequence-based models, binary variable $Y_{ii'}$ is used to establish sequencing relationship between a pair of batches; in time-grid-based models, major events (e.g., start of a batch) are mapped onto time points (T_n) of one or more grids.

In terms of decisions, most models for sequential environments include assignment of batches to units, sequencing between pairs of batches assigned to the same unit, and/or timing of batches; that is, batching decisions are assumed to be given as input. The first models considering batching, assignment, and sequencing decisions simultaneously appeared in the late 2000s. Models for network environments include batching, batch-unit assignment, and timing decisions. Models for general environments, which are typically models developed for network environments augmented with constraints to enforce special material handling constraints, include all three major decisions.

In terms of modelling elements, if batching decisions are given, then problems in sequential environments are typically expressed in terms of batches, stages, and units. On the other hand, problems in network and general environments, are modeled using units, tasks, and materials.

The modeling of time involves three levels (see Figure 1.14):[7]

(1) Selection between sequence-based and time-grid-based approach
(2) If a sequence-based approach is used, then selection between immediate and global sequencing; if a time-grid-based approach is used, then selection between common and multiple time grids
(3) Selection between continuous and discrete representation of time

[7] To describe the three levels, we use some concepts that are formally introduced in subsequent chapters (e.g., sequence-based models). Thus, the discussion, including the concepts illustrated in Figure 1.14, are at high level. The reader can skip the remaining of Section 1.5, without this affecting their understanding.

Scheduling models can then be classified, using the three aforementioned attributes. Broadly speaking, most models for sequential environments (1) consider assignment and sequencing/timing decisions; (2) are represented in terms of units, stages, and batches; and (3) employ either sequence-based or time-grid-based modeling of time. Models for network environments (1) consider all three decisions; (2) are represented in terms of tasks, units, and materials; and (3) are almost exclusively based on time-grid-based modeling of time.

1.6 Book Outline

The remainder of the book is structured as follows. Chapter 2 presents (1) some basic MIP models, which will allow the reader to better understand how binary decision variables can be used to model different types of constraints; and (2) a review of MIP solution methods. The book it then divided into three parts.

Part II (Chapters 3 through 7): Models for some *basic* problems are presented. We start with problems in the simplest environment, the single unit (***Sms***, with $|\mathbf{K}| = 1$ and $|\mathbf{J}_1| = 1$) environment, in Chapter 3, and expand the scope to single-stage multiple-unit environment (***Sms***, with $|\mathbf{K}| = 1$ and $|\mathbf{J}_1| > 1$) in Chapter 4; multistage with multiple parallel units in each stage (***Sms***, with $|\mathbf{K}| > 1$ and at least one k with $|\mathbf{J}_k| > 1$) in Chapter 5; multipurpose facilities (***Smp***) in Chapter 6; and network environment (***N***) in Chapter 7. This presentation is intentionally based on production environments because this facilitates the gradual introduction of new decisions and new types of constraints, leading to increasingly complex models. For example, after introducing the basic decisions of sequencing and timing in Chapter 3, the consideration of multiple units in Chapter 4 naturally leads to the introduction of assignment decisions (which were not considered in Chapter 3). In parallel, the basic processing features are also gradually introduced: release/due times, setups, and changeovers are introduced in Chapter 3; general resource constraints in Chapter 4; and storage constraints in Chapter 5. Problems in network environments with all aforementioned features are then addressed in Chapter 7. Finally, an effort is made to present many alternative formulations for the same problem, with special emphasis placed on the presentation of different approaches for the modeling of time.

Part III (Chapters 8 through 11): Having covered the basic environments and processing features in Part II, Part III expands the scope of scheduling problems. Chapter 8 covers a range of extensions, such as special material transfer and storage constraints, in network environments. Chapter 9 presents how scheduling models for continuous processes can be generated, and Chapter 10 introduces the ideas underlying periodic scheduling. Chapter 11 covers scheduling problems with blending, which lead to mixed-integer nonlinear programming (MINLP) models. Since the focus of Part III is on the modeling of new major features, the presentation adopts a single approach for the modeling of time, namely, a common discrete time grid. This allows us to focus on the novel modeling concepts necessary to address these problems, rather than repeat the ideas that have already been covered in Part II.

Part IV (Chapters 12 through 15): The last part of the book covers some special topics, such as solution methods for problems in sequential (Chapter 12) and network (Chapter 13) environments, real-time scheduling (Chapter 14), and integration of production planning and scheduling (Chapter 15). The goal of this part is to introduce the reader to some general and representative solution approaches rather than review an exhaustive list of approaches. Also, emphasis is placed on the ideas behind the development of the selected methods, so that the interested reader can develop similar methods for other problems, rather than implementation details.

1.7 Notes and Further Reading

(1) The definition of scheduling in Section 1.1.1 is taken from Pinedo [1].

(2) Interestingly, the titles of many early books on the subject include both the words "sequencing" and "scheduling," such as in *Introduction to Sequencing and Scheduling* [2], *Sequencing and Scheduling: An Introduction to the Mathematics of the Job-Shop* [3], and *Principles of Sequencing and Scheduling* [4].

(3) The overall discussion in this chapter borrows concepts introduced in [5], using also some ideas from two review papers [6, 7]. The discussion of the role of scheduling in SC, in Sections 1.1.3 and 1.1.4, is taken from [8, 9]. The SC matrix in Figure 1.2 is modified from [8]. A discussion of inventories policies can be found in Pochet and Wolsey [10].

(4) The two representations of network processes mentioned in Section 1.2.4 were proposed by Pantelides and coworkers: the STN was introduced in [11, 12] and the RTN in [13].

(5) For a complete discussion of processing characteristics (Section 1.3.2) the interested reader in pointed to Mendez et al. [6] and Harjunkoski et al. [7].

(6) The discussion of approaches to scheduling (Section 1.4) follows the presentation in Velez and Maravelias [14]. For additional material on specific methods, the interested reader is pointed to the following sources: priority rules [15], the Carlier–Pinson algorithm for job-shop problems [16], heuristic algorithms [17], constraint programming [18–20], timed automata [21], hybrid methods [20, 22].

(7) For a thorough discussion of mixed-integer programming models, theory, and solution methods, the interested reader is pointed to two seminal books [23, 24].

References

[1] Pinedo M. *Scheduling: Theory, Algorithms, and Systems*. 3rd ed. New York; London: Springer; 2008.

[2] Baker KR. *Introduction to Sequencing and Scheduling*. New York: Wiley; 1974.

[3] French S. *Sequencing and Scheduling: An Introduction to the Mathematics of the Job-Shop*. Chichester, New York: E. Horwood; Wiley; 1982.

[4] Baker, KR, Trietsch, D. *Principles of Sequencing and Scheduling*. Hoboken: Wiley; 2009.
[5] Maravelias CT. General Framework and Modeling Approach Classification for Chemical Production Scheduling. *AIChE J.* 2012;58(6):1812–1828.
[6] Mendez CA, Cerda J, Grossmann IE, Harjunkoski I, Fahl M. State-of-the-Art Review of Optimization Methods for Short-Term Scheduling of Batch Processes. *Comput Chem Eng.* 2006;30(6-7):913–946.
[7] Harjunkoski I, Maravelias CT, Bongers P, Castro PM, Engell S, Grossmann IE, et al. Scope for Industrial Applications of Production Scheduling Models and Solution Methods. *Comput Chem Eng.* 2014;62(0):161–193.
[8] Stadtler H. Supply Chain Management and Advanced Planning – Basics, Overview and Challenges. *Eur J Oper Res.* 2005;163(3):575–588.
[9] Shapiro JF. *Modeling the Supply Chain.* 2nd ed. Belmont: Thomson-Brooks/Cole; 2007.
[10] Pochet Y, Wolsey LA. *Production Planning by Mixed Integer Programming.* New York; Berlin: Springer; 2006.
[11] Kondili E, Pantelides CC, Sargent RWH. A General Algorithm for Short-Term Scheduling of Batch-Operations. 1. Milp Formulation. *Comput Chem Eng.* 1993;17(2):211–227.
[12] Shah N, Pantelides CC, Sargent RWH. A General Algorithm for Short-Term Scheduling of Batch-Operations .2. Computational Issues. *Comput Chem Eng.* 1993;17(2):229–244.
[13] Pantelides CC, editor Unified Frameworks for Optimal Process Planning and Scheduling. *2nd Conference on Foundations of Computer Aided Process Operations*; 1994; Snowmass: CACHE Publications.
[14] Velez S, Maravelias CT. Advances in Mixed-Integer Programming Methods for Chemical Production Scheduling. *Annu Rev Chem Biomol.* 2014;5:97–121.
[15] Haupt R. A Survey of Priority Rule-Based Scheduling. *Or Spektrum.* 1989;11(1):3–16.
[16] Carlier J, Pinson E. An Algorithm for Solving the Job-Shop Problem. *Manage Sci.* 1989;35 (2):164–176.
[17] Adams J, Balas E, Zawack D. The Shifting Bottleneck Procedure for Job Shop Scheduling. *Manage Sci.* 1988;34(3):391–401.
[18] Baptiste P, Le Pape C, Nuijten W. *Constraint-Based Scheduling: Applying Constraint Programming to Scheduling Problems.* Boston: Kluwer Academic; 2001.
[19] Van Hentenryck P, Michel L. *Constraint-Based Local Search.* Cambridge: MIT Press; 2005.
[20] Hooker J. *Logic-Based Methods for Optimization: Combining Optimization and Constraint Satisfaction.* New York: John Wiley & Sons; 2000.
[21] Panek S, Engell S, Subbiah S, Stursberg O. Scheduling of Multi-product Batch Plants Based upon Timed Automata Models. *Comput Chem Eng.* 2008;32(1-2):275–291.
[22] Hooker JN. Logic, Optimization, and Constraint Programming. *INFORMS Journal on Computing.* 2002;14(4):295–321.
[23] Nemhauser GL, Wolsey LA. *Integer and Combinatorial Optimization.* New York: Wiley; 1988.
[24] Wolsey LA. *Integer Programming.* New York: Wiley; 1998.

2 Mixed-Integer Programming

This chapter provides an overview of mixed-integer programming (MIP) modeling and solution methods. In Section 2.1, we present some preliminary concepts on optimization and mixed-integer programming. In Section 2.2, we discuss how binary variables can be used to model features commonly found in optimization problems. In Section 2.3, we present some basic MIP problems and models. Finally, in Section 2.4, we overview the basic approaches to solving MIP models and present some concepts regarding formulation tightness and decomposition methods. Finally, we discuss software tools for modeling and solving MIP models in Section 2.5.

2.1 Preliminaries

2.1.1 General Optimization Problem

The general deterministic optimization problem (P) can be written as

$$\begin{aligned} &\max f(x) \\ &g(x) = 0 \\ &h(x) \leq 0 \end{aligned} \tag{2.1}$$

where $x \in \mathbf{X} \subseteq \mathbb{R}^n, f(x)$ is a scalar function, $g(x)$ is an m-vector function (with $m < n$), and $h(x)$ is an l-vector function. If some of the parameters ξ are not known deterministically, then the solution depends not only on decision variables x but also on the realization of ξ, which leads to an optimization problem under uncertainty. To address such problems, many stochastic optimization methods have been proposed. While scheduling instances have uncertainty, stochastic optimization methods are beyond the scope of this book. Nevertheless, in Chapter 14 we discuss how uncertainty in scheduling instances is addressed via real-time scheduling.

An equality can always be written as $g_1(x) = 0$; e.g., $3x_1 + 2{x_2}^2 = 9 - x_3$ can be written as $g_1(x) = 3x_1 + 2{x_2}^2 + x_3 - 9 = 0$. An inequality can always be written as $g_2(x) \leq 0$; e.g., $\sin x_1 \leq \cos x_2$ can be written as $g_2(x) = \sin x_1 - \cos x_2 \leq 0$.

Next, we introduce some basic concepts:

- *Feasible point*: a point that satisfies all the constraints.
- *Feasible region*: the set of all feasible points.

- At a feasible point x, an inequality constraint $g_i(x) \leq 0$ is said to be *active* (binding) if $g_i(x) = 0$; and inactive (nonbinding) if $g_i(x) < 0$.
- All equality constraints are regarded as active at any feasible point.
- *Active set* at a feasible point x: the set of all constraints that are active at x.
- *Boundary* of the feasible region: the set of feasible points for which at least one inequality is binding.
- *Interior point*: all feasible points that are not on the boundary, that is, those feasible points at which all inequality constraints are nonbinding.

A function is *linear* if it involves a constant weighted sum of variables and a constant; otherwise it is nonlinear. A variable is *continuous* if it can take on any value in a specified interval; for example, $x_1 \in [1, 10]$. A variable is discrete if it is limited to a countable set of values; for example, $x_1 \in \{0, 1, 2, \ldots 10\}$. Depending on the type of functions (f, g, and h) and variables x, we have different types of optimization problems (see also Table 2.1). An optimization model can be any of the following:

- A linear programming (LP) model (or linear program) if the objective function (f) *and all* constraints (g and h) are linear functions in the decision variables x, and all decision variables are continuous.
- A nonlinear programming (NLP) model (or nonlinear program) if the objective function *or any* of the constraints is a nonlinear function in the decision variables x, and all decision variables are continuous.
- A mixed-integer programming (MILP of MIP) model (or mixed-integer program) if the objective function *and all* constraints are linear functions in the decision variables x, and there is at least one integer variable.
- A mixed-integer nonlinear programming (MINLP) model (or simply an MINLP) if the objective function *or any* of the constraints is a nonlinear function of x, and there is at least one integer variable.

Note that problems with only integer variables and nonlinear equations can be reformulated as pure (linear) integer programming (IP) models.

As we discussed in Chapter 1, some of the decisions in scheduling are discrete (e.g., assignment of a batch to a unit), which means that scheduling problems are typically formulated as mixed-integer programming models. In most cases, they can be modeled using linear equations, though processing features such as blending and performance decay lead to nonlinear equations and thus MINLP models. Since the majority of models discussed in this book are linear, our discussion in the remainder of the present chapter will focus on MIP models and solution methods.

2.1.2 General Mixed-Integer Programming Problem

If both the objective function and all constraints are linear and $(\mathbf{J}^{\mathrm{I}}, \mathbf{J}^{\mathrm{C}})$ is a partition of the set $\mathbf{J}$ of variable indices (i.e., $\mathbf{J}^{\mathrm{I}} \cap \mathbf{J}^{\mathrm{C}} = \emptyset$ and $\mathbf{J}^{\mathrm{I}} \cup \mathbf{J}^{\mathrm{C}} = \mathbf{J}$), then the general optimization model in (2.1) can be rewritten as follows:

Table 2.1 Types of optimization problems.

	Constraints	
Variables	Linear	Nonlinear
Continuous	Linear programming (LP)	Nonlinear programming (NLP)
Integer	Integer programming (IP)	
Continuous & integer	Mixed-integer linear programing (MILP or MIP)	Mixed-integer nonlinear programming (MINLP)

$$\begin{gathered} \max c^T x \\ Ax - b = 0 \\ Bx - b \leq 0 \\ x_j \in \mathbb{Z}, j \in \mathbf{J}^{\mathrm{I}}; x_j \in \mathbb{R}, j \in \mathbf{J}^{\mathrm{C}} \end{gathered} \tag{2.2}$$

An inequality can be converted to an equality through the introduction of a nonnegative slack variable; for example, $x_1 + 2x_2 \leq 10$ can be replaced by $x_1 + 2x_2 + s_1 = 10$ with $s_1 \geq 0$. Any free variable $x_1 \in \mathbb{R}$ can be replaced by $\left(x_1^+ - x_1^-\right)$ where x_1^+, x_1^- are nonnegative. Also, every bounded nonnegative integer $x_1 \in \left\{1, 2, \ldots, x_1^U\right\}$ can be replaced by a set of binary variables $\left(x_1^0, x_1^1, \ldots, x_1^U\right)$ such that $x_1 = \sum_{k \in \{0,\,1,\ldots\}} k x_1^k$. In general, a bounded integer $x_1 \in (\mathbb{Z} \cap [y^L, y^U]) = \{y^L, y^L + 1, \ldots, y^U - 1, y^U\}$ with any y^L and y^U can be replaced by a set of $(y^U - y^L + 1)$ binaries such that $x_1 = \sum_{k \in \{y^L,\ldots,\, y^U\}} k x_1^k$. Similarly, any discrete variable $x_1 \in \left\{y^1, y^2, \ldots, y^m\right\}$ can be replaced by m binary variables $x_1 = \sum_{k \in \{1,\ldots,m\}} y^k x_1^k$. Thus, the general MIP model can be rewritten as follows (with $n = n_1 + n_2$):

$$\begin{array}{ll} \max & c^T x + d^T y \\ \text{s.t} & Ax + By \leq b \quad , \\ & y \in \{0,1\}^{n_1}, x \in \mathbb{R}_+^{n_2} \end{array} \tag{2.3}$$

which is the general form we will consider in the present chapter. If $n_1 = 0$, then we have an LP, and if $n_2 = 0$, then we have a pure IP.

2.1.3 Graphs and Networks

Interestingly, many scheduling problems or specific features of scheduling problems can be represented using concepts from graph theory and networks. In this subsection, we review some basic definitions and concepts. We use uppercase letters for graphs and networks (italic), sets (bold italic), and matrices. The element in row i and column j of matrix A is denoted by a_{ij}.

Definition 1. A graph G with n vertices and m edges consists of a vertex set $\boldsymbol{V}(G) = \{v_1, v_2, \ldots v_n\}$ and an edge set $\boldsymbol{E}(G) = \{e_1, e_2, \ldots e_m\}$, where each edge consists of two vertices called its endpoints.

A vertex v is said to be *incident* to an edge $e = \{w, z\}$ if $v = w$ or $v = z$. A vertex v is *adjacent* to vertex w if $\{v, w\} \in \boldsymbol{E}(G)$. A graph is *simple* if it has no *loops* (i.e., no $\{w, z\}$ edges with $w = z$) and no multiple edges between the same two endpoints.

Definition 2. A directed graph or digraph G is a graph where each edge is an ordered pair of vertices; edge $\{w, z\}$ is written as wz with w/z being the tail/head of wz.

Definition 3. A network is a digraph with nonnegative capacity u_e on each edge e.

The definition of a network often includes the condition that the digraph has a distinct source vertex s and a sink vertex t. In this book, we will use Definition 3. Also, following a standard convention, we will use the terms *vertex* and *edge* for graphs, but we will use the terms *node* and *arc*, respectively, for networks. The set of arcs and nodes of network N are denoted as $\boldsymbol{A}(N)$ and $\boldsymbol{N}(N)$.

A graph is depicted as a set of points corresponding to its vertices connected pairwise by curves corresponding to its edges (see Figure 2.1A). Digraphs are shown similarly, but the curves are directed, from the tail to the head of the arc (see Figure 2.2A). A graph (network) G (N) can be represented by its vertex-to-edge (node-to-arc) *incidence* matrix A, where the rows of A correspond to vertices (nodes) of G (N) and its columns correspond to edges (arcs) of G (N). In graphs, $a_{ij} = 1$ if vertex i is incident to edge j, and $a_{ij} = 0$ otherwise (Figure 2.1C); in networks, $a_{ij} = -1$ if node i is the tail of j, $a_{ij} = 1$ if i is the head of j, and $a_{ij} = 0$ otherwise (Figure 2.2C).

Next we define some basic concepts:

- A *subgraph* is obtained from a graph if we delete vertices and/or edges.
- An *induced subgraph* is a subgraph that contains the chosen vertices and the edges between them.

A

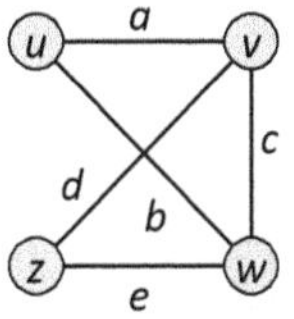

B

$$\begin{aligned}&\text{Graph } G_1 = (\boldsymbol{V}_1, \boldsymbol{E}_1)\\ &\boldsymbol{V}_1 = \boldsymbol{V}(G_1) = \{u, vG, w, z\}\\ &\boldsymbol{E}_1 = \boldsymbol{E}(G_1) = \{(u, v), (u, w), (v, w), (v, z), (w, z)\}\\ &\quad = \{a, b, c, d, e\}\end{aligned}$$

C

$$\begin{array}{c} \\ u \\ v \\ w \\ z \end{array} \begin{array}{c} \begin{array}{ccccc} a & b & c & d & e \end{array} \\ \left[\begin{array}{ccccc} 1 & 1 & & & \\ 1 & & 1 & 1 & \\ & 1 & 1 & & 1 \\ & & & 1 & 1 \end{array}\right] \end{array}$$

Figure 2.1 Graph representation. (A) Example graph G_1. (B) Vertex and edge sets of G_1. (C) Vertex-to-edge incidence matrix of G_1.

A

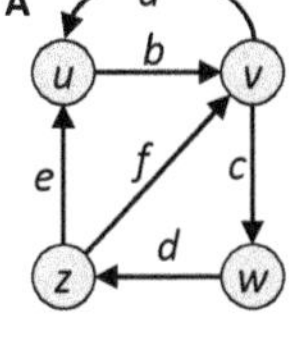

B

$$\begin{aligned}&\text{Digraph } G_2 = (\boldsymbol{V}_2, \boldsymbol{E}_2)\\ &\boldsymbol{V}_2 = \boldsymbol{N}(G_2) = \{u, v, w, z\}\\ &\boldsymbol{E}_2 = \boldsymbol{A}(G_2) = \{(u, v), (v, u), (v, w), (w, z), (z, u), (z, v)\}\\ &\quad = \{a, b, c, d, e, f\}\end{aligned}$$

C

$$\begin{array}{c} \\ u \\ v \\ w \\ z \end{array} \begin{array}{c} \begin{array}{cccccc} a & b & c & d & e & f \end{array} \\ \left[\begin{array}{cccccc} -1 & +1 & & & +1 & \\ +1 & -1 & -1 & 1 & & +1 \\ & & +1 & -1 & & \\ & & & +1 & -1 & -1 \end{array}\right] \end{array}$$

Figure 2.2 Digraph representation. (A) Example digraph G_2. (B) Node and arc sets of G_2. (C) Node-to-arc incidence matrix of G_2.

- A *walk* of length k is a sequence $v_0, e_1, v_1, e_2, \ldots, e_k, v_k$ with $e_k = \{v_{k-1}, v_k\}$ for all i; v_0 and v_k are its endpoints.
- A *trail* is a walk with no repeated edges.
- A *path* is a walk with no repeated vertices.
- A u, w-walk has $v_0 = u$ and $v_k = w$.
- A walk (or trail) is *closed* if it has length $k \geq 1$ and $v_0 = v_k$.
- A *cycle* is a closed trail with $v_0 = v_k$ being the only repeated vertex.

Finally, we define the following:

- A graph is *connected* if there is a path in G between every pair of vertices.
- A graph is *complete* (clique) if it has all edges.
- A graph is *bipartite* if $V(G) = V_1 \cup V_2$ (with $V_1 \cap V_2 = \emptyset$) and there are no edges with both ends on the same subset.
- The degree of vertex v in simple graph G, denoted by $d(v)$, is the number of edges incident with v.
- In digraphs, we have in-degree, $d^-(v)$, which is the number of arcs with v as their head, and out-degree, $d^+(v)$, which is the number of arcs with v as their tail.

2.2 Modeling with Binary Variables

In Section 2.2.1, we present how to model some basic logic conditions using binary variables. In Section 2.2.2, we discuss how to use binary variables to develop piecewise linear approximations of nonlinear functions, and in Section 2.2.3 we discuss the modeling of disjunctions.

2.2.1 Logic Conditions

In many problems, the decision maker has to select one (or more) among multiple alternatives (items) $i \in \mathbf{I}$; for example, a batch has to be assigned to exactly one unit. If $x_i \in \{0, 1\}$ is equal to 1 if item i is selected, then the restriction to select exactly one item can be enforced by

$$\sum_i x_i = 1. \tag{2.4}$$

If at most one item has to be selected, then the equality should be replaced with $\leq$. Assuming that $\mathbf{I}$ is ordered, the set of variables x_i is also called *special ordered set of type 1* (SOS1). Since the selection of up to one item is a condition that arises in many problems, commercial solvers have built-in capability to model and exploit it during branching. Specifically, the user can define variables x_i as SOS1 variables (instead of binary variables) and write (2.4). Note that if the condition to be enforced was $\sum_i x_i = 5.5$ and only one variable x_i can be positive, then declaring x_i variables as SOS1would suffice.

Another common logic condition is that if an item i is selected, then item j should also be selected, that is, if $x_i = 1$, then $x_j = 1$, but if $x_i = 0$, then x_j can be anything. This condition is enforced via

$$x_j \geq x_i. \tag{2.5}$$

If there is equivalence between items i and j (i.e., if i is selected, then j should be selected, and if i is not selected, then j should not be selected), then we have

$$x_i = x_j. \tag{2.6}$$

Binary variables can be used to model discontinuous functions, which, as we will see in subsequent chapters, are common in chemical process operations. For example, the size of a batch should be between a lower and upper bound if the corresponding batch is executed (i.e., a binary is equal to 1) and 0 otherwise. In the simplest case, for parameters x^L/x^U,

$$z = \begin{cases} 0, & \text{if } x = 0 \\ a + bx, & \text{if } x^L \leq x \leq x^U \end{cases} \tag{2.7}$$

can be modeled using $y \in \{0, 1\}$, which is 1 if and only if x is positive,

$$z = ay + bx \tag{2.8}$$

$$x^L y \leq x \leq x^U y. \tag{2.9}$$

Finally, binary variables can be used to *cut off* previously found solutions, thereby allowing us to find the *next-best* solution to a MIP instance. Consider a MIP model with binary variables $x_j, j \in \mathbf{J}^\mathrm{I}$. If $\{\mathbf{B},\mathbf{N}\}$ is a partition of $\mathbf{J}^\mathrm{I}$ (i.e., $\mathbf{B} \cup \mathbf{N} = \mathbf{J}^\mathrm{I}$ and $\mathbf{B} \cap \mathbf{N} = \emptyset$) and, in the current solution, $x_j = 1$ if $j \in \mathbf{B}$ and $x_j = 0$ if $j \in \mathbf{N}$, then the current x_j solution can be cut off via the following *no-good cut*,

$$\sum_{j \in \mathbf{B}} x_j + \sum_{j \in \mathbf{N}} (1 - x_j) \leq |\mathbf{J}^\mathrm{I}| - 1, \tag{2.10}$$

which, since $|\mathbf{J}^\mathrm{I}| - |\mathbf{N}| = |\mathbf{B}|$, can be rewritten as

$$\sum_{j \in \mathbf{B}} x_j - \sum_{j \in \mathbf{N}} x_j \leq |\mathbf{B}| - 1. \tag{2.11}$$

No-good cuts are used in iterative decomposition methods where a relaxation of the original problem is solved at the upper level to provide a candidate solution to the lower-level subproblem that checks for feasibility and optimality. No-good cuts can be used to exclude previously found solutions from the feasible region of the upper-level subproblem so that different solutions can be passed to the lower-level subproblem. Example 2.1 illustrates the use of (2.10) and (2.11).

Example 2.1 No-Good Cuts Consider the IP problem

$$\max \left\{ \frac{1}{1}x_1 + \frac{1}{2}x_2 + \frac{1}{3}x_3 + \frac{1}{4}x_4 + \frac{1}{5}x_5 : \sum_{k=1..5} x_k \leq 3 \right\}.$$

The optimal solution is $x_1 = x_2 = x_3 = 1, x_4 = x_5 = 0$, with optimal objective function value $z = 1\frac{5}{6}$. Equation (2.10) results in the following no-good cut:

$$x_1 + x_2 + x_3 + (1 - x_4) + (1 - x_5) \leq 4.$$

Note that if we substitute the current solution into the preceding equation, the left-hand side becomes equal to 5, so this solution is cut off. Equation (2.11) leads to

$$x_1 + x_2 + x_3 - x_4 - x_5 \leq 2,$$

which also cuts off the current solution.

After adding either of the preceding cuts and reoptimizing, we obtain the next-best solution, $x_1 = x_2 = x_4 = 1,\ x_3 = x_5 = 0$, with an objective function value $z = 1\frac{3}{4}$. To cut off the second-best solution, we can add

$$x_1 + x_2 + x_4 + (1 - x_3) + (1 - x_5) \leq 4$$

or

$$x_1 + x_2 + x_4 - x_3 - x_5 \leq 2.$$

2.2.2 Nonlinear Functions

Binary variables can also be used to approximate nonlinear univariate functions (see Figure 2.3). Let $f(x)$ be the function to be approximated, $x_k, k \in \mathbf{K} = \{0, 1, .. K\}$ the points used for the approximation, and $y_k = f(x_k)$. If we define weights $w_k \in [0, 1]$, then x can be written as

$$x = \sum_k x_k w_k \tag{2.12}$$

and $f(x)$ piecewise linearly approximated by[1]

$$y = \sum_k y_k w_k, \tag{2.13}$$

where

$$\sum_k w_k = 1. \tag{2.14}$$

The condition that up to two w_k can be positive, and that their indices should be adjacent is enforced using binary variables $v_k, k \in \mathbf{K} = \{1, 2, .., K\}$, through, for example,

$$w_{k-1} + w_k \leq v_k, \quad k \in \mathbf{K} \backslash \{0\} = \{1, \ldots, K\}, \tag{2.15}$$

where

$$\sum_k v_k = 1. \tag{2.16}$$

[1] We use $y_k = f\{x_k\}$, that is, the value of $f(x)$ at x_k, for the approximation. Is this necessary? Can you think of cases where using other points would suffice? Or even be better?

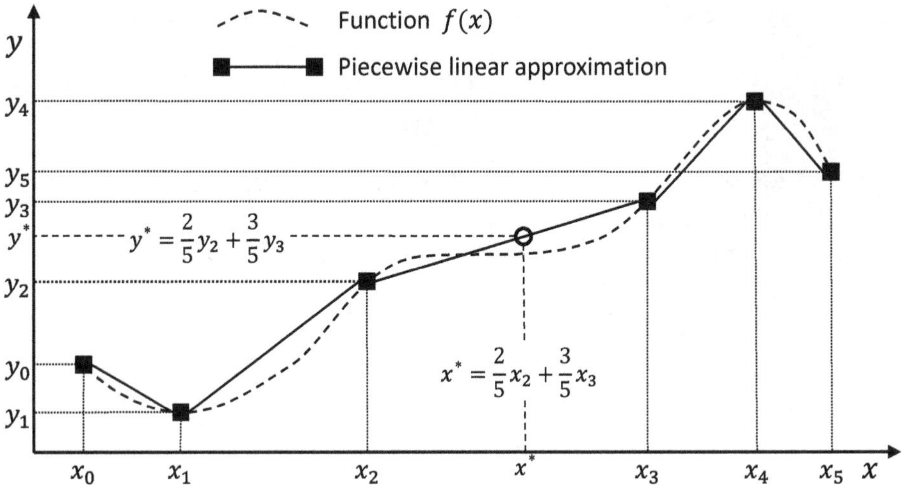

Figure 2.3 Piecewise linear approximation of a nonlinear function; approximation at x^* shown for illustration.

Variable sets that have to satisfy (2.15) and (2.16) are referred to as *special ordered set of type 2* (SOS2). Since the requirement to approximate nonlinear functions using piecewise linear functions is common, modeling languages and commercial solvers allow users to directly define SOS2 variables, which means, in the case of piecewise linear approximation, that it suffices to write (2.12) through (2.14) and define w_k as SOS2 variables. Multivariable functions can also be linearized, at a higher modeling and computational cost, however.

2.2.3 Disjunctions

Finally, binary variables allow the relaxation of constraints and the modeling of disjunctions. If constraint $\sum_i \alpha_i x_i \leq b$ should be enforced only if $y \in \{0, 1\}$ is equal to 1 (i.e., when a logic condition is met), then we can write

$$\sum_i \alpha_i x_i \leq b + M(1 - y), \tag{2.17}$$

where M is a sufficiently large parameter (i.e., the left-hand side [LHS] is smaller than the right-hand side [RHS] for all feasible values of x_i when $y = 0$). Constraints that have the form of (2.17) are often called *big-M* constraints. If $\sum_i \alpha_i x_i = b$ should be enforced when $y = 1$, then we write

$$b - M(1 - y) \leq \sum_i \alpha_i x_i \leq b + M(1 - y). \tag{2.18}$$

More generally, if a set of constraints, $A^1 x \leq b^1$, should be satisfied when Boolean $Y = True$ and another set of constraints, $A^2 x \leq b^2$, should be satisfied when $Y = False$,

$$\begin{bmatrix} Y = True \\ A^1 x \leq b^1 \end{bmatrix} \vee \begin{bmatrix} Y = False \\ A^2 x \leq b^2 \end{bmatrix}, \tag{2.19}$$

then the MIP reformulation consists of the following constraints:

$$A^1 x \leq b^1 + M^1(1 - y) \tag{2.20}$$

$$A^2 x \leq b^2 + M^2 y, \tag{2.21}$$

where $y \in \{0, 1\}$ is equal to 1 if and only if $Y = True$; and M^1 and M^2 are vectors of sufficiently large parameters.

Alternatively, variables x can be dissagregated,

$$x = x^1 + x^2, \tag{2.22}$$

subject to (where $M^{1'}$ and $M^{2'}$ are parameter vectors),

$$x^1 \leq M^{1'} y, \;\; x^2 \leq M^{2'}(1 - y) \tag{2.23}$$

and used to enforce

$$A^1 x^1 \leq b^1 y \tag{2.24}$$

$$A^2 x^2 \leq b^2(1 - y). \tag{2.25}$$

2.3 Basic Integer Programming Problems

In Section 2.3.1, we discuss the *knapsack problem*, which illustrates how binary variables are used to model *selection*. In Section 2.3.2, we present the *assignment* problem, which illustrates, well, how assignment decisions are modeled. The famous *traveling salesman* problem, which is similar but yet so different from the assignment problem, is presented in Section 2.3.3. The *set covering* problem, which illustrates selection subject to suitability constraints is discussed in Section 2.3.4. In Section 2.3.5, we introduce *production planning*, a problem that illustrates multiperiod decision making and will be further discussed in Chapter 15. *Facility location*, another SC problem, is discussed in Section 2.3.6. Finally, in Section 2.3.7, we introduce some basic concepts on networks. In the present section, we use lowercase letters for parameters and uppercase letters for variables.

2.3.1 Knapsack

We are given n items to be packed in a knapsack with capacity w. Each item $i \in \mathbf{I} = \{1, 2, \ldots, n\}$ has value p_i and weight w_i. The goal is to choose the items that fit in the knapsack with the max value.

We introduce $X_i \in \{0, 1\}$ with $X_i = 1$ if item i is chosen. The objective function is

$$\max \sum\nolimits_i p_i X_i \tag{2.26}$$

and the capacity constraint can be written as follows:

$$\sum\nolimits_i w_i X_i \leq w. \tag{2.27}$$

Equation (2.27) is called a *knapsack* constraint or simply a *knapsack*. If there are two constraints (e.g., weight and volume), then two *knapsacks* are required. In general, if there are m resources $j \in \mathbf{J} = \{1, 2, \ldots, m\}$ with capacities w_j, we replace (2.27) with[2]

$$\sum\nolimits_i w_{ij} X_i \leq w_j, \;\; j. \tag{2.28}$$

2.3.2 Assignment

We are given n tasks (e.g., classes, routes), $i \in \mathbf{I} = \{1, 2, \ldots, n\}$; and m resources (e.g., instructors, buses), $j \in \mathbf{J} = \{1, 2, \ldots, n\}$. Each task has to be assigned to a resource, and each resource can carry out only one task. The assignment cost is c_{ij}. The objective is to find the assignment with the minimum total cost.

We define $X_{ij} \in \{0, 1\}$, which is equal to 1 if task i is assigned to resource j. The objective function is

$$\min \sum\nolimits_{i,j} c_{ij} X_{ij}. \tag{2.29}$$

Equation (2.30) enforces that every task is assigned to one resource, while (2.31) enforces that each resource carries out exactly one task.

$$\sum\nolimits_j X_{ij} = 1, \;\; i \tag{2.30}$$

$$\sum\nolimits_i X_{ij} = 1, \;\; j \tag{2.31}$$

Interestingly, if the model is solved as LP (i.e., constraints $X_{ij} \in \{0, 1\}$ are replaced by $X_{ij} \in [0, 1]$), then there is always an optimal solution with integral values.[3]

2.3.3 Traveling Salesman

There are n cities to be visited exactly once by a single traveling salesman. The distance/cost from city i to city j is c_{ij}, with, in general, $c_{ij} \neq c_{ji}$. The objective is to find the sequence of cities that yield the minimum total distance/cost.

We introduce $X_{ij} \in \{0, 1\}$ to represent the selection of arc (i,j) or $i \rightarrow j$. The objective function is then written as

$$\min \sum\nolimits_{i,j} c_{ij} X_{ij}. \tag{2.32}$$

[2] Throughout the book, equation domains will be typically expressed using the corresponding indexes only, that is, the more rigorous $\forall j \in \mathbf{J}$ will be replaced by, simply, j. Subset membership will be used when the equation is expressed for a subset; for example, (2.15) should not be written for $k = 0$ and thus was defined for $k \in \mathbf{K} \backslash \{0\}$.

[3] The assignment problem was studied extensively well before the advent of math programming; e.g., Carl Gustav Jacobi (1890) studied it in the nineteenth century. It was solved effectively by the Hungarian method (Kuhn, 1955) and improved by Munkres (1957).

The condition that each city is visited once can be enforced by requiring that only one arc from another city, (2.33), and to another city, (2.34), should be selected:

$$\sum_i X_{ij} = 1, \ j \tag{2.33}$$

$$\sum_j X_{ij} = 1, \ i. \tag{2.34}$$

Have we seen these two constraints before? Are they sufficient to define a feasible route for the salesman? What type of solution can we get? Answers to these questions are given in Example 2.2.

If $c_{ij} \neq c_{ji}$, then we have an *asymmetric* traveling salesman problem (TSP), which is defined using a digraph, with separate $i \to j$ and $j \to i$ arcs. If $c_{ij} = c_{ji}$ for all i, j, then we have a *symmetric* TSP, which can be defined on a graph using edge variables $X_e \in \{0,1\}, e \in \boldsymbol{E}(G)$. The objective function becomes

$$\min \sum_{e \in \boldsymbol{E}(G)} c_e X_e \tag{2.35}$$

and the condition that each city is visited once can be enforced by

$$\sum_{e \in \boldsymbol{E}(v)} X_e = 2, \quad v \in \boldsymbol{V}(G), \tag{2.36}$$

where $\boldsymbol{E}(v)$ is the set of edges incident to vertex v.

Example 2.2 Traveling Salesman Problem We are given a six-city symmetric TSP and the corresponding costs (see Figure 2.4A). If we solve the IP model that consists of (2.35) and (2.36), we obtain the solution shown in Figure 2.4B, which consists of two *subtours* (closed paths) $\{(u,v),(v,w),(w,u)\}$ and $\{(x,y),(y,z),(z,x)\}$. Solutions with subtours are infeasible because they require multiple salesmen, so additional constraints are necessary to eliminate subtours, that is, (2.36) (and (2.33) and (2.34) for the asymmetric TSP) are not sufficient. How can we systematically generate all such *subtour elimination* constraints? It turns out that this is a daunting task, so we consider an easier question: how can we eliminate (cut off) the solution shown in Figure 2.4B? One option would be to add a no-good cut, but this would

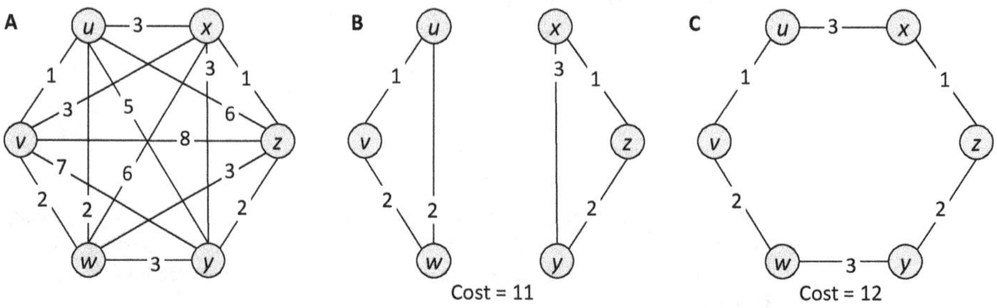

Figure 2.4 Traveling salesman problem. (A) six-city instance with associated costs. (B) Best solution, with two subtours, obtained if no subtour elimination constraints are used. (C) Optimal TSP (single-tour) solution.

require writing a constraint involving 15 variables. (Can you see why?) A more problem-specific approach would be to write constraints that eliminate these specific subtours from any solution:

$$X_{uv} + X_{uw} + X_{vw} \leq 2$$

and

$$X_{xy} + X_{xz} + X_{yz} \leq 2.$$

where we assume that the tail of the single edge between two vertices is the one denoted by the letter that comes first in the alphabet (e.g., $(u, v) \in \boldsymbol{E}(G)$ and $(v, u) \notin \boldsymbol{E}(G)$). The optimal solution is shown in Figure 2.4D. Can you obtain this solution by iteratively solving instances with subtour elimination constraints? How many iterations are required?

2.3.4 Set Covering

We are given n tasks, $i \in \mathbf{I} = \{1, 2, \ldots, n\}$ to be *covered* by m resources, $j \in \mathbf{J} = \{1, 2, \ldots, m\}$, with $m < n$. Each resource j can carry out a subset of tasks $\mathbf{I}_j \subseteq \mathbf{I}$. The task- resource compatibility is represented via parameter $a_{ij} = 1$ if $i \in \mathbf{I}_j$. The cost of a resource is c_j. The objective is to find the set of resources that *covers* all tasks and has the minimum cost.

We introduce $X_j \in \{0, 1\}$ to represent the selection of resource j. The objective function is

$$\min \sum\nolimits_j c_j X_j. \tag{2.37}$$

The constraint that all tasks be covered means that we should select at least one resource with $a_{ij} = 1$, which can be written as follows:

$$\sum\nolimits_j a_{ij} X_j \geq 1, \quad i. \tag{2.38}$$

2.3.5 Production Planning

We are given (1) a planning horizon divided into a set of time periods, $t \in \mathbf{T} = \{1, 2, .., T\}$; (2) a set of products (items), $\mathbf{I}$, with holding cost h_i [\$/(kg·period)], and demand d_{it} [kg], due at the end of time period t; and (3) a resource with production cost c_i [\$/kg], production rate r_i [kg/h], and availability η [h].[4]

[4] The problem we consider is one variant of the famous *lot sizing* problem, which is the backbone of all production planning systems, and has been studied extensively in the literature. Numerous reformulation results and decomposition methods (e.g., column generation, Lagrangian relaxation/decomposition) have been proposed to improve the solution of lot sizing–based production planning problems. To obtain more accurate production targets, the aforementioned formulations have also been extended to include overtime, product substitutes, productivity, and capacity utilization.

We introduce variable $P_{it} \geq 0$ to denote the production of item i during period t, and variable $I_{it} \geq 0$ to denote the inventory of item i during period t. The objective function includes production and inventory costs,

$$\min \sum\nolimits_{i,t} (c_i P_{it} + h_i I_{it}). \tag{2.39}$$

The material balance for item i is written as

$$I_{i,t+1} = I_{it} + P_{it} - d_{it}, \quad i, t, \tag{2.40}$$

where we assume that all the production during period t becomes inventory at the end of the period. The resource capacity constraint is

$$\sum\nolimits_i (1/r_i) P_{it} \leq \eta, \quad t, \tag{2.41}$$

where $(1/r_j)P_{it}$ represents the time allocated to production of item i during period t, and the LHS represents the total production time.

If a setup time σ_i is required if item i is produced in a period and this setup incurs a cost γ_i, then we introduce $X_i \in \{0, 1\}$ to denote a setup for item i during period t. The objective function becomes

$$\min \sum\nolimits_{i,t} (c_i P_{it} + \gamma_i X_{it} + h_i I_{it}), \tag{2.42}$$

the capacity constraint includes setup time,

$$\sum\nolimits_i \left[\left(\frac{1}{r_i} \right) P_{it} + \sigma_i X_{it} \right] \leq \eta, \quad t, \tag{2.43}$$

and the following constraint is required for the activation of the setup binary variable,

$$P_{it} \leq \left(\frac{\eta - \sigma_i}{r_i} \right) X_{it}, \quad i, t, \tag{2.44}$$

where $(\eta - \sigma_i)$ is the maximum time that can be allocated to the production of item i, and thus the ratio in the RHS represents the maximum production of i.

Finally, if unmet demand can be backlogged and met later, at a penalty, β_i [\$/(kg·period)], then we introduce new variables to model backlog, $B_{it} \geq 0$, and shipments, $S_{it} \geq 0$; and modify the objective function

$$\min \sum\nolimits_{i,t} (c_i P_{it} + \gamma_i X_{it} + h_i I_{it} + \beta_i B_{it}), \tag{2.45}$$

and the material balance,

$$I_{i,t+1} = I_{it} + P_{lt} - S_{it}, \quad i, t, \tag{2.46}$$

where shipments and demand are used to calculate backlog as follows:

$$B_{i,t+1} = B_{it} + d_{it} - S_{it}, \quad i, t. \tag{2.47}$$

Note that if we use (2.47) to solve for S_{it}

$$S_{it} = B_{it} - B_{i,t+1} + d_{it}$$

then (2.46) can be rewritten as follows:

$$I_{i,t+1} = I_{it} + P_{it} - (B_{it} - B_{i,t+1} + d_{it}), \quad i, t$$

or

$$(I_{i,t+1} - B_{i,t+1}) = (I_{it} - B_{it}) + P_{it} - d_{it}, \quad i, t. \tag{2.48}$$

Note that if we define a generalized inventory, $\hat{I}_{it} = I_{it} - B_{it}$, which can be negative, to represent backlog, then (2.48) can be written as

$$\hat{I}_{i,t+1} = \hat{I}_{it} + P_{it} - d_{it}, \quad i, t,$$

which has the same form as (2.40) (but cannot be directly used for the calculation of backlog penalties).

2.3.6 Facility Location

Given are n, existing and new, facilities (plants) and m customers. We want to decide what new facilities to build, and how much to produce and ship from each facility to satisfy customer demand. The capacity and cost of setting up facility i is a_i [kg] and f_i [$], respectively. The cost of shipping material from i to j is c_{ij} [$/kg]. The demand of customer $j \in \{1, 2, \ldots, m\}$ is d_j.

We introduce $X_i \in \{0, 1\}$ to denote the installation of facility i and $Y_{ij} \in \mathbb{R}_+$ to denote the amount shipped [kg] from facility i to customer j. The objective function includes the (fixed) facility installation cost plus the (variable) shipping cost:

$$\min \sum_i f_i X_i + \sum_{i,j} c_{ij} Y_{ij}. \tag{2.49}$$

Demand satisfaction can be enforced via

$$\sum_i Y_{ij} = d_j, \quad j. \tag{2.50}$$

If no other constraint is added, then the optimization will yield solutions where material is produced and shipped from a facility that is not installed, so the fixed cost is not paid (can you see why?). To avoid this, we need a constraint enforcing that if $Y_{ij} > 0$, for any j, then $X_i = 1$, which can be accomplished via the following facility capacity constraint (have we seen this constraint type before?):

$$\sum_j Y_{ij} \le a_i X_i, \quad i. \tag{2.51}$$

In other, more realistic variants of the problem, there can be additional features such as production cost, p_i, transportation setup cost, t_{ij}, and transportation capacity s_{ij}. If we introduce $Z_{ij} \in \{0, 1\}$ to model the setup of transportation arc ($i \rightarrow j$), then the objective function becomes

$$\min \sum_i f_i X_i + \sum_i p_i \sum_j Y_{ij} + \sum_{i,j} t_{ij} Z_{ij} + \sum_{i,j} c_{ij} Y_{ij} \tag{2.52}$$

and the following constraint is added to enforce the arc capacity constraint and activate the transportation setup cost:

$$Y_{ij} \leq s_{ij} Z_{ij}, \quad i,j. \tag{2.53}$$

Finally, the problem can be extended to involve multiple periods, $t \in \mathbf{T}$, and multiple products, $k \in \mathbf{K}$.

2.3.7 Network Problems

The three fundamental network problems are the following:

(1) *Shortest path*: Given a network with arc costs (distances), find the shortest path from source node s to all other nodes; it has applications in computer data transmission, vehicle routing, and solution of difference equations.
(2) *Maximum flow*: Given a single-source, single-sink network with arc capacities, find the maximum total flow from source s to sink t; it has applications in uniform parallel machine scheduling and distributed computing on multiprocessor machines.
(3) *Minimum cost flow*: Given a network with arc capacities and costs, find the cheapest way to send a fixed amount of flow from single source s to single sink t; with applications in distribution problems, optimal loading of airplanes, and reconstruction of three-dimensional (3-D) shape from X-ray projections

The mathematical formulation for the solution of problems expressed on a network N often has the form $\{\min c^T x : Ax = b, l \leq x \leq u\}$, where A is the node-to-arc incidence matrix of N. In other words, the matrix used to represent N is also used to express mathematical programming models for problems on N. For example, the minimum cost problem on a network can be expressed as the following LP:

$$\min \sum\nolimits_{i,j} c_{ij} X_{ij} \tag{2.54}$$

$$\sum\nolimits_j X_{ji} - \sum\nolimits_j X_{ij} = b_i, \quad i \tag{2.55}$$

$$0 \leq X_{ij} \leq u_{ij}, \quad i,j, \tag{2.56}$$

where X_{ij} is the flow along arc (i,j); c_{ij} is the unit cost of flow along arc (i,j); b_i is the *demand* of node i, with, typically, $b_s = -f$, $b_t = f$, $b_i = 0$, $i \in \mathbf{I} \backslash \{s,t\}$, where f is the *flow* of the network; and u_{ij} is the capacity of arc (i,j) with $u_{ij} = 0$ if $(i,j) \notin \mathbf{A}(N)$. Note that (2.55) and (2.56) are in the form $Ax = b$ and $l \leq x \leq u$, respectively.

Networks are very important from an optimization standpoint because many problems can be represented as networks and because there are algorithms that address network problems very effectively. Furthermore, the node-to-arc incidence matrix is a special class of the so-called *network matrices* that lead to integral polyhedra; that is, the feasible region $\{x: Ax = b, l \leq x \leq u\}$ has integral vertices if b, l and u are integral.

An interesting extension of the problem in (2.54) through (2.56) is the problem with *generalized flows* when the flow is not preserved along a path. If one unit of flow

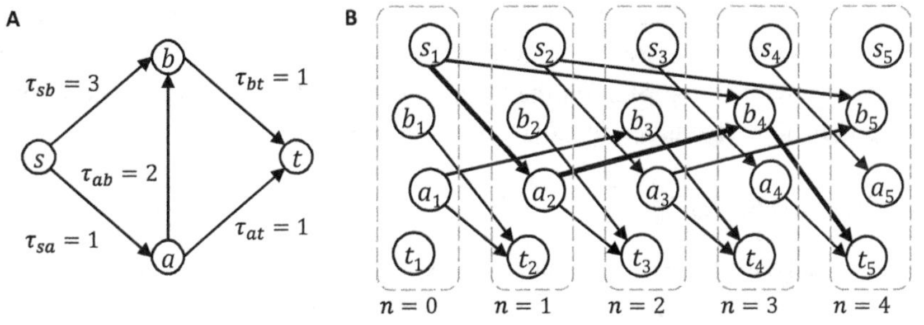

Figure 2.5 Transformation of dynamic network into a time-expanded network. (A) Dynamic network N^D with time delays τ_{ij} shown. (B) Equivalent time-expanded network N^D; path s → a → b → t starting at time 0 shown with bold arcs.

leaving node j through arc (j, i) is converted into μ_{ji} units of flow entering node i, then the flow balance constraint is written as follows:

$$\sum_j \mu_{ji} X_{ji} - \sum_j X_{ij} = b_i, \quad i. \tag{2.57}$$

Note that generalized flows basically introduce multipliers into (2.55). Interestingly, the methods used to address *standard* network problems can be extended to address problems in generalized networks.

Finally, while in many cases the underlying process is *static*, many applications, including production scheduling, require the introduction of the *temporal* dimension. To address this aspect, we use *dynamic* networks, which, in addition to arc capacities, also have arc traversal times, τ_{ij} (delays). A dynamic network N^D can be converted to a *time-expanded* static network N^{TE}, which allows us to use all the methods developed for static networks. Let η be the time horizon we consider, and u_{ij} and τ_{ij} (integer) be the capacity and traversal time of arc (i, j) in N^D. The construction of N^{TE} has the following two steps:

(1) For each node v of N^D, introduce $\eta + 1$ nodes, $v_0, v_1, v_2, \ldots, v_\eta$ in N^{TE}, where node v_n represents node v at time n.

(2) Introduce an arc (v_n, w_l) in N^{TE}, with capacity u_{vw} if (v, w) is an arc in N^D and $\tau_{vw} = l - n$.

A flow along arc (v_n, w_l) in N^{TE} corresponds to a flow from v to w in N^D. The construction of a time-expanded network is illustrated in Figure 2.5.

2.4 Solution Methods

While the discussion of the details of the algorithms used to solve MIP problems is beyond the scope of this book, in this section we give an overview of the two fundamental solution approaches, the branch and bound algorithm in Section 2.4.1 and cutting planes in Section 2.4.2, so that the reader gains an appreciation of the computational issues. Then, to motivate some of the discussion in the subsequent chapters, where alternative MIP formulations are presented for the same problem, we overview the importance of reformulations. Finally, we close this section with an overview of decomposition methods, which will be discussed again, in the context of production scheduling, in Chapters 12 and 13.

2.4.1 Branch-and-Bound Algorithm

The standard (LP-based) branch-and-bound (B&B) algorithm relies on the solution of the linear relaxation of the original MIP problem in (2.3),

$$\begin{aligned} \max \quad & c^T x + d^T y \\ \text{s.t} \quad & Ax + By \leq b \\ & y \in [0,1]^{n_1}, x \in \mathbb{R}_+^{n_2}, \end{aligned} \tag{2.58}$$

where integrality constraints $y \in \{0,1\}^{n_1}$ have been replaced with $y \in [0,1]^{n_1}$. Broadly speaking, the algorithm starts with the solution of (2.58) at the *root node* (node number = 0) to obtain a solution where (at least some of the) binary variables assume fractional values. Next, a binary variable, y_i, with fractional value, y_i^0, at the LP solution is chosen for branching, which leads to two new subproblems added to the list of active subproblems: subproblem 1 generated by adding $y_i \leq 0$ and subproblem 2 by adding $y_i \geq 1$ to the model in (2.58). The problem at the root node is then removed from the list of active subproblems; one of the two new subproblems is selected to be solved next, to yield another fractional solution; and the process is repeated. If a feasible solution, for problem (2.3), is found, then it serves as a lower bound, z^L, on the optimal solution, z^*, of (2.3); that is, if the objective function value of the subproblem in node k, z^k, is smaller than z^L, then subproblem k can be removed from the list of active subproblems (i.e., node k can be *pruned*). The best solution among the active subproblems provides an upper bound, z^U, on z^*. The algorithm terminates when the gap between $z^L$and z^U is closed. Example 2.3 illustrates how B&B is used to solve a simple problem.

Example 2.3 Branch-and-Bound Algorithm Consider the following problem

$$\begin{aligned} \max \quad & 3x_1 + 4x_2 \\ \text{s.t.} \quad & 4x_1 + 3x_2 \leq 10 \\ & x_1 \in \{0,1,2,3\}, x_2 \in \{0,1\}, \end{aligned}$$

where we consider integer variable x_1 to illustrate multiple levels of branching in two dimensions. The feasible integer points, the feasible region of the LP relaxation, and contours of the objective function are shown in Figure 2.6A. The branch and bound tree for the solution of the problem is shown in Figure 2.6B. The solution at the root node $(k = 0)$ is $(x_1^0, x_2^0) = (1.75, 1)$ with $z^0 = 9.25$. Since x_1 is fractional, we branch on it; that is, we generate subproblem (node) 1 by adding $x_1 \leq 1$ and node 2 by adding $x_1 \geq 2$, and discard node 0. The solution at node 1 is $(x_1^1, x_2^1) = (1,1)$, which is integer and thus provides a lower bound (feasible solution) with $z^L = z^1 = 7.0$. The solution at node 2 is $(x_1^2, x_2^2) = (2, 2/3)$ with $z^2 = 8\frac{2}{3}$, which is better than the current lower bound $(z^L = 7.0)$, so we continue branching on fractional variable x_2: we generate node 3 by adding $x_2 \leq 0$ and node 4 by adding $x_2 \geq 1$, and discard node 2. The solution at node 3 is

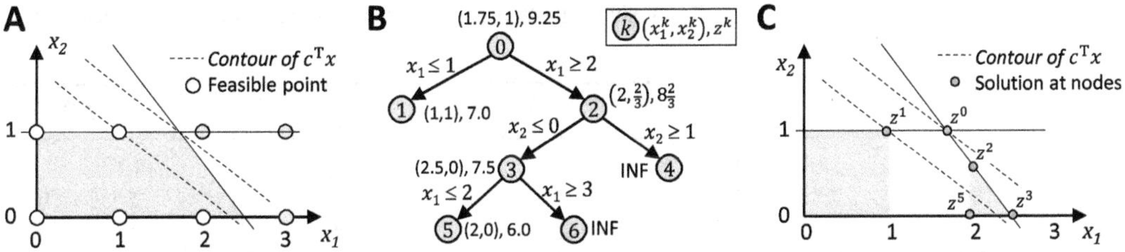

Figure 2.6 Illustration of branch-and-bound algorithm. (A) Integer feasible points, feasible region of LP-relaxation, and objective function contours. (B) Branch-and-bound tree: node number given inside the node along with solution (x_1^k, x_2^k) and objective function value, z^k, outside the node (INF denotes infeasible node); constraints used for branching are given next to arrow representing branching. (C) Representation of the five solutions found during the branch-and-bound search and corresponding objective function values.

$(x_1^3, x_2^3) = (2.5,0)$ with $z^3 = 7.5$, which is better than the current lower bound (7.0), so further branching will be needed. Node 4 is infeasible, so it is pruned. Back to node 3, we branch on x_1: we generate node 5 by adding $x_1 \le 2$ and node 6 by adding $x_1 \ge 3$. The solution of node 5 is $(x_1^5, x_2^5) = (2, 0)$, which is integer feasible, but with $z^5 = 6.0$, which is lower than the current best solution (lower bound) $z^L = z^1 = 7.0$, so node 5 is pruned. Finally, node 6 is infeasible, so it is also pruned. At this point, there are no active nodes, so the optimal solution is the one found at node 1: $(x_1^*, x_2^*) = (1, 1)$ with $z^* = 7.0$.

We note the following:

- The number of binary variables fixed indicates the level of a node in the B&B tree; for example, at the root node (node 0, level 0), no binaries are fixed, while at level 1, after only one branching, one binary is fixed.
- The number of all nodes that can potentially be explored at level $k \in \{0, 1, \ldots, n_1\}$ is 2^k, which means that the maximum number of nodes in a B&B tree is $N^{MAX} = 2^{n_1+1} - 1$ (why is this so?).
- If all binary variables need to be fixed to obtain an integer solution, then solutions will be obtained at nodes at level n_1, which are also called *leaf* nodes.

Example 2.4 illustrates how the computational requirements increase with the number of binary variables.

Example 2.4 Computational Requirements We consider MIP model $(\mathbb{M})$ with $n_1 = 100$ binary variables, and we assume that each LP (at each node) requires 10^{-3} CPU second to be solved. We further assume that we have three solvers available:

(1) Solver 1: explores all combinations of values for the binary variables (brute force).
(2) Solver 2: needs to explore only one out of 10^6 nodes; that is, if the maximum number of nodes in the tree is 10^6 $\left(2^{k+1} - 1 \approx 2^{k+1} = 10^6\right)$, then the solver would find a solution after solving a single LP.
(3) Solver 3: needs to explore only one out of 10^9 nodes;

How long will it take it to solve $(\mathbb{M})$?

Table 2.2 Computational statistics of different instances of model (M) using Solver 3 and a cluster with 1,000 cores.

n_1	N^{MAX}	N^{S3}	T^{CPU}	T^{WC}
20	$2.097 \cdot 10^6$	1	10^{-3}	10^{-6}
100	$2.535 \cdot 10^{30}$	$2.535 \cdot 10^{21}$	$2.535 \cdot 10^{18}$	$2.535 \cdot 10^{15}$
1,000	$2.143 \cdot 10^{301}$	$2.143 \cdot 10^{292}$	$2.143 \cdot 10^{283}$	$2.143 \cdot 10^{274}$

Note: n_1 = number of binary variables, N^{MAX} = maximum number of nodes, N^{S3} = number of nodes explored, T^{CPU} = CPU time in seconds, T^{WC} = wall-clock time in seconds.

The maximum number of nodes in the tree is $N^{MAX} = 10^{100+1} - 1 \approx 2.53 \cdot 10^{30}$. Solver 1 will require $2.53 \cdot 10^{30} \cdot 10^{-3} = 2.53 \cdot 10^{27}$ seconds; Solver 2 will require $2.53 \cdot 10^{30} \cdot 10^{-3} \cdot 10^{-6} = 2.53 \cdot 10^{21}$ seconds; and Solver 3 will require $2.53 \cdot 10^{30} \cdot 10^{-3} \cdot 10^{-9} = 2.53 \cdot 10^{18}$ seconds. Let's further assume that we have a cluster with 1,000 cores and that we can perfectly parallelize Solver 3. The real-time requirements will then be $2.53 \cdot 10^{18} \cdot 10^{-3} = 2.53 \cdot 10^{15}$, which is approximately equal to $8 \cdot 10^7$ years! Table 2.2 illustrates how computational requirements explode with the number of binary variables.

So, does this mean that problems of this size cannot be solved to optimality? Fortunately, modern MIP solvers can do much better than solve one out of 10^9 nodes. Specifically, the enhancement, in terms of the fraction of nodes explored over N^{MAX}, is a function of problem size, that is, the ratio typically decreases as problem size (n_1) increases. Nevertheless, despite major improvements in computational hardware and software, the solution of many MIP models remains hard. As we will discuss in the following subsections, one way to reduce the computational requirements is through the formulation of better (tighter) models.

2.4.2 Cutting Planes

The main idea is to add new constraints, $A^*x + B^*y \leq b^*$, to the formulation so that the LP-relaxation, now given by (2.59), is tighter than (2.58) without cutting off any integer solutions of the original problem, (2.3).

$$
\begin{array}{ll}
\max & c^T x + d^T y \\
\text{s.t} & Ax + By \leq b \\
 & A^* x + B^* y \leq b^* \\
 & y \in [0,1]^{n_1}, x \in \mathbb{R}_+^{n_2}.
\end{array}
\tag{2.59}
$$

Adding constraints makes the LP-relaxation larger, and thus potentially slower to solve, but using a tighter relaxation leads to a smaller B&B tree. Cutting planes can be generated before the B&B search is initiated, during what is called preprocessing, but can also be generated during the B&B search exploiting information regarding the optimal solution(s) of the LP-relaxation of the model. The generation of cutting planes is illustrated through Examples 2.5 to 2.8.

Example 2.5 Impact of Tightening on Branch-and-Bound Search We consider the problem we studied in Example 2.3:

$$\begin{aligned} \max \quad & 3x_1 + 4x_2 \\ \text{s.t.} \quad & 4x_1 + 3x_2 \leq 10 \\ & x_1 \in \{0,1,2,3\}, \; x_2 \in \{0,1\}. \end{aligned}$$

Constraint $4x_1 + 3x_2 \leq 10$ implies $x_1 \leq 2.5$, and since x_1 is integer we can add $x_1 \leq 2$ without removing any integer feasible solution. As shown in Figure 2.7, the B&B search based on the modified formulation explores fewer nodes, four instead of six.

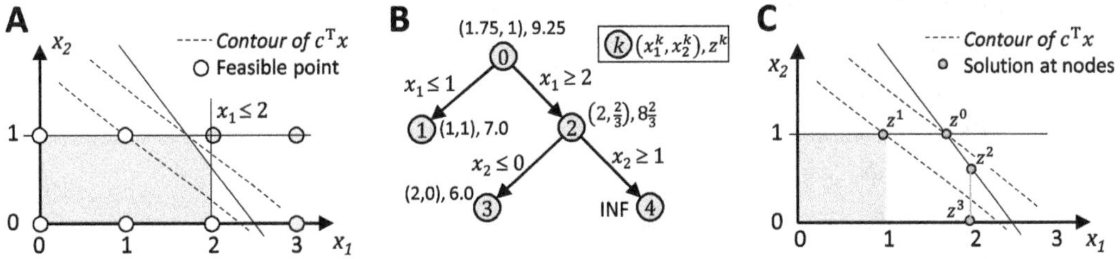

Figure 2.7 Illustration of tightening through cutting plane addition. (A) Integer feasible points, feasible region of LP-relaxation, and objective function contours. (B) Branch-and-bound tree; node number given inside the node along with solution (x_1^k, x_2^k) and objective function value, z^k, outside the node (INF denotes infeasible node); constraints used for branching are given next to the arrow representing branching. (C) Solutions at different nodes.

Example 2.6 Probing Tight cutting planes can also be developed by setting a binary variable to 0 or 1, examining how constraints are modified, and inferring logic conditions based on the modified constraints. This approach, referred to as probing, is typically carried out during preprocessing. To illustrate, consider the following:

$$2x_1 + x_2 + x_3 \geq 1. \tag{2.60}$$

If $x_1 = 1$, then (2.60) becomes strictly redundant, so it can be replaced by

$$x_1 + x_2 + x_3 \geq 1, \tag{2.61}$$

which is tighter. For example, point (0.5, 0, 0) satisfies (2.60) but not (2.61).

Example 2.7 Cover Cuts Consider the following knapsack constraint, where x_1, x_2 and x_3 are binary variables:

$$5x_1 + 5x_2 + 4x_3 \leq 8. \tag{2.62}$$

The sum of the coefficients of x_1 (5) and x_2 (5) is larger than 8, since x_1 and x_2 cannot simultaneously be equal to 1, which implies that the following constraint is valid:

$$x_1 + x_2 \leq 1. \tag{2.63}$$

Similarly, the following inequalities are also valid:

$$x_2 + x_3 \leq 1 \tag{2.64}$$

$$x_1 + x_3 \leq 1. \tag{2.65}$$

Note that the cutting planes in (2.63) through (2.65) cut off points that satisfy (2.62). For example, (2.63) cuts off (1.0, 0.5, 0) and (0.6, 1.0, 0), while (2.64) cuts off (0, 1.0, 0.6) and (0, 0.6, 1.0). However, note the following:

(1) Any one of (2.63) through (2.65) cannot replace (2.62); for example, (1.0, 0.5, 1.0) satisfies (2.63) but not (2.62).
(2) Any combination of two equations cannot replace (2.62); for example, (1.0, 0, 1.0) satisfies (2.63) and (2.64), but does not satisfy (2.62).

Since the sum of any two variable coefficients in (2.62) is larger than 8, the following cutting plane is also valid:

$$x_1 + x_2 + x_3 \leq 1. \tag{2.66}$$

Equation (2.66) is the tightest description of the four feasible integer points satisfying (2.62) (which are the four points?). In fact, along with the no negativity constraints, it defines the convex hull[5] of $F = \left\{x | 5x_1 + 5x_2 + 4x_3 \leq 8, x \in \{0,1\}^3\right\}$, where $x = [x_1, x_2, x_3]^{\mathrm{T}}$, and it can replace (2.62).

In general, if $\mathbf{C}$ is a subset of variables such that $\sum_{i \in \mathbf{C}} w_i$, where w_i is the variable coefficient in the knapsack constraint (see (2.27)), then $\mathbf{C}$ is a *cover* (i.e., covers the knapsack) and the following, the so-called *cover cut*, is a strong valid inequality:

$$\sum_{i \in \mathbf{C}} x_i \leq |\mathbf{C}| - 1. \tag{2.67}$$

Note that (2.63) through (2.66) can be viewed as cover cuts.

Example 2.8 Preprocessing Consider the ith constraint of a MIP model:

$$\sum_j a_{ij} x_j \leq b_i \tag{2.68}$$

If variable coefficients, a_{ij}, are nonnegative, then the following is valid:

$$\sum_j \lfloor a_{ij} \rfloor x_j \leq b_i. \tag{2.69}$$

[5] The convex hull of a set X of points in a Euclidean space is the smallest convex set that contains X.

Further, since the LHS is integer, the following is also valid:

$$\sum_j \lfloor a_{ij} \rfloor x_j \leq \lfloor b_i \rfloor. \tag{2.70}$$

Using this idea, constraints can be replaced by tighter ones. For example, $x_1 + x_2 \leq 1.5$ can be replaced by $x_1 + x_2 \leq 1 = \lfloor 1.5 \rfloor$. However, note that this procedure does not always lead to tighter constraints. For example, from $1.5x_1 + x_2 \leq 2$ we obtain $x_1 + x_2 \leq 2$, which is not tighter.

Modern MIP solvers are very effective in generating cutting planes during preprocessing as well as during the branch-and-bound search. However, based on problem-specific knowledge, modelers are often able to generate effective cutting planes a priori. A potential disadvantage of the a priori addition of cutting planes is that not all of them will be useful, so it may lead to longer solution time for the LP-relaxation. The generation of cutting planes for some scheduling problems will be discussed in Chapter 12.

Cutting planes can also be used, in theory, to solve a MIP model. The idea is that the current solution, x^k, of the LP-relaxation is cut off through the addition of a cutting plane, and a new LP solution, x^{k+1}, is obtained with $z^{k+1} \leq z^k$ for a max problem. The process is repeated until the LP solution is integral. Since the added constraints are valid, the first integral solution of the LP-relaxation augmented by linear valid inequalities is guaranteed to be the optimal solution of the MIP model. While the generation of the cutting planes based on an LP solution is beyond the scope of this book, the idea is illustrated in Figure 2.8. Modern MIP solvers employ a hybrid of branch-and-bound and cutting-plane-based approaches, often referred to as branch-and-cut.

2.4.3 Reformulations

The goal is to generate a tighter formulation via (1) the development of a different problem representation and thus the use of different variables and (tighter) constraints; or (2) the tightening via the reformulation of a subset of constraints (see Example 2.9). To illustrate the main idea, we introduce the following concepts:

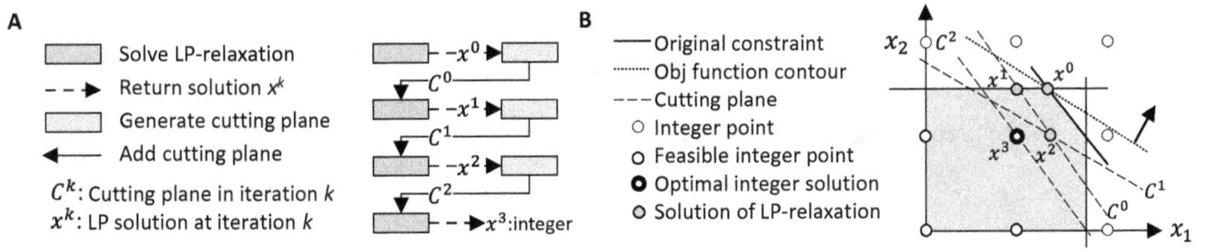

Figure 2.8 Illustration of cutting planes approach to solving a MIP: solution of LP relaxation of original model yields a solution that is used to generate a cutting plane; after adding the cutting plane to the LP-relaxation, the process is repeated. (A) Flow chart of cutting-plane-based algorithm. (B) Geometric interpretation.

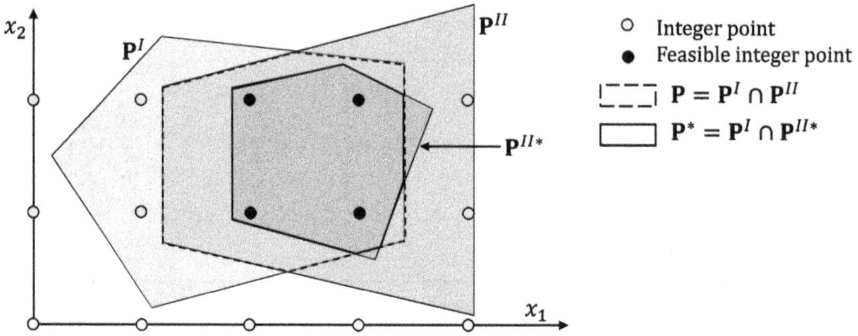

Figure 2.9 Illustration of tightening through reformulation of a subset of constraints. If constraints defining polytope $\mathbf{P}^{II}$ are replaced by constraints yielding polytope $\mathbf{P}^{II^*}$, the resulting formulation, $\mathbf{P}^*$, is tighter. Note that both $\mathbf{P}$ and $\mathbf{P}^*$ contain the same integer points.

- A *polyhedron* $\mathbf{P}$ is the solution set of a finite system of linear inequalities, that is, $\mathbf{P} = \{x | Ax \leq b\}$.
- A polyhedron $\mathbf{P}$ is a (convex) *polytope* (i.e., a bounded polyhedron) if there exist $l, u \in \mathbb{R}^n$ such that $l \leq x \leq u$; that is, $\mathbf{P} = \{x | Ax \leq b, l \leq x \leq u\}$.
- Since a set of inequalities can be converted into a set of equalities by introducing slack variables, and a set of equalities can be converted into a set of inequalities, the definitions of polyhedra and polytopes hold also for systems of linear equalities.
- A (convex) polytope can also be described as the convex hull of a finite number of points. Conversely, the convex hull of a finite number of points is a polytope.
- A vector v of the polyhedron is a *vertex* (extreme point) if and only if it cannot be written as a linear combination of points in $\mathbf{P} \backslash \{v\}$. A convex polytope $\mathbf{P}$ is the convex hull of its vertices.

Consider the following MIP model:

$$\max \left\{c^T x | A^I x \leq b^I, A^{II} x \leq b^{II}, x \in \mathbf{X}\right\} = \max \left\{c^T x | x \in \mathbf{P}^I \cap \mathbf{P}^{II}, x \in \mathbf{X}\right\}, \quad (2.71)$$

where $x \in \mathbf{X}$ includes the integrality constraints. For a reformulation, we derive a tighter formulation for the second set of constraints, that is, we replace $A^{II} x \leq b^{II}$ with $A^{II^*} x \leq b^{II^*}$, where

$$\mathbf{P}^{II^*} = \left\{x | A^{II^*} x \leq b^{II^*}\right\} \subset \left\{x | A^{II} x \leq b^{II}\right\} = \mathbf{P}^{II}.$$

This replacement (reformulation) leads to a tighter formulation (see Figure 2.9):

$$\max \left\{c^T x | x \in \mathbf{P}^I \cap \mathbf{P}^{II*}, x \in \mathbf{X}\right\}. \quad (2.72)$$

Example 2.9 Vertex Packing A *clique* (or complete graph) is a graph in which every pair of vertices is an edge. A *vertex packing* in a graph $G = \{\boldsymbol{V}, \boldsymbol{E}\}$ is a subset $\boldsymbol{U} \subseteq \boldsymbol{V}$ such that no vertices in $\boldsymbol{U}$ are pairwise adjacent. If A is the vertex-to-edge incidence

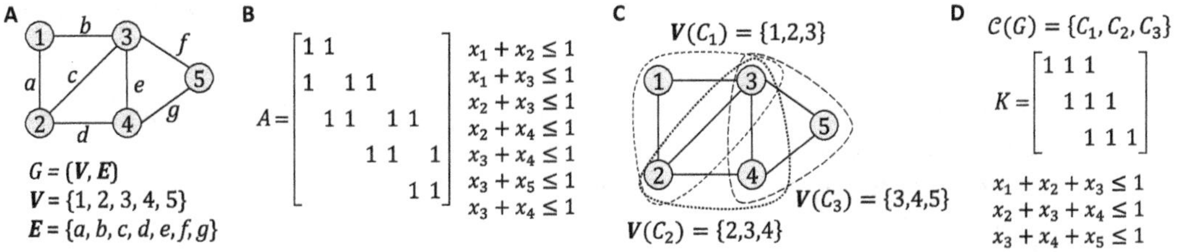

Figure 2.10 Illustration of reformulation for vertex packing problem. (A) Example graph and its vertex and edge sets. (B) Incidence matrix and pairwise constraints. (C) Clicks of the graph. (D) Clique matrix and constraints.

matrix of G, then a packing should satisfy $A^T x \leq \mathbf{1}$, where x is the vector of selection variables and $\mathbf{1}$ is a vector of 1s:

$$\mathbf{P}^{VP} = \{x | A^T x \leq \mathbf{1},\ x \in \{0,1\}^n\} = \{x | x_i + x_j \leq 1, (i,j) \in \boldsymbol{E}(G), x \in \{0,1\}^n\}. \tag{2.73}$$

A tighter reformulation can be obtained if we recognize that a packing cannot have more than one vertex from a clique. Let $\mathcal{C}(G)$ be the set of *maximal* cliques of graph G, that is, cliques that cannot be enlarged by including more vertices. We construct matrix K, whose rows correspond to cliques in $\mathcal{C}(G)$ and columns to vertices in $\boldsymbol{V}$, and coefficient k_{ij} is 1 if vertex j belongs to clique i. The feasible region of the vertex packing problem is then given by

$$\mathbf{P}^{VP-K} = \{x | Kx \leq \mathbf{1},\ x \in \{0,1\}^n\} = \left\{x \middle| \sum\nolimits_{i \in \boldsymbol{V}(C)} x_i \leq 1, C \in \mathcal{C}(G), x \in \{0,1\}^n\right\}, \tag{2.74}$$

where $\boldsymbol{V}(C)$ is the subset of vertices in clique C. Note that solutions that are feasible for the LP-relaxation of $\mathbf{P}^{VP}$ are infeasible for $\mathbf{P}^{VP-K}$.

Figure 2.10 illustrates the development of the reformulation employing graph G with five vertices. The incidence matrix of the graph is shown in Figure 2.10B along with constraints $A^T x \leq \mathbf{1}$. Note that the LP-relaxation of this formulation admits solution $x_1 = x_2 = x_3 = x_4 = x_5 = \frac{1}{2}$. Figure 2.10C shows the three cliques of G and the vertices of each clique, while Figure 2.10D shows the click matrix K and constraints $Kx \leq \mathbf{1}$, defining $\mathbf{P}^{VP-K}$. Note that $x_1 = x_2 = x_3 = x_4 = x_5 = \frac{1}{2}$ is infeasible for the relaxation of $\mathbf{P}^{VP-K}$.

2.4.4 Decomposition Methods

The idea is to decompose the original scheduling problem into smaller subproblems that can be solved faster than the original *full-space* model, even if the subproblems have to be solved multiple times. Broadly speaking, there are two types of decomposition:

(1) Approaches that exploit the *mathematical* structure of the model, such as, Benders and Lagrangian relaxation/decomposition.

(2) Approaches that exploit the *physical* structure of the problem, such as decomposition into assignment and sequencing subproblems.

Next, we outline the basic ideas behind the widely used Benders decomposition and Lagrangian relaxation in the context of scheduling problems.

Benders Decomposition. Consider the following model $(\mathbb{P})$:

$$\begin{array}{lll} \min & c^T x + \sum_j c_j^T y_j & \\ s.t. & A^I x \leq b^I & \text{(I)} \\ & A_j^{II} x + B_j y_j \leq b_j^{II}, \ j \in \mathbf{J} & \text{(II)} \\ & x \in \mathbf{X}, y_j \in \mathbf{Y}_j & \end{array}, \tag{2.75}$$

where variables x describe some systemwide decisions (e.g., assignments of tasks to units), subject to constraints (I); and y_j describe unit specific decisions (e.g., task sequencing between pairs of tasks assigned to the same unit), subject to unit-specific constraints (II) (see Figure 2.11A). If decisions x are fixed, then problem $(\mathbb{P})$ can be decomposed into $|\mathbf{J}|$ independent subproblems. Benders decomposition exploits this model structure: problem $(\mathbb{P})$ is decomposed into a master problem, consisting of constraints (I) and cuts (to be explained next); and the subproblems in constraints (II), which are solved repeatedly for different, fixed values of x. In the first iteration $(k = 1)$, the master problem $(\mathbb{MP}^{k=1})$: $\min\{c^T x | A^I x \leq b^I, x \in \mathbf{X}\}$ is solved to obtain x^1, which is then used to solve subproblems $(\mathbb{SP}_j^{k=1})$: $\min\{c_j^T y_j | B_j y_j \leq b_j^{II} - A_j^{II} x^1, y_j \in \mathbf{Y}_j\}$. If a subproblem $(\mathbb{SP}_j^1)$ is infeasible or the solution is suboptimal, then *feasibility* and *optimality* cuts are generated and added to the master problem, and the process is repeated. If solved to optimality, the solution of $(\mathbb{MP}^k)$ provides a lower bound on the optimal solution of $(\mathbb{P})$, if $c_j^T y_j \geq 0, \forall y_j$ for all subproblems, which is often the case (can you see why?). If all subproblems return a feasible solution at iteration k with objective function value z_j^{k*}, then $c^T x^{k*} + \sum_j z_j^{k*}$ is an upper bound.

If $\mathbf{X}_C^k$ is the set of points satisfying all cuts added up to iteration k, then the pseudocode for Benders decomposition is as follows:

(1) Solve $(\mathbb{MP}^k)$: $\min\{c^T x | A^I x \leq b^I, x \in \mathbf{X} \cap \mathbf{X}_C^k, \}$ to obtain x^k:
(2) Fix x^k and solve subproblems $(\mathbb{SP}_j^k)$: $\min\{c_j^T y_j | B_j y_j \leq b_j^{II} - A_j^{II} x^k, y_j \in \mathbf{Y}_j\}$
(3) If all feasible, stop; if at least one subproblem is infeasible or suboptimal, generate cuts, update $\mathbf{X}_C^k$, and iterate (go to step 1).

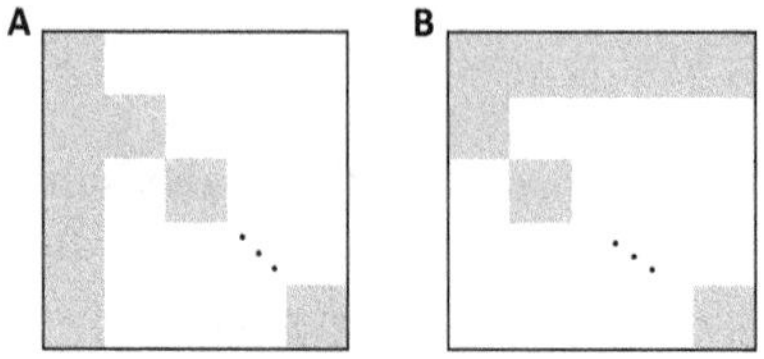

Figure 2.11 Model structures exploited by decomposition methods. (A) Structure of the problem in (2.75). (B) Structure of the problem in (2.76).

If $(\mathbb{P})$ is an LP, then $(\mathbb{P})$ can be solved to optimality using this iterative procedure. While no theoretical guarantees are available if $(\mathbb{P})$ is a MIP model, the method has been shown to be quite effective in yielding high-quality solutions fast. Note that Benders decomposition exploited a structure with *complicating* variables (x), that is, variables that couple the otherwise independent constraints (II).

Lagrangian Relaxation. It is used when complicating constraints are present (see Figure 2.11B). Consider problem $(\mathbb{P})$:

$$\begin{aligned} \min \quad & \sum\nolimits_t c_t^T y_t \\ s.t. \quad & \sum\nolimits_t B_t^I y_t \leq b^I \qquad \text{(I)} \\ & B_t^{II} y_t \leq b_t^{II}, \quad t \in \mathbf{T} \quad \text{(II)} \\ & y_t \in \mathbf{Y}_t \end{aligned}, \tag{2.76}$$

where variables y_t, describing decisions over a planning period, are subject to some linking constraints (I) and period-specific constraints (II). In this case, we can generate an easier problem by *dualizing* the linking constraints (i.e., penalize their violation in the objective function). At iteration k, we solve Lagrangian problem $(\mathbb{LP}^k)$:

$$\begin{aligned} \min \quad & \sum\nolimits_t c_t^T y_t + \left(u^k\right)^T \left(\sum\nolimits_t B_t^I y_t - b^I\right) \\ s.t. \quad & B_t^{II} y_t \leq b_t^{II}, \quad t \in \mathbf{T} \qquad \text{(II)}, \\ & y_t \in \mathbf{Y}_t \end{aligned} \tag{2.77}$$

where u^k is a vector of penalties at iteration k. The problem in (2.77) is a relaxation of (2.76) and thus provides a lower bound. If $(\mathbb{P})$ is an LP, then there is a u^k, determined iteratively using information on constraint violation, for which (2.77) yields the optimal solution of (2.76). If $(\mathbb{P})$ is a MIP, then a similar iterative procedure can still be effective provided that there is method to obtain a feasible solution from an (infeasible) solution of $(\mathbb{LP}^k)$. The pseudocode of the algorithm employing Lagrangian relaxation is as follows:

(1) Dualize (I) and solve $(\mathbb{LP}^k)$: $\max\left\{\sum_t c_t^{\mathrm{T}} y_t + \left(u^k\right)^{\mathrm{T}} \left(\sum_t B_t^I y_t - b^I\right): B_t^{II} y_t \leq b_t^{II}, t \in \mathbf{T}, y_t \in \mathbf{Y}_t\right\}$ to obtain y_t^k.
(2) Find a feasible solution based on current y_t^k.
(3) Update multipliers u^k, and iterate.[6]

2.5 Software Tools

The development and solution of mixed-integer models require, typically, two types of tools, modeling systems and solvers, both of which are described next.[7]

[6] While the updating of multipliers u^k is not trivial, there are numerous methods available in the literature (see the note in Section 2.6).

[7] The information and webpages mentioned in this section are based on the information available when the book was written (2019). Updated information can be found at Cambridge.com under Resources.

2.5.1 Modeling Languages

Models can be developed using a wide range of commercially available high-level modeling systems (i.e., standalone optimization modeling environments), which allow the user to easily build complex, large-scale models as well as change them as necessary. A major advantage of these modeling systems is that they offer an interface with optimization solvers, so the users do not have to do low-level programming (e.g., supplying matrices to solvers and calling subroutines) to use the solvers. Another advantage of these systems is that they can be easily interfaced with spreadsheets, databases, and graphics packages, thereby facilitating the development of new instances, and ultimately the testing and deployment of optimization models. Finally, some of these systems have graphical capabilities, including automatic generation of Gantt charts, thereby allowing easy implementation and testing of prototype applications. Note that many solver vendors also offer modeling languages. Table 2.3 gives the major modeling systems and the corresponding webpages. The magazine *OR/MS Today* often publishes reviews and comparisons of modeling systems.

General computing systems (e.g., Matlab) and programming languages (e.g., Python, C++, Java) can also be used to build models and interfaced with optimization solvers. In addition, some of these systems/languages offer libraries specifically for the development of optimization solutions. The major such systems and libraries are given in Table 2.4.

Table 2.3 Major modeling systems.

System/modeling language	Webpage
AIMMS	www.aimms.com/
AMPL	https://ampl.com/
GAMS	www.gams.com/
LINGO	www.lindo.com/
IBM ILOG CPLEX Optimization Studio[a]	www.ibm.com/products/ilog-cplex-optimization-studio
FICO Xpress Workbench[b]	www.fico.com/en/products/fico-xpress-workbench

[a] Developed by IBM, developer of solver CPLEX.
[b] Developed by FICO, developer of solver FICO Xpress Solver.

Table 2.4 Computing systems/languages and optimization libraries.

System	Library	Webpage
Matlab	Tomlab	https://tomopt.com/tomlab/
Python	Pyomo	www.pyomo.org/
Julia	JuMP	www.juliaopt.org/

Table 2.5 Major commercial mixed-integer linear programming solvers.

System	Webpage
CPLEX	www.ibm.com/analytics/cplex-optimizer
Gurobi	www.gurobi.com/
FICO Xpress Solver	www.fico.com/en/products/fico-xpress-optimization

2.5.2 Solvers

As discussed in the previous subsection, models can be generated using (1) optimization-specific modeling languages, (2) general programming languages, and (3) libraries of general systems. These models are *solved* using optimization solvers. With the exception of Chapter 11, the models presented in this book are mixed-integer linear programming models. The most effective MILP solvers are given in Table 2.5. Information for mixed-integer nonlinear programming solvers can be found under in the webpage of the book under Resources (www.cambridge.org).

Some free and/or open source solvers are available through the *Computational Infrastructure for Operations Research* (COIN-OR) initiative (www.coin-or.org/). Finally, we note that users can use the NEOS Server (https://neos-server.org/neos/) to solve instances online.

2.6 Notes and Further Reading

(1) The origins of mixed-integer programming go back to 1954 when the use of cutting planes and linear programming was proposed as a solution method [1, 2]. In the 1960s, the B&B algorithm was developed, and in 1972, MIP became commercially viable with the LP-based B&B algorithm, which remained the state-of-the-art method until the late 1990s [3, 4]. During this time, extensive research went into cutting planes and other solution methods, but cuts were incorporated into commercial solvers only in the late 1990s, leading to a 50-fold improvement in solution times [5].

(2) Features that have been introduced to commercial solvers leading to significant enhancements include preprocessing routines; node heuristics, which find good integer solutions; and node presolve methods, which further tighten the formulation at a node [5]. Combined with tremendous improvements in hardware and LP algorithms, these advances allow many problems that were previously intractable to be solved in minutes. Modern solvers allow users to change a range of algorithmic settings (e.g., aggressiveness of cut generation, rounds of preprocessing, and frequency of heuristic application) to tailor the solver to their problem. Also, user callbacks allow users to easily develop customized methods.

(3) The reader interested in modeling approaches using mathematical programming is pointed to Williams [6].

(4) Disjunctive programming was pioneered in the 1970s and 1980s by Balas [7, 8]. Generalized disjunctive programming was introduced by Grossmann and coworkers [9, 10].

(5) For the interested reader, Nemhauser and Wolsey [11] and [12] cover advanced mixed-integer modeling techniques and solution methods, including methods for problems expressed in terms of graphs and networks. A more thorough discussion of graphs and the associated problems can be found in [13]. Ahuja et al. [14] discuss networks.

(6) The basics of Benders decomposition and Lagrangian decomposition can be found in [15] and [16], respectively.

(7) A discussion on modeling languages can be found in [17].

(8) More information on MINLP solvers can be found in Belotti et al. [18] and Kronqvist et al. [19].

2.7 Exercises

(1) Formulate mixed-integer linear constraints for the following disjunction, using both big-M and convex-hull formulations: either $x \in [0, 12]$ or $x \in [20, 30]$. What are the tightest (smallest) values of big-M parameters you can use in the reformulations?

(2) Formulate mixed-integer linear constraints for the following disjunction, using both big-M and convex-hull formulations:

$$\begin{bmatrix} 3 \leq x \leq 5 \\ 2 \leq y \leq 6 \end{bmatrix} \vee \begin{bmatrix} 7 \leq x \leq 10 \\ 5 \leq y \leq 8 \end{bmatrix}.$$

Draw the two feasible regions. What are the tightest big-M parameters?

(3) Given is the integer programming problem:

$$\begin{bmatrix} & \max 1.2y_1 + y_2 \\ & y_1 + y_2 \leq 1 \\ st & 0.8y_1 + 1.1y_2 \leq 1 \\ & y_1, y_2 \in \{0, 1\} \end{bmatrix}$$

(a) Plot the contours of the objective function and the feasible region for the case when the binary variables are relaxed as continuous variables $y_1, y_2 \in [0, 1]$.

(b) Determine from inspection the solution of the relaxed problem.

(c) Enumerate the four 0–1 combinations in your plot to find the optimal solution.

(d) Solve the preceding problem with the branch-and-bound method by enumerating the nodes in the tree and solving the LP subproblems with your favorite software tool.

Table 2.6 Data for Exercise 4.

	Production time per unit produced			Production time available per week
	Product 1	Product 2	Product 3	
Plant 1	3 hours	4 hours	2 hours	30 hours
Plant 2	4 hours	6 hours	2 hours	40 hours
Unit profit	5	7	3	(thousands of dollars)
Sales potential	7	5	9	(units per week)

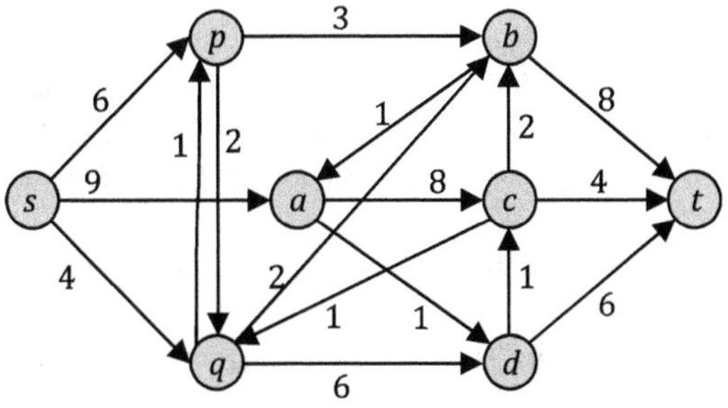

Figure 2.12 Network and associated parameters (arc capacities) for Exercise 7.

(4) The Research and Development (R&D) Division of company XYZ has developed three possible new products. However, to avoid undue diversification of the company"s product line, management has imposed the following restriction: *From the three possible new products, at most two should be chosen to be produced.* Each of these products can be produced in either of two plants. For administrative reasons, management has imposed a second restriction: *Just one of the two plants should be chosen to be the sole producer of the new products.*

The number of hours of production time needed per unit of each product, along with other relevant information, including marketing estimates of the number of units of each product that could be sold per week if it is produced, are given in Table 2.6. The objective is to choose the products, the plant, and the production rates of the chosen products to maximize total profit.

(5) Solve the following MIP by branch-and-bound using an LP code to solve the LP relaxations of the subproblems. Draw the branch-and-bound tree with the objective function value and the solution at every node. Can you do any preprocessing?

$$\left[\begin{array}{rc} & \max 77.9y_1 + 76.8y_2 + 89.6y_3 + 97.1x_1 + 31.3x_2 \\ & 60.9y_1 + 68.9y_2 + 69.0y_3 - 56.9x_1 + 22.5x_2 = 86.5 \\ st & -86.8y_1 + 32.7y_2 + 24.3y_3 + 13.8x_1 - 12.6x_2 \leq 77.3 \\ & 10.9y_1 + 3.6y_2 - 40.8y_3 + 43.9x_1 + 7.1x_2 \leq 82.3 \\ & y_i \in \{0,1\}, i = 1,2,3; x_i \geq 0, i = 1,2 \end{array}\right].$$

(6) Formulate the following combinatorial optimization problems:
 (a) Maximum edge matching
 (b) Minimum edge covering
 (c) Maximum vertex packing
 (d) Minimum vertex covering

 (Hint: Search the web if you need to find the formal problem statements.)

(7) Find a maximum flow for the network shown in Figure 2.12 using the mathematical programming formulation.

References

[1] Dantzig G, Fulkerson R, Johnson S. Solution of a Large-Scale Traveling-Salesman Problem. *Journal of the Operations Research Society of America*. 1954;2(4):393–410.

[2] Gomory RE. Outline of an Algorithm for Integer Solutions to Linear Programs. *Bull Amer Math Soc*. 1958;64:275–278.

[3] Land AH, Doig AG. An Automatic Method of Solving Discrete Programming Problems. *Econometrica*. 1960;28(3):497–520.

[4] Dakin RJ. A Tree-Search Algorithm for Mixed Integer Programming Problems. *Computer Journal*. 1965;8(3):250–253.

[5] Bixby R, Rothberg E. Progress in Computational Mixed Integer Programming – a Look Back from the Other Side of the Tipping Point. *Annals of Operations Research*. 2007;149 (1):37–41.

[6] Williams HP. *Model Building in Mathematical Programming*. 5th ed. Hoboken: Wiley; 2013.

[7] Balas E. Note on Duality in Disjunctive Programming. *Journal of Optimization Theory and Applications*. 1977;21(4):523–528.

[8] Balas E. Disjunctive Programming and a Hierarchy of Relaxations for Discrete Optimization Problems. *Siam J Algebra Discr*. 1985;6(3):466–486.

[9] Raman R, Grossmann IE. Modeling and Computational Techniques for Logic-Based Integer Programming. *Comput Chem Eng*. 1994;18(7):563–578.

[10] Turkay M, Grossmann IE. Disjunctive Programming Techniques for the Optimization of Process Systems with Discontinuous Investment Costs Multiple Size Regions. *Ind Eng Chem Res*. 1996;35(8):2611–2623.

[11] Nemhauser GL, Wolsey LA. *Integer and Combinatorial Optimization*. New York: Wiley; 1988.

[12] Wolsey LA. *Integer Programming*. New York: Wiley; 1998.

[13] West DB. *Introduction to Graph Theory*. 2nd ed. Upper Saddle River: Prentice Hall; 2001.

[14] Ahuja RK, Magnanti TL, Orlin JB. *Network Flows : Theory, Algorithms, and Applications*. Englewood Cliffs: Prentice Hall; 1993.

[15] Benders JF. Partitioning Procedures for Solving Mixed-Variables Programming Problems. *Numerische Mathematik*. 1962;4(1):238–252.

[16] Fisher ML. The Lagrangian Relaxation Method for Solving Integer Programming Problems. *Manage Sci*. 1981;27(1):1–18.

[17] Atamturk A, Savelsbergh MWP. Integer-Programming Software Systems. *Annals of Operations Research*. 2005;140(1):67–124.

[18] Belotti P, Kirches C, Leyffer S, Linderoth J, Luedtke J, Mahajan A. Mixed-Integer Nonlinear Optimization. *Acta Numerica.* 2013;22:1–131.
[19] Kronqvist J, Bernal DE, Lundell A, Grossmann IE. A Review and Comparison of Solvers for Convex MINLP. *Optim. Eng.* 2019;20(2):397–455.

Part II

Basic Methods

3 Single-Unit Environment

This chapter introduces scheduling in the simplest production environment, the single-unit environment. The statements of different problems are presented in Section 3.1. Two different types of MIP sequence-based[1] models are presented in Section 3.2. Section 3.3 discusses models based on a continuous time grid, while Section 3.4 presents models based on a discrete time grid. A number of extensions are discussed in Section 3.5, and we close in Section 3.6 with some general remarks.

Starting in this chapter and throughout the book, we will use the following notation conventions. We use lowercase Latin characters for indices, uppercase Latin bold letters for sets, uppercase Latin characters for variables, Greek letters for parameters, and regular uppercase Latin letters for set elements. All subsets of a set will be denoted by the letter used for the set and a subscript and/or superscript. Specifically, indices are used as subscripts to denote subsets that are index specific (e.g., the subset of units $\mathbf{J}$ in stage k is denoted by $\mathbf{J}_k$); uppercase letters are used as superscripts to further differentiate subsets (e.g., if $\mathbf{J}$ is the set of units, the subset of processing units and storage vessels are denoted by $\mathbf{J}^P$ and $\mathbf{J}^S$, respectively). Parameters and variables may also have superscripts for differentiation; for example, the processing cost of task i is denoted by γ_i^P while the changeover cost between tasks i and i' is denoted by $\gamma_{ii'}^{CH}$.

3.1 Problem Statement

In single-unit environments, batching decisions can be made independently from other decisions. Since there is only one unit available, the minimum number of batches that can meet these orders can be precalculated, leading to a predefined set of batches, some of which may be identical. Thus, we are given a set of batches, $i \in \mathbf{I}$, to be carried out on a single unit, U. The processing time of batch i is denoted by τ_i. In the absence of additional constraints and processing features, this is not an optimization problem. Since all batches will be carried out on the same unit, we observe the following:

- The minimum makespan will be equal to $\sum_i \tau_i$ regardless of the sequencing of batches.

[1] The type of model in this book referred to as *sequence-based* is often referred to, in the PSE literature, as *precedence based*. The reason for this difference is explained in Section 3.7.

- If there is a processing cost, γ_i^P, the total processing cost will be $\sum_i \gamma_i^P$.
- Since there are no due times, lateness, tardiness, and earliness cannot be defined.
- Since the number and size of batches are fixed, the total production and thus profit are fixed.

There are two features that make this problem more interesting:

(1) Each batch is subject to release and due times, ρ_i and ε_i, respectively.
(2) There are sequence-dependent changeover times, $\sigma_{ii'}$, and costs, $\gamma_{ii'}^{CH}$.

Note that setup (i.e., sequence-independent) times and costs are not relevant for single-unit problems because (1) all feasible solutions have exactly the same setup cost (since all batches are assigned to the same unit); and (2) setup times can be simply added to the processing times, so the problem statement remains the same.

In terms of problem classification using the $\alpha/\beta/\gamma$ triplet, we denote the single-unit environment as $\boldsymbol{Sms}$, $|\mathbf{J}| = 1$. In the remainder of the present chapter, we will consider problems with the following:

(1) Release and due dates, that is, $[\boldsymbol{Sms}, |\mathbf{J}| = 1/\rho, \varepsilon/\cdot]$
(2) Release/due times and changeover times, that is, $[\boldsymbol{Sms}, |\mathbf{J}| = 1/\rho, \varepsilon, \sigma/\cdot]$
(3) Release/due times, and changeover times and costs, that is, $[\boldsymbol{Sms}, |\mathbf{J}| = 1/\rho, \varepsilon, \sigma, \gamma^{CH}/\cdot]$

3.2 Sequence-Based Models

3.2.1 Global Sequence Models

The basic idea is to use a binary variable to represent the relative order (sequence) in which a pair of batches is processed, and then employ big-M constraints to enforce a *no-overlap condition* between all pairs of batches. Specifically, binary variable $y_{ii'}$ is equal to 1 if batch i is processed (on unit U) before batch i'. If $S_j \in \mathbb{R}_+$ denotes the starting time of batch i, then the following constraint, often called the disjunctive constraint,[2] ensures that batch i is finished before batch i' starts:[3]

$$S_i + \tau_i \leq S_{i'} + M(1 - Y_{ii'}), \ i, i' \neq i, \tag{3.1}$$

where M is a sufficiently large number. Note that while it would have been easy to define a new variable, E_i, to denote the end of a batch $(E_i = S_i + \tau_i)$, we use $S_i + \tau_i$ instead. Also, since we do not know the value of $Y_{ii'}$ in the optimal solution, (3.1) should be written for each (i, i') pair twice.

[2] In a global sequence model, we want to enforce that given a pair of batches, (i, i'), either i is processed before i' is processed (i.e., i is finished before i' starts) or i' is processed before i is processed, that is, we want to enforce a *disjunction*. This constraint is further discussed in Section 3.5.

[3] Recall that equation domains are expressed using indexes only, unless the equations are defined over subsets.

Equation (3.1) enforces correct batch timing provided that binary variables $Y_{ii'}$ assume *correct* values. To achieve this, we have to ensure that a sequencing relationship is established for every pair of tasks since $Y_{ii'} = 1$ even when batch i' is not processed immediately after i, which is accomplished via

$$Y_{ii'} + Y_{i'i} = 1, \quad i, i' > i. \tag{3.2}$$

Equations (3.1) and (3.2) ensure that there is no overlap between the processing of any two batches, that is, they ensure that unit U processes at most one batch at a time.

For all problem classes, release and due dates can then be readily enforced:

$$S_i \geq \rho_i, \ \ i \tag{3.3}$$

$$S_i + \tau_i \leq \varepsilon_i, \ \ i. \tag{3.4}$$

Equation (3.4) enforces hard due dates (deadlines). If due times can be violated, at a cost, then (3.4) can be replaced with

$$S_i + \tau_i \leq \varepsilon_i + L_i, \quad i, \tag{3.5}$$

where L_i is the lateness/tardiness of batch i. If the objective function is to minimize lateness, then (3.5) will be satisfied as equality; that is, L_i will assume the smallest possible value, which can be negative. If the objective is to minimize tardiness, then we should enforce $L_i \geq 0$, and (3.5) will be satisfied as equality only if $S_i + \tau_i \leq \varepsilon_i$.

If the triangle inequality, $\sigma_{ii''} < \sigma_{ii'} + \tau_{i'} + \sigma_{i'i''}$, holds for all i, i', i'' (see discussion in Section 3.6.1), then changeover times can be addressed through a simple modification of (3.5):

$$S_i + \tau_i + \sigma_{ii'} \leq S_{i'} + M(1 - Y_{ii'}), \ \ i, i' \neq i, \tag{3.6}$$

which enforces that, if batch i is processed before batch i', then the start time of i', the successor, is larger than the finish time $(S_i + \tau_i)$ of i, the predecessor, plus the change-over time from i to i'.

Changeover costs cannot be readily modeled using global sequence models because for a given batch i', potentially many variables $Y_{ii'}$ assume a value of 1.

For makespan, MS, minimization,

$$\min MS, \tag{3.7}$$

we require that makespan is greater than or equal to the end times of all batches,

$$MS \geq S_i + \tau_i, \ \ i. \tag{3.8}$$

For earliness minimization, the general objective function is

$$\min \sum_i \omega_i(\varepsilon_i - (S_i + \tau_i)), \tag{3.9}$$

where ω_i is a weight factor, and we assume that due times are met ((3.4)) so no benefit is obtained from increasing the lateness of batch i in order to decrease earliness of batch i'.

The objective function for total weighted lateness, L^{TOT}, and tardiness, T^{TOT}, minimization is

$$\min \sum\nolimits_i \omega_i L_i \tag{3.10}$$

where

$$L_i \geq (S_i + \tau_i) - \varepsilon_i, \;\; i \tag{3.11}$$

and for tardiness we also require $L_i \geq 0$.

Equtations (3.1) through (3.11) can be combined to develop models for various problems. For example, the model for problem $[\boldsymbol{Sms}, |\mathbf{J}| = 1/\rho, \varepsilon/MS]$ consists of (3.1) through (3.4) and (3.7) through (3.8); while the model for problem $\left[\boldsymbol{Sms}, |\mathbf{J}| = 1/\rho, \varepsilon, \sigma/T^{TOT}\right]$ consists of (3.2) through (3.3), (3.5), (3.6), and (3.10).

3.2.2 Immediate Sequence Models

In an immediate sequence model, binary variable $Y_{ii'} = 1$ if batch i is immediately followed by i'; that is, there is no batch processed between i and i'. The main difference between global and immediate sequence modeling is in the number of binary variables assuming a value of 1 at a feasible solution (see Figure 3.1). In the former, a sequencing relationship is established between all pairs of batches, via (3.2), which means that in any feasible solution there are $|\mathbf{I}|\,(|\mathbf{I}| - 1)/2\; Y_{ii'}$ variables that are equal to 1. In an immediate sequence model, a relationship only between immediate *neighbors* is established, which means that in any feasible solution only $|\mathbf{I}| - 1$ variables are equal to 1. Thus, the main difference between the two approaches lies in the equations that activate variables $Y_{ii'}$. Specifically, (3.2) should be replaced with alternative equations.

Note that all batches have only one immediate predecessor and one immediate successor, except the first batch (which has no predecessor) and the last batch (which has no successor). To exploit this, we introduce two new binary variables:

- $Y_i^F = 1$ if batch i is processed first in unit U.
- $Y_i^L = 1$ if batch i is processed last in unit U.

These should satisfy

$$\sum\nolimits_i Y_i^F = \sum\nolimits_i Y_i^L = 1. \tag{3.12}$$

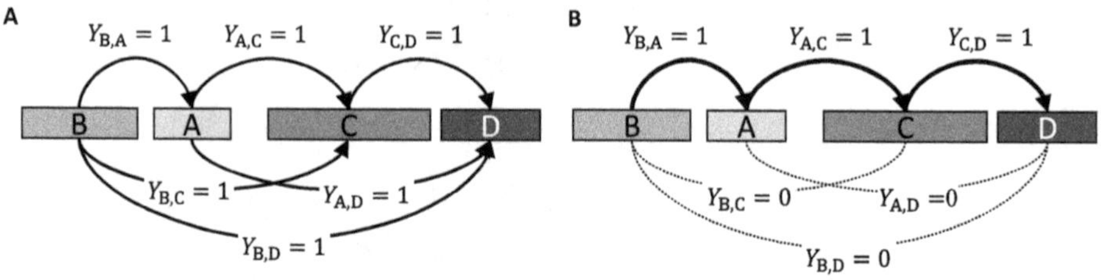

Figure 3.1 Graphic representation of a solution using sequence-based models. (A) Global sequencing. (B) Immediate sequencing (arrows represent binary variables).

The condition on the number of immediate predecessors and successors is then enforced via

$$\sum_{i} Y_{ii'} = 1 - Y^F_{i'}, \; i' \tag{3.13}$$

$$\sum_{i'} Y_{ii'} = 1 - Y^L_{i}, \; i. \tag{3.14}$$

Equation (3.13) enforces that each batch has exactly one immediate predecessor unless it is the first batch, while (3.14) enforces that each batch has exactly one immediate successor unless it is the last. Equations (3.12) through (3.14) can be used to *activate* immediate sequencing binary variables. Once variables $Y_{ii'}$ assume values that can lead to a feasible solution, the equations presented in the previous subsection can be used to enforce the remaining constraints. For example, an immediate sequence-based model for problem $[\boldsymbol{Sms}, |\mathbf{J}| = 1/\rho, \varepsilon/MS]$ consists of (3.1), (3.12) through (3.14) (instead of (3.2)), (3.3), (3.4), (3.7), and (3.8).

An alternative approach exploits the fact that exactly $|\mathbf{I}| - 1$ immediate sequencing variables will be activated in any feasible solution:

$$\sum_{i,i'} Y_{ii'} = |\mathbf{I}| - 1. \tag{3.15}$$

Equation (3.15) coupled with (3.13), (3.14), and (3.1), which can still be used to enforce timing constraints given a set of sequencing relationships, correctly enforces that only one batch is processed at any given time.[4]

In addition, since changeover cost can easily be calculated using $Y_{ii'}$ variables, we can also consider problem $[\boldsymbol{Sms}, |\mathbf{J}| = 1/\cdot/C^{CH}]$ with the objective function

$$\min \sum_{i} \sum_{i'} \gamma^{CH}_{ii'} Y_{ii'}. \tag{3.16}$$

3.3 Models Based on a Continuous Time Grid

The fundamental difference between sequence-based and time-grid-based approaches is that the former do not require any *preconceived* auxiliary structure, while the latter require the definition of one or more time grid(s) onto which the execution of tasks/batches are mapped. As will be discussed in the next chapter, there is a wide range of modeling approaches that can be adopted using time grids. However, in the present chapter we discuss some special cases arising in the simple single-unit environment.

When a continuous grid is adopted, the horizon, η, is divided into $|\mathbf{I}|$ periods of unknown length (see Figure 3.2A). Each period $t \in \mathbf{T}$ starts at time point $t - 1$ and ends at time point t. The timing of point t is denoted by T_t. The horizon starts at $\rho^0 = \min_i\{\rho_i\}$. If deadlines are given, then it ends at $\varepsilon^F = \max_i\{\varepsilon_i\}$; that is, $\eta = \varepsilon^F - \rho^0$. If no deadline information is available, a sufficiently long horizon should be considered. Changing the origin to 0 (i.e., $\rho^0 \to 0$), we can define the horizon from

[4] Can you see why (3.12) is unnecessary? Why does a solution that satisfies (3.13) through (3.15) satisfy (3.12) as well?

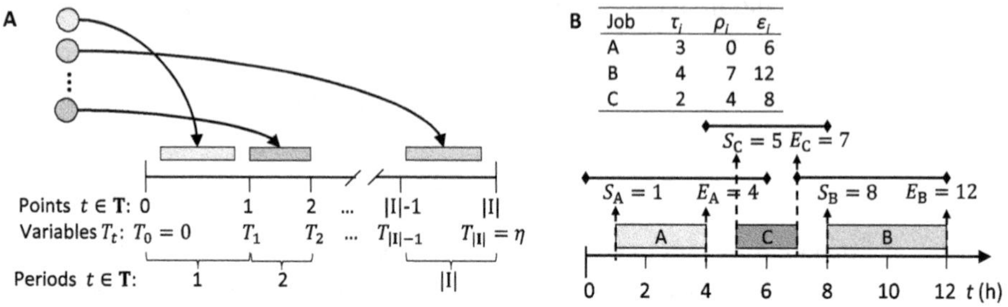

Figure 3.2 Explanation of concepts pertaining to models based on a continuous time grid. (A) Basic ideas: time points and periods, timing variables, and allocation of batches to time periods. (B) Example feasible solution (instance data given in the table).

0 to $\varepsilon^F - \rho^0$. Since the number of batches processed on unit U is known and equal to $|\mathbf{I}|$, and no events other than the start and finish of a batch need to be modeled, a batch can be processed within a time period. Thus, the set of necessary periods is $\mathbf{T} = \{1, 2, \ldots, |\mathbf{I}|\}$, while the set of points is $\mathbf{T}' = \mathbf{T} \cup \{0\} = \{0, 1, 2, \ldots, |\mathbf{I}|\}$.[5] The timing of time points should satisfy the following constraints:[6]

$$T_0 = 0, \quad T_{|\mathbf{T}|} = \eta \tag{3.17}$$

$$T_t \geq T_{t-1}, \quad t. \tag{3.18}$$

The basic idea of single time grid models is the matching between a batch and a time period, often referred to also as the allocation of a batch to a time period. It is accomplished through the introduction of binary variable X_{it}, which is equal to 1 if batch i is allocated to time period t, and the following two constraints

$$\sum\nolimits_t X_{it} = 1, \quad i \tag{3.19}$$

$$\sum\nolimits_i X_{it} = 1, \quad t, \tag{3.20}$$

which enforce that each batch is allocated to exactly one period and each period is used for exactly one batch.

Binary variables X_{it} can then be used to enforce no-overlap between batches (i.e., replace (3.1)). The more intuitive approach is to still employ batch start time variables, S_i, as in the previous subsection and write constraints relating the start and end time, $E_i = S_i + \tau_i$, of a batch with time points, T_t (see Figure 3.2B):

$$S_i \geq T_{t-1} - M(1 - X_{it}), \quad i, t \tag{3.21}$$

$$S_i + \tau_i \leq T_t + M(1 - X_{it}), \quad i, t. \tag{3.22}$$

5 In many papers, the term *time slot*, instead of time period, has been used to describe the setting in which a batch/task starts and finishes within the same period. To streamline the presentation, in this book we will use the term *time period* only.

6 Equations in this section are written for $t \in \mathbf{T} = \{1, ..|\mathbf{I}|\}$, that is, not for $t \in \mathbf{T}'$.

Equations (3.21) and (3.22) enforce that if batch i is allocated to period t, then it is executed within it, where M is a sufficiently large number ($M = \eta$ is valid, though tighter values can be calculated).

Given variables X_{it} satisfying (3.19) and (3.20), (3.21) and (3.22) coupled with (3.17) and (3.18) ensure that unit U process at most one batch at any time. Thus, the first model based on a continuous time grid consists of (3.3), (3.4), or (3.5) and (3.17) through (3.22).

Since start time variables, S_i, are employed, the objective functions introduced in Section 3.2.1 can be used. For example, the model for problem $[\boldsymbol{Sms}, |\mathbf{J}| = 1/\rho, \varepsilon/MS]$ consists of (3.3), (3.4), (3.17) through (3.22), (3.7), and (3.8).

A slightly different model is obtained when S_i is required to be equal to the start of the period to which batch i is allocated, which can be achieved via a double big-M constraint,

$$T_{t-1} - M(1 - X_{it}) \leq S_i \leq T_{t-1} + M(1 - X_{it}), \quad i, t. \tag{3.23}$$

To allow the start time of the first batch to be variable, which may be necessary in say earliness minimization, $T_0 = 0$, in (3.17) should be removed. All remaining equations remain the same. Alternatively, the batch end time, $S_i + \tau_i$, can be enforced to be equal to the end time of the period to which it is allocated, via

$$T_t - M(1 - X_{it}) \leq S_i + \tau_i \leq T_t + M(1 - X_{it}), \quad i, t. \tag{3.24}$$

A different class of single-grid continuous time models is obtained if the batch start time, S_i, is removed and the no-overlap condition is enforced using T_t variables through, for example,

$$T_t \geq T_{t-1} + \sum_i \tau_i X_{it}, \quad t. \tag{3.25}$$

The summation in the RHS of (3.25) represents the processing time of the batch allocated to period t, and the inequality ensures that period t is long enough to accommodate the batch that is allocated to it. Since variable S_i is removed, variables T_t should be used to enforce release and due times, which means that an assumption regarding the placement of a batch within a period is necessary. If we assume that the processing of the batch coincides with the beginning of the period to which it is allocated, then the release and due date constraints can be written as follows:

$$T_{t-1} \geq \rho_i X_{it}, \quad i, t \tag{3.26}$$

$$T_{t-1} + \tau_i X_{it} \leq \varepsilon_i X_{it} + \eta(1 - X_{it}), \quad i, t. \tag{3.27}$$

Note that the assumption that the processing of the batch coincides with the beginning of the period to which it is allocated does not remove degrees of freedom – any feasible schedule can be mapped onto a solution of the model based upon the aforementioned assumption. In the alternative models, with no S_i variables, (3.26) and (3.27) replace (3.3) and (3.4), (3.25) replaces (3.21) and (3.22), and the objective function is expressed without using S_i variables.

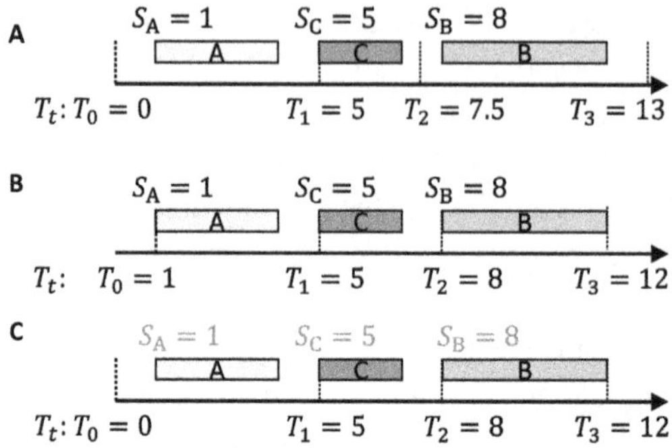

Figure 3.3 Representation of the feasible solution shown in Figure 3.2B using three different sets of assumptions and models. (A) Model based on T_t and S_i variables; a batch can start and end anywhere within a period. (B) Model based on T_t and S_i variables; a batch has to start at the beginning of the period it is allocated to. (C) Model based on T_t variables only; a batch can start anywhere within a period (variables in gray used for illustration, not used in the model).

In general, models based on a continuous time grid can employ different variables and sets of constraints to enforce (1) that a batch starts and ends within the period it is allocated to and (2) release and due time constraints. Consequently, the same schedule can be represented by different variables and, when the same variables are employed, by a different solution vector. Some examples are shown in Figure 3.3.

Accounting for changeovers is, in general, more challenging with time-grid-based models. One approach is to introduce period-specific immediate sequencing binary variables $Y_{ii't}$, which are equal to 1 if the $i \rightarrow i'$ changeover occurs before point t prior to the execution of batch i', and activate them using the batch-period allocation binaries,[7,8]

$$Y_{ii't} \geq X_t + X_{i',t+1} - 1, \quad i, i' \neq i, t \tag{3.28}$$

$$\sum\nolimits_{i'} Y_{ii't} \leq X_{it}, \quad i, t \tag{3.29}$$

$$\sum\nolimits_{i} Y_{ii',t-1} \leq X_{i't}, \quad i', t \tag{3.30}$$

and then use a new timing constraint, whose most intuitive (but not tightest) form is

$$S_i + \tau_i + \sigma_{ii'} \leq T_t + M(1 - Y_{ii't}), \quad i, i' \neq i, t. \tag{3.31}$$

The concept is illustrated in Figure 3.4. Note that sequencing binary variables $Y_{ii't}$ can be defined in multiple ways. For example, $Y_{ii't} = 1$ if the $i \rightarrow i'$ changeover occurs within period t prior to the execution of i', which also occurs in period t. (Can you see the difference? How should (3.28) through (3.31) be modified if this definition is used?)

7 Note that index t in allocation binaries X_{it} denotes periods, while in sequencing binaries $Y_{ii't}$ denotes points.

8 In the single-unit problems we have considered so far, we know how many batches need to be scheduled, and thus all postulated periods have a batch assigned, that is, no period is *empty*. Therefore, all points, except the first and last, should have exactly one changeover (i.e., $\Sigma_{i'} X_{ii't} = 1$). Can you think of a way to directly enforce this equality? Have you seen a similar logic condition enforced before? What can be done differently here?

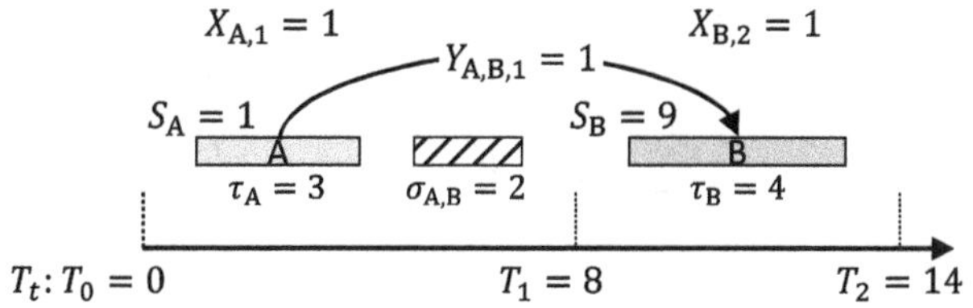

Figure 3.4 Modeling of changeovers using a continuous time grid.

The minimization of changeover cost can then be expressed as follows:

$$\min \sum_{i,i'} \sum_t \gamma_{ii'}^{CH} Y_{ii't}. \tag{3.32}$$

While more effective formulations are available, the modeling of changeover times and costs using models based on a continuous time grid requires the introduction of new binary variables $Y_{ii't}$, which leads to large models, when compared to their immediate-sequence counterparts, which, notably, employ binary variables $Y_{ii'}$ not indexed by t, so they require significantly fewer binary variables.

3.4 Models Based on a Discrete Time Grid

We use index $n \in \mathbf{N} = \{0,1,2, \ldots |\mathbf{N}|\}$ to denote time points. The scheduling horizon is divided into periods $n \in \mathbf{N}' = \mathrm{N} \backslash \{0\} = \{1, 2, \ldots, |\mathbf{N}|\}$ of equal length δ, where period n runs between time points $n - 1$ and n. Note that we use index $t \in \mathbf{T}$ to denote points/periods in continuous time models but index $n \in \mathbf{N}$ in discrete time models. The natural choice for the discretization, δ, is the greatest common factor[9] of all time-related data, that is, processing times, τ_i; changeover times, $\sigma_{ii'}$; and release, ρ_i, and due, ε_i, times. The data used in the model are then obtained by dividing the original parameters by δ,

$$\bar{\tau}_i = {}^{\tau_i}/_{\delta}, \quad \bar{\sigma}_{ii'} = {}^{\sigma_{ii'}}/_{\delta}, \quad \bar{\rho}_i = {}^{\rho_i}/_{\delta}, \quad \bar{\varepsilon}_i = {}^{\varepsilon_i}/_{\delta}.$$

Since this may lead to many time points, especially when changeover times (which are typically significantly smaller than processing times) are considered, a coarser discretization may be necessary. In this case, to ensure feasibility of the solution obtained by a model that employs a coarse grid, time-related parameters should be rounded as follows:

$$\bar{\tau}_i = \lceil {}^{\tau_i}/_{\delta} \rceil, \quad \bar{\sigma}_{ii'} = \lceil {}^{\sigma_{ii'}}/_{\delta} \rceil, \quad \bar{\rho}_i = \lceil {}^{\rho_i}/_{\delta} \rceil, \quad \bar{\varepsilon}_i = \lfloor {}^{\varepsilon_i}/_{\delta} \rfloor. \tag{3.33}$$

(Can you see why some parameters are rounded up and some are rounded down?) The conversion of the actual time-related data, an important step for all models based on discrete time grid(s), is further discussed in Example 3.1.

[9] The use of the term greatest common factor is not rigorous; see Example 3.1 for the details of the required calculations.

We introduce binary variable X_{in}, which is equal to 1 if batch i starts at time point n. The first constraint that should be enforced is that each batch is executed once,

$$\sum_{n} X_{in} = 1, \quad i, \tag{3.34}$$

which is the counterpart of (3.19).

Next, we should enforce the no-overlap restriction, that is, the unit can process only one batch at a time. An intuitive constraint is the counterpart of (3.20),

$$\sum_{i} X_{in} \leq 1, \quad n,$$

where equality has been replaced by inequality since, in general, more time points than batches are employed. While this constraint should be satisfied in any feasible solution (i.e., it is necessary), it is not sufficient. As illustrated in Example 3.2 and Figure 3.5A, a sufficient constraint is the following clique inequality:

$$\sum_{i} \sum_{n'=n-\bar{\tau}_i+1}^{n'=n} X_{in'} \leq 1, \quad n. \tag{3.35}$$

Release times and deadlines are enforced via fixing the binaries outside the allowable window to zero:

$$X_{in} = 0, \qquad i, n < \bar{\rho}_i, n > \bar{\varepsilon}_i - \bar{\tau}_i. \tag{3.36}$$

In general, in discrete time models the timing of an event can be modeled through the multiplication of time points by the corresponding binary variable. For example, the start and end time of a batch can be calculated as follows:

$$S_i = \sum_{n} nX_{in}, \quad E_i = \sum_{n} (n + \bar{\tau}_i) X_{in}.$$

Using this idea, we can enforce constraints for makespan,

$$MS \geq \sum_{n} (n + \bar{\tau}_i) X_{in}, \quad i, \tag{3.37}$$

and also calculate batch earliness, $\bar{\varepsilon}_i - \sum_{n} (n + \bar{\tau}_i) X_{in}$; and lateness, $\sum_{n} (n + \bar{\tau}_i) X_{in} - \bar{\varepsilon}_i$. Thus, the corresponding objective functions become

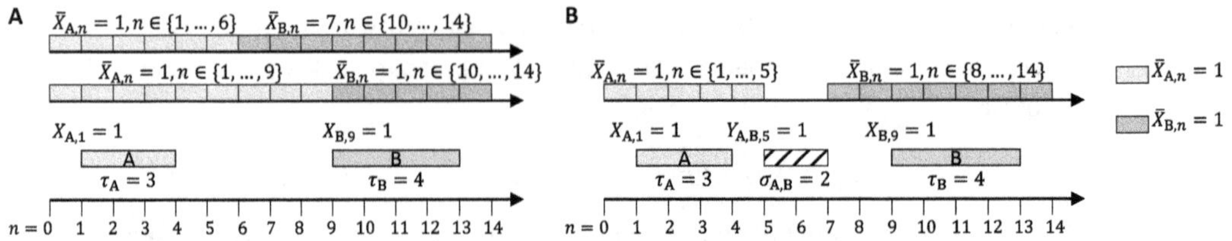

Figure 3.5 Modeling of assignment and changeover time constraints; batch $i = \text{B3}$ is executed at $n = 14$ ($X_{\text{B3},14} = 1$), indicated as diamond; plots show indices of variables included in the constraints. (A) Variables included in the LHS of (3.35): indices (i, n') of included variables are shown as black squares. (B) Variables included in the LHS of (3.43): gray squares indicate variables included in (3.35), and black disks indicate variables included in (3.43). (C) Variables included in (3.44); gray squares denote variables included in (3.35), and black disks denote (i', n') pairs for which (3.44) is expressed.

$$\min MS \tag{3.38}$$

$$\min \sum_i \omega_i \left(\bar{\varepsilon}_i - \sum_n (n + \bar{\tau}_i) X_{in} \right) \tag{3.39}$$

$$\min \sum_i \omega_i \left(\sum_n (n + \bar{\tau}_i) X_{in} - \bar{\varepsilon}_i \right). \tag{3.40}$$

For tardiness minimization, we have to first define and constrain a nonnegative batch-specific tardiness variable ($L_i \geq 0$),

$$L_i \geq \sum_n (n + \bar{\tau}_i) X_{in} - \bar{\varepsilon}_i, \quad i, \tag{3.41}$$

and then use

$$\min \sum_i \omega_i L_i. \tag{3.42}$$

Changeover times can be enforced using binary variables X_{in} via

$$\sum_{i' \neq i} \sum_{n' = n - \bar{\tau}_{i'} - \bar{\sigma}_{i'i} + 1}^{n - \bar{\tau}_{i'}} X_{i'n'} \leq M_i (1 - X_{in}), \quad i, n \tag{3.43}$$

where a valid value for parameter M_i is

$$M_i = \sum_{i' \neq i} \sigma_{i'i}.$$

Equation (3.43) enforces that, if batch i starts at time point n, then no other batch i' can start between points $n - \bar{\tau}_{i'} - \bar{\sigma}_{i'i} + 1$ and $n - \bar{\tau}_{i'}$ (see Figure 3.5B for an illustration). In other words, it enforces a *sufficient separation* between batch i, starting at n, and every other batch i' starting before n.

An alternative approach is to enforce sufficient separation between batch i, starting at n, and each batch i' separately as follows (see Figure 3.5C):

$$X_{in} + X_{i'n'} \leq 1, \quad i, i' \neq i, n, n' \in \{n - \bar{\tau}_{i'}, - \bar{\sigma}_{i'i} + 1, \ldots, n - \bar{\tau}_{i'}\} \tag{3.44}$$

Note that (3.44) is written for every point n, task pair (i, i') , and points n', so it leads to very large formulations.

Equations (3.43) and (3.44) enforce changeover times via binary variables X_{in}, that is, no new variables are introduced. When changeover costs need to be accounted for, however, new variables are needed. While many alternative approaches are available, here we present two based on the following new binary variables:

- $\bar{X}_{in} = 1$ if during period n unit U is setup to carry out batch i; that is, it is in *mode i*.[10]
- $Y_{ii'n} = 1$ if the mode of unit U is changed from batch i to batch i' at time point n.

[10] In general, throughout the book, we will use overbar to distinguish similar variables; overbar will be used, typically, for time-indexed variables when the time index refers to periods as opposed to time points.

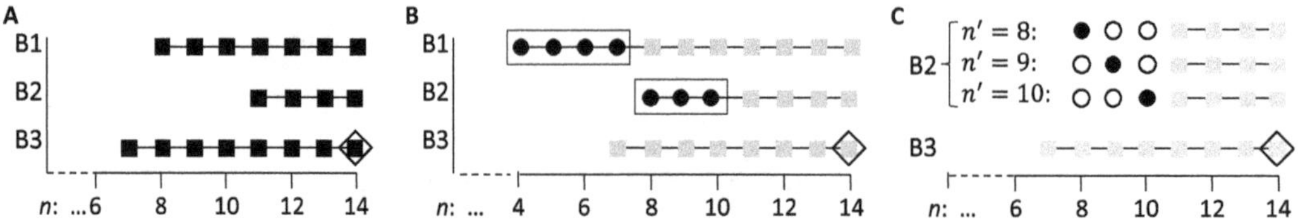

Figure 3.6 Modeling of changeover costs and times using the solution shown in Figure 3.4. (A) Modeling of changeover times using (3.45) through (3.49); positive $\bar{X}_{in}$ variables are shown above the Gantt chart; note that multiple solutions can represent the same schedule (two representative solutions are shown). (B) Modeling of changeover costs and times using (3.50).

The first approach is applicable when there are changeover costs but no changeover times. Variable $\bar{X}_{in}$ should be equal to 1 during the execution of batch i, and it can be 1 also during idle time. The most intuitive and easy to implement set of constraints for correctly activating $Y_{ii'n}$ variables, which are then used to calculate changeover costs, is the following:

$$\sum_i \bar{X}_{in} = 1, \quad n \tag{3.45}$$

$$\sum_{n'=n-\tau_i}^{n-1} X_{in'} \leq \bar{X}_{in}, \quad i, n \tag{3.46}$$

$$\sum_{i'\neq i} Y_{ii'n} \leq \bar{X}_{i,n-1}, \quad i, n \tag{3.47}$$

$$\sum_{i\neq i'} Y_{ii'n} \leq \bar{X}_{i'n}, \quad i', n \tag{3.48}$$

$$Y_{ii'n} \geq \bar{X}_{i,n-1} + \bar{X}_{i'n} - 1, \quad i, i' \neq i, n. \tag{3.49}$$

Equation (3.45) enforces that unit U is always in a mode to carry out a batch, even if it is idle. Equation (3.46) activates $\bar{X}_{in}$ if batch i is at any stage of its execution during period n; note that only one of the binary variables included in the summation can be equal to 1 (why?). Equations (3.47) and (3.48) ensure that a changeover from mode i to mode i' can happen at time n only if the unit was in mode i during period $n-1$ and will be in mode i' during period n. (Have you already seen a similar constraint in the present chapter? What are the similarities and differences?) Finally, (3.49) enforces the changeover from i to i' given the mode of the unit. The concept is illustrated in Figure 3.6A. Note that $Y_{iin} = 0, \forall i$.

The second model accounts for both changeover costs and times. A unit is not in a production mode during the execution of a changeover, which means that (3.45) is not enforced. Instead, we keep track of the mode of a unit through

$$\bar{X}_{in} = \bar{X}_{i,n-1} + \sum_{i'\neq i} Y_{i'i,n-\bar{\sigma}_{i'i}} - \sum_{i'\neq i} Y_{ii'n}, \quad i, n \tag{3.50}$$

which, as we will discuss in Chapter 8, represents, essentially, a *mode balance* over time:[11] a unit is in mode i if it has already been in mode i or a changeover to mode i just

[11] While the RHS of (3.50) appears to include only variables, note that the very first equation, expressed in this case for $n = 1$, includes a parameter, instead of $\bar{X}_{i,n-1}$, which represents the mode of the unit at time zero. This is common in all equations expressing some type of balance over time: the equation for the first period/point will include a parameter representing the *initial condition* of the unit, resource, and so on. For brevity, we will be presenting only the general form.

Table 3.1 Discrete time models for single-stage problems.

Problem	Equations
$[\boldsymbol{Sms}, \lvert\mathbf{J}\rvert = 1/\rho, \varepsilon/MS]$	(3.34), (3.35), (3.36), (3.37), (3.38)
$[\boldsymbol{Sms}, \lvert\mathbf{J}\rvert = 1/\rho, \varepsilon/E_{WGT}]$	(3.34), (3.35), (3.36), (3.39)
$[\boldsymbol{Sms}, \lvert\mathbf{J}\rvert = 1/\rho, \varepsilon, \sigma/MS]^{a}$	(3.34), (3.35), (3.36), (3.43), (3.37), (3.38)
$[\boldsymbol{Sms}, \lvert\mathbf{J}\rvert = 1/\rho, \varepsilon, \sigma, \gamma^{CH}/C^{CH}]$	(3.34), (3.35), (3.36), (3.46), (3.50), (3.51)

[a] Equation (3.44) can be used instead of (3.43).

finished, and there is no changeover from mode i to another mode. Equation (3.50) replaces (3.45) and (3.47) through (3.49). The concept in illustrated in Figure 3.6B.

Using binary variables $Y_{ii'n}$, we can then express the cost minimization objective function:

$$\min \sum\nolimits_{i,i',n} \gamma_{ii'}^{CH} Y_{ii'n}. \tag{3.51}$$

Table 3.1 gives the equations of models based on a discrete time grid for different single-unit problems.

Example 3.1 Conversion of Time-Related Data The goal of this example is to illustrate how time-related data should be processed when formulating models based on a discrete time grid. Regardless of the time units used, time-related parameters $(\tau_i, \sigma_{ii'}, \rho_i, \varepsilon_i)$ are rational numbers; that is, they can be written as a/b, where b is an integer. This means that, using essentially unit conversion, any parameter $\pi_m = a_m/b_m$ can be written as an integer, $\hat{\pi}_m = \pi_m \cdot LCD(\pi_m)$ where $LCD(\pi_m)$ is the least common denominator of all π_m. Thus, to preserve full accuracy, we can use the greatest common factor (GCF) of the converted time-related parameters $(\hat{\tau}_i, \hat{\sigma}_{ii'}, \hat{\rho}_i, \hat{\varepsilon}_i)$ as the time discretization and convert all parameters accordingly: $\bar{\pi}_m = \hat{\pi}_m / GCF(\hat{\pi}_m)$. If π_m are fractional, then the process is equivalent to setting $\delta = 1/LCD(\pi_m)$. If a more coarse discretization is employed, either because full accuracy is unnecessary or leads to computationally expensive models, then one has to first choose the desired discretization, δ, and then use (3.33).

We consider six different instances of a single-unit problem with three batches using (1) two sets of release and due times (RD1 and RD2) and three sets of processing times (PT1, PT2, PT3), given in Table 3.2; and (2) two sets of changeover times (CT1 and CT2), given in Table 3.3. The problem and datasets used for each instance are given in Table 3.4. For clarity, the dataset index is used as superscript; for example, the release time of batch 2 in dataset RD1 is $\rho_2^{\mathrm{RD1}} = 4$.

In instance 1, all time-related parameters are integers, so $\pi_m = \hat{\pi}_m$, and $GCF(\pi_m) = 1$, so we can use $\delta = 1$. In instance 2, π_m is integral again but $GCF(\pi_m) = 2$, so we can use $\delta = 2$, which leads to the updated parameters given in row 3 of Table 3.5. In instance 3, processing times are fractional (see row 4 of Table 3.5), so to calculate the maximum δ that would ensure absolute accuracy, we do the following:

Table 3.2 Release, due, and processing time data.

	Release/due times		Processing times		
Batch	RD1	RD2	PT1	PT2	PT3
1	0/16	0/20	5	6	4.50
2	4/12	4/20	5	4	5.00
3	4/20	4/20	4	4	3.25

Table 3.3 Changeover times $\sigma_{ii'}$; i/i' given in the first column/row.

	CT1			CT2		
	1	2	3	1	2	3
1		0.50	0.75		0.5	1.5
2	1.25		0.50	1.0		1.0
3	1.00	1.5		1.0	2.0	

Table 3.4 Problem type and datasets of each instance.

Instance	Problem	ρ_i^k/ε_i^k	τ_i^k	$\sigma_{ii'}^k$
1	$[\boldsymbol{Sms}, \lvert\mathbf{J}\rvert = 1/\rho, \varepsilon/MS]$	RD1	PT1	
2	$[\boldsymbol{Sms}, \lvert\mathbf{J}\rvert = 1/\rho, \varepsilon/MS]$	RD1	PT2	
3	$[\boldsymbol{Sms}, \lvert\mathbf{J}\rvert = 1/\rho, \varepsilon/MS]$	RD1	PT3	
4	$[\boldsymbol{Sms}, \lvert\mathbf{J}\rvert = 1/\rho, \varepsilon, \sigma/MSS]$	RD1	PT3	CT1
5	$[\boldsymbol{Sms}, \lvert\mathbf{J}\rvert = 1/\rho, \varepsilon, \sigma/MS]$	RD1	PT1	CT2
6	$[\boldsymbol{Sms}, \lvert\mathbf{J}\rvert = 1/\rho, \varepsilon, \gamma^{CH}/C^{CH}]$	RD2	PT3	CT1

(1) Write all parameters as rational numbers (row 5 in Table 3.5).
(2) Calculate $LCD(\pi_m)$, which is equal to 4
(3) Calculate $\hat{\pi}_m = \pi_m \cdot LCD(\pi_m)$ (row 6, in Table 3.5), which are the converted parameters when we use $\delta = 1/LCD(\pi_m) = 0.25$.

If we choose to use $\delta = 0.5$, then the parameters should be updated as described in (3.33), resulting in the values given in row 8 of Table 3.5.

In instance 4, changeover times are integer multiples of 0.25, so $LCD(\pi_m) = 4$, as in instance 3, so we finally obtain $\bar{\pi}_m = 4\pi_m$. Instance 5 is the same as instance 4, except we use PT1 instead of PT3 (so all τ_i are integers) and CT2 instead of CT1 (so changeovers are integer multiples of 0.5). Since data are multiples of 0.5, the standard approach leads to $\delta = 0.5$. However, we observe that in any feasible solution there will be at most one fractional changeover time, since both fractional changeovers are from batch A. Thus, a solution obtained using $\delta = 1$ (i.e., $\sigma_{AB} = 1, \sigma_{AC} = 2$) will be feasible and can be readily *converted* to an exact solution by subtracting 0.5 if batch A is not processed last.

In instance 6, the objective function is the minimization of changeover cost (data not given), which means that the conversion of time-related data does not affect the value of

Table 3.5 Detailed calculations for instances 2 and 3 (In = instance).

In	Parameters	Row	Calculation steps	ρ_A	ρ_B	ρ_C	ε_A	ε_B	ε_C	τ_A	τ_B	τ_C
2	$GCF(\hat{\pi}_m) = 2$	2	$\hat{\pi}_m = \pi_m$	0	4	4	16	12	20	6	4	4
		3	$\bar{\pi}_m = \hat{\pi}_m / GCF(\hat{\pi}_m)$	0	2	2	8	6	10	3	2	2
3		4	π_m	0	4	4	16	12	20	4.5	5.0	3.25
		5	a_m / b_m	0/1	4/1	4/1	16/1	12/1	20/1	9/2	5/1	13/4
	$GCF(\hat{\pi}_m) = 1$	6	$\hat{\pi}_m = \pi_m \cdot LCD(\pi_m)$	0	16	16	64	48	80	18	20	13
	$GCF(\hat{\pi}_m) = 1$	7	$\bar{\pi}_m = \hat{\pi}_m / GCF(\hat{\pi}_m)$	0	16	16	64	48	80	18	20	13
		8	$\bar{\pi}_m$ using $\delta = 0.5$	0	8	8	32	24	40	9	10	7

the optimal solution as long as no feasible solution is cut off. (In general, what does this mean in terms of time data conversion? Can you see why inaccurate time-related data can lead to a suboptimal solution?) The standard approach would lead to $\delta = 0.25$. However, we observe that an upper bound on the total changeover time in any feasible solution is 3, which means that all solutions remain feasible even if a discretization $\delta = 1$ is used.

Example 3.2 is used to illustrate how the assignment constraint, (3.35), and the constraints enforcing changeover times, (3.43) and (3.44), are formulated.

Example 3.2 Assignment and Changeover Time Constraints We consider a toy instance with three batches, ($\mathbf{I} = \{B1, B2, B3\}$), with the processing and changeover times given in Table 3.6. We consider the constraints based on the execution of B3 at $n = 14$, that is, $X_{B3,14} = 1$. To simplify the notation, we define $\mathbf{N}_{in}^{U} = \{n - \bar{\tau}_i + 1, \ldots, n-1, n\}$, which represents the set of indices, for batch i, over which the inner summation in (3.35) should be performed. In other words, (3.35) is rewritten as

$$\sum_i \sum_{n' \in \mathbf{N}_{in}^{U}} X_{in'} \leq 1, \quad n.$$

In terms of the assignment constraint, (3.35), we note that if B3 starts at $n = 14$, then B1 cannot have started at $n \in \mathbf{N}_{B1,14} = \{8,.., 14\}$ because its processing time is 7; and if it had started at any $n \in \mathbf{N}_{B1,14}$, then its execution would continue during period 14, that is, B3 could not start at $n = 14$. Similarly, B2 cannot have started at $n \in \{11,.., 14\}$ because its processing time is 4. Also, note that if B3 starts at 14, then it cannot have started at $n \in \{7,.., 13\}$ because its processing time is 8. Thus, in the general case, we should allow only one batch to start between $n = 14 - \bar{\tau}_i + 1$ and $n = 14$.[12] The (i, n') pairs for which the corresponding $X_{in'}$ variables should be included in the LHS summation of (3.35) are shown as black squares in Figure 3.5A. Note that exactly the same conclusions would have been drawn had we used the execution of batch B1 or B2 at $n = 14$.

Table 3.6 Processing and changeover times of example instance.

		$\bar{\sigma}_{ii'}$ (i = left, i' = top)		
	$\bar{\tau}_{ii'}$	B1	B2	B3
B1	7	-	2	4
B2	4	1	-	3
B3	8	3	2	-

[12] Do you see the relationship between this range and the range used in the summation in the LHS of (3.35)? What happens if you replace 14 with n?

In terms of (3.43), we note that if B3 starts at $n = 14$, then B1 cannot have started at $n \in \{4,5,6,7\}$, in addition to $n \in \{8,.., 14\}$, which was established in the previous paragraph. This is because the changeover from B1 to B3 is 4, which means that even if B1 had started at $n = 14$, its execution would be completed at $n = 11 = 4 + 7$, which in turn means that the changeover to B3 would be completed at $n = 15 = 11 + 4$. Thus, the unit would not be available to start B3 at $n = 14$. Similarly, B2 cannot have started at $n \in \{8,9,10\}$, in addition to $n \in \{11, \ldots, 14\}$ explained in the previous paragraph. Note that no new points must be excluded due to the execution of batch B3 earlier (why?). Thus, in the general case, if batch i is executed at n, then this means that (1) a batch $i' \neq i$ cannot have started between $n - \bar{\tau}_{i'} - \bar{\sigma}_{i'i} + 1$ and $n - \bar{\tau}_{i'}$ (check this range using $n = 14$, $\bar{\tau}_{i'} = \bar{\tau}_{\mathrm{B1}} = 7$, and $\bar{\sigma}_{i'i} = \sigma_{\mathrm{B1,B3}}$); and (2) no additional points should be excluded for batch i.[13] The (i', n') pairs for which the corresponding $X_{i'n'}$ variables should be included in the LHS summation of (3.43), expressed for $i = \mathrm{B3}$ and $n = 14$, are shown as black disks in Figure 3.5B.

Finally, for (3.44), we note that, as in the previous paragraph, if $i = \mathrm{B3}$ starts at $n = 14$, then B2 cannot have started at $n \in \{8, 9, 10\}$, but in this case, this is enforced by three different constraints: $X_{\mathrm{B3},14} + X_{\mathrm{B2},8} \leq 1$, $X_{\mathrm{B3},14} + X_{\mathrm{B2},9} \leq 1$, and $X_{\mathrm{B3},14} + X_{\mathrm{B2},10} \leq 1$. In the general case, for each (i, n) pair, we write $\sum_{i' \neq i} \bar{\sigma}_{i'i}$ constraints (do you see why?). Figure 3.5C shows the points n' (solid black disks) for which (3.44) should be expressed for $(i, n) = (\mathrm{B3}, 14)$ and we consider $i' = \mathrm{B2}$.

3.5 Extensions

In this section, we discuss two extensions: the *prize collection* variant of the single-unit problem and *product families*. In the former, a maximum demand is used to define all batches that can be produced within the horizon, but not all of them can be executed due to unit availability, so one has to select the most profitable ones.[14] In the latter, products belong to product families, and products belonging to different families have sequence-dependent changeovers (costs and/or times), while products within the same family have no or sequence-independent changeover costs and/or times. Both extensions arise in all production environments, but we discuss them in detail in this chapter only.

3.5.1 Prize Collection Problem

If there is no minimum demand, but rather we try to maximize the benefit from carrying (some of) the batches, then we consider a maximization problem that also involves

[13] Do you see why no binary variables for batch i are included in the summation in (3.43)?

[14] The problem takes its name from the *prize collecting* traveling salesman problem (TSP), where a salesman cannot visit all cities, so he has to pick the ones that will maximize the salesman's revenue, that is, find the tour that would maximize the collection of prizes.

batch selection decisions: $Z_i = 1$ if batch i is carried out. The objective function, PC, then for all problems $[\boldsymbol{Sms},\ |\mathbf{J}| = 1/\cdot/PC]$ becomes

$$\max \sum_i \pi_i Z_i, \tag{3.52}$$

where π_i is the *prize* for carrying out batch i.

In global sequence models, binary variable $Y_{ii'}$ can be equal to 1 only if both batches i and i' are selected, a condition that can be enforced via

$$Y_{ii'} \leq Z_i, \quad i, i' \neq i \tag{3.53}$$

$$Y_{i'i} \leq Z_i, \quad i, i' \neq i, \tag{3.54}$$

and if two batches are selected, then a sequence should be established:

$$Y_{ii'} + Y_{i'i} \geq Z_i + Z_{i'}, -1, \quad i, i' > i. \tag{3.55}$$

If a batch is not selected, then it would be logical to set its start and end date to zero and relax the corresponding release time and deadline constraints. Accordingly, start times can be constrained as follows:

$$S_i \leq MZ_i, \quad i, \tag{3.56}$$

and (3.3) and (3.4) are replaced by (3.57) and (3.58), respectively:[15]

$$S_i \geq \rho_i Z_i, \quad i \tag{3.57}$$

$$S_i + \tau_i Z_i \leq \varepsilon_i Z_i, \quad i. \tag{3.58}$$

The global sequence model for problem $[\boldsymbol{Sms},\ |\mathbf{J}| = 1/\rho, \varepsilon/PC]$ consists of (3.1), (3.57), (3.58), (3.53) through (3.55), and (3.52).

In immediate sequence models, (3.12) can be used for the activation of Y_i^F and Y_i^L variables, assuming that at least two batches are selected. The activation of sequencing binary variables can be achieved through the following two constraints, which can be used instead of (3.13) and (3.14):

$$\sum_i Y_{ii'} = Z_{i'} - Y_{i'}^F, \quad i' \tag{3.59}$$

$$\sum_{i'} Y_{ii'} = Z_i - Y_i^L, \quad i \tag{3.60}$$

and

$$\sum_{i,i'} Y_{ii'} = \sum_i Z_i - 1, \tag{3.61}$$

which is the counter part of (3.15). Variables Y_i^F and Y_i^L satisfy

$$Y_i^F + Y_i^L \leq Z_i, \quad i \tag{3.62}$$

[15] Are all equations (3.56) through (3.58) strictly necessary? If (3.56) is omitted, then $S_i > 0$ would be feasible for an unselected batch. But would (3.58) then be sufficient to enforce $S_i = 0$ if $Z_i = 0$? Alternatively, can (3.57) and (3.58) be rewritten so that they correctly enforce release times and deadlines, even if $S_i > 0$ for batches with $Z_i = 0$? Also, if (3.56) is removed, would (3.3) and (3.4) be sufficient?

One immediate sequence model for problem $[\boldsymbol{Sms}, \ |\mathbf{J}| = 1/\rho, \varepsilon/PC]$ then consists of (3.1), (3.3), (3.4), (3.59) through (3.62), and (3.52).

Next, we describe one of the many models that can be developed based on a continuous time grid. Binary variables Z_i are also used to denote the selection of batch i, and binary variable $\hat{Z}_t$ is introduced to denote the allocation of any batch to period t, that is, $\hat{Z}_t = 1$ if a batch is allocated to period t. Since the number of the selected batches is an optimization decision, $|\mathbf{I}|$ time periods should be postulated, but unnecessary periods can be stacked at the end of the horizon,

$$\hat{Z}_t \leq \hat{Z}_{t-1}, \quad t > 1. \tag{3.63}$$

Batch i is allocated to a period only if it is selected,

$$\sum_t X_{it} = Z_i, \quad i. \tag{3.64}$$

and only active periods can be allocated a batch:

$$\sum_i X_{it} = \hat{Z}_t, \quad t. \tag{3.65}$$

In addition to (3.63) through (3.65) which replace (3.19) and (3.20), the model consists of (3.17), (3.18), (3.21), (3.22), and objective function (3.52).[16] Note that variables $\hat{Z}_t$ and (3.63) and (3.65) are not necessary, but lead to a tighter model with fewer equivalent solutions.[17] In general, any model based on a continuous time grid for problem $[\boldsymbol{Sms}, |\mathbf{J}| = 1/\rho, \varepsilon/MS]$ can be converted into a model for problem $[\boldsymbol{Sms}, |\mathbf{J}| = 1/\rho, \varepsilon/PC]$ by (1) introducing variables and equations to account for batch selection (e.g., variables Z_i), and (2) modeling the condition that a batch is allocated to a period if and only if it is selected, which leads to the activation of X_{it} variables, which can then be used to enforce the disjunctive constraint.

Finally, discrete time models can also be modified via the introduction of selection binary variable Z_i, which is then used to constrain the assignment of batches (instead of (3.34)):

$$\sum_n X_{in} = Z_i, \quad i. \tag{3.66}$$

The model based on a discrete time grid for problem $[\boldsymbol{Sms}, |\mathbf{J}| = 1/\rho, \varepsilon/PC]$ consists of (3.66), (3.35), (3.36), and (3.52).

3.5.2 Product Families

Products often belong to product families $f \in \mathbf{F}$. The grouping into families can be based on various criteria, including product similarities, processing similarities, or

[16] Are equations similar to (3.56) through (3.58) necessary?

[17] The term *equivalent solutions* is used to describe sets of solutions that are mathematically different but correspond to the same schedule. Consider the solution shown in Figure 3.3A, where batch starting times are $\boldsymbol{S} = [S_A, S_B, S_C] = [1, 5, 8]$. The shown *scheduling* solution (Gantt chart) can be obtained by a large number of equivalent *mathematical* solutions with the same $[S_A, S_B, S_C] = \boldsymbol{S}$ vector but different $[T_1, T_2, T_3]$ vectors. For example, $\boldsymbol{S}$ is feasible for the following $[T_1, T_2, T_3]$ vectors: $[0, 5, 8], [0, 4, 8], [0, 4.1, 8], [0, 4.2, 8], \ldots, [0, 5, 7], [0, 5, 7.1], \ldots$.

changeover considerations. The goal of this grouping is to lead to computationally tractable optimization models without compromising the quality of solution. Since the number of batches of each product is known, batches can also be assigned to families: the subset of batches of family f is $\mathbf{I}_f$. The transition between batches in different families (i.e., from $i \in \mathbf{I}_f$ to $i' \in \mathbf{I}_{f'}$ with $f \neq f'$) incurs a sequence-dependent changeover time $\sigma_{ff'}$ and/or cost $\gamma^{CH}_{ff'}$, while transitions between a pair of batches in the same family have no changeover or sequence-independent changeover time and/cost (setup).

One approach to modeling product families is to use a model that accounts for changeover times/costs (i.e., models for problems $\boldsymbol{Sms}$, $|\mathbf{J}| = 1/\rho, \varepsilon, \sigma, \gamma/\cdot]$) and simply update changeover parameters as follows:

(1) Products within a family (i.e., $i \in \mathbf{I}_f$ and $i' \in \mathbf{I}_f$ for some f):
- If there is no changeover time/cost, then $\sigma_{ii'} = \gamma^{CH}_{ii'} - 0$.
- If there is setup time, one approach is to still use a model that addresses sequence-dependent times but replace them as follows: $\sigma_{i'i} = \sigma_i, \forall i' \in I_f$, where σ_i is the setup time of $i \in I_f$. If $\sigma_i < \sigma_{f'f}, \forall f, f', i \in I_f$(which is a reasonable assumption), then an alternative approach is to use adjusted processing times $\tau^S_i = \tau_i + \sigma_i$, zero changeovers between batches of the same family, and adjusted changeovers between batches of different families, $\sigma^S_{ii'} = \sigma_{ii'} - \sigma_{i'}$. (Do you see any other adjustments that have to be made? How would you make them?)

(2) Products in different families (i.e., $i \in \mathbf{I}_f$ and $i' \in \mathbf{I}_{f'}$ with $f \neq f'$):
- Sequence-dependent time/cost: $\sigma_{ii'} = \sigma_{ff'}\gamma^{CH}_{ii'} = \gamma^{CH}_{ff'}$.

The treatment of product families using models for problems $[\boldsymbol{Sms}, |\mathbf{J}| = 1/\rho, \varepsilon, \sigma, \gamma/\cdot]$ with adjusted parameters can be effective when immediate-sequence models are employed (why?). However, models can also be simplified in additional ways. For example, in sequence-based models, if changeover times and costs between products of the same families are zero, then (3.1) can be used instead of (3.6) for products in the same family. Similarly, in models based on a continuous time grid, variables $Y_{ii't}$ and (3.28) can be removed for batches i, i' in the same family; the summation in (3.29) can be modified to include batches $i' \in \mathbf{I}_f$ if $i \in \mathbf{I}_f$; and the summation in (3.30) can be over $i \notin \mathbf{I}_f$ where $i' \in \mathbf{I}_f$.

Finally, we note that while the preceding simple approach is easy to implement, it may lead to unnecessarily large formulations, so one may want to develop specific formulations that better exploit product families, especially when the cardinality of $\mathbf{I}_f$ is large. This topic will be discussed again in Chapter 8.

3.6 Remarks

In this section, we discuss some commonly made assumptions (Section 3.6.1), variable fixing (Section 3.6.2), and alternative solution representations and models (Section 3.6.3); present some key points regarding model size (Section 3.6.4); discuss the trade-off between problem-specific and general models (Section 3.6.5); and close, in

Section 3.6.6, with some recommendations for model selection given the characteristic of an instance. All discussed aspects and questions arise in all production environments, but we discuss them in detail in the present chapter only.

3.6.1 Assumptions

In Section 3.2.1, we enforced sequencing between all pairs of batches. This means that a separation larger than σ_{BC} may be enforced between batch B and C if batch A immediately precedes batch B and the following holds true: $\sigma_{\text{AC}} \geq \sigma_{\text{AB}} + \tau_B + \sigma_{\text{BC}}$. Thus, it is assumed that

$$\sigma_{ii''} < \sigma_{ii'}, + \tau_{i'} + \sigma_{i'i''}, \quad i, i', i''. \tag{3.67}$$

Another typical assumption is that

$$\sigma_{ii''} < \sigma_{ii'} + \sigma_{i'i''}, \quad i, i', i''. \tag{3.68}$$

If (3.68) does not hold true but executing a changeover automatically *enforces* the execution of the batch, then this is not an issue. If (3.68) is not satisfied and executing a changeover does not imply the execution of the corresponding batch, then the fastest transition from i to i'', via i' but without executing batch i', will be preferred, leading to an incorrect solution.[18] Note that if (3.68) is satisfied, then (3.67) is as well.

Furthermore, it has been assumed that (1) the processing unit is the only shared resource, that is, no other shared resources, such as labor or electricity, are required for the execution of a task; and (2) that unlimited storage is available for the batch inputs and outputs. As we will see in the following chapters, general resource constraints and storage restrictions can play an important role in scheduling. Specifically, in Chapters 4 and 5 we discuss how general resource and storage constraints, respectively, can impact the quality of a solution, and then discuss modeling approaches to account for them. The approaches we will present in these chapters, in the context of single- and multistage problems, respectively, are readily applicable to single-unit problems that are a special case of the two aforementioned classes of problems.

3.6.2 Variable Fixing

In general, many variables can be fixed prior to solving an optimization model, thus reducing not only the number of variables but also the number of constraints via preprocessing. First, variables can be fixed using release and due time information. In sequence-based models, for example, we can set $Y_{ii'} = 1$ (which implies $Y_{i'i} = 0$) if $\varepsilon_i < \rho_{i'} + \tau_{i'} + \tau_i$ in the presence of deadlines. Similarly, in continuous time models, the same inequality implies that batch i cannot be assigned to the last period (since i' should

[18] Can you think of problems in which the execution of a changeover to batch i' automatically implies the execution of batch i'? Can you think of a problem we have discussed in the present chapter where this may not be the case?

be processed after i), so $X_{i,t=|\mathbf{I}|} = 0$, and i' cannot be assigned to the first time period, so $X_{i,t=1} = 0$. If similar inequalities hold true for batch i and multiple other batches $i', i'', \cdots$, then more binaries can be fixed to zero. In general, fixing binary variables may allow us to either modify (tighten) or remove constraints. Can you identify constraints in the previous sections that can be modified/removed if some sequencing, $Y_{ii'}$, or assignment, X_{it}, variables are fixed?

In discrete time models, we can use release and due times to define the set, $\mathbf{N}_i$, of feasible time points at which batch $i \in \mathbf{I}$ is allowed to start, $\mathbf{N}_i = \{n \in N : \bar{\rho}_i \le n \le \bar{\varepsilon}_i - \bar{\tau}_i\}$; and the set, $\mathbf{I}_n$, of batches that can start at time point n, $\mathbf{I}_n = \{i \in \mathbf{I} : \bar{\rho}_i \le n \le \bar{\varepsilon}_i - \bar{\tau}_i\}$. These two sets can then be used to (1) fix variables to zero (e.g., $X_{in} = 0$ if $n \notin \mathbf{N}_i$); (2) reduce the number of equations (e.g., replace (3.43) with (3.69)); and (3) reduce the number of variables included in the equations (e.g., rewrite (3.45) as in (3.70)):

$$\sum_{i' \neq i} \sum_{n'=n-\bar{\tau}_{i'}-\bar{\sigma}_{ii'}+1}^{n-\bar{\tau}_{i'}} X_{i'n}, \le M_i(1 - X_{in}), \quad i, n \in \mathbf{N}_i \tag{3.69}$$

$$\sum_{i \in \mathbf{I}_n} \bar{X}_{in} = 1, \quad n. \tag{3.70}$$

While the aforedescribed variable fixing can be effective, modern MIP solvers have powerful preprocessing algorithms that in most cases also lead to similar variable fixing and constraint tightening or removal.

3.6.3 Alternative Models

Equations (3.1) and (3.6) enforce that there is no overlap between batches through an inequality that involves, essentially, the finish time of the predecessor and the start time of the successor. The same condition can be enforced using a number of different approaches, as shown in Figure 3.7. Each of these approaches leads to a different sequence-based model. Note that the models can differ, among others, in the following ways:

(1) The exact form of the constraint, enforcing the same inequality (Figure 3.7A vs. Figure 3.7B)
(2) The use of different constraints and variables to enforce the same logic condition – the disjunctive constraint (Figure 3.7A vs. Figure 3.7C vs. Figure 3.7D)
(3) The use of different big-M parameters

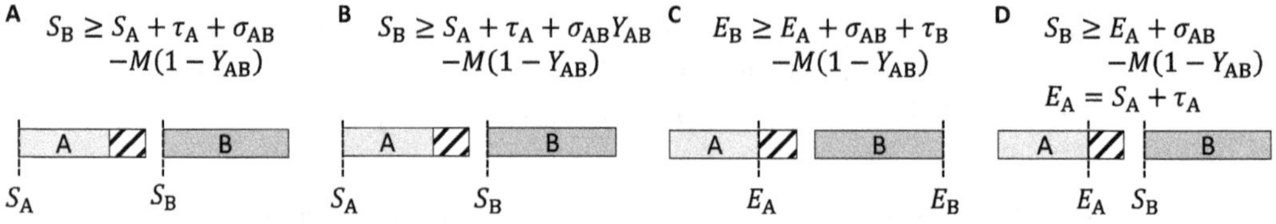

Figure 3.7 Alternative approaches to enforcing no-overlap (disjunctive) constraint.

3.6.4 Model Size

We discuss how the size of the different models increases as instance size increases, so the reader gains an appreciation of this aspect, though we note that, as we will discuss in subsequent chapters, model size is not the only model characteristic determining computational performance. To quantify model size, we will borrow from computer science the concept of the *order of a function*, and specifically the *big O notation,* which describes the limiting behavior of a function as its argument goes to infinity. In our context, we are interested in understanding how the number of variables and constraints of a model increase as the instance increases, in this case, measured by the number of batches, $|\mathbf{I}|$. If $f^V(|\mathbf{I}|)$ is the number of variables of a model as a function of the number of batches, then we say that $f^V(|\mathbf{I}|)$ is of order $|\mathbf{I}|^2$, denoted as $O(|\mathbf{I}|^2)$, if and only if for all sufficiently large values of $|\mathbf{I}|$, the value of $f^V(|\mathbf{I}|)$ is at most a positive constant multiple of $|\mathbf{I}|^2$; that is, $f^V(|\mathbf{I}|) \leq k|\mathbf{I}|^2$.

We consider models for the basic problem, introduced in Section 3.1. Since no additional binary variables are needed to model changeovers in sequence-based models, the number of binary variables $Y_{ii'}$ is always $O(|\mathbf{I}|^2)$. The number of binary variables X_{it} in models based on a continuous time grid is $O(|\mathbf{I}|^2)$ if there are no changeovers, while $O(|\mathbf{I}|^3)$ variables are required if variables $Y_{ii't}$ are used to model changeovers.[19] Discrete time models employ $|\mathbf{I}| \cdot |\mathbf{N}|$ binary variables X_{in}, which is $O(|\mathbf{I}|^2)$ because the number of time periods depends, in general, linearly on the number of batches. The discretization (δ) determines, essentially, the number of periods per batch, which is the processing time, $\bar{\tau}_i$ – see (3.33). If τ_{AVG} is the average processing time, in multiples of δ, then the number of time points will be $(1+\lambda)\tau_{AVG}\,|\mathbf{I}|$, where λ is the ratio of idle time over total processing time, which is not a function of $|\mathbf{I}|$.[20] If fine discretization is employed, then τ_{AVG} increases, but the order remains the same. If changeover costs are present, then the number of variables becomes $O(|\mathbf{I}|^3)$ due to the introduction of variables $Y_{ii'n}$. Table 3.7 gives the number of variables employed by different types of models for different problems and instances of different size.[21]

We make the following observations:

(1) If there are no changeovers (*NC*), then all models are comparable with the exception of the discrete time model, which, if the other models require X variables, requires $(1+\lambda)\tau_{AVG}$X variables. In other words, as expected, the discretization impacts the relative size of the model.

[19] For the basic single-unit problem, defined in Section 3.1, we know that $|\mathbf{I}|$ time periods are necessary ($|\mathbf{I}| = |\mathbf{T}|$), and thus we can easily determine the number of variables indexed by time periods. We note that this is not the case in other problems, discussed in subsequent chapters, where the number of the necessary time periods cannot be easily determined.

[20] Here, we assume that $\tau_{AVG}\,|\mathbf{I}|$ is an estimate of the total processing times, and thus $(1+\lambda)\tau_{AVG}\,|\mathbf{I}|$ is an estimate of the horizon we consider. In the problem defined in Section 3.1, we know that the exact total processing time is $\Sigma_i\bar{\tau}_i$, but this calculation is not possible in other problems where, for example, a batch can be assigned to different units with different processing times.

[21] We assume that the techniques described in Section 3.6.2, to reduce the number of variables, are not applied.

Table 3.7 Number of variables employed by the four types of models, for three problems, and instances of different size.

Model type	Binary variables (*NC*/*CT*/*CC*)	Number of variables[a] (*NC*/*CT*/*CC*) $\lvert\mathbf{I}\rvert = 5$	$\lvert\mathbf{I}\rvert = 20$	$\lvert\mathbf{I}\rvert = 100$
Immediate-sequence	$Y_{ii'}/Y_{ii'}/Y_{ii'}$	20/20/20	380/380/380	9900/9900/9900
Global-sequence	$Y_{ii'}/Y_{ii'}/-$[b]	20/20/ –	380/380/ –	9900/9900/ –
Continuous time	$X_{it}/X_{it}, Y_{ii't}/X_{it}, Y_{ii't}$	25/125/125	$400/8 \cdot 10^3/8 \cdot 10^3$	$10^4/10^6/10^6$
Discrete time[c]	$X_{in}/X_{in}/X_{in}, Y_{ii'n}$	150/150/ 600	$2.4 \cdot 10^2/2.4 \cdot 10^2/$ $4.6 \cdot 10^4$	$6 \cdot 10^4/6 \cdot 10^4/$ $5.94 \cdot 10^6$

Note: Problem abbreviations: *NC*: no changeover time and cost; *CT*: changeover time; *CC*: changeover cost.
[a] Variables Y_{ii}, Y_{iit} and Y_{iin} (denoting changeover from i to itself) can be set to zero.
[b] Global-sequence models cannot readily address problems with changeover costs.
[c] We assume $\lambda = 0.2$ and $\tau_{AVG} = 5$.

(2) In the presence of changeover times (*CT*), the number of binary variables for both sequence-based and discrete-time-based models stays the same, while continuous time models require a new variable, $Y_{ii't}$.

(3) In the presence of changeover costs (*CC*), models based on immediate sequencing and a continuous time grid require as many variables as in the *CT* case, while models based on discrete time grid require new variable, $Y_{ii'n}$.

In terms of constraints, both immediate and global sequence-based models employ $O(|\mathbf{I}|^2)$ constraints. In all sequence-based models, at large $|\mathbf{I}|$, (3.1) or (3.6) dominate (note that (3.2) is roughly half of (3.1)). In models employing a continuous time grid (with $|\mathbf{I}|$ periods), constraints written for all i and t dominate if there are no changeovers, so the number of constraints is $O(|\mathbf{I}|^2)$. If there are changeovers, then (3.28) leads to $O(|\mathbf{I}|^3)$ constraints. In models based on a discrete grid, if there are no changeovers, the number of constraints is $O(|\mathbf{I}|)$ since (3.35) is written for every time point, which depends linearly on the number of batches. If there are changeover times, then the number of constraints becomes $O(|\mathbf{I}|^2)$ if (3.43) is used and $O(|\mathbf{I}|^3)$ if (3.44) is used. Finally, the number of constraints to model changeover costs is $O(|\mathbf{I}|^3)$.

Note that the number of variables and constraints in sequence-based and continuous-grid-based models depends only on the number of batches (i.e., the instance size), whereas the size of a discrete time model depends on the number batches but also the employed discretization, which is a user decision.

3.6.5 Problem-Specific versus General Models

A major question when developing and/or using MIP scheduling models is the trade-off between generality and computational tractability. Since, most MIP models become intractable as their size increases, one has to often exploit the structure of the specific problem at hand in order to develop a model that is computationally effective, albeit applicable to few problems or even few instances of one problem. On the other hand, a general model has the advantage of being applicable to a wider range of problems and/or instances, thus limited development is needed, but can be unnecessarily complex, and thus computationally less effective, for relatively simple problems.

In that respect, models based on a discrete time grid are preferred because, as we will see in Chapters 7 through 10, they remain practically the same for problems in various environments and/or problems with additional processing features, whereas sequence-based models and models employing continuous time grids become increasingly complex as the environment becomes more complex and/or additional features are considered. Specifically, the major advantages of models based on a discrete time grid are the following:

(1) They can be readily extended to account for processing features such as general resource constraints (e.g., labor and utilities); time-varying cost and availability of resources; intermediate deliveries of raw materials and orders of final products; and various storage restrictions.[22]

(2) The extended models that account for the aforementioned features remain linear and are not significantly larger than the "standard" models, with the exception of changeover times and costs.

However, these models are larger than the tailored sequence-based or continuous-time-grid-based models that can be developed for simpler problems (e.g., in single-unit environment) by exploiting structure of these problems.

3.6.6 Recommendations

We present some general guidelines for the selection of a model to address single-unit problems:

(1) Models based on a discrete time grid should be selected only when processing features that render competing models inapplicable or intractable are present. For example, general resource constraints cannot be modeled using sequence-based models, and time-varying resource costs lead to the introduction of bilinear terms, and thus MINLP models, when a continuous time grid is used. The following recommendations are made assuming that there are no general resource constraints.

(2) Immediate sequence-based models are preferable if changeover costs have to be considered, since no additional binary variables are required.

[22] While in the present chapter we assumed that there are no general resource and storage constraints, problems in single-unit environments can be subject to such constraints. For example, limited storage may be available for final products (i.e., the output of a batch).

(3) Immediate or global sequence-based models are preferable if changeover times are considered.

(4) Models based on a continuous time grid are preferred when there are no changeover times/costs.

These guidelines should be used only as a starting point, since there are several other factors that may affect computational performance. The most important are the following:

(1) *Objective function*: The choice of the objective function affects differently the performance of different models.

(2) *Processing/changeover times and desired level of accuracy.* As we discussed in Section 3.6.4, the size of models based on a discrete time grid depends on τ_{AVG} (or $\sum_i \bar{\tau}_i$); for example, if all processing times are integers in $\{2, 3, 4\}$, then a model based on a discrete time grid may be the fastest.

(3) *Release and due times.* As we discussed in Section 3.6.2, release and due times can be used to fix binary variables and, in general, reduce model size. Since the level of preprocessing depends on the type of model, the effect of tight windows will impact different models in different ways. For example, tight release/due dates and multiple rounds of preprocessing can lead to the fixing of multiple $Y_{ii'}$ variables in sequence-based models and X_{in} variables in discrete time models, but it is not obvious how continuous time models would be affected.

(4) *Other instance-specific parameters.* The performance of MIP models depends heavily on the parameters of a given instance. For example, the relative magnitude of processing and changeover times or the uniformity of processing times may affect performance. Thus, different models have to be tested to determine which one is most effective for a particular set of instances. Since most parameters remain the same across all instances for a given facility, this testing has to be performed only once or repeated infrequently (i.e., when the changes in the parameters are substantial).

3.7 Notes and Further Reading

(1) The single-unit problem discussed in this chapter is a major building block for all the problems we will study in the following chapters, rather than a problem often encountered in chemical production facilities. Nevertheless, it is essentially identical to the single machine problem we introduced in Chapter 1, a classic discrete manufacturing problem, and has therefore received a lot of attention in the operations research (OR) community.

(2) Throughout the book, we use three seemingly similar terms: sequential, sequencing (and sequence), and precedence. *Sequential* is used to describe a production environment. *Sequencing* refers to the determination of a sequence in which batches are processed; for example, binaries $Y_{ii'}$ are called sequencing binaries because they describe the relative sequence in which batches i and i' are

processed, that is, they establish a sequencing relationship. Models employing sequencing binary variables are referred to as (global/immediate) sequence-based models or sequence models. *Precedence* is used to describe a given (fixed) constraint, typically referred to as precedence relationship or precedence constraint, regarding the sequence in which two batches have to be processed; for example, batch A should be processed before batch B in all solutions (i.e., A is a predecessor of B). Note that a sequencing (relationship) is determined by the optimization, whereas a precedence relationship is a given constraint.

(3) The use of terms *sequence/sequencing* and *precedence* in the book is consistent with their use in the scheduling literature, but is in disagreement with their use in the process systems engineering literature (including papers of the author of this book), where models employing binary variables $Y_{ii'}$ are referred to as precedence based. However, the use of the term *sequence* and *sequence based*, as opposed to *precedence* and *precedence based*, to describe the modeling based on binary variables $Y_{ii'}$ is correct because, in addition to following the standard scheduling notation, it is also consistent with the use of the term *sequencing* to describe one of the optimization decisions carried out in scheduling (see discussion in Chapter 1).

(4) Equation (3.1), or (3.6), is the basic constraint enforcing sequencing between two batches executed in the same unit. This *disjunctive* constraint, or one of its variants (see discussion in Section 3.6.3), is used, after small modifications, in sequence-based models for problems in all sequential environments. Equation (3.1) is always coupled with constraints for the activation of the *correct* $Y_{ii'}$ binary variables, and it is the way in which this activation is achieved that leads to a wide range of sequence-based models.

(5) A key feature of models based on a continuous time grid, for a wide range of problems, is the allocation of (the start and end times of) batches to time periods and/or points. In that respect, (3.21) and (3.22), or (3.23) and (3.24), are the simplest forms of these constraints. As we will see in subsequent chapters, all models based on continuous time grids employ similar constraints to model this *mapping* of batches onto a time grid.

(6) Equation (3.35) is the basic building block of most models based on a discrete time grid for a wide range of production environments. In fact, it is used, practically unchanged, in models for problems in all production environments to enforce the no-overlap condition in each unit. For example, as we will see in Chapter 7, models for problems in network environments can be generated by simply including (3.35) for all units, and then adding a flow (material) balance constraints for materials and a variable lower/upper bound constraint for batchsizes.

(7) The effective solution of models containing this *clique* constraint (see Example 2.9) has received considerable attention in the context of the single-unit problem [1–5].

(8) Pekny et al. propose an exact parallel algorithm for solving a single-unit problem [6]. Interestingly, they show that the problem can be transformed into a prize collecting TSP.

(9) The modeling of changeovers using a discrete time grid has received considerable attention both in the PSE and OR literature. The constraints for changeover times discussed in Section 3.4 are based on the models of Kondili et al [7] and Shah et al. [8], while the constraints for changeover costs are based on the models proposed by Karmakar and Schrage [9].

(10) Kelly and Zyngier discuss different approaches to modeling changeover times [10]. Velez et al. propose new constraints for the enforcement of changeover times as well as theoretical results regarding their tightness [11]. A review of methods to model setup and changeover times is presented in Wolsey [12]. Velez et al. [11] presented an extensive computational comparison among approaches to model changeover times using discrete time models. A continuous time model was proposed in Lasserre and Queyranne [13].

(11) Different ways to model product families are discussed in Kopanos et al. [14].

(12) A model that accounts for a number of practical constraints was proposed by Lima et al. [15].

3.8 Exercises

(1) Consider a single unit environment consisting of four batches $\mathbf{I} = \{B1, B2, B3, B4\}$ with processing times, release and due times, and changeover times and costs given in Table 3.8.

Using the table's data, formulate and solve the following problems using a sequence-based model:

(a) Makespan minimization without changeovers.

(b) Makespan minimization accounting for changeover times; compare with problem (a).

(c) Earliness minimization with weight factor $\omega_i = \{4, 5, 1, 10\}$.

(d) Makespan minimization with changeovers using the immediate sequence-based model; compare with the solution obtained in (a).

(e) Changeover cost minimization.

Table 3.8 Processing times (τ_i), release (ρ_i) and due (ε_i) times, and changeover times ($\sigma_{ii'}$) and costs ($\gamma_{ii'}$).

			$\sigma_{ii'}/\gamma_{ii'}$ (i = left, i' = top)			
	τ_i	ρ_i/ε_i	B1	B2	B3	B4
B1	2	0/15	-	1/1	2/1	1/1
B2	4	6/15	1/4	-	1/2	1/2
B3	3	5/20	1/1	2/8	-	1/1
B4	5	2/15	1/1	3/1	1/1	-

Table 3.9 Batch processing times (τ_i), release (ρ_i) and due (ε_i) times, weights (ω_i), and prices (π_i).

	B1	B2	B3	B4	B5	B6	B7	B8	B9	B10	B11	B12	B13
τ_i	2	4	5	7	3	4	3	8	5	4	6	7	10
ρ_i	0	0	0	5	5	8	2	10	4	4	14	10	9
ε_i	15	20	10	30	50	50	80	80	80	40	80	60	80
ω_i	5	10	5	5	10	5	5	10	5	5	5	5	10
π_i	6	10	6	8	4	11	5	6	10	3	5	8	12

Table 3.10 Changeover times/costs ($\sigma_{ii'}/\gamma_{ii'}$); i/i' in first column/row.

	F1	F2	F3
F1	–	3/5	4/10
F2	2/5	–	5/10
F3	3/5	3/10	–

(2) Using the data for the instance in Exercise (1), formulate and solve the following problems:
   (a) Makespan minimization without and with changeover times using a grid-based continuous-time model with $\eta = 20$.
   (b) Changeover cost minimization using a grid-based continuous time model.
   (c) Repeat (a) and (b) using a grid-based discrete time model with $\delta = 1$.
   (d) Repeat (c), using $\delta = 2$. Can you explain the infeasibility?

(3) Consider a single-unit system with 13 batches, $\mathbf{I} = \{B1, B2, \ldots, B13\}$, with processing times, release and due times, weight factors, and prize information given in Table 3.9. The batches belong to different product families $\mathbf{F} = \{F1, F2, F3\}$, wherein $\mathbf{I}_{F1} = \{B1, B2, B3, B4\}$, $\mathbf{I}_{F2} = \{B5, B6, B7, B8\}$, and $\mathbf{I}_{F3} = \{B9, B10, B11, B12, B13\}$. The corresponding changeover times and costs are given in Table 3.10. There is also a setup time of 1 for batches B2, B7, and B11. Solve the following instances:
   (a) Makespan minimization accounting for the changeover times.
   (b) Earliness minimization using the given weight factors.
   (c) Changeover cost minimization.
   (d) Tardiness minimization where $\varepsilon_i = 40$ for $i \in \{B7, B8, B9, B11, B13\}$ (instead of $\varepsilon_i = 80$), while ε_i for other batches remains the same.
   (e) Prize collection using the updated ε_i values from (d).

References

[1] Sousa JP, Wolsey LA. A Time Indexed Formulation of Nonpreemptive Single-Machine Scheduling Problems. *Math Program.* 1992;54(3):353–367.

[2] Wolsey LA. Valid Inequalities for 0-1 Knapsacks and MIPs with Generalized Upper Bound Constraints. *Discrete Applied Mathematics.* 1990;29(2–3):251–261.

[3] van den Akker JM, Hurkens CAJ, Savelsbergh MWP. Time-Indexed Formulations for Machine Scheduling Problems: Column Generation. *INFORMS J Comput.* 2000;12(2):111–124.

[4] Sadykov R, Wolsey LA. Integer Programming and Constraint Programming in Solving a Multimachine Assignment Scheduling Problem with Deadlines and Release Dates. *INFORMS J Comput.* 2006;18(2):209–217.

[5] van den Akker JM, van Hoesel CPM, Savelsbergh MWP. A Polyhedral Approach to Single-Machine Scheduling Problems. *Math Program.* 1999;85(3):541–572.

[6] Pekny JF, Miller DL, Mcrae GJ. An Exact Parallel Algorithm for Scheduling When Production Costs Depend on Consecutive System States. *Comput Chem Eng*. 1990;14 (9):1009–1023.
[7] Kondili E, Pantelides CC, Sargent RWH. A General Algorithm for Short-Term Scheduling of Batch-Operations .1. MILP Formulation. *Comput Chem Eng*. 1993;17(2):211–227.
[8] Shah N, Pantelides CC, Sargent RWH. A General Algorithm for Short-Term Scheduling of Batch-Operations .2. Computational Issues. *Comput Chem Eng*. 1993;17(2):229–244.
[9] Karmarkar US, Schrage L. The Deterministic Dynamic Product Cycling Problem. *Oper Res*. 1985;33(2):326–345.
[10] Kelly JD, Zyngier D. An Improved MILP Modeling of Sequence-Dependent Switchovers for Discrete-Time Scheduling Problems. *Ind Eng Chem Res*. 2007;46(14):4964–4973.
[11] Velez S, Dong Y, Maravelias CT. Changeover Formulations for Discrete-Time Mixed-Integer Programming Scheduling Models. *Eur J Oper Res*. 2017;260(3):949–963.
[12] Wolsey LA. MIP modelling of changeovers in production planning and scheduling problems. *Eur J Oper Res*. 1997;99(1):154-65.
[13] Lasserre JB, Queyranne M, editors. Generic Scheduling Polyhedra and a New Mixed-Integer Formulation for Single-Machine Scheduling. Paper from the conference IPCO: Integer Programming and Combinatorial Optimization. 1992.
[14] Kopanos GM, Puigjaner L, Maravelias CT. Production Planning and Scheduling of Parallel Continuous Processes with Product Families. *Ind Eng Chem Res*. 2011;50(3):1369–1378.
[15] Lima RM, Grossmann IE, Jiao Y. Long-Term Scheduling of a Single-Unit Multi-Product Continuous Process to Manufacture High Performance Glass. *Comput Chem Eng*. 2011;35 (3):554–574.

4 Single-Stage Environment

In this chapter, we discuss problems in the single-stage or parallel-units environment. The problem statement is presented in Section 4.1. Three types of models are presented in Section 4.2 (sequence based), Section 4.3 (continuous time grid based), and Section 4.4 (discrete time grid based). In Section 4.5, we present how batching decisions can be handled, and in Section 4.6 we discuss how the three types of models can be extended to handle a new feature, namely, general shared resources. Finally, in Section 4.7 we present extensions on the modeling of general resource constraints using discrete modeling of time.

Building upon the material in the previous chapter, we illustrate how some of the modeling techniques introduced for single-unit problems can be extended to account for multiple units. Our goal is to outline some general ideas that the reader can apply to a wider range of problems. We do not, however, present these extensions for all the model variants presented in the previous chapter. Instead, we focus on (1) problem features that are new, compared to the ones in single-unit problems (i.e., batching decisions and general shared resources); and (2) new modeling techniques that are necessary to account for these features.

4.1 Problem Statement

We first consider the problem where batching decisions have already been made, that is, demand has been converted into batches. In this case, we are given a set, $\mathbf{I}$, of batches and a set, $\mathbf{J}$, of units. Each batch $i \in \mathbf{I}$ has a release, ρ_i, and due, ε_i, time, and has to be carried out in exactly one *compatible* unit $j \in \mathbf{J}_i \subseteq \mathbf{J}$. The set of batches that can be processed on unit j is denoted by $\mathbf{I}_j$. The processing time of batch i on unit j is denoted by τ_{ij} and the processing cost by γ_{ij}^P. The changeover cost/time from batch i to batch i' processed on unit j is denoted by $\gamma_{ii'j}^{CH}/\sigma_{ii'j}$.

In terms of problem classification using the $\alpha/\beta/\gamma$ triplet, we denote the single-stage environment as ***Sms***, $|\mathbf{K}| = 1$, where, as we will see in the next chapter, $\mathbf{K}$ is the set of stages. In the next three sections, we consider the following three problems:

(1) ***Sms***, $|\mathbf{K}| = 1$ with release and due dates, [***Sms***,$|\mathbf{K}| = 1/\rho, \varepsilon/\cdot$].

(2) ***Sms***, $|\mathbf{K}| = 1$ with release/due times, and changeover times, [***Sms***,$|\mathbf{K}| = 1/\rho, \varepsilon, \sigma/\cdot$].

(3) ***Sms***, $|\mathbf{K}| = 1$ with release/due times, and changeover costs, [***Sms***, $|\mathbf{K}| = 1/\rho, \varepsilon, \gamma/\cdot$].

4.2 Sequence-Based Models

The new type of decision, compared to single-unit problems, is the assignment of batches to units. Thus, regardless of the modeling approach, we have to introduce a new binary variable, $X_{ij} \in \{0, 1\}$, which is equal to 1 if batch i is assigned to unit $j \in \mathbf{J}_i$. Clearly, the following constraint should be satisfied,[1]

$$\sum_{j \in \mathbf{J}_i} X_{ij} = 1, \quad i \tag{4.1}$$

If two batches are assigned to the same unit, then the relative order in which they are processed should be determined. We introduce variable $Y_{ii'j} \in \{0, 1\}$, which is equal to 1 if both batches i and i' are assigned to unit j, and batch i is processed before batch i'. As in the single-unit problem, the sequencing relationship can be global or immediate (local). If global, the *activation* of the sequencing binary variables can be achieved via

$$Y_{ii'j} + Y_{i'ij} \geq X_{ij} + X_{i'j} - 1, \quad i, i' > i, j \in \mathbf{J}_i \cap \mathbf{J}_{i'} \tag{4.2}$$

For local sequence models, we introduce binary variables:

- $Y_{ij}^F = 1$ if batch i is processed first in unit j
- $Y_{ij}^L = 1$ if batch i is processed last in unit j

and, assuming that at least one batch is assigned to each unit,[2] enforce

$$\sum_{i,i'} Y_{ii'j} = \sum_i X_{ij} - 1, \quad j \tag{4.3}$$

and[3]

$$\sum_{i'} Y_{i'ij} = X_{ij} - Y_{ij}^F, \quad i, j \tag{4.4}$$

$$\sum_{i'} Y_{ii'j} = X_{ij} - Y_{ij}^L, \quad i, j \tag{4.5}$$

which imply $Y_{ij}^F \leq X_{ij}, Y_{ij}^L \leq X_{ij}$ for all i, j; and also

$$\sum_i Y_{ij}^F = \sum_i Y_{ij}^L = 1, \quad j, \tag{4.6}$$

[1] The summation in the LHS of (4.1) should be over $j \in \mathbf{J}_i$, as shown. However, to keep the presentation simple, in the remainder of the book we will not always use (sub)set membership to define the values of the summation index. When the (subset) membership can be easily inferred, we will often use only the index. It is assumed that only the indices for which the corresponding addends are defined are included. Also, we will never use membership in the superset (e.g., $j \in \mathbf{J}$ will always be replaced by j).

[2] The case where no batches are assigned to a unit, if there is evidence that this can happen in an optimal solution, can be modeled either by introducing a dummy batch, Dj, for each unit (with $\mathbf{J}_{Dj} = \{j\}$ and $\tau_{Dj,j} = 0$) preassigned to unit j; or by introducing a new binary variable, Z_j, which becomes one if at least one batch is assigned to j, and 0 otherwise; and is used to replace 1 in (4.6).

[3] Strictly speaking, (4.4)and (4.5) should be defined for $i, j \in \mathbf{J}_i$ or, equivalently, for $j, i \in \mathbf{I}_j$. However, since, for example, variables X_{ij}, $Y_{i'ij}$ and Y_{ij}^F should not be defined if $j \notin \mathbf{J}_i$ (or $i \notin \mathbf{I}_j$), or they are zero, (4.4) can also be written for i, j. In general and throughout the book, if the subsets over which an equation should be defined are easy to infer, then they will not be explicitly given. For example, $j, i \in \mathbf{I}_j$ will be replaced by j, i.

Note that (4.2), which is written for every unit and pair of compatible batches, enforces one of the two binary variables in the LHS to be 1 if both batches are assigned to this unit; while (4.3), which is written for every unit, enforces exactly $\sum_i X_{ij} - 1$ binaries to be equal to one.

If (4.2) is used, then additional constraints can be added to ensure that a sequencing binary does not become 1 if the batch is not assigned to a unit, but this is not necessary because variables $Y_{ii'j}$ are used to enforce the no-overlap condition on a unit, so the optimization *pushes* variables $Y_{ii'j}$ to zero so that more batches can be carried out in a unit.

If a set of feasible assignments (X_{ij}) and global or immediate (local) sequencing relationships $(Y_{ii'j})$ are available, then all remaining problem restrictions and features can be modeled using the same equations for all sequence-based models. The main constraint blocks are (1) start time disaggregation; (2) enforcement of no-overlap condition; (3) release and due time constraints; and (4) objective function.

Since a batch can be assigned to multiple units, we disaggregate the start time of batch i, S_i, into unit-specific start times, S_{ij},

$$S_i = \sum_j S_{ij}, \quad i \tag{4.7}$$

with

$$S_{ij} \leq MX_{ij}, \quad i,j, \tag{4.8}$$

where M is a sufficiently large parameter.

The disjunctive constraint can then be written as follows:

$$S_{ij} + \tau_{ij} \leq S_{i'j} + M\left(1 - Y_{ii'j}\right), \quad i,i',j \in \mathbf{J}_i \cap \mathbf{J}_{i'} \tag{4.9}$$

As in Chapter 3, we could also define variables $E_i = S_i + \sum_j \tau_{ij} X_{ij}$ (and the corresponding E_{ij} via equations similar to (4.7) and (4.8)) to denote the end of a batch, and use them to express the disjunctive constraint.

For all problem classes, release and due dates can then be readily enforced:

$$S_i \geq \rho_i, \quad i \tag{4.10}$$

$$S_i + \sum_j \tau_{ij} X_{ij} \leq \varepsilon_i, \quad i. \tag{4.11}$$

If due times can be violated, at a cost, then (4.11) is replaced by

$$S_i + \sum_j \tau_{ij} X_{ij} \leq \varepsilon_i + L_i, \quad i, \tag{4.12}$$

where L_i is the lateness/tardiness of batch i. If the objective function is lateness minimization, then $L_i \in \mathbb{R}$ and (4.12) is satisfied as equality; that is, L_i will assume the smallest possible (perhaps negative) value. If the objective is tardiness minimization, then $L_i \in \mathbb{R}_+$ and (4.12) will be satisfied as equality only if $S_i + \sum_j \tau_{ij} X_{ij} \geq \varepsilon_i$.

For makespan, MS, minimization, we require

$$MS \geq S_i + \sum_j \tau_{ij} X_{ij}, \quad i \tag{4.13}$$

and the objective is

$$\min MS \tag{4.14}$$

The earliness minimization objective function is

$$\min \sum_i \omega_i \left(\varepsilon_i - \left(S_i + \sum_j \tau_{ij} X_{ij} \right) \right), \tag{4.15}$$

where ω_i is a weight factor, and we assume that due times are met so no benefit is obtained from increasing the lateness of batch i in order to increase earliness of batch i'.

The objective function for total weighted lateness, L^{TOT}, and tardiness, T^{TOT}, minimization is

$$\min \sum_i \omega_i L_i. \tag{4.16}$$

Single-stage problems are the simplest problems where a cost minimization objective can be defined in the absence of changeover costs,

$$\min \sum_{i,j} \gamma_{ij}^P X_{ij}. \tag{4.17}$$

The constraints presented earlier in this section allow us to develop alternative models for a range of problems. For example, the global sequence model for problem $[\boldsymbol{Sms}, |\mathbf{K}| = 1/\rho, \varepsilon/MS]$ consists of (4.1), (4.2), (4.7) through (4.11), (4.13), and (4.14), whereas one immediate sequence model for the same problem results from replacing (4.2) with (4.3) through (4.5).

Changeover times and costs can be handled using modeling techniques similar to the ones presented in the previous chapter. Assuming that the triangular inequality holds, changeover times can be addressed through a simple modification of (4.9):

$$S_{ij} + \tau_{ij} + \sigma_{ii'j} \leq S_{i'j} + M(1 - Y_{ii'j}), \quad i, i', j \in \mathbf{J}_i \cap \mathbf{J}_{i'}. \tag{4.18}$$

Changeover costs can be addressed using immediate-sequence-based models. The objective function is

$$\min \sum_{i,i'} \sum_j \gamma_{ii'j}^{CH} Y_{ii'j}. \tag{4.19}$$

4.3 Models Based on a Continuous Time Grid

Once a batch is assigned to a unit, there are no constraints *coupling* a unit with the others, which means that for the basic single-stage problems we have considered thus far, (1) independent *unit-specific* time grids can be defined, and (2) a batch can be carried out within a single period/slot of a time grid.[4] Thus, the next step is to define the number of points/periods per unit grid. A simple target number is $\lceil |\mathbf{I}|/|\mathbf{J}| \rceil$. However, since the processing times can be rather different across units, a more conservative approach, with more periods, is often required. Instance-specific data can often be used to

[4] In Section 4.6, we will discuss how these two conditions do not hold true in the presence of general resource constraints that *couple* different units.

determine a good number of periods (see Example 4. 1). For now, we will simply assume that the adopted grids have sufficient number of periods to represent the optimal solution.

The time horizon of each unit is divided into T^j unit-specific periods. If unit j becomes available at $\bar{\varrho}_j$ and deadlines are given, then the start, ϱ_j, and end, ε_j, of the *unit* horizons are given by

$$\varrho_j = \max\left\{\bar{\varrho}_j, \lim_{i \in \mathbf{I}_j}\{\rho_i\}\right\}, \quad \varepsilon_j = \max_{i \in \mathbf{I}_j}\{\varepsilon_i\}. \tag{4.20}$$

If no deadlines are available, then a sufficiently large number should be used for ε_j. Also, the grids can be *moved*, by subtracting $\min_j\{\varrho_j\}$ from all ϱ_j and ε_j, so that the earliest unit start time is zero.

If T^{MAX} is the maximum number of periods over all units $\left(T^{MAX} = \max_j\{T^j\}\right)$, then for simplicity we define superset $\mathbf{T} = \{1, \ldots, T^{MAX}\}$, and the subsets, $\mathbf{T}_j \subseteq \mathbf{T}$, of unit-specific time periods $\mathbf{T}_j = \{1, \ldots, T^j\}$. We introduce variables T_{jt} to denote the timing of point t in the grid of unit j. Each unit period starts at $T_{j,t-1}$ and ends at T_{jt}. The timing of time points should satisfy the following constraints:

$$T_{j,0} = \varrho_j, \quad T_{j,T^j} = \varepsilon_j \tag{4.21}$$

$$T_{jt} \geq T_{j,t-1}, \quad j, t \in \mathbf{T}_j. \tag{4.22}$$

Following the approach presented in the previous section, we can enforce the assignment of each batch to exactly one unit, via (4.1), and then to a time period in that unit,

$$\sum\nolimits_t X_{ijt} = X_{ij}, \quad i, j, \tag{4.23}$$

where $X_{ijt} \in \{0, 1\}$ is equal to 1 if batch i is allocated to time period t of unit j. However, (4.23) can be used to project out variables X_{ij}, so (4.1) and (4.23) can be replaced by

$$\sum\nolimits_{j,t} X_{ijt} = 1, \quad i, \tag{4.24}$$

which enforces that each batch has to be assigned to exactly one unit time period. In addition, at most one batch can be allocated to each period:

$$\sum\nolimits_i X_{ijt} \leq 1, \quad j, t \in \mathbf{T}_j. \tag{4.25}$$

Binary variables X_{ijt} can then be used to enforce no-overlap between batches. One approach is to employ the same disaggregated batch start time variables, S_{ij}, defined through (4.7) and

$$S_{ij} \leq M \sum\nolimits_t X_{ijt}, \quad i, j, \tag{4.26}$$

where, compared to (4.8), we use $\sum_t X_{ijt}$ to project out X_{ij}. Using variables S_{ij}, we write constraints relating the start and end time of a batch with time points, T_{jt}:

$$S_{ij} \geq T_{j,t-1} - \eta\left(1 - X_{ijt}\right), \quad i, j, t \in \mathbf{T}_j \tag{4.27}$$

$$S_{ij} + \tau_{ij} \leq T_{jt} + \eta\left(1 - X_{ijt}\right), \quad i, j, t \in \mathbf{T}_j \tag{4.28}$$

,

where horizon η is used as big-M parameter. Equations (4.27) and (4.28) enforce that if batch i is allocated to a period, then it is executed within that period (with $M = \varepsilon_i$ being a valid value), and since there is no overlap between periods of the same unit, due to (4.22), there is no overlap between the processing of batches assigned to the same unit.

Variables S_i are used to enforce release times via (4.10), and deadlines or due times through

$$S_i + \sum_{j,t} \tau_{ij} X_{ijt} \leq \varepsilon_i, \quad i \tag{4.29}$$

$$S_i + \sum_{j,t} \tau_{ij} X_{ijt} \leq \varepsilon_i + L_i, \quad i, \tag{4.30}$$

where L_i is the lateness/tardiness of batch i. Note, again, how we replace X_{ij} with $\sum_t X_{ijt}$.

Equations (4.14) through (4.16) can be used as objective functions for the minimization of makespan (subject to (4.13)), earliness, and lateness/tardiness, respectively. Equation (4.17), for cost minimization, is modified as follows:

$$\min \sum_{i,j,t} \gamma_{ij}^P X_{ijt}. \tag{4.31}$$

Again, different model variants can be obtained by requiring, for example, S_{ij} to be equal to the start of the period to which batch i is allocated, which can be achieved using techniques shown in the previous chapter.

Also, an alternative model can be obtained if variables S_i and S_{ij} are removed. In this case, (4.21) is used for grid definition, (4.24) through (4.25) to obtain a feasible assignment of batches to unit time periods, and the no-overlap condition is enforced using the following constraint, which also replaces (4.22):

$$T_{jt} \geq T_{j,t-1} + \sum_i \text{þ}\tau_{ij} X_{ijt}, \quad j, t \in \mathbf{T}_j. \tag{4.32}$$

Since variables S_i are removed, we use variables T_{jt} to enforce release and due times. Assuming that the processing of the batch coincides with the beginning of the period to which it is allocated, then the release and due date constraints can be written as follows:

$$T_{j,t-1} \geq \rho_i X_{ijt}, \quad i, j, t \in \mathbf{T}_j \tag{4.33}$$

$$T_{j,t-1} + \tau_{ij} X_{ijt} \leq \varepsilon_i X_{ijt} + \eta\left(1 - X_{ijt}\right), \quad i, j, t \in \mathbf{T}_j. \tag{4.34}$$

The alternative model, with no S_i and S_{ij} variables for problem $[\boldsymbol{Sms}, |\mathbf{K}| = 1/\rho, \varepsilon/MS]$ consists of (4.21), (4.24), (4.25), (4.32) through (4.34), (4.14), and the counterpart of (4.13) expressed without S_i variables.[5] While this model is not tight, especially (4.33) and (4.34), we present it to illustrate how ideas presented for a simpler problem (single unit) can be extended to more complex problems.

To model changeovers, period-specific immediate sequencing binary variables, $Y_{ii'jt}$, can be used following an approach similar to the one presented in the previous chapter:

[5] How would you formulate the counterpart of (4.13) without using S_i variables? Under certain conditions, T_{jt} can be used instead of S_i. How can you use this observation?

$$Y_{ii'jt} \geq X_{ijt} + X_{i'j,t+1} - 1, \quad i, i' \neq i, j \in \mathbf{J}_i \cap \mathbf{J}_{i'}, t \in \mathbf{T}_j \tag{4.35}$$

$$\sum\nolimits_{i'} Y_{ii'jt} \leq X_{ijt}, \quad i, j, t \in \mathbf{T}_j \tag{4.36}$$

$$\sum\nolimits_{i} Y_{ii'j,t-1} \leq X_{i'jt}, \quad i', j, t \in \mathbf{T}_j \tag{4.37}$$

$$S_{ij} + \tau_{ij} + \sigma_{ii'j} \leq T_{jt} + M\left(1 - Y_{ii'jt}\right), \quad i, i', j \in \mathbf{J}_i \cap \mathbf{J}_{i'}, t \in \mathbf{T}_j. \tag{4.38}$$

The minimization of changeover cost can then be expressed as follows:

$$\min \sum\nolimits_{i,i',j} \gamma_{ii'j}^{CH} \sum\nolimits_{t} Y_{ii'jt}. \tag{4.39}$$

Example 4.1 Definition of Unit Time Grids The goal of this example is to illustrate how instance-specific data can impact the number of unit time periods, T^j, necessary to represent a solution. A key consideration is the *load* of a unit in a given solution, which is defined as the total time the unit is processing batches, $\sum_{i,t} \tau_{ij} X_{ijt}$. We also define the average processing time of a unit as $\tau_j^{AVG} = \sum_i \tau_{ij} / |\mathbf{I}|$. We start with some general observations:

(1) The load is expected to be balanced across units if a time-related metric is used. (We will discuss this further, but can you see the reason why?)
(2) Subsets $\mathbf{J}_i$ and $\mathbf{I}_j$ may lead to unbalanced loads; for example, intuitively, if very few batches can be processed on a given unit (i.e., $|\mathbf{I}_j|$ is much smaller than $|\mathbf{I}| / |\mathbf{J}|$), then we expect unit j to have low load. (Can you see when this would not be true?) For now, we will assume $\mathbf{J}_i = \mathbf{J}$ and thus $\mathbf{I}_j = \mathbf{I}$.
(3) If the objective is makespan minimization, then we expect the load to be balanced. (Why? What does the presence of unbalanced units imply?)
(4) In lateness/tardiness minimization problems, if the due times are binding, then we expect unit loads to be balanced.
(5) If the windows are not tight, the objective is tardiness minimization, and a secondary cost minimization objective is used, then the load can be unbalanced. (Why? What is the expected total tardiness if the windows are not tight?)
(6) In earliness minimization, we expect all units to be utilized simultaneously during periods when there are multiple due times. This may not be true if the number of due dates, relative to the number of units, is small and the windows are not tight.
(7) In cost minimization problems, the load of a (cheaper) unit may be significantly higher than the load of another unit.

To further study how instance characteristics impact the selection of T^j, we consider makespan minimization in a process with three units (U1, U2, U3) and eight batches (A, B, ..., H); and study three instances (1, 2, 3) with different processing times, as given in Table 4.1. All release times are zero, there are no due times, and we also assume that $\mathbf{J}_i = \mathbf{J}$ for all i.

In instance 1, processing times are very similar across units; and the average processing time in all units is the same $\left(\tau_j^{AVG} = 3.5\right)$. The processing times in

Table 4.1 Task unit and unit average processing times of the three instances.

	Instance 1		Instance 2		Instance 3	
Unit	$\tau_{Aj}, \tau_{Bj}, \ldots, \tau_{Hj}$	τ_j^{AVG}	$\tau_{Aj}, \tau_{Bj}, \ldots, \tau_{Hj}$	τ_j^{AVG}	$\tau_{Aj}, \tau_{Bj}, \ldots, \tau_{Hj}$	τ_j^{AVG}
U1	3, 3, 3, 3, 4, 4, 4, 4	3.5	3, 3, 3, 3, 5, 5, 5, 5	4.0	3, 3, 3, 8, 8, 8, 8, 7	6.0
U2	4, 4, 4, 4, 3, 3, 3, 3	3.5	4, 4, 4, 4, 5, 5, 5, 5	4.5	6, 6, 6, 6, 4, 6, 4, 6	5.5
U3	4, 3, 4, 3, 4, 3, 4, 3	3.5	4, 4, 4, 4, 5, 5, 5, 5	4.5	8, 7, 8, 4, 8, 4, 6, 3	6.0

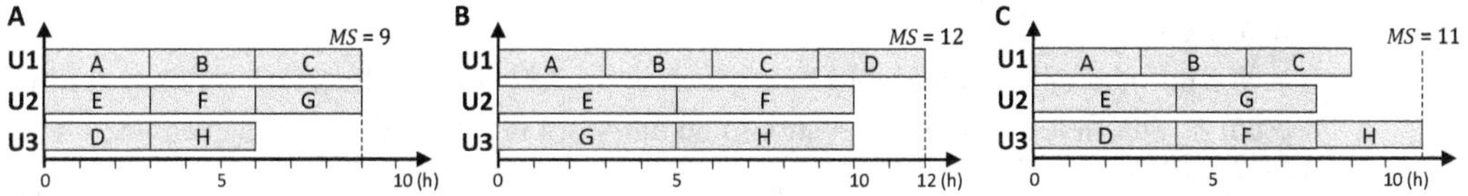

Figure 4.1 Optimal solutions of three instances, showing number of required unit periods. Note that each instance has multiple optimal solutions both in terms of unit–batch assignments and sequencing.

instance 2 are slightly larger than in instance 1, but U1 is still more effective in processing batches A through D and U2 is more effective in processing batches E through H, as in instance 1. In instance 3, processing times are overall larger though they are small for a few unit–batch pairs.

An optimal solution of instance 1 is shown in Figure 4.1A. As expected, the solution requires a maximum of three periods (in U1 and U2), which is equal to the rough estimate $\lceil |\mathbf{I}| / |\mathbf{J}| \rceil = \lceil 8/3 \rceil$. Interestingly, note that if we postulate $T^{U3} = 2$, we cut off one of the multiple optimal solutions of this instance, with assignments: {A, C} → U1, {E, F, G} → U2, {B, D, H} → U3. This assignment may be the optimal one if, for example, cost minimization was considered. An optimal solution to instance 2 is shown in Figure 4.1B. Surprisingly, although the changes in processing times compared to instance 1 are small, the number of periods for U1 increases to 4. Finally, the representation of the optimal solution to instance 3, shown in Figure 4.1C, requires three periods for U1 and U3 despite the fact that based on the values of τ_j^{AVG} these two units appear to be slower than U2, which requires only two periods. This simple example shows that obtaining a good estimate of the number of slots needed to represent the optimal solution(s) is challenging, which means that a conservative estimate should be made.

4.4 Models Based on a Discrete Time Grid

We use index $n \in \mathbf{N} = \{0, 1, 2, \ldots |\mathbf{N}|\}$ to denote the time points between the minimum release date and the maximum due date. The scheduling horizon is divided into periods $n \in \mathbf{N} \backslash \{0\} = \{1, 2, \ldots, \mathbf{N}|\}$ of equal length δ, where period n runs between time points

$n-1$ and n. We assume that, using the techniques described in the previous chapter, we determine δ and then calculate all time-related parameters:

$$\bar{\tau}_{ij} = \lceil \tau_{ij}/\delta \rceil, \quad \bar{\sigma}_{ii'j} = \lceil \sigma_{ii'j}/\delta \rceil, \quad \bar{\rho}_i = \lceil \rho_i/\delta \rceil, \quad \bar{\varepsilon}_i = \lfloor \varepsilon_i/\delta \rfloor. \tag{4.40}$$

Using the ideas presented in Section 4.3, we can define unit-specific grids,

$$\mathbf{N}_j = \left\{ n \middle| \max\left\{\bar{\varrho}_j, \min_{i \in \mathbf{I}_j}\{\bar{\rho}_i\}\right\} \le n \le \max_{i \in \mathbf{I}_j}\{\bar{\varepsilon}_i\} \right\}, \tag{4.41}$$

which can be used to fix binary variables (i.e., $X_{ijn} = 0$ if $n \notin \mathbf{N}_j$), but for simplicity we will use $\mathbf{N}$.

We introduce binary variable X_{ijn}, which is equal to 1 if batch i starts on unit j at time point n. The first constraint that should be enforced is that each batch is executed once,

$$\sum_{j,n} X_{ijn} = 1, \quad i \tag{4.42}$$

Note that, again, $\sum_n X_{ijn}$ can be viewed as a selection binary variable, that is, if $\sum_n X_{ijn} = 1$, then batch i is assigned to unit j $(X_{ij} = 1)$. As in the continuous time model, X_{ij} is projected out. Also, note that $\sum_j X_{ijn}$ can be viewed as equivalent to X_{in}, used in the single-unit model; if $\sum_j X_{ijn} = 1$, then the execution of batch i starts (in some unit) at n.

The processing of at most one batch at any time is enforced using a clique constraint, very similar to the one in the previous chapter,

$$\sum_i \sum_{n'=n-\bar{\tau}_{ij}+1}^{n'=n} X_{ijn'} \le 1, \quad j, n, \tag{4.43}$$

which can be obtained from (3.35) by replacing X_{in} with X_{ijn}, and expressing it for every j and n, instead of every n.

As in the single-unit problem, release and hard due time constraints can be enforced by fixing the binaries that are outside the allowable window to zero,

$$X_{ijn} = 0, \quad i, j, n < \bar{\rho}_i, n > \bar{\varepsilon}_i - \bar{\tau}_{ij}. \tag{4.44}$$

Makespan is constrained as follows:

$$MS \ge \sum_{j,n} (n + \bar{\tau}_{ij}) X_{ijn}, \quad i. \tag{4.45}$$

The weighted earliness, lateness, and tardiness objective functions are expressed, respectively, as follows:

$$\min \sum_i \omega_i \left(\bar{\varepsilon}_i - \sum_{j,n} (n + \bar{\tau}_{ij}) X_{ijn} \right) \tag{4.46}$$

$$\min \sum_i \omega_i \left(\sum_{j,n} (n + \bar{\tau}_{ij}) X_{ijn} - \bar{\varepsilon}_i \right) \tag{4.47}$$

$$\min \sum_i \omega_i L_i, \tag{4.48}$$

with $L_i \in \mathbb{R}_+$ satisfying,

$$L_i \ge \sum_{j,n} (n + \bar{\tau}_{ij}) X_{ijn} - \bar{\varepsilon}_i, \quad i. \tag{4.49}$$

Using the ideas presented in the previous chapter, changeover times can be enforced using binary variables X_{ijn} via

$$\sum_{i' \neq i} \sum_{n'=n-\bar{\tau}_{i'j}-\bar{\sigma}_{i'ij}+1}^{n-\bar{\tau}_{i'j}} X_{i'jn'} \leq M_{ij}\left(1 - X_{ijn}\right), \quad j, i, n \tag{4.50}$$

or

$$X_{ijn} + X_{i'jn'} \leq 1 \quad i, i' \neq i, j \in \mathbf{J}_i \cap \mathbf{J}_{i'}, n, n' \in \left\{n - \bar{\tau}_{i'j} - \bar{\sigma}_{i'ij} + 1, \ldots, n - \bar{\tau}_{i'j}\right\}. \tag{4.51}$$

To model changeover costs ($\gamma_{ii'j}^{CH}$), we introduce two variables:

- $\bar{X}_{ijn} = 1$ if during period n unit j is set up (is in the mode) to carry out batch i.
- $Y_{ii'jn} = 1$ denotes a changeover in the mode of unit j from batch i to batch i' at time point n.

A batch can be carried out only when the unit is in the corresponding mode,[6]

$$\sum_{n'=n-\bar{\tau}_{ij}}^{n-1} X_{ijn'} \leq \bar{X}_{ijn}, \quad j, i, n \tag{4.52}$$

and a unit transitions between modes when a changeover is executed,

$$\bar{X}_{ijn} = \bar{X}_{ij,n-1} + \sum_{i' \neq i} Y_{i'ij,n-\bar{\sigma}_{i'ij}-1} - \sum_{i' \neq i} Y_{ii'j,n-1}, \quad j, i, n \tag{4.53}$$

Note that a unit is not in a mode during the execution of a changeover.

The total cost minimization objective function is

$$\min \sum_{i,j} \left(\gamma_{ij}^{P} \sum_{n} X_{ijn}\right) + \sum_{i,i',j,n} \gamma_{ii'j}^{CH} Y_{ii'jn}. \tag{4.54}$$

4.5 Batching Decisions

If units have different capacities, β_j, then the number of required batches to meet given demand (orders) is not known a priori; that is, batching, assignment, and sequencing decisions have to be optimized simultaneously.[7] Since batches are not fixed, we use index i to denote orders. We are given a set of orders $\mathbf{I}$ and a set of units $\mathbf{J}$. Each order has amount ξ_i due and release/due time ρ_i/ε_i, and the batches toward it can be executed in $j \in \mathbf{J}_i \subseteq \mathbf{J}$. The processing time for a batch of order i in unit j is τ_{ij} and the processing cost is γ_{ij}^{P}. The capacity of unit j is β_j; that is, a batch, of any product, executed in j leads to the production of β_j units of product. Multiple orders for the same product are simply treated as separate orders with different amounts due and release/due times, but the same $\mathbf{J}_i$, and processing times/costs.

[6] Clearly, (4.52) is very similar to (4.43). What is the main difference? Can you see why the variables included in the summation are *shifted* backward by one time point?

[7] In the previous chapter, the number of batches was treated as an optimization decision only when the objective was production/profit maximization, in the prize collection variant. In the present chapter, the number of batches is an optimization decision even when all orders have to be satisfied and the objective function is cost minimization.

4.5.1 Sequence-Based Models

The easiest way to extend the models discussed in Section 4.2 is to postulate a number of *potential* batches, some of which will be chosen, and then enforce the constraints presented in Section 4.2 only for the *selected* batches.

The minimum, λ_i^{MIN}, and maximum, λ_i^{MAX}, number of batches needed to meet order i is

$$\lambda_i^{MIN} = \left\lceil \xi_i / \max_{j \in \mathbf{J}_i} \{\beta_j\} \right\rceil, \quad \lambda_i^{MAX} = \left\lceil \xi_i / \min_{j \in \mathbf{J}_i} \{\beta_j\} \right\rceil, \quad i. \tag{4.55}$$

Thus, $l \in \mathbf{L}_i = \{1, 2, \ldots, \lambda_i^{MAX}\}$ is the set of potential batches for product i, with at least λ_i^{MIN} batches selected. Compared to the models presented in Section 4.2, the major differences are that (1) a batch is now defined by a pair (i, l); and (2) not all batches have to be processed.

We first define a *selection* binary variable Z_{il} that is 1 if batch (i, l) is selected. If it is, then it has to be assigned to a unit:

$$\sum_{j \in \mathbf{J}_i} X_{ilj} = Z_{il}, \quad i, l \in \mathbf{L}_i. \tag{4.56}$$

The demand satisfaction constraint, which *forces* some assignment variables to become 1, then becomes

$$\sum_{l \in \mathbf{L}_i} \sum_j \beta_j X_{ilj} \geq \xi_i, \quad i. \tag{4.57}$$

If variables X_{ilj}, which are the counterparts of X_{ij}, are activated, through (4.56) and (4.57), then the remaining constraints are very similar to the ones presented in Section 4.2. The only changes are that (1) variables are defined for i and $l \in \mathbf{L}_i$ instead of i only (e.g., the sequencing variables are $Y_{ili'l'j}$, instead of $Y_{ii'j}$); and (2) constraints are expressed for i and $l \in \mathbf{L}_i$ instead of i only. For example, (4.2) becomes:

$$Y_{ili'l'j} + Y_{i'l'ilj} \geq X_{ilj} + X_{i'l'j} - 1, \quad i, i' > i, l \in \mathbf{L}_i, l' \in \mathbf{L}_{i'}, j \in \mathbf{J}_i \cap \mathbf{J}_{i'}. \tag{4.58}$$

To reduce the number of equivalent solutions, without loss of generality, we can add

$$Z_{il} \leq Z_{i,l-1}, \quad i, 1 < l \in \mathbf{L}_i. \tag{4.59}$$

Furthermore, since we know that at least λ_i^{MIN} batches should be selected in any feasible solution, we can add

$$Z_{il} = 1, \quad i, l \in \{1, 2, \ldots, \lambda_i^{MIN}\}, \tag{4.60}$$

which can be exploited to develop more effective models. (Can you think of any simplifications that can be made, to other equations, based on (4.60)?)

If the capacity of a unit is product dependent, then we have capacities β_{ij} instead of β_j. Furthermore, if a unit can be operated at less than full capacity, between β_{ij}^{MIN} and β_{ij}^{MAX}, then we introduce variables B_{il} to denote batchsize: if (i, l) is selected, then $B_{il} \in \left[\beta_{ij}^{MIN}, \beta_{ij}^{MAX}\right]$.

In this case, the calculation of λ_i^{MIN} and λ_i^{MAX} is modified as follows (see Example 4.2 for an illustration):

$$\lambda_i^{MIN} = \left\lceil \xi_i / \max_{j \in \mathbf{J}_i} \left\{ \beta_{ij}^{MAX} \right\} \right\rceil, \lambda_i^{MAX} = \left\lceil \xi_i / \min_{j \in \mathbf{J}_i} \left\{ \beta_{ij}^{MIN} \right\} \right\rceil \tag{4.61}$$

and (4.57) is replaced by

$$\sum\nolimits_{l \in \mathbf{L}_i} B_{il} \geq \xi_i, \quad i \tag{4.62}$$

and

$$\sum\nolimits_j \beta_{ij}^{MIN} X_{ilj} \leq B_{il} \leq \sum\nolimits_j \beta_{ij}^{MAX} Z_{ilj}, \quad i, l \in \mathbf{L}_i. \tag{4.63}$$

Equation (4.62) makes B_{il} variables positive, which in turn leads to the activation of some assignment variables X_{ilj} through the first inequality in (4.63), which leads to the selection of batches $(Z_{il} = 1)$ through (4.56). All remaining constraints remain the same.

Example 4.2 Bounding the Number of Selected Batches The goal of this example is to illustrate how parameters λ_i^{MIN} and λ_i^{MAX} are calculated from batchsize parameters. We consider four instances $k \in \{1, 2, 3, 4\}$ in a facility with two units (U1 and U2) producing four products, A, B, C, and D, with demands 5, 10, 15, and 20 units, respectively. In instance 1, batchsizes are fixed (same as unit capacities) and product independent. In instance 2, batchsizes are again the same as the unit capacities, which in this case are product dependent. In instance 3, batchsizes should lie within product-independent lower and upper bounds. In instance 4, batchsizes should lie within product-dependent lower and upper bounds. Data for all instances are given in Table 4.2.

For instance 1, we use (4.55) to calculate λ_i^{MIN} and λ_i^{MAX}. For instance 2, we use (4.55) but replace β_j with β_{ij}. For instance 3, we use (4.61), but replace $\beta_{ij}^{MIN}/\beta_{ij}^{MAX}$ with $\beta_j^{MIN}/\beta_j^{MAX}$. Finally, for instance 4, we use (4.61). (Can you calculate λ_i^{MIN} and λ_i^{MAX} for all orders in all instances?) The results of the calculations are given Table 4.3, along with the argument (ratio) of the function used for the calculation (e.g., $\xi_i / \max_{j \in \mathbf{J}_i} \{\beta_j\}$ in the calculation of λ_i^{MIN} in (4.45)). In bold, are the changes from the previous instance; for example, $\lambda_A^{MAX} = 1$ in instance 1, but becomes $\lambda_A^{MAX} = 2$ in instance 2.

Table 4.2 Unit batchsize capacities for four instances.

	Instance 1	Instance 2				Instance 3	Instance 4			
	β_j	β_{ij}				$\beta_j^{MIN}/\beta_j^{MAX}$	$\beta_{ij}^{MIN}/\beta_{ij}^{MAX}$			
Unit		A	B	C	D		A	B	C	D
U1	5	4	4	5	5	4/5	2/4	3/4	4/5	4/6
U2	10	8	10	12	12	8/10	5/8	6/8	6/10	8/10

Table 4.3 Calculation of parameters $\lambda_i^{MIN}/\lambda_i^{MAX}$ for four instances.

		Instance 1		Instance 2		Instance 3		Instance 4	
Calculation		(4.55)		(4.55), $\beta_j \rightarrow \beta_{ij}$		(4.61), $\beta_{ij}^{MIN} \rightarrow \beta_j^{MIN}$		(4.61)	
i	ξ_i	λ_i^{MIN}	λ_i^{MAX}	λ_i^{MIN}	λ_i^{MAX}	λ_i^{MIN}	λ_i^{MAX}	λ_i^{MIN}	λ_i^{MAX}
A	5	5/10, 1	5/5, 1	5/8, 1	5/4, **2**	5/10, 1	5/4, 2	5/8, 1	5/2, **3**
B	10	10/10, 1	10/5, 2	10/10, 1	10/4, **3**	10/10, 1	10/4, 3	10/8, **2**	10/3, **4**
C	15	15/10, 2	15/5, 3	15/12, 2	15/5, 3	15/10, 2	15/4, **4**	15/10, 2	15/4, 4
D	20	20/10, 2	20/5, 4	20/12, 2	20/5, 4	20/10, 2	20/4, **5**	20/10, 2	20/4, 5

4.5.2 Model Based on a Continuous Time Grid

We use the same basic idea: batch selection is modeled with $Z_{il} \in \{0, 1\}$ and the selected batches are assigned to a unit using (4.64), which is the counterpart of (4.24),

$$\sum_{j,t} X_{iljt} = Z_{il}, \quad i, l \in \mathbf{L}_i, \tag{4.64}$$

where X_{ijt} is replaced by X_{iljt}. Variables S_i and S_{ij} are defined for (i, l) instead of i, and the following equations are expressed for (i, l) instead of i: start time disaggregation, (4.7) and (4.26); release/due time constraints, (4.10) and (4.29); and batch-period time matching, (4.27) and (4.28). Variables T_{jt} and (4.21) and (4.22) remain the same. Changeovers and variable batchsizes can be modeled using the modifications presented in the previous subsection.

4.5.3 Model Based on a Discrete Time Grid

Following the same idea, we can introduce $Z_{il} \in \{0, 1\}$ and activate them via (4.57). Since a selected batch should be assigned to a unit and time point, we have $Z_{il} = \sum_{j \in \mathbf{J}_i, n} X_{iljn}$. However, since sequencing is achieved via the mapping of the starting times onto the grid, there is no need to differentiate between batches of the same product, which means that variables Z_{il} can be removed altogether and the selection of the necessary number of batches can be achieved via

$$\sum_{j \in \mathbf{J}_i, n} \beta_j X_{ijn} \geq \xi_i, \quad i \tag{4.65}$$

instead of (4.42). Since no new types of variables are used and the correct number of batches and assignments are ensured through (4.65), all remaining variables and constraints remain the same. Thus, the model based on a discrete time grid consists of (4.65), (4.43), (4.44), and an objective function. In other words, the model for problems with batching decisions remains almost the same.[8]

[8] In Chapter 3 (Section 3.6.5), we discussed how models based on a discrete time grid can be readily extended to account for various processing feature and constraints. We just covered one such

If variable batchsizes should be modeled, then (4.65) is replaced by

$$\sum\nolimits_{j \in \mathbf{J}_i, n} B_{ijn} \geq \xi_i, \quad i \tag{4.66}$$

and

$$\beta_{ij}^{MIN} X_{ijn} \leq B_{ijn} \leq \beta_{ij}^{MAX} X_{ijn}, \quad i,j,n. \tag{4.67}$$

Interestingly, the two preceding constraints are sufficient to model, with small modifications, batching decisions in all production environments.[9] Note that although no batch *labeling* (using index l) is necessary, batches are unique. No new variables nor constraints are necessary to model changeovers.

4.6 General Shared Resources

So far, we have assumed that the only type of shared resources are units and that each batch requires only one type of shared resource. In many problems, this is not true: the execution of a batch requires, simultaneously, multiple types of resources; for example, labor, electricity, and cooling water.[10] In general, shared resources can be classified as renewable and nonrenewable resources. The former are *used* during the execution of a batch and *freed* after a batch is finished. For example, in the case of cooling (power), a certain load (which is subtracted from the available cooling capacity of the facility) is necessary during the execution of a batch. Nonrenewable resources are *consumed* by a batch, that is, such resources do not become available after the batch is finished; for example, a promoter, necessary for the initiation of a polymerization batch, can be viewed as a nonrenewable resource. In Chapter 7, we will discuss how the treatment of materials as nonrenewable resources allows us to model all scheduling-relevant entities (units, materials, utilities, etc.) as resources, and thus represent a scheduling problem as a time-indexed resource allocation problem. In the present chapter, we will focus on renewable resources. Also, for simplicity, in the remainder of this section we will consider problems with no changeovers.

4.6.1 Preliminaries

At a second level, resources can be classified as discrete (e.g., labor) and continuous (e.g., cooling load). A special type of discrete resource is a *unary* resource, that is, a resource for which the demand (requirement) is one unit and its capacity is also one.

case – batching decisions were handled with almost no increase in model complexity. Can a sequence-based model remain simple when batching decisions are considered? Can you formulate a sequence-based model without index l? What about continuous time grid-based models?

[9] This is another advantage, in terms of modeling, of discrete time models: the basic constraints are the same across models developed to address problems in different environments.

[10] In the present chapter, we introduce resource requirements of batches and describe the interactions between batches and resources. However, the same concepts are applicable in more general problems, defined in terms of *tasks,* as we will see in Chapter 7.

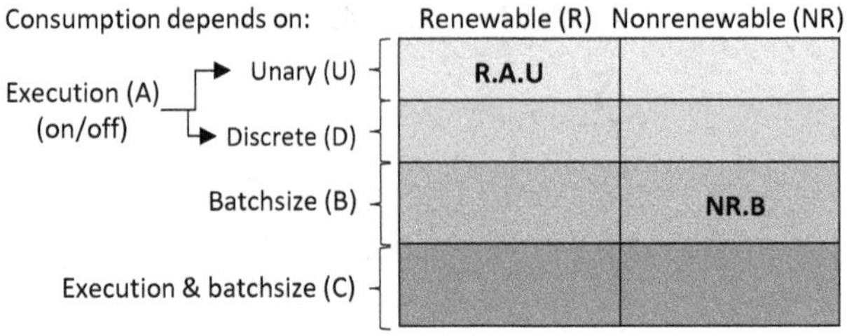

Figure 4.2 Resource type classification. Resources can be renewable (R) or nonrenewable (NR). Resource consumption can depend on the execution of a batch (type A), its batchsize (type B), or both (type C). If of type A, then a resource can be further classified as unary (U) or discrete (D). Two examples are shown.

Interestingly, an equipment unit can be viewed as a unary resource. While this is a classification that has been used in the literature, we note that given a sufficient *discretization* level, all resources can be viewed as discrete (especially when batchsizes are fixed[11]), so we will instead classify them as resources the demand for which depends on (1) the execution (on/off) of a batch (type A); (2) the size (continuous variable) of a batch (type B); and (3) both the execution and batch size (type C). If type A, then the resource can be further classified as unary (e.g., unit, labor) and discrete (e.g., pumps needed during the execution of a batch to load/withdraw material). Also, if the batchsize is fixed, then the resource requirements can be precalculated, so resources can be treated as type A.

An example of a type A resource is labor: one supervisor is required regardless of the batchsize. An agent used to maintain the necessary pH is an example of type B – it changes (proportionally) with the amount processed (batchsize). Heating/cooling loads are often resources of type C since some load will be used to heat/cool the unit, independently of the batchsize, and some of it will be used to heat/cool the processed material. The proposed classification is shown in Figure 4.2.

If φ_{im} is the fixed requirement of resource m by batch i and ψ_{im} is the variable resource requirement (per unit of batchsize B_i), then the resource use, R_{im}, during the execution of a batch is given by

$$R_{im} = \varphi_{im} + \psi_{im} B_i \tag{4.68}$$

Note that for resources of type A, $\psi_{im} = 0$; for resources of type B, $\varphi_{im} = 0$; while both parameters are positive for resources of type C. The interaction between batches and different types of resources is illustrated in Example 4.3.

[11] If batchsizes are fixed, then the demand of a batch for a resource is a known constant, which can be converted to an integer. If the resource demand depends on a variable batchsize, then the resource demand becomes a continuous variable.

We note that a resource can be classified differently based on the batch that requires it. For example, the batch of a product produced in small units may always require one pump for loading/unloading, which means that a pump can be viewed as a unary resource; whereas the batch of a different product, produced in the same facility but in larger units with variable batchsizes may require a number of pumps that is an (integer) function of its batchsize, which means that pumps should be treated collectively as a discrete resource. Thus, it is more accurate to say that a resource–batch pair is classified according to the scheme shown in Figure 4.2 rather than the resource.

Example 4.3 Resource Use/Consumption We consider three instances to illustrate the concepts of (non)renewable resources (Instance 1, Figure 4.3A), resource types A-C (Instance 2, Figure 4.3B), and the relationship between resource consumption and availability (Instance 3, Figure 4.3C). We use the batches and the corresponding resource requirements given in Table 4.4. Resource capacities, denoted by χ_m, are also given.

Table 4.4 Resource classification, resource capacity, and batch resource requirements for three instances.

	Instance 1		Instance 2		Instance 3
Resource	R1	R2	R1	R2	R1
Classification	NR.A[a]	R.A	R.B	R.C	R.A
Capacity (χ_m)	3	4	3	2	4.5
	φ^S_{im}	φ_{im}	ψ_{im}	φ_{im}/ψ_{im}	φ_{im}
B1	2		0.3	0.5/0.1	1
B2		2			2
B3					2

[a] Parameter φ^S_{im} is used, instead of φ_{im}, because consumption of renewable resource, at the start of the batch, is permanent.

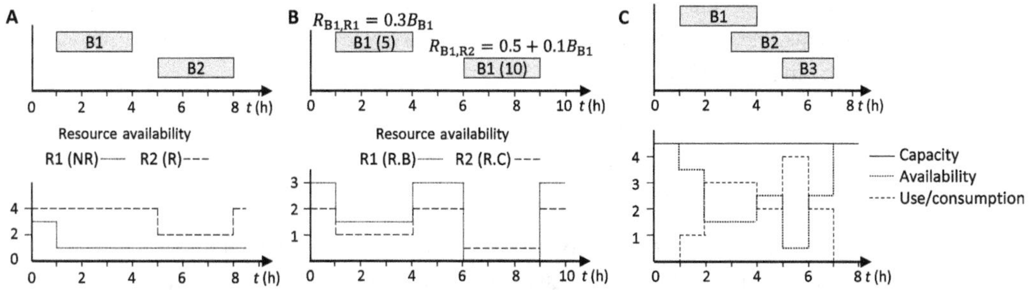

Figure 4.3 Resource availability for different types of resources. (A) Nonrenewable versus renewable. (B) Type B versus type C. (C) Relationship between resource use/consumption and availability.

Figure 4.3A shows the decrease in the availability of a nonrenewable resource R1 due to the execution of batch B1, and the reduced availability of renewable resource R2 during the execution of B2. Figure 4.3B shows how the availability of two types of renewable resources is impacted by the execution of two batches of the same product with two different batchsizes. Finally, Figure 4.3C shows that the availability of a resource during a time period is equal to its capacity minus its use (consumption) during the same period.

We close this subsection with a note on terminology. So far, we have used the term *consumption* for nonrenewable resources because of the *permanent* nature of the consumption, and the term *use* for renewable resources because the consumption is temporary – the availability increases once the task is finished (the resource is *returned* or *produced*). Since a renewable resource is also (temporarily) consumed and we will not further discuss nonrenewable resources, we will henceforth use the term *consumption*, during the execution of a batch, for renewable resources as well. We will also use the terms *engagement* and *release* for the (instantaneous) beginning and end, respectively, of resource consumption.

4.6.2 Sequence-Based Models

A key attribute of single-stage problems is that batches assigned to different units are not interacting with each other, which means that batch sequencing is unit specific. Thus, in sequence-based models sequencing variables ($Y_{ii'j}$) and (4.2), or (4.3), and (4.9) are defined for a single unit. If batches assigned to different units require the same resource, then monitoring resource consumption across units is necessary, which means that the parallel units are now *coupled*. In general, sequence-based models are not well suited to address these problems unless the additional shared resources are unary. This is the case we will consider in this subsection. We will further consider one type of resource (e.g., labor) but with multiple units (e.g., workers).

If $m \in \mathbf{M}$ is the set of resource units (e.g., workers), and $m \in \mathbf{M}_i$ the subset of resource units capable of performing batch i, then resource constraints can be enforced by simply using the same equations, which have been used to enforce batch-unit assignments and batch sequencing, for each unary resource separately. First, we define the following variables:

- $X_{im} \in \{0, 1\}$: = 1 if batch i is assigned to resource unit m.
- $Y_{ii'm} \in \{0, 1\}$: = 1 if resource unit m is used to perform batch i before batch i'.
- $S_{im} \in \mathbb{R}_+$: disaggregated start time.

If $\mathbf{I}^L$ is the set of batches requiring one unit of resource, then, for a global sequence model, we add the following equations (which are the counterparts of (4.1), (4.2), and (4.7) through (4.9)):

$$\sum\nolimits_{m \in \mathbf{M}_i} X_{im} = 1, \quad i \in \mathbf{I}^L \tag{4.69}$$

$$Y_{ii'm} + Y_{i'im} \geq X_{im} + X_{i'm} - 1 \quad i, i' > i, m \in \mathbf{M}_i \cap \mathbf{M}_{i'} \tag{4.70}$$

$$S_i = \sum_{m \in \mathbf{M}_i} S_{im}, \quad i \tag{4.71}$$

$$S_{im} \leq MX_{im}, \quad i \in \mathbf{I}^L, m \in \mathbf{M}_i \tag{4.72}$$

$$S_{im} + \sum_j \tau_{ij} X_{ij} \leq S_{i'm} + M(1 - Y_{ii'm}), \quad i, i', m \in \mathbf{M}_i \cap \mathbf{M}_{i'}, \tag{4.73}$$

where the summation $\sum_j \tau_{ij} X_{ij}$ in (4.73) is used to calculate the processing time, since the equation is not written for a unit.[12] All remaining equations remain the same. In other words, batches are assigned to all unary resources, sequencing should satisfy all resource types, and the timing subject to independent sequencing should satisfy release and due times. Local sequence models can be developed using the same ideas.

The case where multiple types of unary resources are required by batches (e.g., labor with skill set A and labor with skill set B) is modeled by introducing the variables and constraints presented in this subsection for $m^A \in \mathbf{M}^A$ and $m^B \in \mathbf{M}^B$ (assuming $\mathbf{M}^A \cap \mathbf{M}^B = \emptyset$).

4.6.3 Models Based on a Common Continuous Time Grid

The special case of unary resources can be treated in a manner similar to the one described in the previous subsection:

(1) Define resource-unit-specific grids, using equations similar to (4.21) and (4.22).
(2) Assign batches to resource unit slots, using equations similar to (4.24) and (4.25).
(3) Disaggregate start time variables with respect to all resources, using equations similar to (4.7) and (4.26).
(4) Enforce that batches are carried out within resource unit slots, as in (4.27) and (4.28), using the disaggregated start time variables.

Note that each batch has unique S_i, to which all disaggregated variables are related, so release and due times are enforced through (4.10) and (4.29), respectively.

The general case, where batches assigned to different units consume the same resource, cannot be readily handled by models based on unit-specific grids (see Figure 4.4). Thus, we present a model based on a common time grid[13] (see Figure 4.4C). While a wide range of models is available, here we present one model based on the following two binary variables:

- $X^S_{ijt} = 1$ if batch i starts on unit j at time point t.
- $X^F_{ijt} = 1$ if batch i finishes on unit j at time point t.

[12] Can you think of a different (new) parameter that can be used, instead of τ_{ij}, in (4.73) so no summation is necessary, as in (4.9)?

[13] If no two batches start or finish at the same time, then the number of required time points by a single grid model is $2|\mathbf{I}|$, which can be significantly higher than T^{MAX}, the maximum number of time points in a model employing unit-specific grids. Since all variables and constraints are defined for every t, the model can potentially be significantly larger than models employing unit-specific grids.

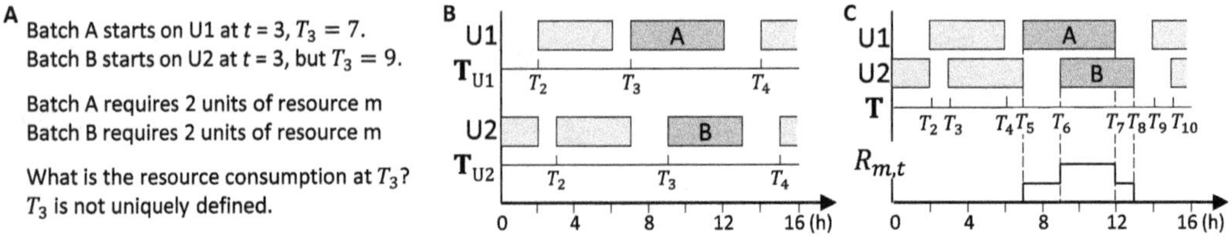

Figure 4.4 Illustration of resource consumption monitoring. (A) Description of instance. (B) Depiction of solution representation using unit-specific grids; T_3 corresponds to $t = 7$ in U1 but $t = 9$ in U2. (C) Representation of the same solution using a common time grid; the start of A and B now correspond to different time points.

Note that, as opposed to variables X_{ijt} in Section 4.3, variables X_{ijt}^S and X_{ijt}^F match (the start/finish of) a batch with time points, not periods.

Fixed Batching Decisions. A batch should start and finish at a time point (replaces (4.24)):

$$\sum_{j,t} X_{ijt}^S = \sum_{j,t} X_{ijt}^F = 1, \quad i. \tag{4.74}$$

In a given unit, at most one batch can start/finish at a point (replaces (4.25)):

$$\sum_i X_{ijt}^S \leq 1, \quad \sum_i X_{ijt}^F \leq 1 \quad j,t. \tag{4.75}$$

The difference, compared to the models presented in Section 4.3, is that (4.75) is not sufficient to enforce that no other batch will start (or finish) between the start and end of batch i because a batch can span more than one time period. In other words, if $X_{ijt}^S = 1$ and $X_{ijt'}^F = 1$, then (4.75) cannot prevent $X_{i'jt''}^S = 1$ for $t < t'' < t'$. This can be accomplished via

$$\sum_i \sum_{t' \leq t} \left(X_{ijt'}^S - X_{ijt'}^F \right) \leq 1, \quad j,t, \tag{4.76}$$

which basically enforces that the number of batches that have started on unit j through time point t can be at most one more than the batches that have finished on unit j through t. Equation (4.26) along with the following constraints, which replace (4.27) and (4.28), enforce that the start/end time of a batch is mapped, exactly, onto time point t, (i.e., time T_t) if $X_{ijt}^S / X_{ijt}^F = 1$:

$$T_t - M\left(1 - X_{ijt}^S\right) \leq S_{ij} \leq T_t + M\left(1 - X_{ijt}^S\right), \quad i,j,t \tag{4.77}$$

$$T_t - M\left(1 - X_{ijt}^F\right) \leq S_{ij} + \tau_{ij} \leq T_t + M\left(1 - X_{ijt}^F\right), \quad i,j,t \tag{4.78}$$

An alternative approach is to allow a batch to finish before the time point it is mapped to, that is, the batch end time falls within period t rather than at time point t

$$T_{t-1} - M\left(1 - X_{ijt}^F\right) \leq S_{ij} + \tau_{ij} \leq T_t + M\left(1 - X_{ijt}^F\right), \quad i,j,t \tag{4.79}$$

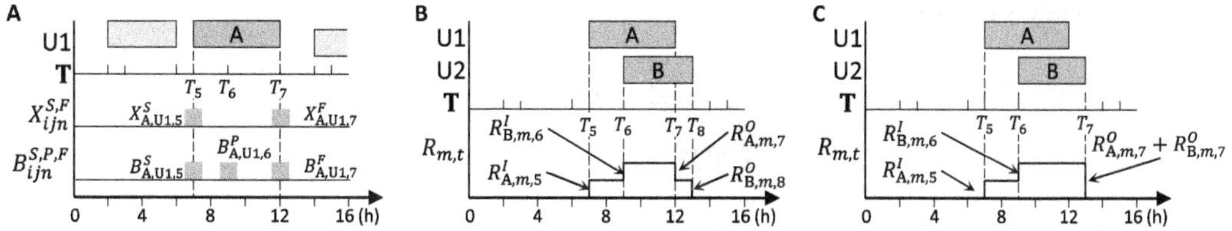

Figure 4.5 Illustration of single, common time grid modeling. (A) Mapping of batch start/finish onto the common grid through (4.77)and (4.78); and calculation of batchsizes from (4.83) through (4.86). (B) Calculation of resource engagement and release for batches A and B shown in Figure 4.4. (C) Alternative representation of the solution shown in B, using a time grid with 8, instead of 7, periods, and using (4.79); resource consumption is assumed to be constant during period 7, and thus overestimated during [12,13].

The preceding equation, which is used instead of (4.78), leads to an overestimation of resource consumption during the time between the actual batch end time and the corresponding time point, but this overestimation does not cut off solutions because no batches are allowed to start within a period (see Figure 4.5).

Using binaries X_{ijt}^S/X_{ijt}^F, we can calculate the amount of resource m engaged (R_{imt}^I)/ released (R_{imt}^O) by batch i starting/ending at time point t,

$$R_{imt}^I = \varphi_{im}\sum_j X_{ijt}^S + \psi_{im}\sum_j \beta_i X_{ijt}^S = \psi_{im}^T \sum_j X_{ijt}^S, \quad i,m,t \tag{4.80}$$

$$R_{imt}^O = \varphi_{im}\sum_j X_{ijt}^F + \psi_{im}\sum_j \beta_i X_{ijt}^S = \psi_{im}^T \sum_j X_{ijt}^F, \quad i,m,t, \tag{4.81}$$

which can then be used to express the total resource consumption, R_{mt}, during period t:

$$R_{mt} = R_{m,t-1} + \sum_i R_{im,t-1}^I - \sum_i R_{im,t-1}^O \leq \chi_m, \quad m,t, \tag{4.82}$$

where β_i is the fixed batchsize of batch i and χ_m is the *capacity* of resource m. Note that we can calculate resource requirements $\psi_{im}^T = \varphi_{im} + \psi_{im}$ because batchsizes are fixed.

In summary, the need to account for resources, consumed by batches carried out in different units, led to (1) the adoption of a common grid, with $|\mathbf{T}|$ larger than T^{MAX}; (2) the replacement of assignment binary variable X_{ijt} with variables X_{ijt}^S and X_{ijt}^F; (3) the modification of the grid-batch matching constraints; and (4) the introduction of variables and equations to calculate and constrain total resource consumption. The objective function would also have to be modified accordingly.

Batching Decisions. If batching decisions are considered, say due to unequal unit capacities, the following batchsize variables are introduced:

- B_{ijt}^S: batchsize of batch i starting on unit j at time point t.
- B_{ijt}^F: batchsize of batch i finishing on unit j at time point t.
- B_{ijt}^P: batchsize of batch i being processed on unit j during time period t and does not finish at time point t.

They should satisfy

$$\sum_{j\in \mathbf{J}_i,t} B^S_{ijt} \geq \xi_i, \quad i \tag{4.83}$$

$$\beta^{MIN}_{ij} X^S_{ijt} \leq B^S_{ijt} \leq \beta^{MAX}_{ij} X^S_{ijt}, \quad i,j,t \tag{4.84}$$

$$\beta^{MIN}_{ij} X^F_{ijt} \leq B^F_{ijt} \leq \beta^{MAX}_{ij} X^F_{ijt}, \quad i,j,t \tag{4.85}$$

and

$$B^S_{ij,t-1} + B^P_{ij,t-1} = B^P_{ijt} + B^F_{ijt}, \quad i,j,t, \tag{4.86}$$

which is a batchsize *balance* constraint ensuring that the starting (B^S_{ijt}) and finishing ($B^F_{ijt'}$) batchsize variables for the same batch are equal.

Equtation (4.74) is replaced by

$$\sum_{j,t} X^S_{ijt} = \sum_{j,t} X^F_{ijt}, \quad i, \tag{4.87}$$

while (4.75) through (4.78) remain the same.[14] Finally, R^I_{imt} and R^O_{imt} are calculated as follows (instead of (4.80) and (4.81)):

$$R^I_{imt} = \varphi_{im} \sum_j X^S_{ijt} + \psi_{im} \sum_j B^S_{ijt}, \quad i,m,t \tag{4.88}$$

$$R^O_{imt} = \varphi_{im} \sum_j X^F_{ijt} + \psi_{im} \sum_j B^F_{ijt}, \quad i,m,t \tag{4.89}$$

These are used to calculate and bound resource consumption using (4.82).

Figure 4.5 illustrates the basic ideas used to model resource consumption using a common continuous time grid. Note that variables R^I_{imt}/R^O_{imt} denote instantaneous engagement/release of resource (at time point t), while variable R_{mt} denotes resource consumption during period t.

4.6.4 Models Based on a Discrete Time Grid

One of the major advantages of models based on a (common) discrete time grid is that all types of resources can be readily modeled by simply using variables X_{ijn} (and B_{ijn}, if applicable) to calculate resource consumption regardless of the other characteristics of the problem. Specifically, R^I_{imn} and R^O_{imn} are calculated using the following two equations

$$R^I_{imn} = \varphi_{im} \sum_j X_{ijn} + \psi_{im} \sum_j B_{ijn}, \quad i,m,n \tag{4.90}$$

$$R^O_{imn} = \varphi_{im} \sum_j X_{ij,n-\bar{\tau}_{ij}} + \psi_{im} \sum_j B_{ij,n-\bar{\tau}_{ij}}, \quad i,m,n \tag{4.91}$$

[14] In the present chapter, we assume that the processing time is independent of batchsize. If this is not the case, then (4.78) should be modified: the end time $(S_{ij} + \tau_{ij})$ should be calculated based on a variable processing time. This feature will be discussed in Chapter 7.

and then plugged into

$$R_{mn} = R_{m,n-1} + \sum_i R^I_{im,n-1} - \sum_i R^O_{im,n-1} \leq \chi_m, \quad m,n \tag{4.92}$$

to calculate and bound the total resource consumption. If batchsizes are fixed, then $\psi_{im}\sum_j B_{ijn} = \psi_{im}\beta_i\sum_j X_{ijn}$ in (4.90), and $\psi_{im}\sum_j B_{ijn-\tau_{ij}} = \psi_{im}\beta_i\sum_j X_{ijn-\tau_{ij}}$ in (4.91), so (4.92) becomes

$$R_{mn} = R_{m,n-1} + \sum_{i,j}\psi^T_{im}X_{ij,n-1} - \sum_{i,j}\psi^T_{im}X_{ij,n-\bar{\tau}_{ij}-1} \leq \chi_m, \quad m,n, \tag{4.93}$$

where $\psi^T_{im} = \varphi_{im} + \psi_{im}\beta_i$; and (4.93) replaces (4.90) through (4.92).

Alternatively, recognizing that the consumption of resource m by a task i during period n will be given by (4.68) if batch i starts between $n - \bar{\tau}_{ij}$ and $n - 1$, we can write the following resource constraint,

$$\sum_{i,j}\sum_{n'=n-\bar{\tau}_{ij}}^{n'=n-1}\left(\varphi_{im}X_{ijn'} + \psi_{im}B_{ijn'}\right) \leq \chi_m, \quad m,n, \tag{4.94}$$

which, interestingly, is a generalization of the assignment constraint (4.43), which, in light of the discussion in this section, can now be viewed as representing a unary resource constraint. Equation (4.43) is obtained from (4.94) by setting $\varphi_{im} = 1$, $\psi_{im} = 0$, and $\chi_m = 1$. Note that the consumption of unary resources does not depend on batchsize (i.e., $\psi_{im} = 0$) and that the definition of unary resources given in Section 4.6.1 directly implies $\varphi_{im} = 1$ and $\chi_m = 1$.

The remaining constraints of a model based on a single discrete time grid are used to enforce the following:

(1) Unit-batch assignment in problems with fixed batching, (4.42)
(2) No overlap between batches processed on the same unit, present in all models, (4.43)
(3) Batch selection in problems with batching decisions but fixed unit-specific batchsizes, (4.65)
(4) Batch selection in problems with variable batchsizes, (4.66) and (4.67)

They all remain unchanged. Thus, the modeling of all types of resources can be achieved by, essentially, adding constraints (4.90) through (4.92) or (4.94).

In addition to enforcing general resource constraints, models based on a discrete time grid can be used to calculate resource consumption costs. If γ^{RES}_m is the resource unit cost,[15] then the total resource cost is $\sum_{m,n}\gamma^{RES}_m\delta R_{mn}$, which can be added to the cost minimization objective function,

$$\min \sum_{i,j}\left(\gamma^P_{ij}\sum_n X_{ijn}\right) + \sum_{i,i',j,n}\sigma_{ii'j}Z_{ii'jn} + \sum_{m,n}\gamma^{RES}_m\delta R_{mn}. \tag{4.95}$$

[15] Variable R_{mn} typically represents a rate. For example, if m is heating, then R_{mn} represents the power [W] (heat per unit time) required during period n. In this case, γ^{RES}_m would represent the cost per unit of energy [\$/J]. The (heating) energy [J] consumed during period n is $R_{mn}\,[\mathrm{J/sec}] \cdot \delta\,[\mathrm{sec}]$ and the corresponding cost [\$] is γ^{RES}_m [\$/J]δR_{mn} [J].

4.7 General Shared Resources: Extensions

4.7.1 Time-Varying Resource Capacity and Cost

The available resources may vary over time; for example, the number of workers available during the night may be smaller than during the day and, consequently, the capacity of other resources such as cleaning-in-place can also be lower during certain periods. Scheduled unit (e.g., boiler) maintenance can also lead to time-varying resource capacity. Similarly, resource cost may also vary over time; for example, electricity cost is different between day and night.

To handle these two features using continuous time models, a large number of new binary variables and constraints are necessary to map the (known) timing of resource capacity/cost changes onto the (variable) time points of the grid, thereby leading to computationally expensive models. Both features, however, can be readily handled using models based on discrete modeling of time by simply using time-indexed capacity (χ_{mn} instead of χ_m) and cost (γ_{mn}^{RES} instead of γ_m^{RES}) parameters.

The first step is the calculation of χ_{mn} and γ_{mn}^{RES} from the given data. To ensure feasibility, if resource capacity increases, from χ_m^1 to χ_m^2, between point $n-1$ and n (to be precise, during $((n-1)\delta, n\delta]$), then we set $\chi_{m,n-2} = \chi_{m,n-1} = \chi_{mn} = \chi_m^1$ and $\chi_{m,n+1} = \chi_{m,n+2} = \cdots = \chi_m^2$ (see Figure 4.6A). If the capacity decreases, then the new value is assigned to the interval during which the change happens. (Can you see why?) For cost, to be conservative, the higher cost can be assigned to the interval during which the change occurs, though an exact calculation is also possible.[16] Once parameters χ_{mn}

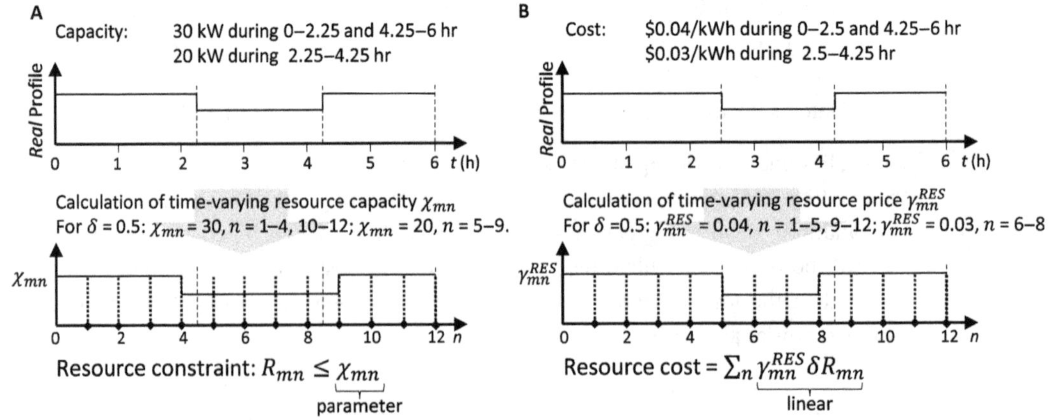

Figure 4.6 Modeling of time-varying resource capacity and cost using a discrete time model with $\delta = 0.5$ h. Calculation of parameters χ_{mn} (A) and γ_{mn}^{RES} (B) from real data.

[16] If the change, from γ_m^1 to γ_m^2, occurs δ' time units after time point $n-1$ (i.e., at time $(n-1)\delta + \delta'$), then the resource cost during period n is $\gamma_{mn}^{RES} = \delta'\gamma_m^1 + (\delta - \delta')\gamma_m^2$.

and γ_{mn}^{RES} are calculated, they can be simply used in (4.92) through (4.94) and (4.95), respectively. No new variables or equations need to be introduced.

4.7.2 Varying Resource Consumption during Batch Execution

In general, a resource may be engaged or released at the start, end, or during the execution of a batch. For example, labor may be needed only during the beginning (for loading) and end (for unloading) of a batch, but not in between. Similarly, a utility may be necessary during certain phases of a batch. This feature can be readily modeled using discrete modeling of time because the time at which a resource is engaged or released relative to the beginning of a batch can be readily associated with a point of the grid.

Extending the concepts of fixed and variable resource consumption, we define parameters φ_{ims} and ψ_{ims} to denote the fixed and proportional, respectively, engagement (φ_{ims}, $\psi_{ims} > 0$) and release (φ_{ims}, $\psi_{ims} < 0$) of resource m, s periods after the start of batch i (note that φ_{ims} and ψ_{ims} have to be calculated based on the chosen discretization δ). The net change in consumption (engagement minus release), R_{imn}^{NET}, of resource m by batch i during period n is calculated as follows:

$$R_{imn}^{NET} = \sum_j \sum_{s=0}^{\bar{\tau}_i} \left(\varphi_{ims} X_{ij,n-s} + \psi_{ims} X_{ij,n-s} \right), \quad i, m, n. \tag{4.96}$$

Note that (4.96) replaces both (4.90) and (4.91) because parameters φ_{ims} and ψ_{ims} account for engagement and release. The total resource consumption is then given and bounded by

$$R_{mn} = R_{m,n-1} + \sum_i R_{im,n-1}^{NET} \leq \chi_{mn}, \quad m, n. \tag{4.97}$$

The modeling ideas are illustrated in Example 4.4.

Example 4.4 Varying Resource Consumption during Batch Execution We consider a discrete grid with $\delta = 1$ and a batch that has processing time $\bar{\tau}_i = 6$ and requires the following:

(1) One unit of labor ($m = L$) during loading (0–1 h) and unloading (5–6 h)

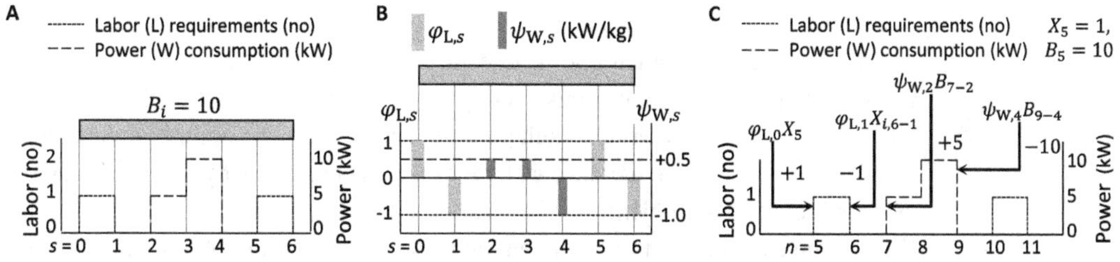

Figure 4.7 Modeling of varying resource consumption during batch execution. (A) Sample resource consumption during batch with batchsize of 10 kg. (B) Resource consumption parameters. (C) Resource consumption calculations for batch starting at $n = 5$. (Index i, for simplicity, from panels B and C).

(2) Electricity ($m = W$, kW/kg): 0.5 kw/kg from 2 to 3 h, and 1.0 kW/kg from 3 to 4 h

Figure 4.7A shows the resource consumption during the execution of batch B1 with batchsize equal to 10 kg – note that index s is used in the time axis. Figure 4.7B shows the parameters φ_{ims} and ψ_{ims}, calculated for $\delta = 1$, used to model the aforementioned resource requirements. Figure 4.7C shows resource consumption (R_{mn}) and some of the relevant calculations of R_{imn}^{NET} when batch i starts at $n = 5$ with batchsize equal to 10 ($X_{i,5} = 1$, $B_{i,5} = 10$).

4.8 Notes and Further Reading

(1) The extensions discussed in the previous chapter (prize collection and product families) can be addressed using the same ideas. Interestingly, the approach to prize collection has parallels with the batching problem: the number of batches is a decision variable in both. Also, the remarks in the previous chapter regarding commonly made assumptions and alternative solution representations hold true for the single-stage environment. In terms of variable domains, we will discuss some general methods in the next chapter, which are also applicable to problems in the single-stage environment since the latter are a special case of the multistage environment.

(2) The modeling of changeover times shows how an approach employed in a simple problem (single unit) can be extended to address a more general problem (single stage). Specifically, the idea was applied by defining all variables over and expressing all constraints for all j. In all approaches, changeover constraints are unit specific, that is, they are enforced using variables defined for a specific unit. Thus, regardless of the environment, and the features leading to the assignment of batches (or, as we will see later, tasks) to units, changeover constraints in all environments can be enforced using the techniques presented in the present chapter. The equations used to model changeover times in sequence-based ((4.18)) and continuous grid-based ((4.35) through (4.38)) models can be applied, practically unchanged to problems in multistage (Chapter 5) and multipurpose (Chapter 6) environments. Sequence-based models cannot be extended to address problems in network environments, while the modeling of changeovers using continuous-grid-based models in network environments is more complex because a common grid is necessary (see discussion in Chapter 7). Finally, the modeling of changeover times using discrete time modes ((4.50) or (4.51)) remains the same in multistage, multipurpose, and network environments. Similarly, the ideas presented in the present chapter for the modeling of changeover costs can be extended to address problems in more complex environments.

(3) We also discussed how to model general resource constraints. While the environment can be more complex, the calculation of resource consumption remains the same: (1) the per-batch resource consumption has a fixed and a batch-size-dependent component and (2) batches carried out in different units contribute to the total resource consumption. Thus, the equations presented in the present

chapter are applicable to all other environments. A generalization of the concept of *resource* will be discussed in Chapter 7.

(4) In terms of formulation size, as discussed in the previous chapter, an advantage of sequence-based models is that no additional binary variables are needed to model changeovers, but these models cannot be readily extended to account for other features (e.g., general shared resources). Models based on unit-specific continuous time grids can be effective if no changeovers are present and especially when the $|\mathbf{I}| / |\mathbf{J}|$ ratio is low or when good estimates of the number of slots per units can be obtained. Modeling based on a discrete time grid can potentially lead to large models, but the resulting models can be extended to account for many other features at almost no computational cost. The modeling of shared resources, in particular, can be performed effectively only through discrete time models.

(5) A review of algorithms for parallel machine, preemptive and nonpreemptive, scheduling can be found in Blazewicz [1].

(6) Pinto and Grossmann proposed a continuous-time-grid-based model that uses preordering constraints (heuristic rules) to estimate the number of periods [2]. Cerda and coworkers proposed a series of sequence-based models [3, 4]. Lim and Karimi proposed a model that employs unit-specific time grids [5]. Castro and Grossmann presented a series of models based on discrete and continuous time grids [6].

(7) Mendez et al. proposed a two-stage method for batching and scheduling [4], whereas Marchetti et al. proposed a single model for simultaneous batch and scheduling [7]. The models that were developed for simultaneous batch and scheduling in multistage environments are, clearly, applicable to problems in single-stage environments. Such models are discussed in Section 5.6.

(8) Velez et al. presented different formulations for single-unit and single-stage problems, including ones for simultaneous batch and scheduling, based on a discrete time grid, with special emphasis on the modeling of changeover times [8].

(9) An elegant derivation of models for single-unit and single-stage problems, using disjunctive programming, can be found in Castro and Grossmann [9].

(10) The ideas used in Section 4.6.3 for the modeling of resource consumption using a single continuous time grid are adopted from Maravelias and Grossmann [10], who developed methods for problems in network environments. Similarly, the ideas in Section 4.6.4 are adopted from Kondili et al. [11] and Pantelides [12].

4.9 Exercises

(1) Consider a single-stage environment consisting of two units, $\mathbf{J} = \{U1, U2\}$, and four batches, $\mathbf{I} = \{B1, B2, B3, B4\}$. The processing times and costs for the compatible units, along with release and due times, are given in Table 4.5, while the changeover times and costs are given in Table 4.6.

Using the preceding data, formulate and solve the following problems using a sequence-based model:

Table 4.5 Processing times (τ_{ij}) and costs (γ_{ij}^{P}); and release (ρ_i) and due (ε_i) times.

Batch	$\tau_{ij}/\gamma_{ij}^{P}$ U1	U2	ρ_i/ε_i
B1	2/5	-	0/15
B2	2/10	3/5	2/15
B3	3/10	4/5	3/20
B4	-	5/10	4/12

Table 4.6 Changeover times ($\sigma_{ii'j}$) and costs ($\gamma_{ii'j}^{CH}$).

	$\sigma_{ii'j}/\gamma_{ii'j}^{CH}$ (i = left, i' = top)			
j = U1	B1	B2	B3	B4
B1	-	0/0	0/0	-
B2	1/5	-	1/5	-
B3	1/5	2/10	-	-
j = U2				
B2	-	-	1/5	1/5
B3	-	2/10	-	1/5
B4	-	1/5	1/5	-

(a) Makespan minimization without changeovers (global sequence model)
(b) Makespan minimization without changeovers using an immediate sequence model; compare with (a) in terms of model size and computational performance
(c) Processing cost minimization
(d) Makespan minimization with changeover times
(e) Changeover cost minimization
(f) Total (processing and changeover) cost minimization

(2) Using the data for the system in Exercise 1, formulate models to solve the following instances using both continuous and discrete time (with $\delta = 1$) grid-based models:
(a) Makespan minimization without changeovers
(b) Makespan minimization with changeover times
(c) Processing cost minimization without changeovers
(d) Changeover cost minimization
(e) Total cost minimization

(3) Consider a single-stage environment consisting of three units, $\mathbf{J} = \{U1, U2, U3\}$, and 13 batches, $\mathbf{I} = \{B1, B2, \ldots, B13\}$. The batch processing times and costs for the compatible units, release and due times, and weight factors are given in

Table 4.7 Batch processing times (τ_i) and costs (γ_{ij}^P), release (ρ_i) and due (ε_i) times, and weights (ω_i).

		B1	B2	B3	B4	B5	B6	B7	B8	B9	B10	B11	B12	B13
τ_i/γ_{ij}^P	U1	2/5	4/5	5/10	7/5	3/15	4/10					6/5	7/15	10/10
	U2	3/5	5/5	3/5	6/10			3/10	8/5	5/5	4/10			
	U3	4/10	3/5						6/5	7/5	4/5			
ρ_i/ε_i		0/15	0/20	0/10	5/30	5/50	8/50	2/80	10/80	4/80	4/40	14/80	10/60	9/80
ω_i		5	10	5	5	10	5	5	10	5	5	5	5	10

Table 4.8 Changeover times/costs $(\sigma_{ii'j}/\gamma^{CH}_{ii'j})$.

(i,i')	$j=$ U1	(i,i')	$j=$ U2	(i,i')	$j=$ U3
(B1,B2)	2/20	(B1,B2)	3/6	(B1,B2)	1/16
(B2,B1)	3/16	(B2,B1)	2/8	(B2,B1)	3/12
(B2,B3)	3/18	(B2,B3)	5/10	(B2,B8)	4/10
(B2,B4)	4/16	(B2,B4)	3/12	(B2,B9)	3/14
(B2,B5)	3/18	(B2,B7)	5/8	(B2,B10)	4/12
(B2,B6)	2/20	(B2,B8)	4/10	(B8,B2)	2/6

Table 4.9 Batch processing times/costs (τ_i/γ^P_{ij}), order amounts (ξ_i), and resource consumption coefficients (φ_{im}/ψ_{im}).

	$\tau_{i,\mathrm{U1}}/\gamma^P_{i,\mathrm{U1}}$	$\tau_{i,\mathrm{U2}}/\gamma^P_{i,\mathrm{U2}}$	ξ_i	$\varphi_{i,\mathrm{M1}}/\psi_{i,\mathrm{M1}}$	$\varphi_{i,\mathrm{M2}}/\psi_{i,\mathrm{M2}}$
B1	2/5		20		
B2	2/10	3/5	15		0/1
B3	3/10	4/5	5	1/0	0/1
B4		5/10	10	1/0	

Table 4.10 Changeover times/costs $(\sigma_{ii'j}/\gamma^{CH}_{ii'j})$.

(i,i')	$j=$ U1	(i,i')	$j=$ U2
(B1,B2)	1/5	(B2,B3)	1/6
(B2,B3)	2/5	(B2,B4)	2/4
(B3,B1)	1/8	(B3,B2)	2/10
(B3,B2)	2/10	(B3,B4)	3/5
		(B4,B2)	2/6
		(B4,B3)	1/5

Table 4.7. The nonzero changeover times/costs for batches in different units are given in Table 4.8. Solve the following instances:

(a) Makespan minimization with changeover times
(b) Earliness minimization with changeover times
(c) Processing cost minimization
(d) Changeover cost minimization
(e) Total cost minimization

(4) Consider a single-stage environment consisting of four batches, $\mathbf{I}=\{\mathrm{B1},\mathrm{B2},\mathrm{B3},\mathrm{B4}\}$; three units, $\mathbf{J}=\{\mathrm{U1},\mathrm{U2},\mathrm{U3}\}$ with variable batchsizes; and two resources, $\mathbf{M}=\{\mathrm{M1},\mathrm{M2}\}$. The minimum and maximum batch sizes for all batches in all units are five and 10, respectively (i.e., $\beta^{MIN}_{ij}=\beta^{MIN}_j=5$, $\beta^{MAX}_{ij}=\beta^{MAX}_j=10$). Resource M1 is a unary resource and M2 has capacity $\chi_{\mathrm{M2}}=10$. The batch processing times and costs for the compatible units, order amounts due at 20 hours, and resource requirements are given in Table 4.9. The

nonzero changeover times/costs in different units are given in Table 4.10. Solve the following instances using time-grid-based models:

(a) Makespan minimization
(b) Processing cost minimization

References

[1] Blazewicz J, Dror M, Weglarz J. Mathematical-Programming Formulations for Machine Scheduling – a Survey. *Eur J Oper Res.* 1991;51(3):283–300.

[2] Pinto JM, Grossmann IE. A Continuous-Time Mixed-Integer Linear-Programming Model for Short-Term Scheduling of Multistage Batch Plants. *Ind Eng Chem Res.* 1995;34(9):3037–3051.

[3] Cerda J, Henning GP, Grossmann IE. A Mixed-Integer Linear Programming Model for Short-Term Scheduling of Single-Stage Multiproduct Batch Plants with Parallel Lines. *Ind Eng Chem Res.* 1997;36(5):1695–1707.

[4] Mendez CA, Henning GP, Cerda J. Optimal Scheduling of Batch Plants Satisfying Multiple Product Orders with Different Due-Dates. *Comput Chem Eng.* 2000;24(9-10):2223–2245.

[5] Lim M-F, Karimi IA. A Slot-Based Formulation for Single-Stage Multiproduct Batch Plants with Multiple Orders per Product. *Ind Eng Chem Res.* 2003;42(9):1914–1924.

[6] Castro PA, Grossmann IE. An Efficient MILP Model for the Short-Term Scheduling of Single Stage Batch Plants. *Comput Chem Eng.* 2006;30(6–7):1003–1018.

[7] Marchetti PA, Mendez CA, Cerda J. Mixed-Integer Linear Programming Monolithic Formulations for Lot-Sizing and Scheduling of Single-Stage Batch Facilities. *Ind Eng Chem Res.* 2010;49(14):6482–6498.

[8] Velez S, Dong YC, Maravelias CT. Changeover Formulations for Discrete-Time Mixed-Integer Programming Scheduling Models. *Eur J Oper Res.* 2017;260(3):949–963.

[9] Castro PM, Grossmann IE. Generalized Disjunctive Programming as a Systematic Modeling Framework to Derive Scheduling Formulations. *Ind Eng Chem Res.* 2012;51(16):5781–5792.

[10] Maravelias CT, Grossmann IE. New General Continuous-Time State-Task Network Formulation for Short-Term Scheduling of Multipurpose Batch Plants. *Ind Eng Chem Res.* 2003;42(13):3056–3074.

[11] Kondili E, Pantelides CC, Sargent RWH. A General Algorithm for Short-Term Scheduling of Batch-Operations .1. MILP Formulation. *Comput Chem Eng.* 1993;17(2):211–227.

[12] Pantelides CC, editor, Unified Frameworks for Optimal Process Planning and Scheduling. 2nd Conference on Foundations of Computer Aided Process Operations; 1994 1994; Snowmass, CO: CACHE Publications.

5 Multistage Environment

In this chapter, we discuss scheduling in multistage environments. The problem statement is presented in Section 5.1 and three types of models are presented in Section 5.2 (sequence-based), Section 5.3 (continuous grid-based), and Section 5.4 (discrete grid-based). In Section 5.5, we introduce an important new feature, namely, storage constraints. Again, we build upon the material covered in the previous chapters to model assignment and sequencing decisions, as well as other constraints such as release and due times.

5.1 Problem Statement

We consider the problem with fixed batching decisions. The facility consists of processing stages $k \in \mathbf{K}$, and each stage has units $j \in \mathbf{J}_k$ with $\cup_k \mathbf{J}_k = \mathbf{J}$ and $\mathbf{J}_k \cap \mathbf{J}_{k'} = \emptyset$ for all k, k'; that is, each unit belongs to only one stage (see Figure 5.1). Further, we are given a set, $\mathbf{I}$, of batches that have to be processed on exactly one unit in each stage, where the processing of a batch in stage $k+1$ can start only after its processing in stage k is completed. This new type of constraint, compared to the single-stage problem, will henceforth be referred to as *precedence relation*, *precedence constraint*, or simply *precedence*. Set $\mathbf{J}_{ik}$ is the subset of units in stage k suitable for processing batch i, and $\mathbf{I}_j$ is the set of batches that can be carried out in unit j. Each batch $i \in \mathbf{I}$ has a release, ρ_i, and due, ε_i, time. The processing time of batch i on unit j is denoted by τ_{ij} and the processing cost by γ_{ij}^P. The changeover time/cost from batch i to batch i' processed on unit j is denoted by $\sigma_{ii'j}/\gamma_{ii'j}^{CH}$. We denote the multistage environment as $\boldsymbol{Sms}$, $|\mathbf{K}| > 1$. Since the modeling of changeover times and costs can be achieved using the techniques presented in the previous chapters, we will focus on problem class $[\boldsymbol{Sms}, |\mathbf{K}| > 1/\rho, \varepsilon/\cdot]$.

Assuming that unit j becomes available at $\bar{\varrho}_j$, the start, ϱ_j, and end, ε_j, of the *unit horizons* are given by

$$\varrho_j = \max\left\{\bar{\varrho}_j, \min_{i \in \mathbf{I}_j}\left\{\rho_i + \sum\nolimits_{k'<k} \min_{j' \in \mathbf{J}_{ik'}} \tau_{ij'}\right\}\right\}, \quad k, j \in \mathbf{J}_k \tag{5.1}$$

$$\varepsilon_j = \max_{i \in \mathbf{I}_j}\left\{\varepsilon_i - \sum\nolimits_{k'>k} \min_{j' \in \mathbf{J}_{ik'}} \tau_{ij'}\right\}, \quad k, j \in \mathbf{J}_k. \tag{5.2}$$

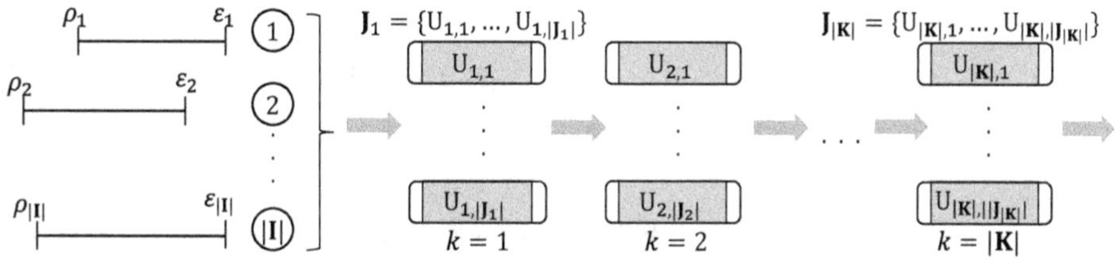

Figure 5.1 Representation of scheduling problem in multistage environment.

The summation in the RHS of (5.1) represents the fastest time a batch can be processed in the first $k-1$ stages, and the entire term in the outer min function represents the earliest time a batch can be available for processing in stage k. Similarly, the summation in (5.2) represents the shortest time it would be required for a batch that was just completed in stage k to go through all remaining $(k' > k)$ stages; and the term inside the max function is the latest time the processing of any batch in stage k can be completed while still meeting its deadline. If no deadlines are available, then a sufficiently large number should be used for ε_j. The unit windows calculated through (5.1) and (5.2) can be used to bound and/or variables in all models.

5.2 Sequence-Based Models

A batch has to be assigned to a unit in each stage. This is achieved through variable $X_{ij} \in \{0,1\}$, which is equal to 1 if batch i is assigned to unit $j \in \mathbf{J}_{i,k}$,

$$\sum\nolimits_{j \in \mathbf{J}_{i,k}} X_{ij} = 1, \quad i, k, \tag{5.3}$$

which is the same as (4.1), but it is expressed for every batch and stage.

If two batches are assigned to the same unit, then the relative sequence in which they are processed should be determined. We introduce $Y_{ii'j} \in \{0,1\}$, which is equal to 1 if both batches i and i' are assigned to unit j, and batch i is processed before batch i'. Note that the stage information is *hidden* in index j. A global sequencing relationship is activated via

$$Y_{ii'j} + Y_{i'ij} \geq X_{ij} + X_{i'j} - 1, \quad i, i' > i, k, j \in \mathbf{J}_{ik} \cap \mathbf{J}_{i'k}. \tag{5.4}$$

For local sequencing, we can use

$$\sum\nolimits_{i,i'} Y_{ii'j} = \sum\nolimits_{i} X_{ij} - 1, \quad j \tag{5.5}$$

or, following the ideas presented in the previous chapter, define two new binary variables, Y_{ij}^F (=1 if batch i is processed first in unit j) and Y_{ij}^L (=1 if batch i is processed last in unit j), and assuming that at least one batch is assigned to each unit, enforce

$$\sum\nolimits_{i} Y_{ij}^F = \sum\nolimits_{i} Y_{ij}^L = 1, \quad j \tag{5.6}$$

and[1]

$$\sum_{i'} Y_{i'ij} = X_{ij} - Y_{ij}^F, \quad i,j \tag{5.7}$$

$$\sum_{i'} Y_{ii'j} = X_{ij} - Y_i^L, \quad i,j. \tag{5.8}$$

Next, we define the start time of batch i in stage k, S_{ik}, which is disaggregated into unit-specific start times, S_{ij},

$$S_{ik} = \sum_{j \in \mathbf{J}_{ik}} S_{ij}, \quad i,k \tag{5.9}$$

with

$$S_{ij} \le MX_{ij}, \quad i,k,j \in \mathbf{J}_{ik} \tag{5.10}$$

where M is a sufficiently large parameter. Based on (5.1), we can also bound $S_{ik} \ge \min_{j \in \mathbf{J}_{ik}} \{\varrho_j\}$.

Start times, S_{ik}, are then used to enforce a timing relationship between the processing of the same batch in consecutive stages (precedence constraint),

$$S_{ik} + \sum_{j \in \mathbf{J}_{ik}} \tau_{ij} X_{ij} \le S_{i,k+1}, \quad i,k < |\mathbf{K}|. \tag{5.11}$$

The disjunctive constraint is written as follows:

$$S_{ij} + \tau_{ij} \le S_{i'j} + M(1 - Y_{ii'j}), \quad i,i',k,j \in \mathbf{J}_{ik} \cap \mathbf{J}_{i'k}. \tag{5.12}$$

If changeover times are present, then (5.12) is replaced by

$$S_{ij} + \tau_{ij} + \sigma_{ii'j} \le S_{i'j} + M(1 - Y_{ii'j}), \quad i,i',k,j \in \mathbf{J}_{ik} \cap \mathbf{J}_{i'k}. \tag{5.13}$$

For all problem classes, release and due dates can readily be enforced:

$$S_{i,k=1} = S_{i,1} \ge \rho_i, \quad i \tag{5.14}$$

$$S_{i,|\mathbf{K}|} + \sum_{j \in \mathbf{J}_{i,|\mathbf{K}|}} \tau_{ij} X_{ij} \le \varepsilon_i, \quad i \tag{5.15}$$

If due times can be violated, then we use

$$S_{i,|\mathbf{K}|} + \sum_{j \in \mathbf{J}_{i,|\mathbf{K}|}} \tau_{ij} X_{ij} \le \varepsilon_i + L_i, \quad i, \tag{5.16}$$

where L_i is the lateness/tardiness of batch i. If the objective function is lateness minimization, then $L_i \in \mathbb{R}$, whereas if the objective is tardiness minimization, then $L_i \in \mathbb{R}_+$.

For makespan, MS, minimization, it suffices to enforce

$$MS \ge S_{i,|\mathbf{K}|} + \sum_{j \in \mathbf{J}_{i,|\mathbf{K}|}} \tau_{ij} X_{ij}, \quad i \tag{5.17}$$

and the objective is

$$\min MS. \tag{5.18}$$

[1] Recall that we intentionally use i,j for the equation domain instead of $i,j \in \mathbf{J}_i$.

For earliness minimization and total weighted lateness/tardiness minimization, we use (5.19) and (5.20), respectively:

$$\min \sum_i \omega_i \left(\varepsilon_i - \left(S_{i,|\mathbf{K}|} + \sum_{j \in \mathbf{J}_{i,|\mathbf{K}|}} \tau_{ij} X_{ij} \right) \right) \tag{5.19}$$

$$\min \sum_i \omega_i L_i. \tag{5.20}$$

For cost minimization, we use

$$\min \sum_{i,j} \gamma_{ij}^P X_{ij} + \sum_{i,i'} \sum_j \gamma_{ii'j}^{CH} Y_{ii'j}. \tag{5.21}$$

The constraints presented in this section can be combined to address different problems. To give an example, the global sequence model for problem $[\boldsymbol{Sms}, |\mathbf{K}| = 1/\rho, \varepsilon/MS]$ consists of (5.3), (5.4), (5.9) through (5.12), (5.14), (5.15), (5.17), and (5.18).

5.3 Models Based on a Continuous Time Grid

Similarly to the single-stage problem, there are no constraints *linking* a unit with other units in the same stage, but unlike the single-stage problem, units are not completely decoupled because of the precedence constraints between consecutive stages. Nevertheless, enforcing the precedence constraints does not require the adoption of a common time grid. Thus, the model we present employs unit-specific grids where, in addition, a batch can be carried out within a single period of a time grid.[2]

The next step is to define the number of time periods, T^j, for each unit. A simple estimate for the units in stage k is $\lceil |\mathbf{I}| / |\mathbf{J}_k| \rceil$ (note that units in different stages can have different number of periods), though a more systematic approach, using the ideas presented in the previous chapter, can be followed. Also, the grids can be *moved* so that the earliest unit start is zero, by subtracting $\min_j\{\varrho_j\}$ from all ϱ_j and ε_j.

If $T^{MAX} = \max_j\{T^j\}$, we define superset $\mathbf{T} = \{1, \ldots, T^{MAX}\}$, and the subsets, $\mathbf{T}_j = \{1, \ldots, T^j\} \subseteq \mathbf{T}$, for each unit-specific grid. We introduce variables T_{jt} to denote the timing of point t in the grid of unit j, where each unit period starts at $T_{j,t-1}$ and ends at T_{jt}. We enforce the following:

$$T_{j,0} = \varrho_j, \quad T_{j,T^j} = \varepsilon_j \tag{5.22}$$

$$T_{jt} \geq T_{j,t-1}, \quad j, t \in \mathbf{T}_j. \tag{5.23}$$

We introduce $X_{ijt} \in \{0, 1\}$ to denote the assignment of batch i to time period t of unit j. Each batch has to be assigned to one unit in each stage and to one period of the selected unit,

[2] Recall that unit-specific grids cannot be employed when general shared resources, other than processing units, are present. In that case, a common grid is needed, and a batch should be allowed to span more than one time period. Since we will not consider general shared resources in the present chapter, we discuss models that employ unit-specific grids because they are smaller.

$$\sum_{j \in \mathbf{J}_k} \sum_{t \in \mathbf{T}_j} X_{ijt} = 1, \quad i, k, \tag{5.24}$$

which is similar to the constraint used in the previous chapter after projecting out binary X_{ij} using $X_{ij} = \sum_{t \in \mathbf{T}_j} X_{ijt}$. In addition, at most one batch can be allocated to each unit period:

$$\sum_i X_{ijt} \le 1, \quad j, t \in \mathbf{T}_j. \tag{5.25}$$

Equivalent solutions can be removed through the addition of the following:

$$\sum_i X_{ijt} \le \sum_i X_{ij,t-1}, \quad j, \ t \in \mathbf{T}_j \backslash \{1\}. \tag{5.26}$$

One approach to enforce precedence constraints and release/due time constraints is to employ, as in Section 5.2, variables S_{ik} and S_{ij}, defined through (5.9) and subject to

$$S_{ij} \le M \sum_t X_{ijt}, \quad i, j. \tag{5.27}$$

If batch i is allocated to a period, then it is executed within that period:

$$S_{ij} \ge T_{j,t-1} - M(1 - X_{ijt}), \quad i, j, t \in \mathbf{T}_j \tag{5.28}$$

$$S_{ij} + \tau_{ij} \le T_{jt} + M(1 - X_{ijt}), \quad i, j, t \in \mathrm{T}_j. \tag{5.29}$$

Then the precedence constraint for a given batch in consecutive stages is enforced through (5.11). Note that in (5.27) we use $\sum_t X_{ijt}$ instead of X_{ij}, which is used in (5.10); and that (5.28) and (5.29) are identical with the corresponding constraints in the previous chapter.

Release time, deadline, and due time constraints can be enforced via (5.14), (5.30), and (5.31) respectively,

$$S_{i,|\mathbf{K}|} + \sum_{j \in \mathbf{J}_{i,|\mathbf{K}|}} \sum_t \tau_{ij} X_{ijt} \le \varepsilon_i, \quad i \tag{5.30}$$

$$S_{i,|\mathbf{K}|} + \sum_{j \in \mathbf{J}_{i,|\mathbf{K}|}} \sum_t \tau_{ij} X_{ijt} \le \varepsilon_i + L_i, \quad i. \tag{5.31}$$

Equation (5.18) can be used for the minimization of makespan subject to

$$MS \ge S_{i,|\mathbf{K}|} + \sum_{j \in \mathbf{J}_{i,|\mathbf{K}|}} \sum_t \tau_{ij} X_{ijt}, \quad i. \tag{5.32}$$

Equations (5.19) and (5.20) can be used for earliness and lateness/tardiness minimization, respectively. The modeling of changeovers requires the introduction of time-indexed binary variables $Y_{ii'jt}$ to denote immediate sequencing relationship and equations similar to the ones presented in the previous chapter. The minimization of total production cost is then written as

$$\min \sum_{i,j} \gamma_{ij}^P \sum_t X_{ijt} + \sum_{i,i',j} \gamma_{ii'j}^{CH} \sum_t Y_{ii'jt}. \tag{5.33}$$

As in the previous chapter, different models can be formulated by enforcing different restrictions with respect to S_{ik} and T_{jt}.

5.4 Models Based on a Discrete Time Grid

We use index $n \in \mathbf{N} = \{0, 1, 2, \dots |\mathbf{N}|\}$ to define the time points (between the minimum release and the maximum due date) and periods $n \in \mathbf{N}\backslash\{0\} = \{1, 2, \dots, |\mathbf{N}|\}$ of equal length δ, where period n starts at point $n-1$ and ends at n. Time-related data are calculated as follows:

$$\overline{\tau_{ij}} = \lceil {}^{\tau_{ij}}/_{\delta} \rceil, \ \bar{\sigma}_{ii'j} = \lceil {}^{\sigma_{ii'j}}/_{\delta} \rceil, \ \bar{\rho}_i = \lceil {}^{\rho_i}/_{\delta} \rceil, \ \overline{\varepsilon_i} = \lfloor {}^{\varepsilon_i}/_{\delta} \rfloor. \tag{5.34}$$

We introduce $X_{ijn} \in \{0, 1\}$ to represent the beginning of processing batch i in unit j at time point n.[3] Release time constraints and deadlines can be enforced by fixing the *early* and *late* binaries to zero:

$$X_{ijn} = 0, \quad i,j,n < \bar{\rho}_i; \ X_{ijn} = 0, \quad i,j,n > \bar{\varepsilon}_i - \bar{\tau}_{ij}. \tag{5.35}$$

Each batch has to be processed on exactly one unit in each stage and assigned to one time point,

$$\sum\nolimits_{j \in \mathbf{J}_{ik}} \sum\nolimits_n X_{ijn} = 1, \quad i,k. \tag{5.36}$$

The no-overlap condition is enforced using the, by now familiar, clique constraint:

$$\sum\nolimits_i \sum\nolimits_{n'=n-\bar{\tau}_{ij}+1}^{n'=n} X_{ijn'} \leq 1, \quad j,n. \tag{5.37}$$

Next, we enforce the precedence constraints. We present three approaches to accomplish this.

Batch-Stage Balance. In the first approach, we define $U_{ikn} \in \{0, 1\}$ to denote the *availability* of batch i to undergo processing in stage $k+1$ at the end of period n.[4] In other words, $U_{ikn} = 1$ if the processing of batch i in stage k is completed by $n-1$ but its processing in stage $k+1$ has not started through and including point $n-1$, resulting in batch i remaining idle during period n. It is defined for $k \in \{0, 1, \dots |\boldsymbol{K}|\}$ with $U_{i,\,k=0,n}$ (parameters) representing the *initial conditions*, and $U_{i,k=|\mathbf{K}|,n} = 1$ representing that batch i has been processed in all stages by point n. Variable U_{ikn} is calculated as follows:

$$U_{ik,n+1} = U_{ikn} + \sum\nolimits_{j \in \mathbf{J}_{ik}} X_{ij,n-\bar{\tau}_{ij}} - \sum\nolimits_{j \in \mathbf{J}_{i,k+1}} X_{ijn}, \quad i,k > 1, n < |\mathbf{N}|. \tag{5.38}$$

Note that U_{ikn} can be defined as continuous variable, $U_{ikn} \in [0, 1]$, and still assume binary values at all feasible solutions. (Can you see why?) Release date constraints can alternatively be enforced by using a dummy stage $k = 0$ and setting,

$$U_{i,k,0} = 0, k > 0; \quad U_{i,0,n} = 0, i, n < \bar{\rho}_i; \quad U_{i,1,\bar{\rho}_i} = 1, \quad i. \tag{5.39}$$

[3] An alternative would be to define X_{ikn} to denote the start of processing of batch i in stage k and then enforce $\sum_{j \in \mathbf{J}_{ik}} X_{ijn} = X_{ikn}$. In the formulation presented here, we have essentially used this equation to project out X_{ikn}.

[4] As we will see in Chapter 7, variable U_{ikn} can be viewed as the *inventory* of the feed required for the processing of batch i in stage k and, in that regard, (5.38) can be viewed as a material balance over time.

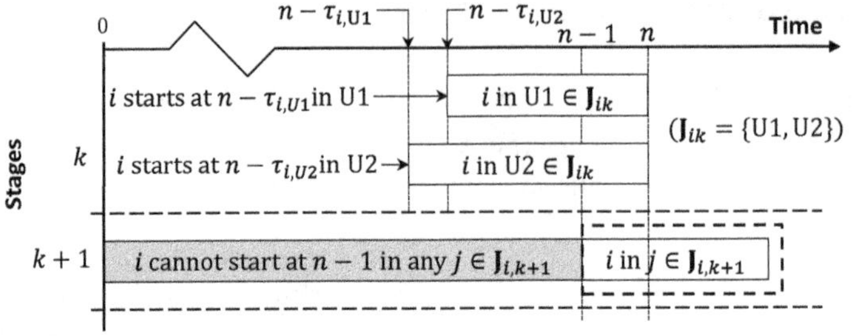

Figure 5.2 Enforcement of precedence constraints using the disaggregated start and finish batch-stage times.

Deadline constraints can be enforced by setting

$$U_{i,|\mathbf{K}|,\bar{\varepsilon}_i+1} = 1, \quad i.$$

Aggregated Start and Finish Batch-Stage Time.[5] In the second approach, we employ X_{ijn} variables to directly enforce precedences. The processing of batch i in stage k starts at $\sum_{j\in\mathbf{J}_{ik}}\sum_n nX_{ijn}$ and ends at $\sum_{j\in\mathbf{J}_{ik}}\sum_n (n+\bar{\tau}_j)X_{ijn}$. Thus, precedence constraints can be enforced as follows:

$$\sum_{j\in\mathbf{J}_{ik}}\sum_n nX_{ijn} \geq \sum_{j\in\mathbf{J}_{i,k-1}}\sum_n (n+\bar{\tau}_{ij})X_{ijn}, \quad i, k > 1. \tag{5.40}$$

Equation (5.40) is written once for each batch and stage, and thus leads to small models. Note that, following the logic of (5.40), we can enforce release and due date constraints by adding

$$\sum_{j\in\mathbf{J}_{i,1}}\sum_n nX_{ijn} \geq \bar{\rho}_i; \; \sum_{j\in\mathbf{J}_{i,|\mathbf{K}|}}\sum_n (n+\bar{\tau}_{ij})X_{ijn} \leq \bar{\varepsilon}_i, \tag{5.41}$$

though this is not necessary.

Disaggregated Start and Finish Batch-Stage Time.[6] The third approach also employs X_{ijn} variables but leads to one constraint per batch, stage, and time point. For batch i at stage k, the precedence constraint is expressed as the following logic condition: *if the processing of batch i at stage k finishes at or after n, then its processing at stage* $k+1$ *cannot have started before n, and vice versa.* This condition can be enforced as follows (see Figure 5.2):

$$\sum_{j\in\mathbf{J}_{ik}}\sum_{n'\geq n} X_{ij,n'-\bar{\tau}_{ij}} + \sum_{j\in\mathbf{J}_{i,k+1}}\sum_{n'<n} X_{ijn'} \leq 1, \quad i, k \geq 1, n. \tag{5.42}$$

Compared to (5.40), (5.42) leads to larger (why?) but tighter MIP models.

[5] The modeling of start and finish times using binaries X_{ijn} directly builds upon a MIP formulation for the resource-constrained project scheduling problem (RCPSP) (see note 4 in Section 5.6).

[6] This approach is based on a generalization of the disaggregated constraints used to enforce precedence relations in the resource-constrained project scheduling problem (see discussion in Section 5.6).

Objective functions are independent of the approach used for the modeling of precedence constraints. Makespan is subject to

$$MS \geq \sum_{j \in \mathbf{J}_{i,|\mathbf{K}|}} \sum_{n} (n + \bar{\tau}_{ij}) X_{ijn}, \quad i. \tag{5.43}$$

The weighted earliness, lateness, and tardiness are expressed as follows:

$$\min \sum_{i} \omega_i \left(\bar{\varepsilon}_i - \sum_{j \in \mathbf{J}_{i,|\mathbf{K}|}} \sum_{n} (n + \bar{\tau}_{ij}) X_{ijn} \right) \tag{5.44}$$

$$\min \sum_{i} \omega_i \left(\sum_{j \in \mathbf{J}_{i,|\mathbf{K}|}} \sum_{n} (n + \bar{\tau}_{ij}) X_{ijn} - \bar{\varepsilon}_i \right) \tag{5.45}$$

$$\min \sum_{i} \omega_i L_i \tag{5.46}$$

with $L_i \in \mathbb{R}_+$ subject to,

$$L_i \geq \sum_{j \in \mathbf{J}_{i,|\mathbf{K}|}} \sum_{n} (n + \bar{\tau}_{ij}) X_{ijn} - \bar{\varepsilon}_i, \quad i. \tag{5.47}$$

Finally, changeover times and costs can be modeled using the ideas presented in the previous chapter.

5.5 Storage Constraints

So far, we have assumed that there are enough storage vessels of appropriate size to store all materials consumed and produced by the processing of batches in all stages. Consequently, there are no limitations in the time elapsed between the processing of a batch between two consecutive stages. In practice, however, limited storage vessels may be available, which means that to obtain an *implementable* schedule, storage constraints should be included in the model. In fact, the presence of storage constraints is one of the attributes that often differentiates chemical production from discrete manufacturing. Accordingly, the goal of this section is to introduce all the different types of storage constraints, and then present how to model the major types using a representative (sequence-based) model. The modeling of storage constraints will be discussed again in Chapters 7 and 8.

5.5.1 Preliminaries

Chemical manufacturing involves the handling of fluids (as opposed to discrete parts), which often leads to different types of storage, and, more generally, material handling restrictions. While the majority of the research papers in the field have focused on the modeling of storage constraints in network production environments,[7] the no

[7] Interestingly, although models for scheduling in sequential environments predate the models for scheduling in network environments, storage constraints were first addressed in the context of network environments. This was, partly, due to the fact that, as we will see in Chapter 7, the representations developed for network environments enabled the seamless modeling of (some types of) storage constraints, whereas the modeling of storage constraints in sequential environments requires more substantial model modifications.

splitting/mixing restrictions present in sequential environments result in more complex storage constraints, which will be the topic of this section.

A fundamental assumption in most approaches developed to address storage in network production environments is that multiple batches of the same material can be mixed in the same storage vessel, which means that the storage of each material can be modeled using a single vessel. This further implies that in order to enforce storage capacity constraints, we have to simply upper bound the inventory level in the single vessel.

The treatment of storage in sequential facilities is inherently different because different batches of the same product (which lead to the production and consumption of the same intermediate materials) cannot be mixed. Thus, the number of available vessels is also relevant and needs to be considered. In addition, if batching decisions are considered, the size of the batches are not known a priori, which means that the size of the vessels have to be also taken into account. Thus, the storage policy in multistage processes is determined not only by the size of the storage vessels but also their relative number with respect to the number of batches. If enough vessels of sufficient size are available, then we can assume that we are operating under *unlimited* storage policy; otherwise, we are under *limited* storage, and storage constraints become important. Another aspect is the time (the material produced by) a batch can be stored in a vessel, that is, the *storage time*. Storage time is relevant in sectors, such as food and pharmaceuticals, which involve intermediate products with limited shelf life. Interestingly, the time a batch can wait in a processing unit after its completion (henceforth referred to as *waiting time*) becomes also relevant in the presence of storage constraints because waiting in a processing unit can be used instead of actual storage in a vessel if the next processing stage is the bottleneck of the entire system and there is a limited number of vessels.

In summary, in the general case a storage *policy* is based on both *capacity* and *timing* constraints. The former depend on both the *number* and the *size* of storage vessels, while the latter depend on the *waiting* and *storage* times (see Figure 5.3). Note that, in this section, we use the term *size* to describe the capacity of an individual storage vessel, and the term *capacity* to describe a feature of the entire process, which depends on both the number and size of all vessels. Also, we use the term *waiting* time to describe the time a batch is held in a processing unit, while we reserve the term *storage* time for the time spent in a storage vessel.

Figure 5.4 shows a classification of the different storage policies. The number and size of vessels determine whether we have unlimited (US), limited (LS), or no storage

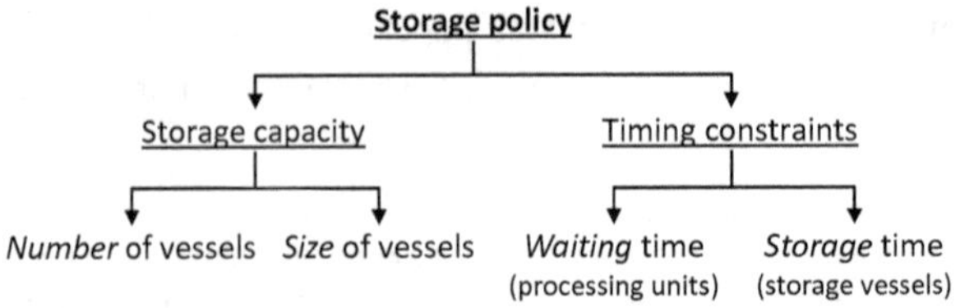

Figure 5.3 Elements of storage policies.

Timing constraints: waiting and storage time		Storage capacity: number and size of storage vessels		
		Unlimited (US)	Limited (LS)	No storage (NS)
	Unlimited (UT)	US/UT *Traditional approaches*	LS/UT • Vessel resource constraints • Modeling of transfers • Modeling of waiting & storage times	NS/UT* • Account for transfers (i.e., enforce storage bypass) • Modeling of waiting times
	Limited (LT)	US/LT • Modeling & bounding of waiting & storage times	LS/LT • Vessel resource constraints • Account for transfers • Modeling & bounding of waiting & storage times	NS/LT* • Account for transfers (i.e., enforce storage bypass) • Modeling & bounding of waiting times

* UT and LT refer to waiting time only; since there are no storage vessels, all storage times are zero

Figure 5.4 General classification of storage policies based on storage capacity and timing constraints.

(NS). The waiting and storage time constraints determine whether we have unlimited time (UT) or limited time (LT) storage/waiting time policies. Each box in Figure 5.4 corresponds to a different *case*, represented by a pair β^C/β^T, where β^C and β^T refer to storage capacity and timing, respectively, constraints. We note the following:

(1) In the US/UT case, it can be assumed that a batch can always be stored as soon as its processing in a given stage is completed, and that it can remain stored for unlimited time. This is the problem we have considered so far.

(2) In the US/LT case, it is necessary to model waiting and storage times and introduce constraints that bound them.

(3) In the presence of limited storage, storage vessels become scarce resources, and hence the assignment and sequencing of competing batches to storage vessels should be considered. We should also account for the transfer of a batch from a processing unit to a storage vessel or a processing unit in the next stage, and consequently the waiting/storage times, even if they are not bounded.

(4) When no storage is available, there is no need to model batch-vessel assignments. However, unlike the US/UT case, a batch should be directly transferred from one stage to the next with no *idle* time. No-storage automatically implies zero-wait *storage* time, so zero-wait is relevant only for the waiting time in a processing unit. We classify zero-wait as a special case of the limited waiting time.

5.5.2 Problem Statement

Before we give a formal problem statement, we note the following:

(1) To express resource constraints appropriately, it is necessary to explicitly account for the time a batch is transferred from a processing unit to a storage vessel and vice versa. Waiting time in processing units can be used to achieve optimal solutions. Waiting and storage times should therefore be modeled and bounded in the limited time case.

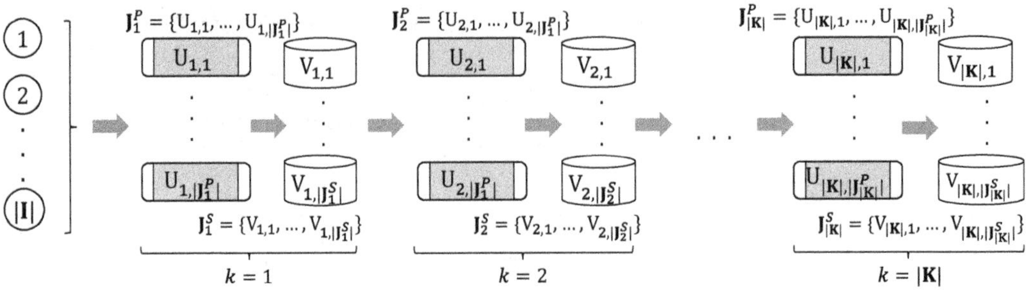

Figure 5.5 Structure and major subsets for scheduling in multistage environments under storage constraints.

(2) If batching decisions are not given, then the (number and) size of batches are variables, which means that the number and size of storage vessels should be taken into account. If batching decisions are known, then batchsizes are given, which means that we can predetermine which vessels are appropriate for storing batches at the different stages, which further means that it is not necessary to consider batchsizes and vessel sizes. In the present chapter, we consider the problem where batching decisions are given.

The facility consists of processing stages $k \in \mathbf{K}$. Each stage has processing units $j \in \mathbf{J}_k^P$ with $\cup_k \mathbf{J}_k^P = \mathbf{J}^P$ and $\mathbf{J}_k^P \cap \mathbf{J}_{k'}^P = \emptyset$ for all k, k'; and in addition, each stage has storage vessels $j \in \mathbf{J}_k^S$, where the output of stage k can be stored, with $\cup_k \mathbf{J}_k^S = \mathbf{J}^S$ and $\mathbf{J}_k^S \cap \mathbf{J}_{k'}^S = \emptyset$ for all k, k' (see Figure 5.5). We assume unlimited storage for raw materials (inputs to first stage); if raw material storage is limited, then we introduce a dummy $k = 0$ stage that contains only storage vessels. Each batch $i \in \mathbf{I}$ has to be processed on exactly one compatible unit, $j \in \mathbf{J}_{ik}^P$, in each stage, and can be stored in a compatible vessel, $j \in \mathbf{J}_{ik}^S$, before its processing in stage $k + 1$ starts. Set $\mathbf{I}_j$ is the set of batches that can be carried out in unit j. Each batch has a release time, ρ_i, and due time, ε_i, and the processing time of batch i on unit j is denoted by τ_{ij}. The changeover time in processing units and storage vessels is denoted by $\sigma_{ii'j}^P$ and $\sigma_{ii'j}^S$, respectively.

We will use the classification in Figure 5.4 as a road map for the modeling of storage constraints. First, in Section 5.5.3, we will present the model for the general (LS/LT) case (highlighted in Figure 5.4), and then, in Section 5.5.4, we will discuss how this model can be modified to address all other cases.

5.5.3 Basic Sequence-Based Model

We introduce three types of binary variables :

- X_{ij}^P, which is equal to 1 if batch i is assigned to processing unit $j \in \mathbf{J}_i^P$
- X_{ij}^S, which is 1 if batch i is stored in storage vessel $j \in \mathbf{J}_i^S$

- X_{ik}^{T}, which is equal to 1 if batch i is directly transferred (i.e., without storage) from stage k to stage $k+1$.

We also introduce *start time* variables $S_{ik}^{P}/S_{ik}^{S} \geq 0$ to denote the start of processing/storage of batch i in stage k, and $S_{ik}^{T} \geq 0$ to denote the time at which batch i is transferred from stage k to stage $k+1$. The transfer is assumed to occur instantaneously, that is, to have zero duration. Finally, we introduce *duration* variables T_{ij}^{W} and T_{ij}^{S} to denote the time batch i is stored in $j \in \mathbf{J}_{i}^{P}$ after its execution (*waiting* time) and the time it is stored in $j \in \mathbf{J}_{i}^{S}$ (*storage* time), respectively.[8]

A batch has to be assigned to a processing unit in each stage:

$$\sum\nolimits_{j \in \mathbf{J}_{ik}^{P}} X_{ij}^{P} = 1, \quad i, k. \tag{5.48}$$

A batch can be stored, in a vessel, or directly transferred to the next stage:

$$\sum\nolimits_{j \in \mathbf{J}_{ik}^{S}} X_{ij}^{S} + X_{ik}^{T} = 1, \quad i, k. \tag{5.49}$$

The relative order in which two batches assigned to the same unit are processed is modeled through variable $Y_{ii'j}^{P} \in \{0, 1\}$ (global sequencing) activated via,

$$Y_{ii'j}^{P} + Y_{i'ij}^{P} \geq X_{ij}^{P} + X_{i'j}^{P} - 1, \quad i, i' > i, k, j \in \mathbf{J}_{ik}^{P} \cap \mathbf{J}_{i'k}^{P} \tag{5.50}$$

Similarly, if two batches are stored in the same vessel, then there should be no overlap:

$$Y_{ii'j}^{S} + Y_{i'ij}^{S} \geq X_{ij}^{S} + X_{i'j}^{S} - 1, \quad i, i' > i, k, j \in \mathbf{J}_{ik}^{S} \cap \mathbf{J}_{i'k}^{S}. \tag{5.51}$$

For local sequencing binary variables, we can use the techniques discussed in Section 5.2.

Next, the start time of processing, S_{ik}^{P}, and storage, S_{ik}^{S}, of batch i in stage k are disaggregated into unit-specific times, S_{ij}^{P} and S_{ij}^{S}, respectively:

$$S_{ik}^{P} = \sum\nolimits_{j \in \mathbf{J}_{ik}^{P}} S_{ij}^{P}, \quad i, k \tag{5.52}$$

$$S_{ik}^{S} = \sum\nolimits_{j \in \mathbf{J}_{ik}^{S}} S_{ij}^{S}, \quad i, k \tag{5.53}$$

subject to

$$S_{ij}^{P} \leq M_{j}^{P} X_{ij}^{P}, \quad i, k, j \in \mathbf{J}_{ik}^{P}; S_{ij}^{S} \leq M_{j}^{S} X_{ij}^{S}, \quad i, k, j \in \mathbf{J}_{ik}^{S} \tag{5.54}$$

while S_{ik}^{T} should also satisfy

$$S_{ik}^{T} \leq M_{j}^{T} X_{ik}^{T}, \quad i, k, \tag{5.55}$$

where M_{j}^{P}, M_{j}^{S}, and M_{j}^{T} are sufficiently large parameters that can be calculated using parameters ε_{j} calculated in (5.2).

[8] Note that S variables (S_{ik}^{P}, S_{ik}^{S}, and S_{ik}^{T}) represent the absolute time (i.e., the time elapsed from the beginning of the scheduling horizon) of an event, while, in the present subsection, T variables (T_{ij}^{W} and T_{ij}^{S}) represent duration of an activity. Recall that, up to this subsection, T variables had been used only in grid-based models, to denote the timing (absolute time) of time points (e.g., variables T_{jt} in Section 5.3).

The disjunctive constraint has to now be written for both processing units,

$$S_{ij}^P + \tau_{ij} + T_{ij}^W \leq S_{i'j}^P + M\left(1 - Y_{ii'j}^P\right), \; i, i', k, j \in \mathbf{J}_{ik}^P \cap \mathbf{J}_{i'k}^P \tag{5.56}$$

and storage vessels,

$$S_{ij}^S + T_{ij}^S \leq S_{i'j}^S + M\left(1 - Y_{ii'j}^S\right), \; i, i', k, j \in \mathbf{J}_{ik}^S \cap \mathbf{J}_{i'k}^S, \tag{5.57}$$

where T_{ij}^W and T_{ij}^S should satisfy

$$T_{ij}^W \leq \omega_i^{MAX} X_{ij}^P, \quad i, k, j \in \mathbf{J}_{ik}^P; \quad T_{ij}^S \leq \varsigma_i^{MAX} X_{ij}^S, \quad i, k, j \in \mathbf{J}_{ik}^S, \tag{5.58}$$

where ω_i^{MAX} and ς_i^{MAX} are the maximum time a batch can remain in a processing unit and storage vessel, respectively.

The start time (S_{ik}^P, S_{ik}^S, and S_{ik}^T) and duration (T_{ij}^W and T_{ij}^S) variables are coupled as follows. The start of processing and storing in the same stage is described by

$$S_{ik}^S + S_{ik}^T = S_{ik}^P + \sum\nolimits_{j \in \mathbf{J}_{ik}^P} \left(\tau_{ij} X_{ij} + T_{ij}^W\right), \quad i, k, \tag{5.59}$$

where at most one of S_{ik}^S and S_{ik}^T can be positive due to (5.49), and then (5.54) and (5.55). The precedence between two consecutive stages is expressed as follows:

$$S_{i,k+1}^P = \left(S_{i,k}^S + \sum\nolimits_{j \in \mathbf{J}_{ik}^S} T_{ij}^S\right) + S_{i,k}^T, \quad i, k < |\mathbf{K}|, \tag{5.60}$$

where the summation in the parentheses represents the transfer time after intermediate storage. Figure 5.6 illustrates how the new variables ($X_{ik}^T, X_{ij}^S, S_{ik}^S, S_{ik}^T, T_{ij}^S, T_{ik}^W$) are used to model storage constraints using a two-stage facility with one storage vessel in stage 1.

Release date constraints can then be readily enforced through (5.14) with $S_{i,1}$ replaced by $S_{i,1}^P$. If a batch can be shipped to the customer any time before its due time, then storage in the last stage doesn't need to be modeled (i.e., $\mathbf{J}_{|\mathbf{K}|}^S = \emptyset$ and the corresponding variables and constraints are not defined) and deadlines can be enforced using (5.15) with $S_{i,|\mathbf{K}|}$ replaced by $S_{i,|\mathbf{K}|}^P$. If a batch must be shipped within an order

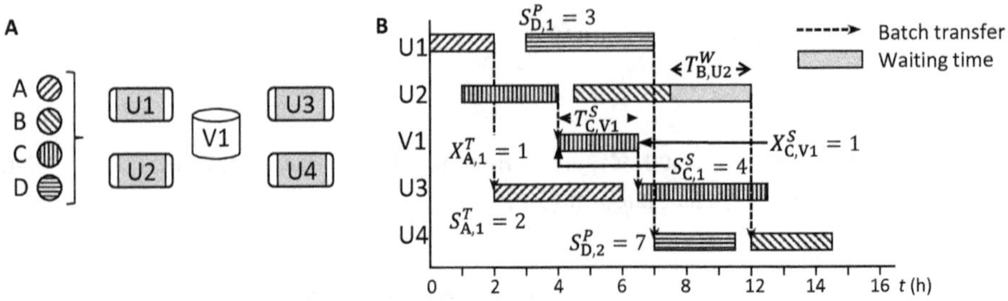

Figure 5.6 Modeling of storage constraints. (A) Illustrative instance (batches and production environment). (B) Gantt chart with values of select variables.

satisfaction window, $[\varepsilon_i^L, \varepsilon_i^U]$, then storage in the final stage should be considered, and the shipment should occur within the window:

$$\varepsilon_i^L \leq \left(S_{i,|\mathbf{K}|}^S + \sum\nolimits_{j \in \mathbf{J}_{i,|\mathbf{K}|}^S} T_{ij}^S\right) + S_{i,|\mathbf{K}|}^T \leq \varepsilon_i^U, \quad i. \tag{5.61}$$

Lateness and tardiness can be modeled either through a constraint similar to (5.16) or by replacing ε_i^U with $\varepsilon_i^U + L_i$ in (5.61).

If changeover times need to be taken into account, then (5.56) and (5.57) should be replaced, respectively, by

$$S_{ij}^P + \tau_{ij} + T_{ij}^W + \sigma_{ii'j}^P \leq S_{i'j}^P + M\left(1 - Y_{ii'j}^P\right), \quad i, i', k, j \in \mathbf{J}_{ik}^P \cap \mathbf{J}_{i'k}^P \tag{5.62}$$

and

$$S_{ij}^S + T_{ij}^S + \sigma_{ii'j}^S \leq S_{i'j}^S + M\left(1 - Y_{ii'j}^S\right), \quad i, i', k, j \in \mathbf{J}_{ik}^S \cap \mathbf{J}_{i'k}^S. \tag{5.63}$$

5.5.4 Modifications and Extensions

Next, we describe how the special cases shown in Figure 5.4 can be modeled.

Case US/UT. If there are no storage and timing constraints, then the model is reduced to the model presented in Section 5.2.

Case US/LT. We start from the model presented in Section 5.2. If we assume that the total maximum time a batch can wait in a unit or be stored in a vessel is equal to ς_i^{MAX}, then we can assume that the batch is transferred to a vessel (or directly to the next stage), that is, it does not wait in a processing unit. Consequently, the time a batch remains stored between its processing in stages k and $k+1$ is equal to $S_{i,k+1} - (S_{ik} + \sum_{j \in \mathbf{J}_{ik}} \tau_{ij} X_{ij})$, and we simply enforce

$$S_{i,k+1} - \left(S_{ik} + \sum\nolimits_{j \in \mathbf{J}_{ik}} \tau_{ij} X_{ij}\right) \leq \varsigma_i^{MAX}, \quad i, k < |\mathbf{K}|. \tag{5.64}$$

If a batch can wait in a processing unit and then be stored in a vessel and the timing constraints are independent from each other, then we (1) introduce variable T_{ij}^W and bound it via $T_{ij}^W \leq \omega_{ij}^{MAX} X_{ij}$; (2) use (5.65), instead of (5.12), to enforce sequencing in a unit; and (3) bound storage time via (5.66):

$$S_{ij} + \tau_{ij} + T_{ij}^W \leq S_{i'j} + M(1 - Y_{ii'j}), \quad i, i', k, j \in \mathbf{J}_{ik} \cap \mathbf{J}_{i'k} \tag{5.65}$$

$$S_{i,k+1} - \left\{S_{ik} + \sum\nolimits_{j \in \mathbf{J}_{ik}} \left(\tau_{ij} X_{ij} + T_{ij}^W\right)\right\} \leq \varsigma_i^{MAX}, \quad i, k < |\mathbf{K}|. \tag{5.66}$$

Case LS/UT. Starting from the model presented in Section 5.5.3, we simply replace (5.58) with

$$T_{ij}^W \leq MX_{ij}^P, i, k, j \in \mathbf{J}_{ik}^P; T_{ij}^S \leq MX_{ij}^S, \quad i, k, j \in \mathbf{J}_{ik}^S. \tag{5.67}$$

Case NS/UT. Since there are no storage vessels, the waiting time in processing units is unlimited, but there is no storage time. Starting from the model presented in Section 5.2, we (1) introduce variable T_{ij}^{W} subject to $T_{ij}^{W} \leq MX_{ij}, i, k, j \in \mathbf{J}_{ik}$; (2) use (5.65) to enforce sequencing in a unit; and (3) enforce no storage through

$$S_{i,k+1} = S_{ik} + \sum\nolimits_{j \in \mathbf{J}_{ik}} \left(\tau_{ij} X_{ij} + T_{ij}^{W} \right), \quad i, k < |\mathbf{K}|. \tag{5.68}$$

Case NS/LT. Starting from the model used in the NS/UT case, we replace $T_{ij}^{W} \leq MX_{ij}, i, k, j \in \mathbf{J}_{ik}$ with $T_{ij}^{W} \leq \omega_i^{MAX} X_{ij}, i, k, j \in \mathbf{J}_{ik}$.

5.6 Notes and Further Reading

(1) Early examples of sequence-based formulations include the models proposed by Méndez et al. [1] and Gupta and Karimi [2].

(2) Continuous-time-grid-based formulations for the scheduling of multiproduct, multistage batch plants with fixed number of batches have been proposed by Pinto and Grossmann [3, 4], Lamba and Karimi [5], Castro and Grossmann [6], and Castro et al. [7].

(3) The development of discrete time models for multistage problems is more recent; Castro and Grossmann presented some models [6], while the presentation in Section 5.4 is based on the models proposed by Merchan et al. [8].

(4) The aggregated MIP model for the resource-constrained project scheduling problem (RCPSP), which was modified in Section 5.4 to model the precedence between consecutive stages was proposed by Pritker et al. [9]. The disaggregated model was proposed by Christofides et al. [10].

(5) In terms of storage, Ku and Karimi [11] and Kim et al. [12] addressed storage policies in multistage processes with one unit per stage, while Kim et al. [13] studied scheduling under finite intermediate storage policy in more general multipurpose processes (this will be studied in the next chapter) with one unit per stage. Mendez and Cerda [14] and Wu and He [15] considered storage constraints in sequential environments with multiple units per stage.

(6) Karimi and coworkers proposed various methods to address specific problems in multistage environments [16 through 18].

(7) Simultaneous batching, assignment, and sequencing/timing decisions in general multistage environments were studied by Prasad and Maravelias [19] and Sundaramoorthy and Maravelias [20]. Sundaramoorthy and Maravelias [21] extended these models to account for storage, while Sundaramoorthy et al. [22] considered simultaneous batching, assignment, and sequencing/timing decisions under storage and utility constraints. Baumann and Trautmann also proposed a model that accounts for multiple process features simultaneously [23].

(8) Merchan et al. [8] proposed improved models for two extensions: (1) product diversification, where the batches of different products are identical in early stages and gradually become distinguishable as they move through the stages; and (2) multiple orders for the same products.

(9) To address larger instances, Harjunkoski and Grossmann [24] and Maravelias [25] studied decomposition methods that consider the assignment and sequencing subproblems independently. Neumann et al. [26] and Roe et al. [27] have also developed algorithms that exploit the structure of the problem.

(10) In terms of MIP solution methods, Merchan et al. [8] proposed a range of solution methods for discrete time models, including (1) tightening constraints based on fixed and variable stage-specific batch windows, (2) reformulations, (3) priority-based branching, and (4) a two-stage method to improve the accuracy of solutions obtained by discrete time models.

5.7 Exercises

(1) Consider a multistage environment with three stages, each one having two units ($\mathbf{J}_1 = \{U1, U2\}$, $\mathbf{J}_2 = \{U3, U4\}$, $\mathbf{J}_3 = \{U5, U6\}$) and eight batches. The processing times and costs as well as release and due times are given in Table 5.1.

Using the data in Table 5.1, formulate and solve the following instances using three models: (1) a sequence-based model (of your choice), (2) a grid-based continuous time model, and (3) a discrete time model:

(a) Makespan minimization
(b) Processing cost minimization
(c) Earliness minimization

(2) Consider the multistage environment with three stages from Exercise (1), but with the modified release and due times given in Table 5.2.

Using the preceding data, formulate and solve the following instances using three models: (1) a sequence-based model (of your choice), (2) a grid-based continuous time model, and (3) a discrete time model:

(a) Lateness minimization
(b) Tardiness minimization

Table 5.1 Processing times, τ_{ij} [h]; processing costs, γ_{ij}^P [\$]; and release, ρ_i, and due, ε_i, times [h].

	τ_{ij} / γ_{ij}^P							
	U1	U2	U3	U4	U5	U6	ρ_i	ε_i
B1	6/10	6/10	8/10	8/10	5/10	4/10	0	48
B2	6/15	8/10	8/10	6/15	8/10	6/15	3	48
B3	8/15	10/10	10/10	8/15	2/15	4/10	6	48
B4	6/10	4/15	4/15	6/10	2/10	2/10	0	48
B5	10/10	4/15	4/15	8/10	2/10	2/10	4	48
B6	6/10	6/10	6/10	6/10	6/10	6/10	2	48
B7	8/10	4/15	2/15	6/10	4/10	2/15	3	48
B8	10/10	8/15	4/15	6/10	8/10	4/15	10	48

Table 5.2 Batch release, ρ_i, and due, ε_i, times [h].

	B1	B2	B3	B4	B5	B6	B7	B8
ρ_i	0	3	6	0	4	2	3	10
ε_i	24	24	24	24	24	24	24	24

Table 5.3 Processing times, τ_{ij} [h]; processing costs, γ_{ij}^P [\$]; and release, ρ_i, and due, ε_i, times [h].

	τ_{ij}/γ_{ij}^P									
	U1	U2	U3	U4	U5	U6	U7	U8	ρ_i	ε_i
B1	7/35	9/71	9/63	6/66	4/94	8/12	3/66	2/34	0	48
B2	9/78	10/18	9/60	7/54	5/87	8/87	7/64	6/69	0	48
B3	4/97	6/82	5/63	10/48	7/75	4/83	8/65	8/37	0	48
B4	4/86	2/89	8/16	6/83	7/92	10/48	6/62	6/28	0	48

Table 5.4 Changeover times, $\sigma_{ii'j}$ [h]/costs, $\gamma_{ii'j}^{CH}$ [\$].

	U1				U2			
	B1	B2	B3	B4	B1	B2	B3	B4
B1	–	4/10	2/15	3/15	–	3/10	2/10	2/15
B2	4/10	–	2/15	3/10	4/10	–	2/15	4/10
B3	3/10	2/15	–	4/10	2/15	3/10	–	4/10
B4	2/15	3/10	3/15	–	2/15	3/10	4/10	–

(3) Consider a multistage environment with four stages, each one having two units ($\mathbf{J}_1 = \{U1, U2\}$, $\mathbf{J}_2 = \{U3, U4\}$, $\mathbf{J}_3 = \{U5, U6\}$, $\mathbf{J}_4 = \{U7, U8\}$) and four batches. The processing and changeover times and costs as well as release and due times are given in Tables 5.3 and 5.4.

Using the data in Tables 5.3 and 5.4, formulate and solve the following instances using three models: (1) a sequence-based model (of your choice), (2) a grid-based continuous time model, and (3) a discrete time model:

(a) Makespan minimization without changeover times
(b) Makespan minimization with changeover times
(c) Changeover cost minimization
(d) Total cost minimization

(4) Consider a multistage environment with two stages with one processing unit per stage, one vessel for intermediate storage (i.e., between the two stages), and two batches. If a batch is not immediately processed in the next stage, it should be stored. All processing data, including changeover times and costs, are given in Tables 5.5 and 5.6.

Table 5.5 Processing times, τ_{ij} [h]; processing costs, γ_{ij}^{P} [\$]; and release, ρ_i, and due, ε_i, times [h].

Batch	U1	U2	ρ_i	ε_i
A	3/4	7/2	0	19
B	5/5	1/2	3	10

Table 5.6 Changeover times, $\sigma_{ii'j}$ [h]/costs, $\gamma_{ii'j}^{CH}$ [\$].

		$\sigma_{ii'j}/\gamma_{ii'j}^{CH}$	
From	To	U1	U2
A	B	1/1	2/2
B	A	1/1	1/1

Formulate a sequence-based model (of your choice) to solve the following instances:

(a) Makespan minimization with and without changeover times
(b) Changeover cost minimization with and without changeover times

References

[1] Mendez CA, Henning GP, Cerda J. An MILP Continuous-Time Approach to Short-Term Scheduling of Resource-Constrained Multistage Flowshop Batch Facilities. *Comput Chem Eng*. 2001;25(4–6):701–711.

[2] Gupta S, Karimi IA. An Improved MILP Formulation for Scheduling Multiproduct, Multistage Batch Plants. *Ind Eng Chem Res*. 2003;42(11):2365–2380.

[3] Pinto JM, Grossmann IE. A Continuous-Time Mixed-Integer Linear-Programming Model for Short-Term Scheduling of Multistage Batch Plants. *Ind Eng Chem Res*. 1995;34 (9):3037–3051.

[4] Pinto JM, Grossmann IE. An Alternate MILP Model for Short-Term Scheduling of Batch Plants with Preordering Constraints. *Ind Eng Chem Res*. 1996;35(1):338–342.

[5] Lamba N, Karimi IA. Scheduling Parallel Production Lines with Resource Constraints. 1. Model Formulation. *Ind Eng Chem Res*. 2002;41(4):779–789.

[6] Castro PM, Grossmann IE. New Continuous-Time MILP Model for the Short-Term Scheduling of Multistage Batch Plants. *Ind Eng Chem Res*. 2005;44(24):9175–9190.

[7] Castro PM, Grossmann IE, Novais AQ. Two New Continuous-Time Models for the Scheduling of Multistage Batch Plants with Sequence Dependent Changeovers. *Ind Eng Chem Res*. 2006;45(18):6210–6226.

[8] Merchan AF, Lee H, Maravelias CT. Discrete-Time Mixed-Integer Programming Models and Solution Methods for Production Scheduling in Multistage Facilities. *Comput Chem Eng*. 2016;94:387–410.

[9] Pritsker AAB, Waiters LJ, Wolfe PM. Multiproject Scheduling with Limited Resources: A Zero-One Programming Approach. *Manage Sci.* 1969;16(1):93–108.

[10] Christofides N, Alvarez-Valdes R, Tamarit JM. Project Scheduling with Resource Constraints – a Branch and Bound Approach. *Eur J Oper Res.* 1987;29(3):262–273.

[11] Ku HM, Karimi IA. Scheduling in Serial Multiproduct Batch Processes with Finite Interstage Storage – a Mixed Integer Linear Program Formulation. *Ind Eng Chem Res.* 1988;27(10):1840–1848.

[12] Kim M, Jung JH, Lee IB. Optimal Scheduling of Multiproduct Batch Processes for Various Intermediate Storage Policies. *Ind Eng Chem Res.* 1996;35(11):4058–4066.

[13] Kim SB, Lee HK, Lee IB, Lee ES, Lee B. Scheduling of Non-sequential Multipurpose Batch Processes under Finite Intermediate Storage Policy. *Comput Chem Eng.* 2000;24(2–7):1603–1610.

[14] Mendez CA, Cerda J. An MILP Continuous-Time Framework for Short-Term Scheduling of Multipurpose Batch Processes under Different Operation Strategies. *Optimization and Engineering*. 2003;4(1–2):7–22.

[15] Wu JY, He XR. A New Model for Scheduling of Batch Process with Mixed Intermediate Storage Policies. *Journal of the Chinese Institute of Chemical Engineers.* 2004;35 (3):381–387.

[16] Gupta S, Karimi IA. Scheduling a Two-Stage Multiproduct Process with Limited Product Shelf Life in Intermediate Storage. *Ind Eng Chem Res.* 2003;42(3):490–508.

[17] Liu Y, Karimi IA. Novel Continuous-Time Formulations for Scheduling Multi-Stage Batch Plants with Identical Parallel Units. *Comput Chem Eng.* 2007;31(12):1671–1693.

[18] Liu Y, Karimi IA. Scheduling Multistage Batch Plants with Parallel Units and No Interstage Storage. *Comput Chem Eng.* 2008;32(4–5):671–693.

[19] Prasad P, Maravelias CT. Batch Selection, Assignment and Sequencing in Multi-Stage Multi-Product Processes. *Comput Chem Eng.* 2008;32(6):1106–1119.

[20] Sundaramoorthy A, Maravelias CT. Simultaneous Batching and Scheduling in Multistage Multiproduct Processes. *Ind Eng Chem Res.* 2008;47(5):1546–1555.

[21] Sundaramoorthy A, Maravelias CT. Modeling of Storage in Batching and Scheduling of Multistage Processes. *Ind Eng Chem Res.* 2008;47(17):6648–6660.

[22] Sundaramoorthy A, Maravelias CT, Prasad P. Scheduling of Multistage Batch Processes under Utility Constraints. *Ind Eng Chem Res.* 2009;48(13):6050–6058.

[23] Baumann P, Trautmann N. A Continuous-Time MILP Model for Short-Term Scheduling of Make-and-Pack Production Processes. *International Journal of Production Research.* 2013;51(6):1707–1727.

[24] Harjunkoski I, Grossmann IE. Decomposition Techniques for Multistage Scheduling Problems Using Mixed-Integer and Constraint Programming Methods. *Comput Chem Eng.* 2002;26(11):1533–1552.

[25] Maravelias CT. A Decomposition Framework for the Scheduling of Single- and Multi-Stage Processes. *Comput Chem Eng.* 2006;30(3):407–420.

[26] Neumann K, Schwindt C, Trautmann N. Advanced Production Scheduling for Batch Plants in Process Industries. *Or Spectrum.* 2002;24(3):251–279.

[27] Roe B, Papageorgiou LG, Shah N. A Hybrid MILP/CLP Algorithm for Multipurpose Batch Process Scheduling. *Comput Chem Eng.* 2005;29(6):1277–1291.

6 Multipurpose Environment

In this chapter, we discuss scheduling in multipurpose environments,[1] which are the most general sequential environments. As in single- and multistage facilities, the general problem is posed in terms of facility (e.g., number and capacity of units) and product (e.g., processing times) data, as well as raw material and resource availability (e.g., batch release times), and product demand (e.g., due times). If the batching problem is solved independently, however, then the problem can be expressed in terms of batches instead of products.[2] This is the problem that we will study in the present chapter. Models for the simultaneous batching and scheduling can be formulated using the ideas presented in Chapter 4. General shared resources and storage policies can be modeled using the techniques presented in Chapters 4 and 5, respectively.

The main differences between multistage and multipurpose environments are that (1) the sequence of stages is batch-specific, and (2) a processing unit may belong to different and/or multiple stages. The key concepts in representing problems in multipurpose environments are the introduction of (1) batch routings and (2) batch-unit suitability that depends on batches and stages. Once batch routings and batch-unit suitability are determined, then modeling techniques similar to the ones presented in the previous chapter, for both sequence-based and time-grid-based models, can be employed. To limit the discussion of similar ideas, we choose to present the constraints of a sequence-based (Section 6.2) and a continuous time grid-based (Section 6.3) model for a special case of the problem, and then present a model based on a discrete time grid for the most general case (Section 6.4).

6.1 Problem Statement

Since we consider fixed batching decisions, the problem is expressed in terms of batches, not products. Batches cannot be mixed or split and have to go through multiple

[1] The term *multipurpose* has also been used in the PSE literature to describe *network* environments. The term *multipurpose* reflects the suitability of units to perform different tasks, which means that they can be present in both *sequential* and *network* environments. In this book, we will reserve the term to refer to sequential environments only.

[2] If the batching problem is solved and, in addition, there are no utility and storage constraints, then problems in multipurpose facilities are practically equivalent to the flexible job-shop problem introduced in Chapter 1.

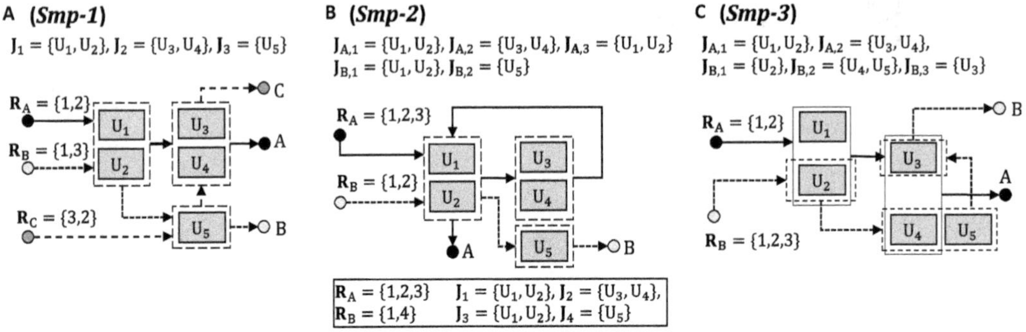

Figure 6.1 Representation of three types of multipurpose environments.

stages, as in multistage environments, but the sequence of stages is batch specific, and a unit may belong to different and/or multiple stages depending on the batch.

We are given a set of batches $i \in \mathbf{I}$ and units $j \in \mathbf{J}$. The routing of batch i is $\mathbf{R}_i$. The simplest case arises when units belong to the same work center (Figure 6.1A). If work centers are viewed as independent stages, then this case is modeled by defining routings $\mathbf{R}_i$ and the subset of units belonging to stage k, $\mathbf{J}_k$. A first extension arises when we have routings with reentries; that is, the same batch goes through some units (or stages) multiple times. To model reentries, we add the same stage more than once in the routing of the batch; that is, $\mathbf{R}_i$ is an ordered multiset (Figure 6.1B). To facilitate modelling, however, and to also account for the case where the reentering batch undergoes a different operation, and thus has a different processing time, we do the following:

(1) Treat the *repeated* stage as a new stage in the routing $\mathbf{R}_i$, and represent processing times in terms of stages.[3]

(2) Use subsets $\mathbf{J}_{i,k}$, where the stage number and the membership in a stage is batch dependent.

An alternative representation is to introduce distinct stages and use unique batch-independent subsets $\mathbf{J}_k$ (see Figure 6.1B and following discussion). The most general case results when units belong to different stages depending on the batch they process (Figure 6.1C), which can be represented using subsets $\mathbf{J}_{i,k}$. Note that all sequential facilities can be viewed as special cases of multipurpose with reentries, and unit suitability that is batch and stage specific. In other words, any problem in sequential facilities can be stated in terms of product routings, $\mathbf{R}_i$, and batch/stage-unit suitability information, $\mathbf{J}_{i,k}$.

[3] To illustrate, consider an environment with two stages, K1 and K2, with one unit per stage (U1 in K1 and U2 in K2), and batch P, which has to go through three operations O1 in K1 → O2 in K2 → O3 in K1. This problem can be modeled by defining routing $\mathbf{R}_i = \{1, 2, 3\}$ with $\mathbf{J}_{P,1} = \{U1\}$, $\mathbf{J}_{P,2} = \{U2\}$ and $\mathbf{J}_{P,3} = \{U1\}$, and the associated processing times $\tau_{P,1}$, $\tau_{P,2}$, and $\tau_{P,3}$ (note that we do not use $\tau_{P,U1}$ because two different operations are carried out in U1).

In summary, there are three types of multipurpose environment as ***Smp-X***, where $X \in \{1, 2, 3\}$:

(1) ***Smp-1***: unique partition of units to stages and unit-stage compatibility ($\mathbf{J}_{i,k} = \mathbf{J}_k$, $\forall i, k \in \mathbf{R}_i$); no reentries

(2) ***Smp-2***: unique partition of units to stages and unit-stage compatibility; with reentries

(3) ***Smp-3***: batch- and stage-specific unit-stage compatibility; with reentries

Since the modeling of changeover times and costs can be achieved using the techniques presented in the previous chapters, we will focus on problem $[\mathbf{Smp\text{-}X}/\rho, \varepsilon/\cdot]$.

Figure 6.1 illustrates the representations of the three aforementioned environments using sets $\mathbf{R}_i$, $\mathbf{J}_k$, and $\mathbf{J}_{i,k}$. Note that the numbering of stages in ***Smp-2*** and ***Smp-3*** is arbitrary; for example, in Figure 6.1B, we could have used $\mathbf{R}_B = \{1, 3\}$ and $\mathbf{J}_{B,3} = \{U_5\}$. The alternative representation of ***Smp-2*** employs distinct stages and unique $\mathbf{J}_k$; for example, U_5, which is used only by batch B for its second *step,* is now considered to be stage 4 of the facility, because stage 2, with $\mathbf{J}_2 = \{U_3, U_4\}$, and stage 3, with $\mathbf{J}_3 = \{U_1, U_2\}$ (for the reentry of A), are already defined. The sets for this alternative representation are given in the box at the bottom of Figure 6.1B.

In addition to $\mathbf{R}_i$, $\mathbf{J}_k$, and $\mathbf{J}_{i,k}$, we define the following:

- $\mathbf{J}_i = \cup_{k \in \mathbf{R}_i} \mathbf{J}_{i,k}$: the set of units batch i can be processed in
- $\mathbf{I}_j = \{i | j \in \mathbf{J}_i\}$: the set of batches that can be processed in unit j, possibly more than once

Furthermore, since the routings do not follow numeric sequence, we define the following functions:

- $f(i)$: first stage in the routing of batch i
- $l(i)$: last stage in the routing of batch i
- $s(i, k)$: the successor stage of k in $\mathbf{R}_i$

We also definte the following sets:

- $\mathbf{K}_{i,k}^{-}$: the set of stages preceding stage k in the routing of batch i
- $\mathbf{K}_{i,k}^{+}$: the set of stages succeeding stage k in the routing of batch i

For example, if $\mathbf{R}_i = \{1, 4, 2, 3, 5\}$, then $f(i) = 1$, $l(i) = 5$, $s(i, 4) = 2$, $\mathbf{K}_{i,3}^{-} = \{1, 4, 2\}$, $\mathbf{K}_{i,2}^{+} = \{3, 5\}$. The aforementioned defined functions and sets are illustrated in Figure 6.2 using the facility shown in Figure 6.1C as example, but with the routing of batch A represented as $\mathbf{R}_A = \{1, 3\}$ instead of $\mathbf{R}_A = \{1, 2\}$.

Each batch has a release time, ρ_i, and due time, ε_i. The processing time of batch i on unit j is denoted by τ_{ij} and the processing cost by γ_{ij}^{P}. If there are reentries, then processing times and costs are denoted by τ_{ijk} and γ_{ijk}^{P}, respectively.

Unit windows $[\varrho_j, \varepsilon_j]$ can be calculated from batch release and due time information (see discussion in Section 5.1). Assuming that $\bar{\varrho}_j = 0, \forall j$, we obtain

$\mathbf{J}_{A,1} = \{U_1, U_2\}, \mathbf{J}_{A,3} = \{U_3, U_4\},$
$\mathbf{J}_{B,1} = \{U_2\}, \mathbf{J}_{B,2} = \{U_4, U_5\}, \mathbf{J}_{B,3} = \{U_3\}$

$\mathbf{J}_A = \{U_1, U_2, U_3, U_4\}, \mathbf{J}_B = \{U_2, U_4, U_5, U_3\}$

$\mathbf{I}_{U1} = \{A\}, \mathbf{I}_{U2} = \{A, B\}, \mathbf{I}_{U3} = \{A, B\},$
$\mathbf{I}_{U4} = \{A, B\}, \mathbf{I}_{U5} = \{B\}$

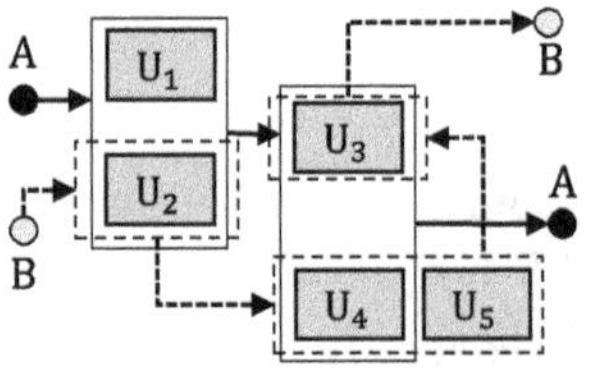

$\mathbf{R}_A = \{1,3\}, \mathbf{R}_B = \{1,2,3\}$

$f(A) = 1, l(A) = 3$
$s(A, 1) = 3, s(B, 1) = 2$

$\mathbf{K}^-_{A,1} = \emptyset, \mathbf{K}^-_{A,3} = \{1\},$
$\mathbf{K}^+_{A,1} = \{3\}, \mathbf{K}^+_{A,3} = \emptyset$

Figure 6.2 Illustration of functions and additional sets used for modeling multipurpose environments.

$$\varrho_j = \max_{i \in \mathbf{I}_k} \left\{ \rho_i + \sum\nolimits_{k' \in \mathbf{K}^-_{i,k}} \min_{j' \in \mathbf{J}_{k'}} \tau_{ij'k'} \right\}, \quad k, j \in \mathbf{J}_k \tag{6.1}$$

$$\varepsilon_j = \max_{i \in \mathbf{I}_k} \left\{ \varepsilon_i - \sum\nolimits_{k' \in \mathbf{K}^+_{i,k}} \min_{j' \in \mathbf{J}_{k'}} \tau_{ij'k'} \right\}, \quad k, j \in \mathbf{J}_k, \tag{6.2}$$

where $\mathbf{I}_k$ is the subset of batches that have to be processed in stage k. In ***Smp-1*** and if the alternative representation of ***Smp-2*** is used, then $\mathbf{I}_k$ is defined as $\mathbf{I}_k = \cup_{j \in \mathbf{J}_k} \mathbf{I}_j$. (Can you see why the second representation of ***Smp-2*** is required? How would you define $\mathbf{I}_k$ in the ***Smp-3*** environment?)

6.2 Sequence-Based Model

We present a sequence-based model for $[$***Smp-1***$/\rho, \varepsilon/\cdot]$. As in multistage environments, a batch i has to be assigned to a unit in each stage in routing $\mathbf{R}_i$. Since in ***Smp-1*** a batch will not be processed in a unit more than once, we introduce variable $X_{ij} \in \{0, 1\}$, which is equal to 1 if batch i is assigned to unit $j \in \mathbf{J}_{i,k} = \mathbf{J}_k$, for $k \in \mathbf{R}_i$:[4]

$$\sum\nolimits_{j \in \mathbf{J}_k} X_{ij} = 1, \quad i, k \in \mathbf{R}_i. \tag{6.3}$$

If two batches are assigned to the same unit, then the sequence in which they are processed should be determined. Since each unit belongs to a single stage and we have no reentries, each batch will be carried out in a unit at most once, so there is no need to track the stage in which unit j belongs.[5] We introduce $Y_{ii'j} \in \{0, 1\}$, which is equal to 1 if both batches i and i' are assigned to unit j, and batch i is processed before batch i'. For global sequencing, we have

$$Y_{ii'j} + Y_{i'ij} \geq X_{ij} + X_{i'j} - 1, \quad i, i' > i, k \in \mathbf{R}_i \cap \mathbf{R}_{i'}, j \in \mathbf{J}_k. \tag{6.4}$$

For local sequencing, we can use the ideas presented in the previous chapters.

[4] For problems in ***Smp-2*** and ***Smp-3*** environments, variables $X_{ikj} \in \{0, 1\}$ are introduced, where index k specifies the stage in the routing $\mathbf{R}_i$ where batch i can be processed by unit j. In these cases, the LHS of the equation that would replace (6.3) includes a summation of X_{ikj} and the summation domain is $j \in \mathbf{J}_{i,k}$ instead of $j \in \mathbf{J}_k$.

[5] If we have reentries, then the sequencing should be performed between (k, i) pairs; that is, sequencing binaries $Y_{iki'k'j}$ are employed and activated, using a constraint similar to (6.4), by X_{ikj} binary variables. Additionally, the domain of (6.4) has to be modified accordingly.

The disjunctive constraint is written as follows:

$$S_{ij} + \tau_{ij} \leq S_{i'j} + M\left(1 - Y_{ii'j}\right), \quad i, i', k \in \mathbf{R}_i \cap \mathbf{R}_{i'}, j \in \mathbf{J}_k. \tag{6.5}$$

Next, we define the start time S_{ik} of batch i in $k \in \mathbf{R}_i$, which is disaggregated into unit-specific S_{ij}:[6]

$$S_{ik} = \sum\nolimits_{j \in \mathbf{J}_k} S_{ij}, \quad i, k \in \mathbf{R}_i, \tag{6.6}$$

subject to

$$S_{ij} \leq MX_{ij}, \quad i, k \in \mathbf{R}_i, \ j \in \mathbf{J}_k, \tag{6.7}$$

where M is a sufficiently large number.

Start times, S_{ik}, are then used to enforce the precedence constraint between consecutive stages (i.e., for batch i, between stage $k \in \mathbf{R}_i$ and $s(i,k)$):

$$S_{i,s(i,k)} \geq S_{ik} + \sum\nolimits_{j \in \mathbf{J}_k} \tau_{ij} X_{ij}, \quad i, k \in \mathbf{R}_i \backslash l(i). \tag{6.8}$$

Release and due dates can be readily enforced:

$$S_{i,f(i)} \geq \rho_i, \ i \tag{6.9}$$

$$S_{i,l(i)} + \sum\nolimits_{j \in \mathbf{J}_{l(i)}} \tau_{ij} X_{ij} \leq \varepsilon_i + L_i, \ i, \tag{6.10}$$

where L_i is the lateness ($L_i \in \mathbb{R}$) or tardiness ($L_i \in \mathbb{R}_+$) of batch i (see previous chapters).

For makespan, MS, minimization, we enforce

$$MS \geq S_{i,l(i)} + \sum\nolimits_{j \in \mathbf{J}_{l(i)}} \tau_{ij} X_{ij} \tag{6.11}$$

and the objective is

$$\min MS. \tag{6.12}$$

The objective function for total weighted lateness and tardiness minimization is

$$\min \sum\nolimits_i \omega_i L_i \tag{6.13}$$

and for cost minimization, we use

$$\min \sum\nolimits_{i,j} \gamma_{ij}^P X_{ij}. \tag{6.14}$$

6.3 Model Based on a Continuous Time Grid

We study, again, problems $[\textbf{\textit{Smp-1}}/\rho, \varepsilon/\cdot]$, that is, problems with no shared utilities. If shared utilities are present, then a common grid is necessary and the processing of a batch can span multiple time periods. If no shared utilities are present, then

[6] If reentries are present, then we have to define S_{ikj} instead of S_{ij}, and sum, in the RHS of (6.6), over $j \in \mathbf{J}_{i,k}$.

unit-specific grids can be used, in conjunction with precedence relations between consecutive stages.

A simple estimate of the number of periods for a unit in stage k is $T^j = \lceil |\mathbf{I}_k|/|\mathbf{J}_k| \rceil$, while a more accurate estimate can be determined using the ideas presented in Chapter 4. We define sets $\mathbf{T}_j = \{1, \ldots, T^j\}$ for each unit-specific grid, introduce variables T_{jt} to denote the timing of point t, and assume that period t spans $[T_{j,t-1}, T_{jt}]$:

$$T_{j,0} = \varrho_j, \quad T_{j,T^j} = \varepsilon_j \tag{6.15}$$

$$T_{jt} \geq T_{j,t-1}, \quad j, t \in \mathbf{T}_j. \tag{6.16}$$

The allocation of batch i to $t \in \mathbf{T}_j$ is modeled with $X_{ijt} \in \{0, 1\}$. Each batch has to be allocated to one unit in each stage in $\mathbf{R}_i$ and to one period of the selected unit:[7]

$$\sum\nolimits_{j \in \mathbf{J}_k} \sum\nolimits_{t \in \mathbf{T}_j} X_{ijt} = 1, \quad i, k \in \mathbf{R}_i. \tag{6.17}$$

In addition, at most one batch can be allocated to each unit period:

$$\sum\nolimits_{i \in \mathbf{I}_j} X_{ijt} \leq 1, \quad j, t \in \mathbf{T}_j. \tag{6.18}$$

Equivalent solutions can be removed through

$$\sum\nolimits_i X_{ijt} \leq \sum\nolimits_i X_{ij,t-1}, \quad j, t \in \mathbf{T}_j \backslash \{1\}. \tag{6.19}$$

We introduce start times, S_{ik}, and their disaggregated counterparts, S_{ij}, defined through (6.6) and subject to

$$S_{ij} \leq M \sum\nolimits_t X_{ijt}, \quad i, j. \tag{6.20}$$

If a batch is allocated to period t, then

$$S_{ij} \geq T_{j,t-1} - M(1 - X_{ijt}), \quad i, j \in \mathbf{J}_i, t \in \mathbf{T}_j \tag{6.21}$$

$$S_{ij} + \tau_{ij} \leq T_{jt} + M(1 - X_{ijt}), \quad i, j \in \mathbf{J}_i, t \in \mathbf{T}_j. \tag{6.22}$$

Precedence constraints and release times are enforced using (6.8) and (6.9), respectively, and due times using

$$S_{i,l(i)} + \sum\nolimits_{j \in \mathbf{J}_{l(i)}} \tau_{ij} \sum\nolimits_t X_{ijt} \leq \varepsilon_i + L_i, \quad i. \tag{6.23}$$

For makespan minimization, we use (6.12) subject to

$$MS \geq S_{i,l(i)} + \sum\nolimits_{j \in \mathbf{J}_{l(i)}} \tau_{ij} \sum\nolimits_t X_{ijt}, \quad i. \tag{6.24}$$

[7] Note that the only difference between (6.17) and its counterpart in Chapter 5 is that it is written for $k \in \mathbf{R}_i$ instead of k. In the models for problems in environments ***Smp-2*** and ***Smp-3***, we would have to replace $\mathbf{J}_k$ in the LHS of (6.17) with $\mathbf{J}_{i,k}$, something that has to be done in all equations as we move from models for ***Smp-1*** to models for ***Smp-2*** or ***Smp-3***.

Equations (6.13) and (6.25) are used for lateness/tardiness and cost minimization, respectively:

$$\min \sum_{i,j} \gamma_{ij}^{P} \sum_{t} X_{ijt}. \tag{6.25}$$

6.4 Models Based on a Discrete Time Grid

We consider the problem in the general multipurpose environment (***Smp-3***). We use index $n \in \mathbf{N} = \{0,1,2,\ldots |\mathbf{N}|\}$ to define time points and periods $n \in \mathbf{N}\backslash\{0\} = \{1,2,\ldots,\mathbf{N}\}$ of length δ, where period n starts at point $n-1$ and ends at n. Time-related data are approximated, using the approaches discussed in the previous chapters to obtain $\bar{\rho}_i$, $\bar{\varepsilon}_i$ (deadlines) and $\bar{\tau}_{ijk}$. Batch- and stage-specific earliest start times, ρ_{ik}, and latest finish times, ε_{ik}, can be calculated as follows:[8]

$$\rho_{ik} = \bar{\rho}_i + \sum_{k' \in \mathbf{K}_{ik}^-} \min_{j \in \mathbf{J}_{k'}} \bar{\tau}_{ijk'}, \quad i,k \in \mathbf{R}_i \tag{6.26}$$

$$\varepsilon_{ik} = \bar{\varepsilon}_i - \sum_{k' \in \mathbf{K}_{ik}^+} \min_{j \in \mathbf{J}_{k'}} \bar{\tau}_{ijk'}, \quad i,k \in \mathbf{R}_i. \tag{6.27}$$

Parameters ρ_{ik} and ε_{ik} are used to define two subsets:[9]

- $\mathbf{N}_{ijk}$: time points at which the processing of batch i in $k \in \mathbf{R}_i$ can start in unit j:

$$\mathbf{N}_{ijk} = \left\{ n \in \mathbf{N} | \, \varepsilon_{ik} \le n \le \varepsilon_{ik} - \bar{\tau}_{ijk} \right\}, \quad i,k \in \mathbf{R}_i, j \in \mathbf{J}_{ik} \tag{6.28}$$

- $\mathbf{J}_{ikn}$: units in which the processing of batch i in $k \in \mathbf{R}_i$ can start at n:

$$\mathbf{J}_{ikn} = \left\{ j \in \mathbf{J}_k | \, \varepsilon_{ik} \le n \le \varepsilon_{ik} - \bar{\tau}_{ijk} \right\}, \quad i \in \mathbf{I}, k \in \mathbf{R}_i, n \in \mathbf{N} \tag{6.29}$$

These sets can be used to implicitly enforce release times and deadlines, by setting binary variables to zero, and also reduce the number of binary variables included in various constraints (see, for example, (6.30) and (6.31)).

Variables $X_{ijkn} \in \{0,1\}$ are introduced to model the start of processing of batch i on stage $k \in \mathbf{R}_i$ and unit $j \in \mathbf{J}_{ik}$ at time point n. Each batch has to be processed on exactly one unit in each stage and assigned to start at exactly one time point:

$$\sum_{j \in \mathbf{J}_{i,k}} \sum_{n \in \mathbf{N}_{ijk}} X_{ijkn} = 1, \quad i,k \in \mathbf{R}_i. \tag{6.30}$$

The no-overlap condition is enforced using the clique constraint:

$$\sum_{i} \sum_{k \in \mathbf{K}_{ij}} \sum_{n' \in \mathbf{N}_{ijkn}^{C} \cap \mathbf{N}_{ijk}} X_{ijkn'} \le 1, \quad j,n, \tag{6.31}$$

[8] We use ρ_{ik} and ε_{ik} to denote the earliest start and latest finish times because they can be viewed as generalizations of the batch release time (ρ_i) and deadline (ε_i), respectively.

[9] We introduce these subsets here, and use them in (6.30) through (6.32), to show how a basic form of preprocessing can be carried out and exploited by the user.

where $\mathbf{K}_{ij} = \{k \in \mathbf{R}_i | j \in \mathbf{J}_{ik}\}$ is the set of stages in $\mathbf{R}_i$ in which batch i can be processed in unit j; and, to simplify the summation domain, we introduce $\mathbf{N}^C_{ijkn} = \{n - \bar{\tau}_{ijk} + 1, \ldots, n\}$, which includes all the points at which the processing of batch i in stage $k \in \mathbf{R}_i$ and unit $j \in \mathbf{J}_{ik}$ could start and still be running at time point n.[10]

The precedence constraints between consecutive stages in the routing of a batch can be enforced in different ways (see discussion in Section 5.4). Here, we present the one based on *batch balances*. We define U_{ikn} to denote the *availability* of batch i to undergo processing in the successor stage, $s(i,k)$, that is, $U_{ikn} = 1$ if the processing of batch i in stage k is completed by $n - 1$ but its processing in stage $s(i,k)$ has not started by $n - 1$:

$$U_{ik,n+1} = U_{ikn} + \sum_{j \in \mathbf{J}_{ik}} X_{ijk,n-\bar{\tau}_{ijk}} - \sum_{j \in \mathbf{J}_{i,s(i,k)}} X_{ijkn}, \quad i,k \in \{\{0\} \cup \mathbf{R}_i\} \backslash \{l(i)\}, n. \tag{6.32}$$

For makespan, *MS*, minimization, the objective function is (6.12) subject to

$$MS \geq \sum_{j \in \mathbf{J}_{i,l(i)}} \sum_n \left(n + \bar{\tau}_{ij,l(i)}\right) X_{ij,l(i),n}, \quad i. \tag{6.33}$$

The weighted earliness, lateness, and tardiness are expressed by (6.34), (6.35), and (6.36), respectively:

$$\min \sum_i \omega_i \left\{ \bar{\varepsilon}_i - \sum_{j \in \mathbf{J}_{i,l(i)}} \sum_n \left(n + \bar{\tau}_{ij,l(i)}\right) X_{ij,l(i),n} \right\} \tag{6.34}$$

$$\min \sum_i \omega_i \left\{ \sum_{j \in \mathbf{J}_{i,l(i)}} \sum_n \left(n + \bar{\tau}_{ij,l(i)}\right) X_{ij,l(i),n} - \bar{\varepsilon}_i \right\} \tag{6.35}$$

$$\min \sum_i \omega_i L_i, \tag{6.36}$$

where $L_i \in \mathbb{R}_+$ in (6.36) is subject to

$$L_i \geq \sum_{j \in \mathbf{J}_{i,l(i)}} \sum_n \left(n + \bar{\tau}_{ij,l(i)}\right) X_{ij,l(i),n} - \bar{\varepsilon}_i, \quad i. \tag{6.37}$$

6.5 Notes and Further Reading

(1) Most mathematical programming models for multipurpose facilities have appeared in the operations research literature [1–5].

(2) The representation of problems in multipurpose environments based on batch routings ($\mathbf{R}_i$) and batch- and stage-dependent unit compatibility ($\mathbf{J}_{i,k}$) is taken from [6].

(3) The discrete time model discussed in Section 6.4 is based on the model in [7].

(4) Within the PSE community, a completion time algorithm was proposed for scheduling in multipurpose batch processes with reentry [8], and continuous time MIP models that consider various storage policies [9] and material transfer operations [10] were developed.

[10] The reader has seen essentially the same assignment constraint many times, starting in Chapter 3. In the remainder of the book, we will use, directly, a set ($\mathbf{N}^C_{ijkn}$) to define the summation domain over times points.

(5) Batching decisions can be modeled using the techniques discussed in Chapter 4. Two alternative models for the simultaneous batching and scheduling in multipurpose environments were proposed in [11].

(6) A model for the general problem in multipurpose environments with simultaneous batching and scheduling, as well as storage constraints and stage-dependent batchsizes, was presented in [11].

6.6 Exercises

(1) Consider a multipurpose environment with five units and three batches, where each batch goes through two batch-dependent stages. The processing times and costs as well as release and due times are given in Table 6.1.

Using the data in Table 6.1, solve the following instances using three models: (1) a sequence-based model (of your choice), (2) a grid-based continuous time model, and (3) a discrete time model (with $\delta = 1$):

(a) Makespan minimization
(b) Repeat (a) using a discrete time model with $\delta = 2$
(c) Processing cost minimization

(2) Consider a multipurpose environment with five units and six batches, where each batch goes through two or three stages. The processing times and costs as well as release and due times are given in Table 6.2.

Using the data in Table 6.2, solve the following instances using three models: (1) a sequence-based model (of your choice), (2) a grid-based continuous time model, and (3) a discrete time model (with $\delta = 1$):

(a) Makespan minimization
(b) Processing cost minimization
(c) Lateness and tardiness minimization

Table 6.1 Routing information, $\mathbf{J}_{ik}$; processing times, τ_{ij} [h]/costs, γ_{ij}^P [$]; and release, ρ_i, and due, ε_i, times [h].

Batch	Routing	U1	U2	U3	U4	U5	ρ_i	ε_i
A	$\mathbf{J}_{A,1} = \{U1, U2\}$	3/50	4/40				0	12
	$\mathbf{J}_{A,2} = \{U3, U4\}$			4/40	5/30			
B	$\mathbf{J}_{B,1} = \{U1, U2\}$	3/50	4/40				3	12
	$\mathbf{J}_{B,2} = \{U5\}$					2/60		
C	$\mathbf{J}_{C,1} = \{U5\}$					3/50	0	12
	$\mathbf{J}_{C,2} = \{U3, U4\}$			4/40	5/30			

Table 6.2 Routing information, $\mathbf{J}_{ik}$; processing times, τ_{ij} [h]/costs, γ_{ij}^P [$]; and release, ρ_i, and due, ε_i, times [h].

Batch	Routing	U1	U2	U3	U4	U5	ρ_i	ε_i
A	$\mathbf{J}_{A,1} = \{U1, U2\}$	3/50	4/40				0	16
	$\mathbf{J}_{A,2} = \{U3, U4\}$			4/40	5/30			
B	$\mathbf{J}_{B,1} = \{U1, U2\}$	3/50	4/40				3	16
	$\mathbf{J}_{B,2} = \{U5\}$					2/60		
C	$\mathbf{J}_{C,1} = \{U5\}$					3/50	0	16
	$\mathbf{J}_{C,2} = \{U3, U4\}$			4/40	5/30			
D	$\mathbf{J}_{D,1} = \{U1, U2\}$	5/40	6/30				6	16
	$\mathbf{J}_{D,2} = \{U3, U4\}$			6/30	4/50			
	$\mathbf{J}_{D,3} = \{U5\}$					4/50		
E	$\mathbf{J}_{E,1} = \{U1, U2\}$	3/60	4/50				0	16
	$\mathbf{J}_{E,2} = \{U5\}$					2/70		
F	$\mathbf{J}_{F,1} = \{U5\}$					4/50	7	16
	$\mathbf{J}_{F,2} = \{U3, U4\}$			4/50	4/40			
	$\mathbf{J}_{F,2} = \{U1, U2\}$	7/20	7/20					

References

[1] Gomes MC, Barbosa-Povoa AP, Novais AQ. Reactive Scheduling in a Make-to-Order Flexible Job Shop with Re-entrant Process and Assembly: A Mathematical Programming Approach. *Int. J. Prod. Res.* 2013;51(17):5120–5141.

[2] Gomes MC, Barbosa-Povoa AP, Novais AQ. Optimal Scheduling for Flexible Job Shop Operation. *Int. J. Prod. Res.* 2005;43(11):2323–2353.

[3] Fattahi P, Jolai F, Arkat J. Flexible Job Shop Scheduling with Overlapping in Operations. *Appl Math Model.* 2009;33(7):3076–3087.

[4] Fattahi P, Mehrabad MS, Jolai F. Mathematical Modeling and Heuristic Approaches to Flexible Job Shop Scheduling Problems. *J Intell Manuf.* 2007;18(3):331–342.

[5] Demir Y, Isleyen SK. Evaluation of Mathematical Models for Flexible Job-Shop Scheduling Problems. *Appl Math Model.* 2013;37(3):977–988.

[6] Maravelias CT. General Framework and Modeling Approach Classification for Chemical Production Scheduling. *AIChE J.* 2012;58(6):1812–1828.

[7] Lee H, Maravelias CT. Discrete-Time Mixed-Integer Programming Models for Short-Term Scheduling in Multipurpose Environments. *Comput Chem Eng.* 2017;107:171–183.

[8] Kim YJ, Kim MH, Jung JH. Optimal Scheduling of the Single Line Multi-Purpose Batch Process with Re-Circulation Products. *J. Chem. Eng. Japan.* 2002;35(2):117–130.

[9] Mendez CA, Cerda J. An MILP Continuous-Time Framework for Short-Term Scheduling of Multipurpose Batch Processes Under Different Operation Strategies. *Optim. Eng.* 2003;4 (1-2):7–22.

[10] Ferrer-Nadal S, Capon-Garcia E, Mendez CA, Puigjaner L. Material Transfer Operations in Batch Scheduling. A Critical Modeling Issue. *Ind Eng Chem Res.* 2008;47(20):7721–7732.

[11] Lee H, Maravelias CT. Mixed-Integer Programming Models for Simultaneous Batching and Scheduling in Multipurpose Batch Plants. *Comput Chem Eng.* 2017;106:621–644.

7 Network Environment: Basics

In this chapter, we discuss scheduling in network environments, which, as mentioned in Chapter 1, are unique in the process industries. We will study the *basic* features of this class of problems, including tasks consuming and producing multiple materials, shared utilities, and time-varying utility cost and capacity. In the next chapter, we will consider additional features, including material consumption and production during batch execution, material transfer activities, and unit deterioration.

Problems in sequential environments can be expressed using the *traditional* representation based on batches (orders), stages (operations), and units (machines). This representation is insufficient for problems in network environments. Thus, we start in Section 7.1, with two frameworks for problem representation and the corresponding problem statements. In Section 7.2, we present two models based on a common discrete time grid, and in Section 7.3 we present a model based on a common continuous time grid.

7.1 Problem Representation

We first note that the notion of a batch of a product going through different stages is irrelevant in this environment because there is no requirement for a batch to maintain its identity.[1] Similarly, processing stages are not typically used because products can be produced in different ways, which further implies that batching decisions have to be made. Since batches and stages are not defined, the representation used for problems in sequential facilities is inadequate for the representation of problems in network environments. Accordingly, we will use the concept of *task*, instead of batches and stages, to express problems in these environments. A task is defined in terms of input (consumed) and output (produced) materials and the corresponding conversion coefficients, and it can be performed by a subset of units (task-unit suitability). In this context, we will use

[1] In traditional scheduling notation, a task is a single activity to be scheduled. When modeling sequential environments, we used the term *batch* to represent the entities that move intact through the various processing stages (equivalent to jobs in discrete manufacturing problems). Hence, in that context, one batch is associated with multiple tasks that have to be scheduled. In network environments, to stay consistent with the original publications on network environments, we use the term *task* to represent something different: a task is a type of activity rather than a specific execution.

the term *batch* to denote different executions of the same task, that is, a schedule may include multiple batches of the same task. In addition, we will use the concept of *materials*, consumed and produced by tasks. Finally, we have *shared utilities* required by tasks. In the present chapter, we assume that utility (resource) consumption during the execution of a task remains constant. The extension to varying resource consumption during the execution of a task can be addressed using the ideas presented in Section 4.7.2.

A production environment is classified as *network* when there are no restrictions in the way all input and output materials are handled; that is, multiple batches of the same task can be mixed or material produced by a single batch can be consumed by multiple batches. Furthermore, as we have seen in Section 1.2.2, network environments are not necessarily defined based on the structure of the facility. In the absence of other processing restrictions (e.g., unit connectivity, transfer equipment), materials in a network environment can flow freely from storage vessels (suitable for their storage) to processing units where they are consumed, and from processing units where they are produced to storage vessels. In the present chapter, we will use, interchangeably, the terms *network facility* and *network process* to refer to facilities that can be classified as network environments. We will also use the term *network processing* to denote processing in a network facility.

An example network environment is shown in Figure 7.1. The facility consists of three processing units and six storage vessels, and produces two final products (P1 and P2) from two raw materials (RM1 and RM2) through four tasks, which require shared utilities (steam and cooling water). The basic elements (tasks, processing units and storage vessels, materials, and utilities) along with all the associated information are given in Figure 7.1A. The structure of the facility is shown in Figure 7.1B, with connections between units representing potential material transfers. Note that Figure 7.1B is not a process flow diagram (PFD):

(1) The process is not continuous, so the connections represent distinct, most often modeled as instantaneous material transfers [kg] rather than flows [kg/h].
(2) Not all connections will necessarily be *active* in a solution, and different connections are active at different times.
(3) Units and vessels are not equivalent to unit operations.[2] A unit may carry out different tasks at different times and a vessel may store different materials.

The two most widely used representations of network processes, the state-task network (STN) and resource-task network (RTN), rely on the consumption/production of materials (modeled either as states in STN or resources in RTN) by tasks. STN- and RTN-type approaches have been used to address problems with a wide range of additional characteristics, some of which will be studied in the next chapter (e.g.,

[2] Note that the term *operation* has been used differently in discrete manufacturing scheduling (see Chapter 1) and in process simulation, or chemical engineering in general. In the former, it denotes a processing stage, while in the latter it is used as *unit operation* to describe the continuous processing in a unit.

A

Basic elements
Tasks: T1, T2, T3, T4
Processing units: R1, R2, R3
Storage units: V1, V2, V3, V4, V5, V6
Materials: RM1, RM2, IN1, IN2, IN3, P1, P2
Utilities: hot steam (HS), cooling water (CW)

Task conversions
T1: 0.8 RM1 + 0.2 IN1 → IN3
T2: RM2 → 0.3 IN1 + 0.7 IN2
T3: IN3 → P1
T4: 0.6 IN2 + 0.4 IN3 → P2

Task – unit suitability
T1 and T2 carried out on R1 or R2
T3 and T4 carried out on R3

Material – vessel compatibility
RM1 in V1; RM2 in V2;
IN2 in V3; IN2 and IN3 in V4;
P1 in V5; P2 in V6

Utility requirements
T1 and T3 require HS
T2 and T4 require CW

B

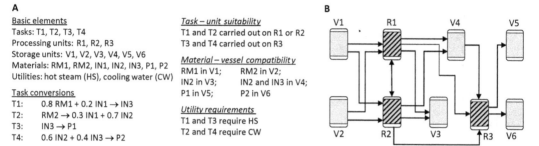

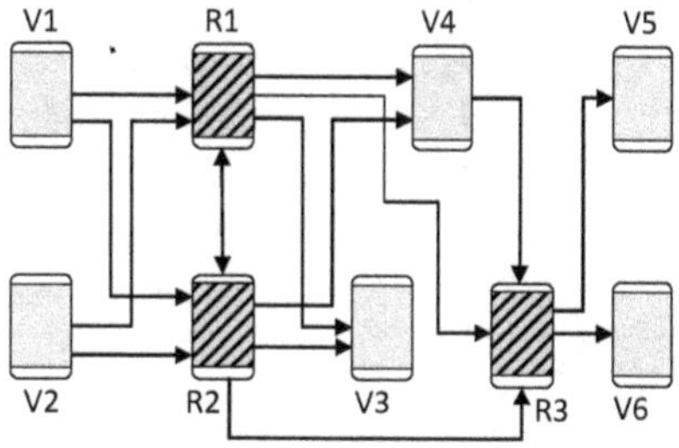

Figure 7.1 Example problem in network production environment. (A) Sets and information necessary to define the problem. (B) Structure of the facility.

resource-constrained material transfers). However, we note that these characteristics are not specific to network processing.

7.1.1 State-Task Network

The STN representation is based on the following basic concepts:

(1) States (materials) include feeds, intermediates, and final products, and are represented by circles.
(2) Tasks are activities that produce and consume materials and are represented by rectangles.
(3) Units are unary resources for the execution of tasks; multiple units may be suitable for the execution of a single task, and a unit may be suitable to carry out multiple tasks.
(4) Shared utilities are required by tasks through their execution.

We will use the term *material*, rather than the originally proposed *state*, because the latter is also used to describe the *system state*, a concept we will use when we discuss real-time scheduling (Chapter 14).

In the graphical representation, material (circles) and task (rectangles) nodes are connected via arcs (*streams*) denoting feasible material transfers. If a task consumes a material, there is an arc from the material to the task; if it produces it, then there is an arc from the task to the material. Processing units and shared utilities are represented implicitly through mappings (sets) and corresponding parameters, as described next. In the problem variant we consider in the present chapter, each material is assumed to have a dedicated storage vessel or it cannot be stored, so storage vessels are not explicitly considered.

The processing facility is defined in terms of the following sets, subsets, and parameters:

Indices/Sets

$i \in \mathbf{I}$ Tasks
$j \in \mathbf{J}$ Processing units
$k \in \mathbf{K}$ Materials
$l \in \mathbf{L}$ Utilities

Subsets

$\mathbf{I}_k^+/\mathbf{I}_k^-$	Tasks producing/consuming material k
$\mathbf{I}_j$	Tasks that can be carried out on unit j
$\mathbf{I}_l$	Tasks requiring utility l
$\mathbf{J}_i$	Processing units that can process task i
$\mathbf{K}_i^+/\mathbf{K}_i^-$	Materials produced/consumed by task i

Parameters

$\beta_j^{MIN}/\beta_j^{MAX}$	Minimum/maximum capacity of unit j
$\gamma_{ij}^F/\gamma_{ij}^V$	Fixed/variable cost for carrying out task i in unit j
π_k	Price of material k
ρ_{ik}	Conversion coefficient of material k produced (>0) or consumed (<0) by task i[3]
τ_i	Processing time of task i
φ_{il}/ψ_{il}	Fixed/variable requirement of task i for utility l
χ_k^M	Capacity of storage vessel dedicated to material k
χ_l^{UT}	Capacity of utility l

Note that we use: $i \in \mathbf{I}$ to denote tasks, instead of batches as in previous chapters; $k \in \mathbf{K}$, to represent materials, instead of stages (not defined in network environments); and $l \in \mathbf{L}$, instead of $m \in \mathbf{M}$, to denote shared utilities.[4]

The STN representation of the facility introduced in Figure 7.1 is given in Figure 7.2. The graphic representation is given in Figure 7.1A, while the major sets ($\mathbf{I}$, $\mathbf{J}$, $\mathbf{K}$, $\mathbf{L}$),

A

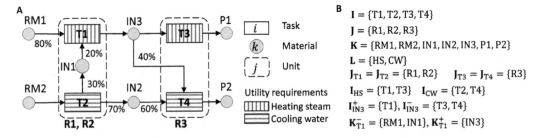

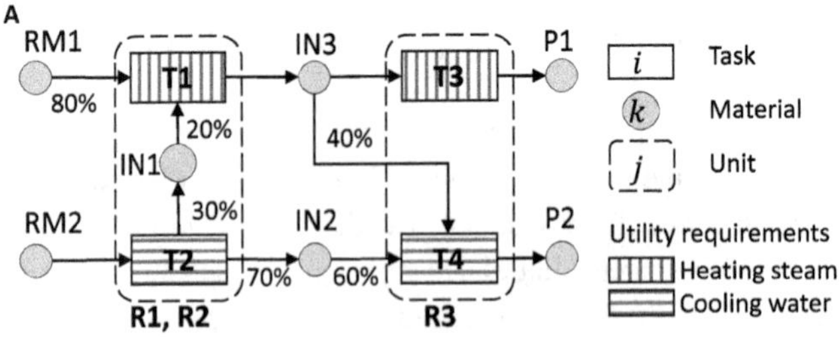

B

$\mathbf{I} = \{T1, T2, T3, T4\}$

$\mathbf{J} = \{R1, R2, R3\}$

$\mathbf{K} = \{RM1, RM2, IN1, IN2, IN3, P1, P2\}$

$\mathbf{L} = \{HS, CW\}$

$\mathbf{J}_{T1} = \mathbf{J}_{T2} = \{R1, R2\}$ $\mathbf{J}_{T3} = \mathbf{J}_{T4} = \{R3\}$

$\mathbf{I}_{HS} = \{T1, T3\}$ $\mathbf{I}_{CW} = \{T2, T4\}$

$\mathbf{I}_{IN3}^+ = \{T1\}$, $\mathbf{I}_{IN3}^- = \{T3, T4\}$

$\mathbf{K}_{T1}^- = \{RM1, IN1\}$, $\mathbf{K}_{T1}^+ = \{IN3\}$

Figure 7.2 State-task network representation of problem introduced in Figure 7.1. (A) Graphical representation. (B) Major sets and subsets, and representative minor subsets.

[3] This is one of the few times where the same letter is used to represent two rather different parameters. In Chapters 3 through 6, we used ρ_i to denote the release time of batch i; and in Section 6.4 we used ρ_{ik} to denote the earliest start time for the processing of batch i in stage k. Starting in the present chapter, we will no longer be using ρ_i/ρ_{ik} to denote these time-related parameters, with the exception of Chapter 12, where we will discuss solution methods for sequential environments.

[4] Recall that when discussing sequential environments, we used $l \in \mathbf{L}$ to represent, product/order batches, as pair (i, l), in problems where batching decisions were considered (see discussion in Section 4.5).

subsets ($\mathbf{J}_i$, $\mathbf{I}_i$), and some representative $\mathbf{I}_k^+$, $\mathbf{I}_k^-$, $\mathbf{K}_i^+$, and $\mathbf{K}_i^-$ subsets are given in Figure 7.1B. Note that the information contained in subsets $\mathbf{I}_j$ and $\mathbf{I}_l$ is also represented in Figure 7.2A via task inclusion inside the unit shape and task shading, respectively. Subsets $\mathbf{I}_k^+$, $\mathbf{I}_k^-$, $\mathbf{K}_i^+$, and $\mathbf{K}_i^-$ can also be inferred from edge-node incidence: subset $\mathbf{I}_k^+$ ($\mathbf{I}_k^-$) includes the tails (heads) of the incoming (outgoing) arcs to (from) the node representing material k. Similarly, $\mathbf{K}_i^+$ ($\mathbf{K}_i^-$) includes the heads (tails) of the outgoing (incoming) arcs from (into) the node representing task i.

7.1.2 Resource-Task Network

As in STN, we assume that each material has a dedicated storage vessel or it cannot be stored, so storage vessels are not explicitly considered. The major difference between STN and the RTN representations is that in the latter, materials and units are also treated as resources. Shared utilities are also, naturally, treated as resources, so the only necessary modeling entities in RTN are tasks and resources.[5]

Recall that in Chapter 4, where we discussed only renewable resources, we used the term resource *consumption* to represent the constant consumption of a renewable resource during the execution of batch; that is, consumption was viewed as a rate. Further, we used the terms *engagement* and *release* for the (instantaneous) beginning and end, respectively, of (renewable) resource consumption. In the present chapter, we consider both renewable (e.g., units and utilities) and nonrenewable (e.g., materials) resources. Thus, to unify the notation, we will use the terms *consumption* and *production* only, for both types. The *consumption rate* of a renewable resource is modeled as an instantaneous consumption at the start of the task and an instantaneous production (of equal amount) at its end. Input materials are treated as resources consumed at the start of a task (and never produced back by the same task), and output materials as resources produced at the end of the task.

The sets, subsets, and parameters used in the RTN representation are the following:

Indices/Sets

$i \in \mathbf{I}$	Tasks
$r \in \mathbf{R}$	Resources

Subsets

$\mathbf{I}_r^+/\mathbf{I}_r^-$	Tasks producing/consuming resource r
$\mathbf{I}_r$	Tasks interacting with resource r; $\mathbf{I}_r = \mathbf{I}_r^+ \cup \mathbf{I}_r^-$
$\mathbf{R}^U/\mathbf{R}^M/\mathbf{R}^{UT}$	Unit/material/utility resources
$\mathbf{R}_i^+/\mathbf{R}_i^-$	Resources produced/consumed by task i
$\mathbf{R}_i$	Resources whose availability is affected by the execution of task i, $\mathbf{R}_i = \mathbf{R}_i^+ \cup \mathbf{R}_i^-$

[5] In terms of the resource types introduced in Chapter 4, units are unary, renewable, based on execution (R.A.U); materials are nonrenewable, based on batchsize (NR.B); and utilities are renewable, based on both execution and batchsize (NR.C).

Parameters

$\beta_{ir}^{MIN}/\beta_{ir}^{MAX}$	Minimum/maximum extent of task i when executed on resource $r \in \mathbf{R}^U$
γ_i^F/γ_i^V	Fixed/variable cost for carrying out task i
τ_i	Processing time for task i
$\varphi_{ir}^S/\varphi_{ir}^E$	Fixed net consumption of resource r by task i at the start/end of task i
ψ_{ir}^S/ψ_{ir}^E	Variable net consumption of resource r by task i at the start/end of task i
χ_r	Capacity of resource r

Coefficients $\beta_{ir}^{MIN}/\beta_{ir}^{MAX}$ (defined for unit resources only) are the counterparts of $\beta_j^{MIN}/\beta_j^{MAX}$ in the STN representation. If task i requires unit resource r, then its batchsize, which in the RTN representation is called its *extent*, should be between β_{ir}^{MIN} and β_{ir}^{MAX}. Coefficients $\varphi_{ir}^S/\varphi_{ir}^E$ and ψ_{ir}^S/ψ_{ir}^E are termed task–resource interaction coefficients because they determine the amount of resource r consumed (via φ_{ir}^S and ψ_{ir}^S) and produced (φ_{ir}^E and ψ_{ir}^E) at the start and end, respectively, of task i.[6]

The modeling of task–resource interactions is illustrated in Figure 7.3. Note that the consumption/production based on $\varphi_{ir}^S/\varphi_{ir}^E$ does not depend on the extent (size) of the task; for example, if $i = \text{T1}$ is carried out on resource unit $r = \text{U1}$, then $\varphi_{\text{T1,U1}}^S = -1$ (U1 is consumed when T1 starts) and $\varphi_{\text{T1,U1}}^E = 1$ (U1 is produced when T1 ends). On the other hand, the consumption/production based on ψ_{ir}^S/ψ_{ir}^E is a (linear) function of its extent; for example, if $i = \text{T1}$ consumes 0.7 kg of $r = A$ and 0.3 kg of $r = B$ and produces 0.8 kg of C for every kg of its batchsize, then $\psi_{\text{T1,A}}^S = -0.7$ and $\psi_{\text{T1,B}}^S = -0.3$ (A and B are consumed when T1 starts), and $\psi_{\text{T1,C}}^E = 0.8$ (C is produced when T1 ends).

An implication of the modeling of the execution of a task on a unit as consumption and production of a specific unit resource is that if a task can be executed on more than one unit, then the task should be modeled as multiple tasks, each one consuming and producing a different resource unit. For example, if task T1 in Figure 7.3 can be assigned to three units, R1, R2, and R3, then three tasks, T1-R1, T1-R2, and T1-R3, will be needed in the RTN representation of the process with the corresponding parameters (e.g., $\varphi_{\text{T1-R1,R1}}^S = -1$ and $\varphi_{\text{T1-R1,R1}}^E = 1$ for T1-R1).

An exception to the approach we described in the previous paragraph, and an interesting feature of RTN, is that if two distinct resources, r and r', (1) can carry out exactly the same tasks (i.e., $\mathbf{I}_r = \mathbf{I}_{r'}$) and (2) have identical interactions with all tasks in $\mathbf{I}_r$ (i.e., $\beta_{ir}^{MIN} = \beta_{ir'}^{MIN}$, $\beta_{ir}^{MAX} = \beta_{ir'}^{MAX}$, $\varphi_{ir}^S = \varphi_{ir'}^S$, $\varphi_{ir}^E = \varphi_{ir'}^E$, and $\psi_{ir}^S = \psi_{ir'}^S$, $\forall i \in \mathbf{I}_r = \mathbf{I}_{r'}$), then they can be modeled as a single resource with capacity equal to 2. More generally, m identical unary resources can be lumped into a single resource with capacity $\chi_r = m$.

[6] In Chapter 4, we defined only φ_{im} and ψ_{im} coefficients (recall that we used $m \in \mathbf{M}$ for resources). This was because (renewable) resource consumptions and production were identical at the start and end of the task, and thus the same two parameters were sufficient to calculate both resource consumption and production (see, for example, (4.88) and (4.89) and (4.90) and (4.91)). Since this is not the case with nonrenewable resources, in the present chapter, we have to define two pairs of coefficients, one for the start (consumption) and one of the end (production) of the task.

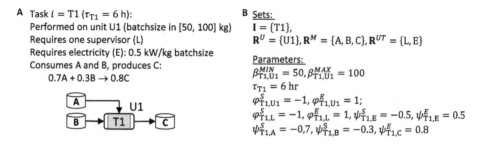

A Task $i = \mathrm{T1}$ ($\tau_{\mathrm{T1}} = 6$ h):
Performed on unit U1 (batchsize in [50, 100] kg)
Requires one supervisor (L)
Requires electricity (E): 0.5 kW/kg batchsize
Consumes A and B, produces C:
0.7A + 0.3B → 0.8C

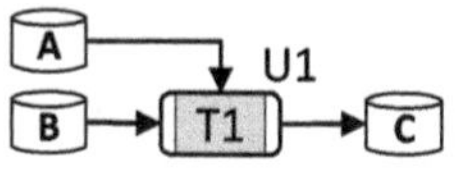

B Sets:

$\mathbf{I} = \{\mathrm{T1}\}$,
$\mathbf{R}^U = \{\mathrm{U1}\}, \mathbf{R}^M = \{\mathrm{A, B, C}\}, \mathbf{R}^{UT} = \{\mathrm{L, E}\}$

Parameters:

$\beta^{MIN}_{\mathrm{T1,U1}} = 50, \beta^{MAX}_{\mathrm{T1,U1}} = 100$
$\tau_{\mathrm{T1}} = 6\ \mathrm{hr}$
$\varphi^S_{\mathrm{T1,U1}} = -1, \varphi^E_{\mathrm{T1,U1}} = 1;$
$\varphi^S_{\mathrm{T1,L}} = -1, \varphi^E_{\mathrm{T1,L}} = 1, \psi^S_{\mathrm{T1,E}} = -0.5, \psi^E_{\mathrm{T1,E}} = 0.5$
$\psi^S_{\mathrm{T1,A}} = -0{,}7, \psi^S_{\mathrm{T1,B}} = -0.3, \psi^E_{\mathrm{T1,C}} = 0.8$

Figure 7.3 Modeling of units and materials as renewable and nonrenewable resources, respectively, in the RTN representation. (A) Task recipe. (B) Corresponding parameters in the RTN representation.

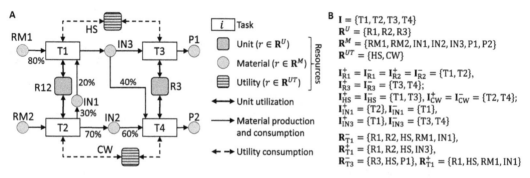

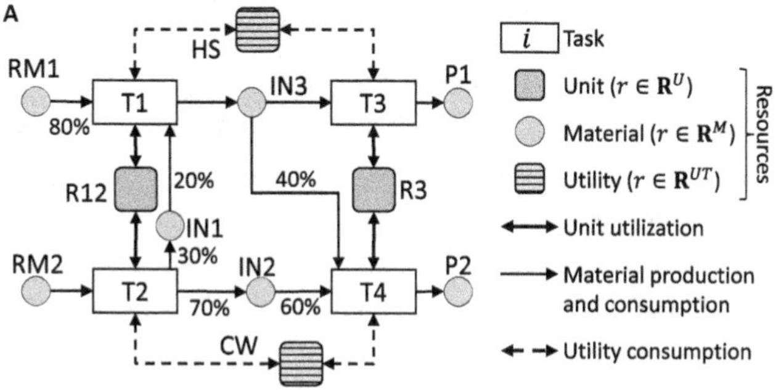

B
$\mathbf{I} = \{\mathrm{T1, T2, T3, T4}\}$
$\mathbf{R}^U = \{\mathrm{R1, R2, R3}\}$
$\mathbf{R}^M = \{\mathrm{RM1, RM2, IN1, IN2, IN3, P1, P2}\}$
$\mathbf{R}^{UT} = \{\mathrm{HS, CW}\}$

$\mathbf{I}^+_{\mathrm{R1}} = \mathbf{I}^-_{\mathrm{R1}} = \mathbf{I}^+_{\mathrm{R2}} = \mathbf{I}^-_{\mathrm{R2}} = \{\mathrm{T1, T2}\}$,
$\mathbf{I}^+_{\mathrm{R3}} = \mathbf{I}^-_{\mathrm{R3}} = \{\mathrm{T3, T4}\}$;
$\mathbf{I}^+_{\mathrm{HS}} = \mathbf{I}^-_{\mathrm{HS}} = \{\mathrm{T1, T3}\}, \mathbf{I}^+_{\mathrm{CW}} = \mathbf{I}^-_{\mathrm{CW}} = \{\mathrm{T2, T4}\}$;
$\mathbf{I}^+_{\mathrm{IN1}} = \{\mathrm{T2}\}, \mathbf{I}^-_{\mathrm{IN1}} = \{\mathrm{T1}\}$,
$\mathbf{I}^+_{\mathrm{IN3}} = \{\mathrm{T1}\}, \mathbf{I}^-_{\mathrm{IN3}} = \{\mathrm{T3, T4}\}$

$\mathbf{R}^-_{\mathrm{T1}} = \{\mathrm{R1, R2, HS, RM1, IN1}\}$,
$\mathbf{R}^+_{\mathrm{T1}} = \{\mathrm{R1, R2, HS, IN3}\}$,
$\mathbf{R}^-_{\mathrm{T3}} = \{\mathrm{R3, HS, P1}\}, \mathbf{R}^+_{\mathrm{T1}} = \{\mathrm{R1, HS, RM1, IN1}\}$

Figure 7.4 RTN representation of the facility shown in Figure 7.1. (A) Graphical representation based on four types of nodes (tasks, unit resources, material resources, and utility resources) and three types of arcs denoting resource consumption by the three types of resources. (B) Tasks and resource sets used to define the problem, along with representative subsets.

Additionally, in this case, no new tasks need to be defined. For example, if units R1, R2, and R3 in the case discussed in the previous paragraph are identical, then a new resource RL is defined with $\chi_{r=\mathrm{RL}} = 3$ and only task T1 is used. Note that this resource *lumping* cannot be performed when the resources are identical, but different tasks can be assigned to these resources. For example, if task T2 can be assigned to R1 and R2 only and T3 to R2 and R3 only, then the three resources cannot be lumped, which further means that T2 and T3 should be modeled as four separate tasks (T2-R1, T2-R2, T3-R2, and T3-R3).

The RTN graphical representation of the facility introduced in Figure 7.1 is shown in Figure 7.4. It is assumed that reactors R1 and R2 are identical, and thus are modeled as a single resource, R12, with capacity 2 and tasks T1 and T2 are not split.

7.2 Models Based on a Discrete Time Grid

As in the previous chapters, we use index n to define time points $n \in \mathbf{N} = (\{0,1,2, \ldots |\mathbf{N}|\})$ and periods $(n \in \mathbf{N} \backslash \{0\})$ of length δ, with period n starting at point $n-1$ and ending at n. We further define the timing of point n at $t_n = n\delta$. In both STN- and RTN-based models, processing times are approximated as $\bar{\tau}_{ij} = \lceil \tau_{ij}/\delta \rceil$.

In addition to the sets and parameters used to define the production facility, an instance may also have associated material shipments, as well as time-varying utility capacity and cost. Since the modeling of these, practically common and important, features requires a nontrivial calculation of new parameters, Section 7.2.1 is devoted to the conversion of the raw data regarding shipments and time-varying utility capacity and cost into parameters that can be readily used in the STN-based MIP model.[7] The conversion for RTN-based models is similar and thus omitted. The STN-based formulation is presented in Section 7.2.2 and the RTN-based formulation in Section 7.2.3. Finally, in Section 7.2.4 we offer a more in-depth discussion of the modeling of backlogs and lost sales.

7.2.1 Intermediate Shipments and Time-Varying Utility Capacity and Pricing

Intermediate shipments and time-varying utility capacity and cost can be given in terms of an externally defined time grid with points $t \in \mathbf{T}$, with material shipments occurring at $t \in \mathbf{T}^{MS} \subseteq \mathbf{T}$ and utility capacity and cost (price) changes occurring at $t \in \mathbf{T}^{UC} \subseteq \mathbf{T}$ and $t \in \mathbf{T}^{UP} \subseteq \mathbf{T}$.[8] Further, shipments are defined in terms of (1) the associated material, (2) timing (corresponding to a $t \in \mathbf{T}^{MS}$), and (3) amount. In other words, we are given a set of shipments, $o \in \mathbf{O}$, and each shipment has an associated material, $\bar{k}(o)$ (resource in RTN), timing t_o, and amount $\tilde{\xi}_o$ (>0 for deliveries and <0 for orders). Also, $\mathbf{O}_k$ is the subset of shipments of material k (i.e., $\mathbf{O}_k = \{o | \bar{k}(o) = k\}$, $\mathbf{T}_k^+/\mathbf{T}_k^-$ is the subset of time points where a delivery/order of material k occurs, and $\mathbf{T}_k^{MS} = \mathbf{T}_k^+ \cup \mathbf{T}_k^-$. For time-varying utility capacity, we assume that we are given (1) the timing t_t of the points $t \in \mathbf{T}^{UA}$ at which a change occurs; (2) the utility whose capacity changes, $\bar{l}(t)$; and (3) the *new* capacity value, $\tilde{\chi}_{lt}^{UT}$, with $\tilde{\chi}_{l,t=0}^{UT}$ being the capacity at the beginning of the horizon. Also, $\mathbf{T}_l^{UC}$ is the subset of points where a change in the capacity of utility l occurs, including the beginning of the horizon ($t = 0$). Similarly, for utility cost changes we are given the timing of the points $t \in \mathbf{T}^{UP}$; the mapping $\bar{l}(t)$; the new cost, $\tilde{\gamma}_{lo}^{UT}$; and subset $\mathbf{T}_l^{UP}$.[9]

7 A detailed treatment of intermediate shipments and time-varying utility capacity and cost is presented in this subsection to highlight that, in a practical setting, data may not be necessarily available in a suitable form and that calculating the parameters necessary to fully define an instance is not always straightforward. Accordingly, we illustrate, at a high level, how *raw* data are converted into parameters that (1) are compatible with the chosen problem representation and (2) can be directly used in the corresponding MIP model. Nevertheless, since data gathering and handling is beyond the scope of this book, in the remainder of the book, after the present section, we will assume that model parameters are readily available. The only exception will be Chapter 14, where real-time scheduling is discussed.

8 In the present chapter, we use index t to denote the points of both (1) the externally defined (fixed) grid used to represent material shipments and time-varying utility capacity and cost in Section 7.2.1 and (2) the (variable) time grid used in the continuous time models discussed in Section 7.3.

9 We use t_o and t_t to denote the known (parameter) timing of shipments and time points of the externally defined grid, respectively. This is one of the few times we use Latin letters for parameters. We chose to do so because we have already used T_t to denote the (variable) timing of time points of time grids. We still use lowercase because t_o and t_t are parameters.

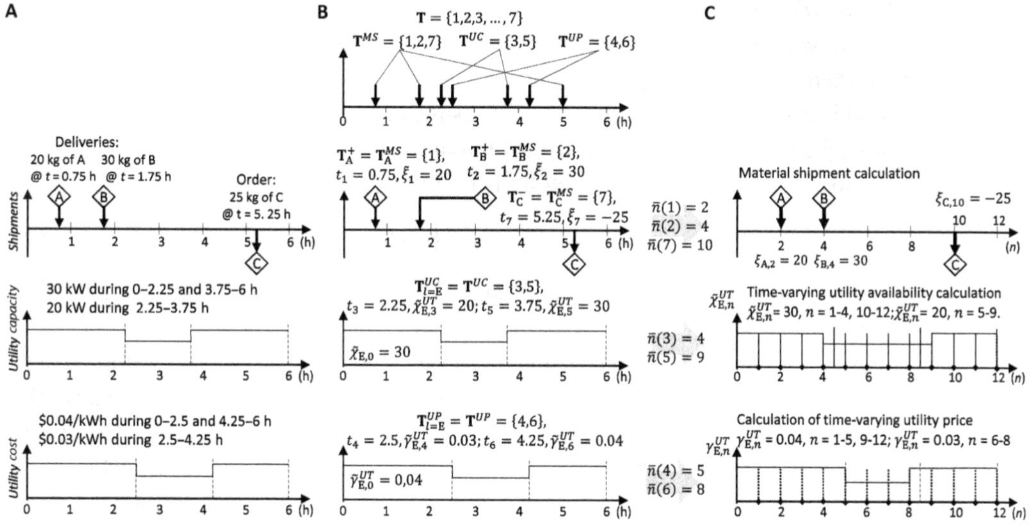

Figure 7.5 Modeling of intermediate shipments and time-varying utility capacity and cost. (A) Original raw data. (B) Auxiliary time grid and associated subsets and parameters. (C) Parameter calculation for model based on grid with length of period δ = 0.5 h.

First, we map points $t \in \mathbf{T}$ onto the grid we will use to formulate the MIP model. Let $\bar{n}(t)$ be the mapping of point $t \in \mathbf{T}$ onto $n \in \mathbf{N}$. To obtain feasible solutions, we follow a conservative approach (see Figure 7.5):

(1) Deliveries are mapped onto the *next* time point; that is, if $t_o \in (t_{n-1}, t_n]$, then delivery o is mapped onto point n (i.e., $\bar{n}(t_o) = n$); $\mathbf{O}_{kn}^{+}$ is the set of deliveries of material k mapped onto n (note that multiple deliveries, originally scheduled at different times, may be mapped onto the same point n).

(2) Orders are mapped onto the *previous* time point; that is, if $t_o \in [t_n, t_{n+1})$, then delivery o is mapped onto point n (i.e., $\bar{n}(t_o) = n$); $\mathbf{O}_{kn}^{-}$ is the set of orders of material k mapped onto n.

(3) Points $t \in \mathbf{T}^{UC}$, where resource capacity increases, are mapped onto the *next* time point of the model grid, and points where capacity decreases are mapped onto the *previous* point.

(4) Changes in utility pricing do not impact feasibility, so they can be mapped to either the previous or next time point; here, we assume that points $t \in \mathbf{T}^{UP}$ are mapped to the closest grid point of the model.

The total delivery of material k at point n, ξ_{kn}, is calculated as follows:

$$\xi_{kn} = \sum\nolimits_{o \in \mathbf{O}_{kn}} \tilde{\xi}_o, \tag{7.1}$$

where $\mathbf{O}_{kn} = \mathbf{O}_{kn}^{+} \cup \mathbf{O}_{kn}^{-}$.

For the capacity of utility l during period n, χ_{ln}^{UT}, we calculate

$$\chi_{ln}^{UT} = \tilde{\chi}_{ln}^{UT}, t \in \mathbf{T}_l^{UC}, n \in \left\{\bar{n}(t) + 1, \ldots, \bar{n}(\bar{t}^{+}(t))\right\}, \tag{7.2}$$

where $\bar{t}^+(t)$ is the next time point in the grid defined for changes in the capacity of l (i.e., the *next* point in $\mathbf{T}_l^{UC}$), which means that $\bar{n}(\bar{t}^+(t))$ represents the time point in the model grid where the next change in the capacity of utility l occurs. Note that the first affected period is $\bar{n}(t)+1$, which starts at $\bar{n}(t)$. If t is the last point in $\mathbf{T}_l^{UC}$, then $\bar{n}(\bar{t}^+(t)) = |\mathbf{N}|$, that is, the capacity remains unchanged through the end of the horizon.

Similarly, the cost of utility l during period n, γ_{ln}^{UT}, is calculated as follows:

$$\gamma_{ln}^{UT} = \tilde{\gamma}_{ln}^{UT}, t \in \mathbf{T}_l^{UP}, n \in \{\bar{n}(t), \ldots, \bar{n}(\bar{t}^+(t))\}, \tag{7.3}$$

in which $\bar{t}^+(t)$ is the next time point in $\mathbf{T}_l^{UP}$, and if t is the last point in $\mathbf{T}_l^{UP}$, then $\bar{n}(\bar{t}^+(t)) = |\mathbf{N}|$.

Figure 7.5 illustrates the calculation of the parameters for the modeling of intermediate shipments and time-varying resource capacity and cost in STN-based models. The externally defined time grid ($t \in \mathbf{T}$), including all subgrids, is shown at the top of panel B. For the RTN-based formulation, we calculate, using very similar ideas, net increase of (nonrenewable) resources, ξ_{rn}; and time-varying capacity, χ_{rn}, and cost, γ_{rn}^{RES}, of (renewable) resources.

7.2.2 STN-Based Models

We introduce variable $X_{ijn} \in \{0,1\}$ to denote the start of the execution of a batch of task i on unit j at time point n. Only one batch can be executed at any time in a given unit:

$$\sum\nolimits_{i \in \mathbf{I}_j} \sum\nolimits_{n' \in \mathbf{N}_{in}^U} X_{ijn'} \leq 1, \quad j, n, \tag{7.4}$$

where $\mathbf{N}_{in}^U = \{n - \bar{\tau}_i + 1, \ldots, n\}$ represents the points at which, if a batch of task i had started, then no other batch can be assigned to start on j at point n.

The batchsize of a batch of task i that started on unit j at point n is denoted by $B_{ijn} \in \mathbb{R}_+$ and is subject to unit capacity constraints:

$$\beta_j^{MIN} X_{ijn} \leq B_{ijn} \leq \beta_j^{MAX} X_{ijn}, \quad i, j, n. \tag{7.5}$$

Variable $I_{kn} \in \mathbb{R}_+$ denotes the inventory level of material i during period n, defined as follows:

$$I_{k,n+1} = I_{kn} + \sum\nolimits_{i \in \mathbf{I}_k^+} \sum\nolimits_{j \in \mathbf{J}_i} \rho_{ik} B_{ij,n-\bar{\tau}_{ij}} + \sum\nolimits_{i \in \mathbf{I}_k^-} \sum\nolimits_{j \in \mathbf{J}_i} \rho_{ik} B_{ijn} + \xi_{kn} - S_{kn} \leq \chi_k^M, \quad k, n, \tag{7.6}$$

where ξ_{kn} is the net material shipment at time point n, calculated by (7.1); χ_k^M is the capacity of the storage vessel dedicated to material k; and S_{kn} are the sales, in addition to fixed orders, of material k at time point n. Sales are allowed only for products and can be further bounded so they can be positive only at predefined time points. Equation (7.6) expresses a material balance over time: the inventory during period $n+1$ is equal to the inventory during the previous time period, plus the material produced by batches that finished at time point n (the beginning of period $n+1$), minus the material consumed by batches that started at point n, plus the net material shipment at time point n, minus sales. For $n = 0$, we use $I_{k,n=0} = \iota_k^0$, where ι_k^0 is the given initial inventory of material k.

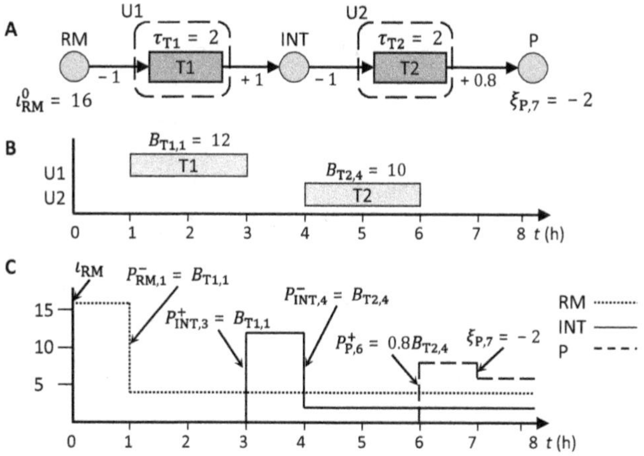

Figure 7.6 Illustration of material balance. (A) STN representation of illustrative example: conversion coefficients given under corresponding arcs; initial RM inventory and order for product P are also given; grid with δ = 1 is assumed. (B) Gantt chart of illustrative solution along with corresponding batchsize variables (index j is omitted for simplicity). (C) Inventory profiles and terms of (7.6) that lead to changes.

Note that we choose to keep track of inventory during a period, rather than a time point, because two inventory levels can be calculated at a point. An example illustrating how inventory level is calculated through (7.6) is shown in Figure 7.6, in which, to simplify the representation of inventories changes, we use

$$P^+_{kn} = \sum\nolimits_{i\in \mathbf{I}^+_k}\sum\nolimits_j \rho_{ik}B_{ij,n-\tau_{ij}}; P^-_{kn} = -\sum\nolimits_{i\in \mathbf{I}^-_k}\sum\nolimits_j \rho_{ik}B_{ijn} \tag{7.7}$$

to represent the total production and consumption of material k at time point n.

The total use (rate) of utility l during period n is denoted by R_{ln}, and defined and bounded as follows:

$$R_{l,n+1} = R_{ln} + \sum\nolimits_{\substack{i\in \mathbf{I}_l\\ j\in \mathbf{J}_i}} \left(\varphi_{il}X_{ijn} + \psi_{il}B_{ijn}\right) - \sum\nolimits_{\substack{i\in \mathbf{I}_l\\ j\in \mathbf{J}_i}} \left(\varphi_{il}X_{ij,n-\bar{\tau}_{ij}} + \psi_{il}B_{ij,n-\bar{\tau}_{ij}}\right) \le \chi^{UT}_{l,n+1},\ \ l,n, \tag{7.8}$$

where χ^{UT}_{ln} is the time-varying resource capacity calculated in (7.2). Similar to (7.6), (7.8) represents a *resource balance*: the use during the current period is equal to the use during the previous period, plus *new* consumption by tasks that just started (at time point n, the beginning of period $n+1$), minus production by tasks that just finished (at time point n).

An alternative way to enforce utility constraints, without introducing variable R_{ln}, is through the following constraint, in which the LHS represents use during period n,

$$\sum\nolimits_{i,j\in \mathbf{J}_i}\sum\nolimits_{n'\in \mathbf{N}^C_{i,n-1}} \left(\varphi_{ij}X_{ijn'} + \psi_{il}B_{ijn'}\right) \le \chi^{UT}_{ln},\ \ l,n, \tag{7.9}$$

where $\mathbf{N}^C_{i,n-1} = \left\{n - \bar{\tau}_{ij}, \ldots, n-1\right\}$. As noted in Chapter 4, (7.9) is a generalization of (7.4): the latter can be obtained from the former by setting $\psi_{il} = 0$ (a unit is occupied regardless of batchsize), and $\varphi_{il} = 1$ and $\chi_{ln}^{UT} = 1$ (unit is a unary resource). Interestingly, if a unit is unavailable during a period due to, say, maintenance, then the capacity of that unit can be calculated using the methods described in Section 7.2.1, and χ_{ln}^{UT} can be set to zero during the maintenance period. Also, note that (7.8) and (7.9) are the counterparts of (4.92) and (4.94).

The general objective function, profit maximization, includes revenue from sales, C^{REV}, and the following costs: processing, C^{PR}; utility, C^{UT}; changeover, C^{CH}; backlog, C^{BCK}; and lost sales, C^{LS}:

$$\max\left(C^{REV} - C^{PR} - C^{UT} - C^{CH} - C^{BCK} - C^{LS}\right) \tag{7.10}$$

with

$$C^{REV} = \sum_{k} \pi_k \sum_{n} S_{kn} \tag{7.11}$$

$$C^{PR} = \sum_{i,j} \left\{ \gamma_{ij}^F \sum_n X_{ijn} + \gamma_{ij}^V \sum_n B_{ijn} \right\} \tag{7.12}$$

$$C^{UT} = \sum_{l,n} \gamma_{ln}^{UT} R_{ln} \tag{7.13}$$

and the changeover cost C^{CH} is modeled and calculated using binary variables $Y_{ii'jn}$, as described in previous chapters. Since this is a short-term problem, we do not consider inventory costs.

Backlogs or lost sales are introduced when demand cannot be met on time, that is, no schedule exists that produces enough material to meet ξ_{kn} in (7.6). Backlogs are used when demand can be met later, after it is due, at a penalty; so, backlog costs incur until demand is met. Lost sales are used when the demand is *lost*, thus lost sales cost incurs once, at the time the unmet demand was due. Let π_k^{BCK} [\$/(kg·h)] and π_k^{LS} [\$/kg] be the unit cost of backlogs and lost sales, respectively. To simplify the discussion, we consider (7.6) only for products for which the net shipments are negative ($\xi_{ik} < 0$), that is, there is net demand for them. We further assume that the cost of not meeting demand far exceeds the revenue from sales (i.e., $\pi_k^{LS} > \pi_k^{BCK}\delta \gg \pi_k$), so the optimization will never favor additional sales ($S_{kn} > 0$) when there is unmet demand.

Using P_{kn}^+ and P_{kn}^-, as defined in (7.7), we rewrite (7.6) as follows:

$$I_{k,n+1} = I_{kn} + P_{kn}^+ - P_{kn}^- - V_{kn} - S_{kn} \le \chi_k^M, \quad k, n, \tag{7.14}$$

where $V_{kn} \in \mathbb{R}_+$ represent actual shipments toward order satisfaction, and orders ($\xi_{ik} < 0$) have been replaced by $-V_{kn}$.

Backlogs. A balance of the backlogged demand, I_{kn}^B, is maintained over time so that it is met later:

$$I_{k,n+1}^B = I_{kn}^B - (\xi_{kn} + V_{kn}), \quad k, n. \tag{7.15}$$

Following the logic of the material balance, (7.15) states that the backlog in period $n+1$ is equal to the backlog in the previous period, plus the demand due ($-\xi_{kn} > 0$) at the beginning of period $n+1$, minus the shipments toward this or previously backlogged demand. Note that when $V_{kn} > -\xi_{kn}$, the backlog is reduced, as expected;

whereas the backlog increases when shipments cannot keep up with demand ($V_{kn} < -\xi_{kn}$). The backlog cost is then calculated as follows:

$$C^{BCK} = \sum_k \pi_k^{BCK} \delta \sum_n I_{kn}^B. \tag{7.16}$$

Lost Sales. The balance of lost sales, I_{kn}^{LS}, which is used along with (7.14), and instead of (7.15), is expressed as follows:

$$I_{kn}^{LS} = -\xi_{kn} - V_{kn}, \quad k, n, \tag{7.17}$$

and is used to calculate

$$C^{LS} = \sum_{k,n} \pi_k^{LS} I_{kn}^{LS}. \tag{7.18}$$

Modeling backlogs (or lost sales) is important for two reasons. First, there are many instances where it is indeed impossible to meet all demand on time, so backlogs allow the user to find the next best solution. Interestingly, minimizing backlog cost is equivalent to minimizing weighted lateness in sequential environments with fixed batching decisions. (Can you see the equivalence? How should weights be defined in sequential environments for fixed batchsize β_i to be consistent with π_k^{BCK}?) Second, there are feasible but hard instances for which no feasible solution can be found within reasonable computational time. In this case, the introduction of backlogs allows us to find a solution, even though it may not satisfy all demand on time, and it also enhances the solution of the instance because it provides a lower bound (for a maximization problem) that can be found fast and used for pruning.

7.2.3 RTN-Based Models

As discussed in Section 7.1.2, the only two entities used in RTN are tasks and resources. Conceptually, tasks in STN and RTN are the same, though duplicate tasks may have to be introduced in RTN, while resources in RTN are used to model units, materials, and utilities. In addition to the parameters introduced in Section 7.1.2, we are also given parameters describing delivery of material resources, and time-varying utility resource capacity and cost, calculated using the approaches presented in Section 7.2.1:

- ξ_{rn}^M: net addition of (nonrenewable) material resource $r \in \mathbf{R}^M$ at time point n
- χ_{rn}: capacity of (renewable) resource $r \in \mathbf{R}^{UT}$ during period n
- γ_{rn}^{RES}: cost of resource $r \in \mathbf{R}^{UT}$ during period n

For $r \in \mathbf{R}^{UT}$, we can either precalculate resource capacity parameters, χ_{rn}, using the ideas presented in Section 7.2.1, or calculate net capacity changes, ξ_{rn}^{UT}, which can then be used in the resource balances.

The two major variables in RTN are $X_{in} \in \{0, 1\}$, denoting the start of the execution of a batch i at time point n; and $R_{rn} \in \mathbb{R}^+$, denoting the availability of resource r during period n.[10] In addition, when a task is carried out, then it has an associated *extent*, $B_{in} \in \mathbb{R}^+$, which in the STN-based formulation would be its batchsize.

A unit resource $r \in \mathbf{R}^U$ is available during a period if it was already available and no batch of task $i \in \mathbf{I}_r$ started at the beginning of the current period, or if a task assigned to it finished at the beginning of the current period:

$$R_{r,n+1} = R_{rn} + \sum\nolimits_{i \in \mathbf{I}_r} X_{i,n-\bar{\tau}_{ij}} - \sum\nolimits_{i \in \mathbf{I}_r} X_{in}, \quad r \in \mathbf{R}^U, n. \tag{7.19}$$

Note that (7.19), which is essentially a resource balance similar to (7.8), enforces the same constraint as (7.4): no more than one task can be executed on a unit at any time. Unit maintenance can be modeled by external unit consumption ($\xi_{rn}^U = -1$) and production ($\xi_{rn'}^U = 1, n' > n$), added to the RHS of (7.19).

If a task is carried out, then its extent is subject to

$$\beta_{ir}^{MIN} X_{in} \leq B_{in} \leq \beta_{ir}^{MAX} X_{in}, \quad i, r \in \mathbf{R}_i \cap \mathbf{R}^U, n. \tag{7.20}$$

The availability of a material resource $r \in \mathbf{R}^M$ is defined similarly to the material balance in (7.6):

$$R_{r,n+1} = R_{rn} + \sum\nolimits_{i \in \mathbf{I}_r^+} \psi_{ir}^E B_{i,n-\bar{\tau}_i} + \sum\nolimits_{i \in \mathbf{I}_r^-} \psi_{ir}^S B_{in} + \xi_{rn}^M - S_{rn} \leq \chi_r^M, \quad r \in \mathbf{R}^M, n, \tag{7.21}$$

where S_{rn} are the additional sales of material resource r and χ_r^M is the capacity of the dedicated storage tank. Note that in RTN each task can be assigned to only one $r \in \mathbf{R}^U$, so a summation over $r \in \mathbf{R}^U \cap \mathbf{R}_i$ (which would be equivalent to $j \in \mathbf{J}_i$) is removed.

Finally, the availability of a utility resource, $r \in \mathbf{R}^{UT}$, is expressed as follows:

$$R_{r,n+1} = R_{rn} + \sum\nolimits_{i \in \mathbf{I}_r^+} \left(\varphi_{ir}^E X_{i,n-\bar{\tau}_i} + \psi_{ir}^E B_{i,n-\bar{\tau}_i}\right) - \sum\nolimits_{i \in \mathbf{I}_r^-} \left(\varphi_{ir}^S X_{in} + \psi_{ir}^S B_{in}\right) + \xi_{rn}^{UT}, \; r \in \mathbf{R}^{UT}, n. \tag{7.22}$$

Note that since $R_{rn} \geq 0$, consumption would never exceed capacity because this would mean that the RHS of (7.22) would become negative.

Interestingly, balances (7.19), (7.21), and (7.22) can be written in the following general form:

$$R_{r,n+1} = R_{rn} + \sum\nolimits_{i \in \mathbf{I}_r^+} \left(\varphi_{ir}^E X_{i,n-\bar{\tau}_i} + \psi_{ir}^E B_{i,n-\bar{\tau}_i}\right) + \sum\nolimits_{i \in \mathbf{I}_r^-} \left(\varphi_{ir}^S X_{in} + \psi_{ir}^S B_{in}\right) + \xi_{rn} + S_{rn}, \tag{7.23}$$

in which, typically:

- For $r \in \mathbf{R}^U$, we have $\mathbf{I}_r^+ = \mathbf{I}_r^- = \mathbf{I}_r$; $\varphi_{ir}^S = -1$, $\varphi_{ir}^E = 1$, for $i \in \mathbf{I}_r$; $\varphi_{ir}^S = \varphi_{ir}^E = 0$, for $i \notin \mathbf{I}_r$; $\psi_{ir}^S = \psi_{ir}^E = 0$, for all i; $S_{rn} = 0$; and $\xi_{rn} = 0$, if maintenance is not considered.

[10] Recall that in Chapter 4 and Section 7.2.2, we modeled resource constraints using variables denoting resource consumption (study Figure 4.3C to refresh your memory). In the present chapter, we choose to define a variable for resource availability in order to make the reader aware of an alternative approach. Also, when modeling materials as resources, keeping track of the available amount of materials is more intuitive.

- For $r \in \mathbf{R}^M$, we have $\varphi_{ir}^S = \varphi_{ir}^E = 0$ for all i ; $\psi_{ir}^S = 0$, if $i \notin \mathbf{I}_r^-$; $\psi_{ir}^S < 0$ if $i \in \mathbf{I}_r^-$; $\psi_{ir}^E = 0$ if $i \notin \mathbf{I}_r^+$; $\psi_{ir}^E > 0$ if $i \in \mathbf{I}_r^+$; and both ξ_{rn} and S_{rn} can be nonzero.
- For $r \in \mathbf{R}^{UT}$, we have $\mathbf{I}_r^+ = \mathbf{I}_r^- = \mathbf{I}_r$ and all coefficients can be nonzero for $i \in \mathbf{I}_r$; we typically have $-\varphi_{ir}^S = \varphi_{ir}^E > 0$ and $-\psi_{ir}^S = \psi_{ir}^E > 0$; ξ_{rn} can be nonzero; and $S_{rn} = 0$.

As in the STN-based model, the general objective function includes revenue from sales, C^{REV}, and various cost components:

$$\max \left(C^{REV} - C^{PR} - C^{UT} - C^{CH} - C^{BCK} - C^{LS} \right) \tag{7.24}$$

with

$$C^{REV} = \sum_{r \in \mathbf{R}^M} \sum_n \pi_r S_{rn} \tag{7.25}$$

$$C^{PR} = \sum_i \left\{ \gamma_i^F \sum_n X_{in} + \gamma_i^V \sum_n B_{in} \right\} \tag{7.26}$$

$$C^{UT} = \sum_{r \in \mathbf{R}^{UT}} \sum_n \gamma_{rn}^{RES} (\chi_{rn} - R_{rn}). \tag{7.27}$$

Note, again, that we do not sum over unit resources in (7.26) because each task can be assigned to a single $r \in \mathbf{R}^U$. Also, the resource utilization is equal to $(\chi_{rn} - R_{rn})$. Backlog and lost sales costs can be modeled using ideas similar to the ones presented in the previous subsection.

7.2.4 Interpretation of Backlogs and Lost Sales

While (7.15) can be used in conjunction with (7.14) to replace (7.6) when backlogs need to be modeled, it can also be used to project out shipment variables,

$$V_{kn} = I_{kn}^B - I_{k,n+1}^B - \xi_{kn}, \quad k, n, \tag{7.28}$$

which can then be used to replace V_{kn} in (7.14):

$$I_{k,n+1} = I_{kn} + P_{kn}^+ - P_{kn}^- - \left(I_{kn}^B - I_{k,n+1}^B - \xi_{kn} \right) - S_{kn} \leq \chi_k^M, \quad k, n. \tag{7.29}$$

After rearrangement, we obtain

$$I_{k,n+1} - I_{k,n+1}^B = I_{kn} - I_{kn}^B + P_{kn}^+ - P_{kn}^- + \xi_{kn} - S_{kn} \leq \chi_k^M, \quad k, n. \tag{7.30}$$

Equation (7.30), which can be used instead of (7.14) to account for backlogs, is interesting because it illustrates how backlog can be viewed as *negative inventory* as well as the concept of generalized inventory. First, note that the only difference between (7.30) and (7.14) is that traditional inventory variables I_{kn} have been replaced by $(I_{kn} - I_{kn}^B)$, which can be negative. In other words, (7.30) allows us to keep track of the actual inventory (when $I_{kn} \geq 0$ and $I_{kn}^B = 0$) or the backlog (when $I_{kn} = 0$ and $I_{kn}^B > 0$), which is equivalent to having negative inventory, since $I_{kn} - I_{kn}^B < 0$. Note that it is suboptimal to have simultaneously $I_{kn} > 0$ and $I_{kn}^B > 0$. (Why?) Based on this interpretation, and if we replace I_{kn} by I_{kn}^+ (positive = actual inventory) and I_{kn}^B by I_{kn}^-

(negative inventory = backlog), we can define the generalized inventory $I^G_{kn} = I^+_{kn} - I^-_{kn}$ and rewrite (7.30):

$$\overbrace{I^+_{k,n+1} - I^-_{k,n+1}}^{I^G_{k,n+1}} = \overbrace{I^+_{kn} - I^-_{kn}}^{I^G_{kn}} + P^+_{kn} - P^-_{kn} + \xi_{kn} - S_{kn} \leq \chi^M_k, \quad k, n. \tag{7.31}$$

When lost sales need to be modeled, (7.17) is used to calculate V_{kn},

$$V_{kn} = -\xi_{kn} - I^{LS}_{kn}, \quad k, n, \tag{7.32}$$

and project it out from (7.14):

$$I_{k,n+1} = I_{kn} + P^+_{kn} - P^-_{kn} - \left(-\xi_{kn} - I^{LS}_{kn}\right) - S_{kn} \leq \chi^M_k, \quad k, n \tag{7.33}$$

or

$$I_{k,n+1} = I_{kn} + P^+_{kn} - P^-_{kn} + \left(\xi_{kn} + I^{LS}_{kn}\right) - S_{kn} \leq \chi^M_k, \quad k, n. \tag{7.34}$$

Equation (7.34) offers another interesting interpretation of unmet demand. First, note that the only difference compared to (7.6) is that $\xi_{kn} < 0$ has been replaced by $\left(\xi_{kn} + I^{LS}_{kn}\right)$. Unmet demand occurs when the RHS of (7.6) is negative (and thus the equation becomes infeasible) because ξ_{kn} is so negative that current inventory (I_{kn}) and production $(P^+_{kn} = \sum_{i \in \mathbf{I}^+_k} \sum_{i \in \mathbf{I}_j} \rho_{ik} B_{ij,n-\tau_{ij}})$ cannot meet it. By introducing I^{LS}_{kn}, we essentially add a slack variable: the RHS can always remain nonnegative by simply increasing I^{LS}_{kn} (and paying a lost sales penalty).

7.3 Models Based on a Common Continuous Time Grid

As discussed in previous chapters, the modeling of shared resources, which couple different units, requires a common, across units, time reference. In addition to shared resources, the modeling of multiple tasks producing or consuming the same material can be facilitated by a single common grid. Thus, while a wide range of models have been proposed in the literature, we present, in Section 7.3.1, a model based on the STN representation and a common grid. Extensions to this model, including shared utilities and intermediate material shipments will be discussed in Section 7.3.2. Finally, general ideas behind the development of alternative models are discussed in Section 7.3.3.

7.3.1 Basic Model

The scheduling horizon, η, is divided into variable length periods $t \in \mathbf{T} = \{1, 2, \ldots, |\mathbf{T}|\}$, defined by $|\mathbf{T}| + 1$ time points, also referred to as simply *points*, $t \in \{0, 1, 2, \ldots, |\mathbf{T}|\}$ where period t starts at point $t - 1$ and ends at point t.

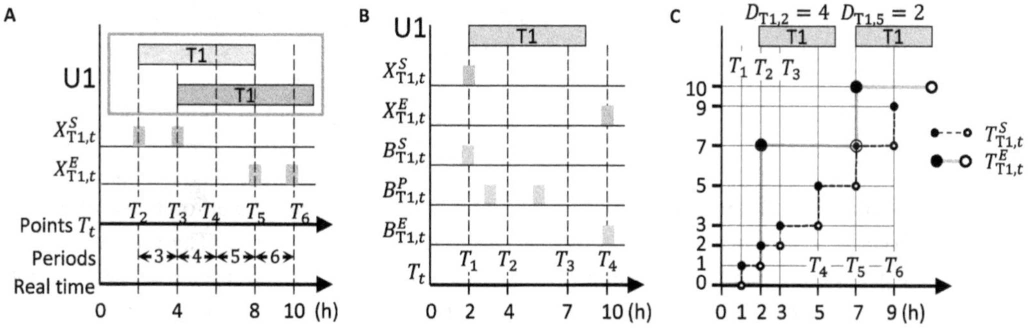

Figure 7.7 Major modeling features of continuous time model. (A) Shown solution satisfies (7.37) through (7.39), but not (7.40). (B) Activation of batchsize variables for a batch starting at time point 2 and ending during period 5; note that batchsize variables would be identical if the batch was ending exactly at point 5. (C) Example of evolution of timing variables.

To be precise, we define period t as the interval $(T_{t-1}, T_t]$. The timing of time point t is denoted by $T_t \in \mathbb{R}_+$ with

$$T_0 = 0, \quad T_{|\mathbf{T}|} = \eta \tag{7.35}$$

$$T_t \geq T_{t-1}, \quad t. \tag{7.36}$$

Batch-Unit Assignments. When a discrete grid is adopted and processing times are fixed, the point at which a batch ends is uniquely determined by the start variables (X_{ijn} or X_{in}). This is not the case when the length of periods is variable. Thus, we introduce two variables to map a batch onto the grid[11] (see Figure 7.7A):

- $X^S_{ijt} \in \{0, 1\}$, which is 1 if a batch of task i starts on unit j exactly at point t (i.e., at time T_t)
- $X^E_{ijt} \in \{0, 1\}$, which is 1 if a batch of task i ends on unit j during period t (i.e., in $(T_{t-1}, T_t]$)

Note that, based on the preceding definitions, the start of a batch has to coincide with a time point, while this is not necessary for its end. Obviously, only one batch can start on unit at a time point,

$$\sum\nolimits_{i \in \mathbf{I}_j} X^S_{ijt} \leq 1, \quad j, t, \tag{7.37}$$

which implies that only one batch can end within a time period (why?):

$$\sum\nolimits_{i \in \mathbf{I}_j} X^E_{ijt} \leq 1, \quad j, t. \tag{7.38}$$

[11] We use the term *allocate* (or *assign*) to describe a matching between an event and a set element or between two set elements; for example, a batch is allocated to a time period in the sequence-based models for problems in sequential environments. We use the term *map* to describe a timing relation: for example, the time a batch starts is mapped onto (is equal to) the time, T_t, time point t occurs.

Further, if we assume that no batch starts before the beginning of the horizon ($T_0 = 0$), and that no batch will start but not end within the scheduling horizon, we have[12]

$$\sum_t X_{ijt}^S = \sum_t X_{ijt}^E, \quad i,j \tag{7.39}$$

with $X_{ij,t=0}^E = 0$ and $X_{ij,t=|\mathbf{T}|}^S = 0$ for all i and j.

While (7.37) through (7.39) are valid, they are not sufficient to enforce that only one batch can be carried out on a unit at any time (see Figure 7.7A). The unit utilization constraint, which is the counterpart of (7.4), enforces that the number of batches that have started on a unit through time T_t can be at most one more than the batches that have finished in the same unit:

$$\sum_{i \in \mathbf{I}_j} \sum_{t' \le t} \left(X_{ijt'}^S - X_{ijt'}^E \right) \le 1, \quad j,t. \tag{7.40}$$

Note that the summation includes all previous points because it is unknown how many periods the execution of a batch spans.

Batching. The following variables are introduced to model batchsizes (see Figure 7.7B). :

- $B_{ijt}^S \in \mathbb{R}_+$: batchsize of a batch of task i that starts on unit j at point t (i.e., at T_t)
- $B_{ijt}^E \in \mathbb{R}_+$: batchsize of a batch of task i that ends on unit j during period t (i.e., in $(T_{t-1}, T_t]$)
- $B_{ijt}^P \in \mathbb{R}_+$: batchsize of a batch of task i that is being processed on unit j during the entire period t (i.e., starts at or before T_{t-1} and ends after T_t)

Note that B_{ijt}^S and B_{ijt}^E are defined at time points, whereas B_{ijt}^P is defined over a period. Also, if $B_{ijt}^S > 0$ (i.e., a batch starts at point t), then $B_{ijt}^P = 0$.

Obviously, variables B_{ijt}^S, B_{ijt}^E, and B_{ijt}^P should be positive and satisfy unit capacity constraints only when the corresponding batch starts at T_t, ends in $(T_{t-1}, T_t]$, and is being processed during $(T_{t-1}, T_t]$, respectively:

$$\beta_i^{MIN} X_{ijt}^S \le B_{ijt}^S \le \beta_i^{MAX} X_{ijt}^S, \quad i,j,t \tag{7.41}$$

$$\beta_i^{MIN} X_{ijt}^E \le B_{ijt}^E \le \beta_i^{MAX} X_{ijt}^E, \quad i,j,t \tag{7.42}$$

$$\beta_i^{MIN} \left(\sum_{t'<t} X_{ijt'}^S - \sum_{t' \le t} X_{ijt'}^E \right) \le B_{ijt}^P \le \beta_i^{MAX} \left(\sum_{t'<t} X_{ijt'}^S - \sum_{t' \le t} X_{ijt'}^E \right), \quad i,j,t. \tag{7.43}$$

Note that $(\sum_{t'<t} X_{ijt'}^S - \sum_{t' \le t} X_{ijt'}^E) = 1$ if the number of batches of task i that have started on unit j by T_{t-1} and do not end by T_t is one, which is equivalent to saying that a batch of task i is being processed on unit j during period t. Equations (7.41) through (7.43), which are the counterpart of (7.5), couple batchsize variables with the

[12] While these two assumptions appear to be reasonable, they are not valid when scheduling is repeatedly executed to react to new information and disturbances. This recursive solution of scheduling problems, termed *real-time scheduling*, will be discussed in Chapter 14.

corresponding assignment decisions but do not enforce equality among them. To accomplish that, we add

$$B^S_{ij,t-1} + B^P_{ij,t-1} = B^P_{ijt} + B^E_{ijt}, \quad i,j,t, \tag{7.44}$$

which can be viewed as a *batchsize balance*. Because at most one variable can be positive in the LHS and RHS of the equation (why?), it enforces the following:

- If a batch starts at T_{t-1} and does not finish by T_t, which means that it is being processed during $(T_{t-1}, T_t]$, then $B^S_{ij,t-1} = B^P_{ijt}$.
- If a batch starts at T_{t-1} and finishes by T_t, which means that it is not being processed during $(T_{t-1}, T_t]$, then $B^S_{ij,t-1} = B^E_{ijt}$.
- If a batch is being processed during period $t-1$ and does not finish by T_t, which means that it is being processed during $(T_{t-1}, T_t]$, then $B^P_{ij,t-1} = B^P_{ijt}$.
- If a batch is being processed during period $t-1$ and finishes by T_t, which means that it is not being processed during $(T_{t-1}, T_t]$, then $B^P_{ij,t-1} = B^E_{ijt}$.

Mapping onto the Time Grid. An advantage of having variables X^E_{ijt} to map the end of a batch onto the time grid is that variable processing times can be readily modeled. Variable processing times appear when, for example, material transfer to and from a unit is slow, so the total processing time of the corresponding task becomes a function of batchsize. If we assume that the processing time has a fixed τ^F_{ij} [h] and variable τ^V_{ij} [h/kg] component, which is a reasonable assumption, then the processing time (duration), $D_{ijt} \in \mathbb{R}_+$, of a batch of task i that starts on unit j at point t is given by

$$D_{ijt} = \tau^F_{ij} X^S_{ijt} + \tau^V_{ij} B^S_{ijt}, \quad i,j,t. \tag{7.45}$$

Next, we use variable D_{ijt} to couple the start time of a batch with its end time. To accomplish this, we introduce nonnegative continuous variables T^S_{ijt}/T^E_{ijt} to denote the start (end) time of a batch of task i that is executed on unit j and starts at T_t (finishes in $(T_{t-1}, T_t]$). Variables T^S_{ijt}/T^E_{ijt} are *unusual* in that they are positive even when the corresponding binary variables are zero; for example, we have $T^S_{\mathrm{T1,U1},1} > 0$ even if $X^S_{\mathrm{T1,U1},1} = 0$ (see Figure 7.7C). Specifically, T^S_{ijt} is always equal to T_t,

$$T^S_{ijt} = T_t, \quad i,j,t, \tag{7.46}$$

and thus can be projected out, but we use it to facilitate the presentation. Variable T^E_{ijt} is nondecreasing in time:

$$T^E_{ijt} \geq T^E_{ij,t-1} + D_{ijt}, \quad i,j,t. \tag{7.47}$$

It remains unchanged if no task starts,

$$T^E_{ijt} \leq T^E_{ij,t-1} + \eta X^S_{ijt}, \quad i,j,t, \tag{7.48}$$

and should, in addition, satisfy

$$T^E_{ij,t-1} \leq T_t + \eta\left(1 - X^E_{ijt}\right), \quad i,j,t, \tag{7.49}$$

that is, it cannot exceed T_t if the batch ends in $(T_{t-1}, T_t]$. Note that the end time of a batch ending in period t is represented by $T^E_{ij,t-1}$ rather than T^E_{ijt}, which, if another batch starts at T_t (i.e., $X^S_{ijt} = 1$), it will increase to represent the end time of the second batch (see Figure 7.7C).

The coupling of start and end times is achieved through

$$T^S_{ijt} + D_{ijt} - \eta\left(1 - X^S_{ijt}\right) \le T^E_{ijt} \le T^S_{ijt} + D_{ijt} + \eta\left(1 - X^S_{ijt}\right), \quad i,j,t, \tag{7.50}$$

which enforces a *jump* in T^E_{ijt} from its previous value $(T^E_{ij,t-1})$, representing the end time of the last batch of i executed on j, to the end time of the batch starting at $T_t = T^S_{ijt}$.

Figure 7.7C shows how variables T^S_{ijt} and T^S_{ijt} *evolve* given two batches of task T1, the first starting at time point 2 (corresponding to $T_2 = 2$); and the second starting at time point 5 ($T_5 = 7$). First, note that T^S_{ijt} (dashed line) is identical to T_t at all points (see how black solid circles lie on the 45^0 line). Second, note that T^E_{ijt} (gray thick line) increase only at time points 2 and 5, when a batch starts.

Material Balance. If I_{kn} is the inventory level of material k during period t, the material balance, without shipments, can be expressed as follows:

$$I_{k,t+1} = I_{kt} + \sum\nolimits_{i\in\mathbf{I}^+_k}\sum\nolimits_{j\in\mathbf{J}_i}\rho_{ik}B^E_{ijt} + \sum\nolimits_{i\in\mathbf{I}^-_k}\sum\nolimits_{j\in\mathbf{J}_i}\rho_{ik}B^S_{ijt} - S_{kt} \le \chi_k, \quad k,t, \tag{7.51}$$

where S_{kt} are the sales of material k at point t. Note that the transfer of material produced by a batch finishing in $(T_{t-1}, T_t]$ (i.e., $X^E_{ijt} = 1$ and $B^E_{ijt} > 0$) is assumed to occur at T_t, which, if the same material is consumed at T_t, may lead to a violation of storage vessel capacity, χ_k, between the time the batch is actually finished and T_t (see discussion in Section 7.3.3).

Objective Function. The maximization of profit can be written as follows:

$$\max \sum\nolimits_k \pi_k \sum\nolimits_t S_{kt} - \sum\nolimits_{i,j}\left\{\gamma^F_{ij}\sum\nolimits_t X^S_{ijt} + \gamma^V_{ij}\sum\nolimits_t B^S_{ijt}\right\}. \tag{7.52}$$

Tightening Constraints. The following constraints, though not necessary, enhance the solution of the model:

$$\sum\nolimits_{i\in\mathbf{I}_j}\sum\nolimits_t D_{ijt} \le \eta, \quad j \tag{7.53}$$

$$\sum\nolimits_{i\in\mathbf{I}_j}\sum\nolimits_{t'\in\{t,..,\,|\mathbf{T}|\,-1\}} D_{ijt'} \le \eta - T_t, \quad j,t \tag{7.54}$$

$$\sum\nolimits_{i\in\mathbf{I}_j}\sum\nolimits_{t'\le t}\left(\tau^F_j X^E_{ijt'} + \tau^V_{ij} B^E_{jt'}\right) \le T_t, \quad j,t. \tag{7.55}$$

Equation (7.53) enforces that the sum of processing times of the batches assigned to a unit cannot exceed the scheduling horizon. Equation (7.54) enforces that the sum of processing times of the batches starting on a unit at or after time T_t cannot exceed the remaining time, $\eta - T_t$; and (7.55) is its counterpart for batches finishing through time T_t.

Finally, the domains of the continuous nonnegative variables are as follows:

$$\begin{aligned} &T_t \in [0,\eta], \ t; T^S_{ijt} \in [0,\eta], \ T^E_{ijt} \in [0,\eta], \ D_{ijt} \in \left[0, \tau^F_{ij} + \tau^V_{ij}\beta^{MAX}_{ij}\right], \ i, \ j, \ t; \\ &B^s_{ijt} \in \left[0, \beta^{MAX}_{ij}\right], \ B^P_{ijt} \in \left[0, \beta^{MAX}_{ij}\right], \ B^E_{ijt} \in \left[0, \beta^{MAX}_{ij}\right], \ i, \ j, \ t; I_{kt} \in [0, \chi_k]. \end{aligned} \tag{7.56}$$

7.3.2 Extensions

For the modeling of shared utilities, we adopt the approach discussed in Section 4.6.3. We introduce variables $R^I_{ilt} \in \mathbb{R}_+$ and $R^O_{ilt} \in \mathbb{R}_+$ to denote the amount of utility l consumed and produced, respectively, by a batch of task i at T_t:

$$R^I_{ilt} = \varphi_{il} \sum_j X^S_{ijt} + \psi_{il} \sum_j B^S_{ijt}, \quad i, l, t \tag{7.57}$$

$$R^O_{imt} = \varphi_{il} \sum_j X^E_{ijt} + \psi_{il} \sum_j B^E_{ijt}, \quad i, l, t. \tag{7.58}$$

In (7.58), we assume that if a task finishes in $(T_{t-1}, T_t]$ (i.e., $X^E_{ijt} = 1$), the utility is produced at T_t, which leads to an underestimation of its availability between the end the batch and T_t, which, however, can be readily addressed, as we will discuss in Section 7.3.3. Variables R^I_{ilt} and R^O_{ilt} are then used to calculate and upper bound use during period t, R_{lt}:

$$R_{lt} = R_{l,t-1} + \sum_i R^I_{il,t-1} - \sum_i R^O_{il,t-1} \le \chi_l, \quad l, t. \tag{7.59}$$

The modeling of intermediate material shipments in network environments using models based on a continuous time grid is challenging because the time a shipment takes place cannot be preassigned to a time point and thus readily included in the corresponding material balance as in discrete time models. As will be shown in Example 7.1, the assignment of orders to points prior to optimization is, in general, impossible because, among other reasons, materials can be produced by different tasks, executed on different units with different processing times and processing costs. Furthermore, unlike sequence-based models for problems in sequential environments,[13] batches cannot be preassigned to shipments/orders, which means that keeping track of the lateness of a specific shipment is nontrivial.

The approach we follow here is to introduce variables $Z_{ot} \in \{0, 1\}$ to *assign* shipments to time points. If we assume that each order, of ξ_o units due at t_o (see Section 7.2.1) is met on time by a single shipment, we can then allocate it to one time point,

$$\sum_t Z_{ot} = 1, \quad o. \tag{7.60}$$

Note that more than one shipment can be assigned to the same point.

Next, variables T_t are disaggregated into T_{to} and $\tilde{T}_{to}$:

$$T_t = T_{to} + \tilde{T}_{to}, \quad t, o. \tag{7.61}$$

If shipment o is assigned to point t, then

$$T_{to} = t_o Z_{ot}, \quad t, o, \tag{7.62}$$

[13] Recall that the models we discussed in Chapter 4, employed batches or orders ($i \in \mathbf{I}$) as the basic modeling entity, even in the presence of batching decisions; i.e., practically all variables were indexed by i or (i, l). Thus, introducing variables to keep track of the earliness or lateness of an order could be readily done.

whereas in all other cases (i.e., shipment o occurs at $t' \neq t$), dummy variable $\tilde{T}_{to}$ is nonnegative:

$$\tilde{T}_{to} \leq \eta(1 - Z_{ot}), \quad t, o. \tag{7.63}$$

Through (7.61) – (7.63) the time, t_o, of a shipment is mapped onto exactly one time, T_t, of the grid.

Next, shipment variables $V_{kt} \in \mathbb{R}_+$ are introduced to model the net shipment of material k at time T_t. If the material associated with shipment o is k (i.e., $\bar{k}(o) = k$), then

$$V_{kt} = \tilde{\xi}_o Z_{ot}, \quad o, t, k = \bar{k}(o). \tag{7.64}$$

Finally, the material balance, which replaces (7.51), becomes

$$I_{k,t+1} = I_{kt} + \sum\nolimits_{i \in \mathbf{I}_k^+} \sum\nolimits_{j \in \mathbf{J}_i} \rho_{ik} B_{ijt}^F + \sum\nolimits_{i \in \mathbf{I}_k^-} \sum\nolimits_{j \in \mathbf{J}_i} \rho_{ik} B_{ijt}^S + V_{kt} - S_{kt} \leq \chi_k, \quad k, t. \tag{7.65}$$

The extended model that accounts for shared utilities and intermediate material shipments consists of (7.35) through (7.50) and (7.52) through (7.65).

Example 7.1 Mapping of Shipment Times to Time Points The goal of this example is to illustrate the challenge of mapping shipments to time points when a continuous grid is used. We consider a system with two units – U1, which is small but fast, and U2, which is large but slow – and one order for 8 kg of P1 due after four hours. The STN representation of the facility along with processing data are shown in Figure 7.8A. Note that all tasks for the production of P2 and P3 last more than four hours.

Figures 7.8B and 7.8C show the Gantt charts and the inventory profiles of material P1 in two solutions over the first five hours, which is a fraction of the horizon considered. In the solution shown in Figure 7.8B (solution 1), P1 is produced using U1, which means that two batches are needed to meet the order, and thus they would have to start at time points 0 and 1 (i.e., $T_0 = 0$, $T_1 = 2$). Consequently, the shipment occurs at the third time point, at $T_2 = 4$. In the solution shown in Figure 7.8C (solution 2), P1 is produced using U2, which means that the demand is met using one batch, which starts

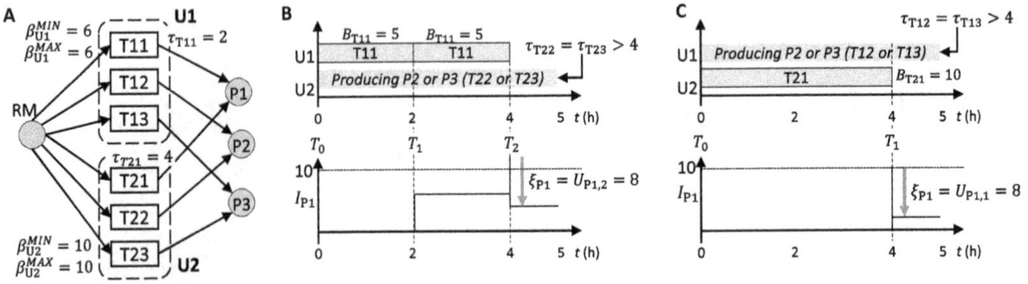

Figure 7.8 Allocation of shipments to time points. (A) Example STN and process parameters (amounts are in kg and times in h). (B) Solution 1: shipment occurs at time point 2. (C) Solution 2: shipment occurs at time point 1.

at $T_0 = 0$, and demand is satisfied at the second time point, at $T_1 = 4$. Note that the optimal assignment of tasks to units, and therefore the mapping of due times to points, will depend on the demand for P1 after four hours as well as the processing costs of all tasks, and thus it cannot be predetermined.

The model presented in the present subsection can be extended, though nontrivially, to account for backlogs and lost sales. (Can you think how you would develop such extension? Can you see any challenges?)

7.3.3 Remarks

The development of models based on continuous time grids has been an area of active research in the 1990s and 2000s, leading to a wide range of models. The vast majority of these models adopt an STN or RTN representation and consider all three decisions: batching, batch-unit assignment, and timing. The main differentiating attributes across them are the following (see Figure 7.9):

(1) The adoption of a common versus unit-specific time grids
(2) The assumptions regarding the mapping of the start and end of a batch onto grid times
(3) The type of variables used to map the execution of batches onto the time grid

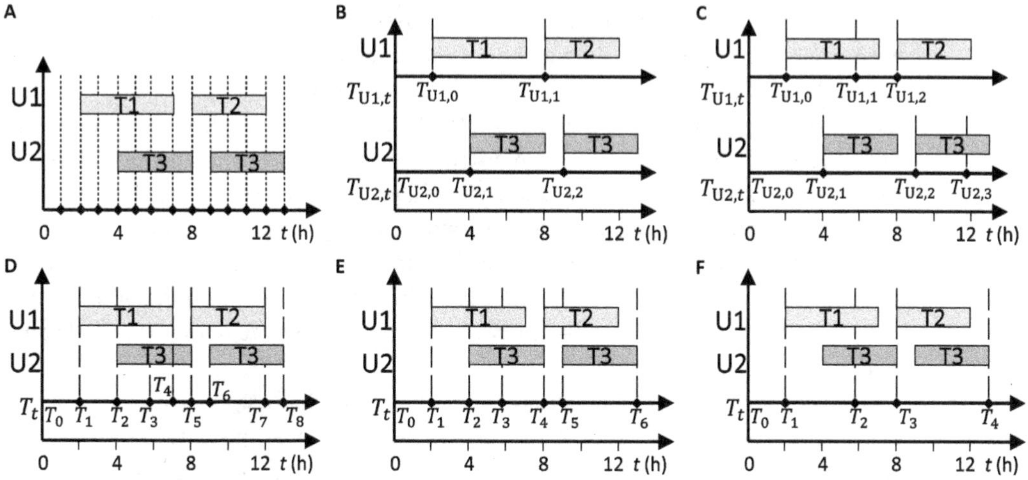

Figure 7.9 Illustration of different types of time grids and approaches for mapping batches onto grid points/periods; each panel shows how the same solution (Gant chart at the top of each panel) is mapped onto grid(s). (A) Discrete common time grid, with $\delta = 1$ h. (B) Continuous unit-specific grids: a batch should start at a time point, can end at or before a time point, and spans one period (adopted typically in models for sequential environments). (C) Continuous unit-specific grids: a batch should start at a time point, end at or before a time point, and can span multiple periods (adopted in models for network environments). (D) Continuous common grid: a batch should start and end at a time point, and can span multiple periods. (E) Continuous common grid: a batch should start at a time point, can end at/within a point/period, and can span multiple periods. (F) Continuous common grid: a batch can start and end either at a point or within a period, and can span multiple periods.

Next, we discuss these attributes in more detail, so through this discussion the reader gains an appreciation of the different approaches and ideas regarding the formulation of different continuous time models.

Common versus. Unit-Specific Time Grids. The adoption of unit-specific grids leads to smaller models because the number of time points needed to represent a given solution is smaller than or equal to the number of points needed when a common grid is adopted, as shown in Figure 7.9 (points in panels B and C are fewer than points used in panels D through F). However, the modeling of features such as (1) consumption and production of multiple materials by a single task, (2) production or consumption of the same material by multiple tasks, and (3) shared utilities[14] is challenging when multiple grids are employed. The challenge stems from the need to *synchronize* different grids, which are now coupled through tasks (e.g., consuming the same material) and shared utilities. In essence, the synchronization is necessary because different units impact the same material and utility balances. The approach to address this challenge is to use more points for each grid; and to also add constraints coupling the grids of the units impacting the same material or utility. The use of additional time points and introduction of coupling constraints (which would be unnecessary in a model based on a common grid) lead to models that are comparable in size and, potentially, more complex than models based on a common grid.

Mapping of Batch Start and End Times onto Grid Times. For the discussion here we assume that a single grid – common across units, materials, and utilities – is used. As already discussed, the same solution (Gantt chart) can be mapped onto different grids depending on the assumptions used for the mapping. The early STN- and RTN-based continuous time models were based on the assumption that both the start and the end of a batch should coincide exactly with a time point. Recognizing that the processing time of a batch can often be artificially prolonged by simply not removing the material after its completion, researchers developed models in which the requirement for the mapping of the end of a batch is relaxed: the end of a batch is mapped onto time, T_t, even though it ends during period $(T_{t-1}, T_t]$, as already discussed in Section 7.3.1. In other words, if the absolute time, $\hat{T}^F_{it}$, a batch of task actually ends satisfies, $T_{t-1} < \hat{T}^F_{it} \leq T_t$, then the end time is considered to be equal to T_t. This leads to an underestimation of the inventory of the output materials and utility availability during $(\hat{T}^F_{it}, T_t)$.

There are two potential shortcomings with this approach. First, if the output material (s) cannot be temporarily stored in the unit where a batch just ended, then the material will have to be moved to the dedicated storage tank earlier, at $\hat{T}^F_{it}$, rather than T_t. The model can still be used, but it may lead to an underestimation of the inventory during $[\hat{T}^F_{it}, T_t)$, which may lead to infeasibility because of limited storage capacity (see Figure 7.10). If this is a concern, then the tasks whose processing time cannot be

[14] Note that the first two features are not present, and the third feature has not been traditionally considered in problems in sequential environments. This explains why models based on unit-specific grids have been studied extensively for, and are often the most effective to address, problems in sequential environments.

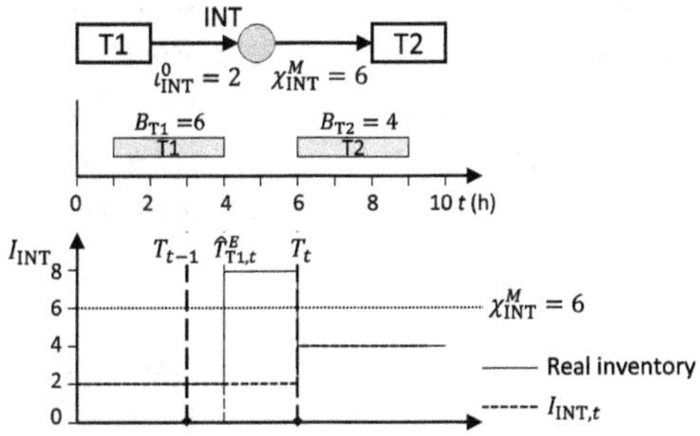

Figure 7.10 Inventory calculation when a batch is allowed to end within a period. Units are not considered to simplify notation. Starting inventory of INT is 2 kg, vessel capacity is 6 kg, and the batchsizes of batches of T1 and T2 are 6 and 4 kg, respectively. If INT is removed immediately upon completion of T1, its inventory would increase to 8 (2 + 6), which is higher than the storage capacity, before it decreases to 4 (8 – 4).

artificially prolonged can be modeled as *no-unit-storage* tasks,[15] $i \in \mathbf{I}^{NUS} \subseteq \mathbf{I}$, and be required to end at exactly a time point, rather than within a period. This can be accomplished by changing the mapping constraints. For example, in the model presented in Section 7.3.1, the following constraint is added:

$$T^E_{ij,t-1} \geq T_t - \eta\left(1 - X^E_{ijt}\right), \quad i \in \mathbf{I}^{NUS}, j, t, \tag{7.66}$$

which, combined with (7.49), enforces that if $X^E_{ijt} = 1$ then a batch of task i ends exactly at T_t.

Second, it would seem that the underestimation of inventory and utility availability may lead to suboptimal solutions because opportunities to start a batch, which consumes the material whose inventory is underestimated or requires the utility whose availability is underestimated, are lost. This is not the case because any solution where a batch starts exactly at the time when another batch ends can be represented using additional time points, thereby allowing batches to finish exactly at a time point if doing so would lead to a better solution. This point is further explained in Example 7.2.

Finally, we note that models have been proposed where both the actual start and the actual end of a batch can occur within a period. Specifically, assuming that $\hat{T}^S_{it}/\hat{T}^E_{it}$ are the actual start/end times of a batch of task i, if $X^S_{ijt} = 1$, then $\hat{T}^S_{it} \in [T_t, T_{t+1})$, and if $X^E_{ijt} = 1$, then $\hat{T}^E_{it} \in (T_{t-1}, T_t]$ (see Figure 7.9F).

[15] Note that there are differences between *no-unit-storage* (NUS) tasks and no intermediate storage (NIS) and zero wait (ZW) storage policies. First, in the context of network environments, NIS and ZW refer to policies applied to materials, not tasks. In NIS policy, there is no dedicated storage vessel for a material, but the material can be stored in the unit it was produced. However, in ZW policy, there is no dedicated storage vessel and the material cannot be stored in the unit it was produced – it has to be transferred out of it immediately. Second, the materials produced by a NUS task may have dedicated storage vessels but have to be immediately transferred to them. Interestingly, materials produced by NUS tasks and have no dedicated vessels (NIS policy) correspond to materials for which ZW policy is applicable. The modeling of different storage constraints and activities in network environments will be discussed in the next chapter.

Example 7.2 Alternative Solution Representation via Time Point Addition This example illustrates how the underestimation of inventory can be addressed by adding a point to the grid. We consider the network shown in Figure 7.11A, where we assume that we have fixed batch sizes (β_{ij}), no initial IN1 and IN2 inventories, and unlimited storage for all materials. Fixed batchsizes and processing times are given above each box representing a task. Figure 7.11B shows part of the solution obtained if the end of the batch of T2 is not assigned to a point: inventory of IN2 is underestimated between 3 and 4 h, and so a batch of task T4 cannot start at 3 h. Figure 7.11C shows the an alternative solution that can be obtained if an additional point is used: the new point coincides with the end of batch of T2, so the calculated and actual inventory of IN2 are the same and a batch of task T4 starts at 3 h.

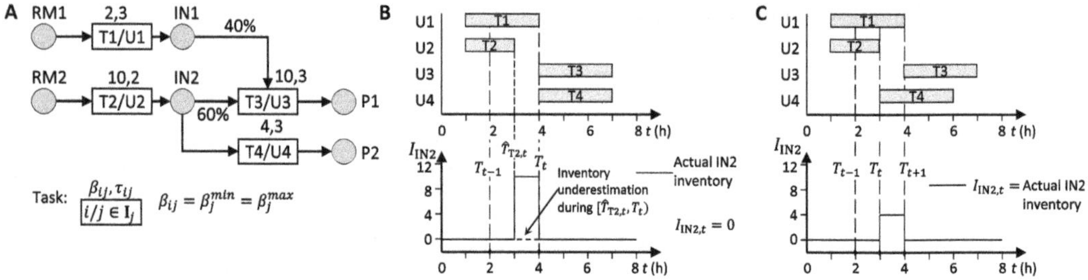

Figure 7.11 Handling of inventory underestimation through time point addition. (A) Example network. (B) Part of the solution obtained if the end of batch of T2 is not assigned to a point. (C) Alternative solution obtained if an additional point is used.

Variables Used for Mapping Batches onto Time Grid. One source of model complexity stems from the fact that the point at or period within which a batch ends is not uniquely determined by the point at which the same batch starts. Thus, additional variables and constraints, as compared to models based on a discrete time grid, are needed to achieve the coupling between batch start and end times as well as the mapping of the end of batch onto the grid. There are, in general, two approaches.

In the first, which was adopted in the model presented in Sections 7.3.1 and 7.3.2, separate binary variables for the allocation of the start and end of a batch to time points are used. Note that a third variable is often also employed to denote the continuing processing of a batch during a period. For example, in the model presented in Section 7.3.1, we could have introduced X_{ijt}^P, to denote the processing of a batch of task i on unit j during period t (i.e., started before T_{t-1} and will end after T_t),

$$X_{ijt}^P = \sum_{t'<t} X_{ijt'}^S - \sum_{t' \leq t} X_{ijt'}^E, \tag{7.67}$$

which would then have allowed us to express (7.43) as follows:

$$\beta_{ij}^{MIN} X_{ijt}^P \leq B_{ijt}^P \leq \beta_{ij}^{MAX} X_{ijt}^P, \quad i,j,t \tag{7.68}$$

as well as write additional tightening constraints.

In the second approach, a single variable $X_{itt'} \in \{0,1\}$ is used to denote the execution of a batch of task i, starting at time point t and ending at point t' or within period t', where we drop index j for simplicity. For example, if a batch of task i starts at T_2 and ends at T_5, then $X_{i,2,3} = X_{i,2,4} = 0$, $X_{i,2,5} = 1$, and $X_{i,2,t'} = 0$ for $t' > 5$. Clearly, $X_{itt'}$ is defined for $t' > t$, which leads to the introduction of $|\mathbf{T}|(|\mathbf{T}| - 1)/2$ binary variables for each i (or for each $(i,j), j \in \mathbf{J}_i$ pair, if variables $X_{ijtt'}$ are considered). If it is assumed that a batch cannot span more than a certain number of periods, which is a reasonable assumption, then the number of $X_{itt'}$ variables can be reduced. A variant of the second approach is to introduce variables $\bar{X}_{itt'} \in \{0,1\}$ to denote the start of a batch of task i at point t and its continuing processing during period t'. If a batch starts at point t, then, by definition, $\bar{X}_{i,t,t+1} = 1$, so $\bar{X}_{i,t,t+1}$ does need to be defined. Thus, if a batch of task i starts at T_2 and ends at T_5, then $\bar{X}_{i,2,4} = \bar{X}_{i,2,5} = 1$, and $\bar{X}_{i,2,t'} = 0$ for $t' > 5$. Variables $X_{itt'}$ and/or $\bar{X}_{itt'}$ are used, typically through variable upper bound constraints, to activate continuous nonnegative variables for the modeling of material balances.

7.4 Notes and Further Reading

(1) The STN representation and a first MILP model accounting for a range of processing features, based on a discrete grid, was proposed by Kondili et al. [1]. An improved STN-based model employing a discrete time grid was proposed by Shah et al. [2]. The RTN representation and the first RTN-based MIP model were proposed in Pantelides [3]. The presentation in Section 7.1 is based on the three aforementioned papers, though the notation is changed to maintain consistency throughout this book. Also, the term *material* instead of *state* has been adopted. Other changes include the introduction of additional constraints to make the presentation easier to follow, and a more systematic discussion of intermediate shipments and time-varying utility capacity and cost.

(2) Pantelides [3] discussed in detail varying resource consumption during the execution of a task. This subject has already been covered in Chapter 4, so it is not discussed here.

(3) Approaches based on a discrete time grid were developed in the 1990s to address a range of problems [4–6].

(4) Soon after the STN and RTN representations were developed, the PSE community focused on developing models based on one or more continuous time grids. The motivation was that a given solution can be represented by far fewer time points if their timing is an optimization decision, rather than predetermined. Therefore, if a continuous grid is employed, the corresponding model will be significantly smaller and thus easier to solve.

(5) The first continuous time models, developed between 1995 and 2000, adopted a common time grid [7–11], and some of them were mixed-integer nonlinear programming (MINLP) models. Some new modeling ideas were proposed in early 2000s, including the model upon which the presentation in Sections 7.3.1

and 7.3.2 is based [12–14]. The model where batches can start and end within periods (see Figure 7.9F) were proposed in [15, 16].

(6) The first models based on unit-specific continuous time grids were developed in the late 1990s and early 2000s [17–20]. A series of extensions have been proposed since then to address the challenges discussed in Section 7.3.3 [21, 22]. A wide range of improved models, based however for the most part on the ideas in the aforementioned models, has appeared in the literature since 2005 [23].

(7) A model based on a mixed-time representation has also been proposed [24]. It employs a discrete time grid but accounts for variable processing times, which means that the assignment of the end of a batch is not uniquely defined by its start.

(8) An alternative framework for facility representation, the so-called *Unit-Operation-Port-State Superstructure* (UOPSS), was proposed and used to develop various solution techniques by Kelly and coworkers [25, 26].

(9) Using ideas from STN, Maravelias and coworkers proposed models for problems in general production environments (see discussion in Section 1.2.5) [27, 28].

(10) Both STN- and RTN-based models rely on the consumption and production of materials to enforce precedence relations between batches of different tasks. In that respect, they fall into the category of *material-based* models, as opposed to *batch-based* models used for the scheduling in sequential environments.

(11) An extensive discussion of computational aspects for models for problems in network environments can be found in Sundaramoorthy and Maravelias [29].

(12) The development of solution methods for material-based models received attention in the 2010s [30] and will be discussed in more detail in Chapter 13.

7.5 Exercises

(1) Consider a network environment with four units, nine materials, and five tasks. All processing data are given in Tables 7.1 through 7.4.

(a) Based on the parameters in Table 7.1, identify sets $\mathbf{I}_j$ and $\mathbf{J}_i$.

(b) Based on the parameters in Table 7.3, identify sets $\mathbf{I}_k^+$, $\mathbf{I}_k^-$, $\mathbf{K}_i^+$, and $\mathbf{K}_i^-$.

(c) Draw the STN and RTN representations of this facility.

Table 7.1 Fixed [h], τ_{ij}^F, and variable [h/kg], τ_{ij}^V, processing time coefficients; fixed [$], γ_{ij}^F and variable [$/kg], γ_{ij}^V, processing costs.

	Processing time coefficients				Processing costs			
Task	U1	U2	U3	U4	U1	U2	U3	U4
T1	1/0.01				10/0.1			
T2		2/0.01	2/0.01			40/0.3	20/0.2	
T3		2/0.01	2/0.01			50/0.3	30/0.2	
T4		1/0.01	1/0.01			50/0.3	30/0.2	
T5				2/0.01				30/0.2

Table 7.2 Unit capacities, β_j^{MAX} [kg]; $\beta_{ij}^{MIN} = 0, \forall i, j$.

	U1	U2	U3	U4
Capacity	100	80	50	200

Table 7.3 Conversion coefficients, ρ_{ik}.

	F1	F2	F3	I1	I2	I3	I4	P1	P2
T1	–1			1					
T2		–0.5	–0.5		1				
T3				–0.4	–0.6	0.6		0.4	
T4			–0.2			–0.8	1		
T5							–1		0.9

Table 7.4 Material-related data: price [\$/kg], π_k; initial inventory [kg], ι_k^0; storage capacity [kg], χ_k^M; and demand [kg] at 24, $\dot{\xi}_k^{24}$, and 48, $\dot{\xi}_k^{48}$, h.

	F1	F2	F3	I1	I2	I3	I4	P1	P2
π_k								2	3
ι_k^0	500	500	500						
χ_k^M	500	500	500					500	500
$\dot{\xi}_k^{24}$								100	200
$\dot{\xi}_k^{48}$								200	250

Using the data for this system, generate three models:

(d) STN-based discrete time, with $\tau_{ij} = \tau_{ij}^F$ and $\tau_{ij} = \tau_{ij}^F + \tau_{ij}^V \beta_j^{MAX}$, using $\delta = 1$.

(e) RTN-based discrete time, with $\tau_{ij} = \tau_{ij}^F$ and $\tau_{ij} = \tau_{ij}^F + \tau_{ij}^V \beta_j^{MAX}$, using $\delta = 1$.

(f) STN-based continuous time (common grid).

Using the preceding models, solve the following instances:

(g) Profit maximization, subject to given demand, for a horizon of 48 h. Can you see the impact of using $\tau_{ij} = \tau_{ij}^F$ versus $\tau_{ij} = \tau_{ij}^F + \tau_{ij}^V \beta_j^{MAX}$?

(h) Profit maximization, subject to all demand ($\dot{\xi}_k^{24} + \dot{\xi}_k^{48}$) met at the end of the 48 h horizon.

(i) Cost minimization.

A number of papers describe how discrete time models can be modified to account (approximately) for variable processing times (see, for example, [1] and [29]).

(j) Formulate an STN-based discrete time model that accounts for variable processing times.

Table 7.5 Processing times, τ_{ij}, [h]/costs [\$], γ_{ij}^F.

	U1	U2	U3	U4	U5
T1	5/120				
H	1/0				
T2		2/45			
T3			2/70		
T4				2/60	2/55
T5				2/90	2/80
T6				2/75	2/50

Table 7.6 Unit capacities, $\beta_j^{MIN}/\beta_j^{MAX}$ [kg].

	U1	U2	U3	U4	U5
Capacity	0/240	13/40	23/70	17/50	15/45

Table 7.7 Material-related data: price, π_k [\$/kg]; initial inventory, ι_k^0 [kg]; storage capacity, χ_k^M [kg]; and demand [kg] at 24, $\dot{\xi}_k^{24}$, 48, $\dot{\xi}_k^{48}$, and 72 $\dot{\xi}_k^{72}$ h.

	F1	F2	F3	F4	I1	I2	P1	P2	P3	B
π_k							2	2.5	3	1
ι_k^0	1,500	1,500	1,500	1,500						
χ_k^M	1,500	1,500	1,500	1,500		75	1,500	1,500	1,500	1,500
$\dot{\xi}_k^{24}$							100	50	150	30
$\dot{\xi}_k^{48}$							100	50	150	30
$\dot{\xi}_{ksi}^{72}$							100	50	150	30

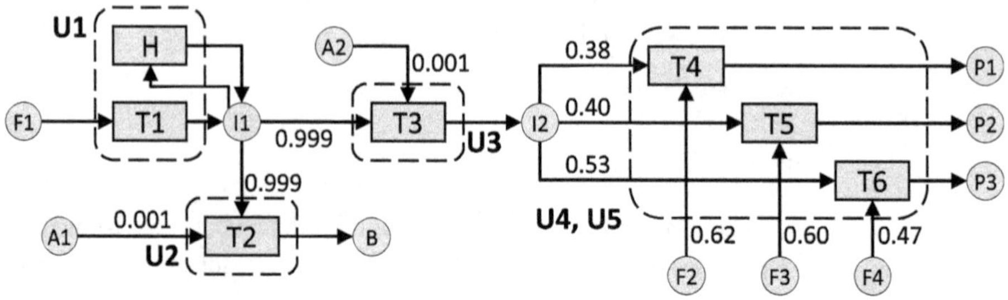

Figure 7.12 STN representation of the system considered in Exercise 2; conversion coefficients not shown are either –1 or 1.

Table 7.8 Processing times, τ_{ij}, [h]/costs γ_{ij}^F [$].

	T1	T2	T3	T4	T5	T6	T7	T8	T9	T10
U1	7.75/ 20			4.25						
U2		2.75/ 10	3.25/ 10							
U3		5.50/ 10	6.50/ 10							
U4					6.00/ 20	9.50/ 20				
U5							7.50/ 40		6.00/ 20	
U6								8.00/ 20		5.25/ 30

Table 7.9 Unit capacities, $\beta_j^{MIN}/\beta_j^{MAX}$ [10^2 kg].

	U1	U2	U3	U4	U5	U6
Capacity	3/10	3/8	3/8	3/10	3/6	3/8

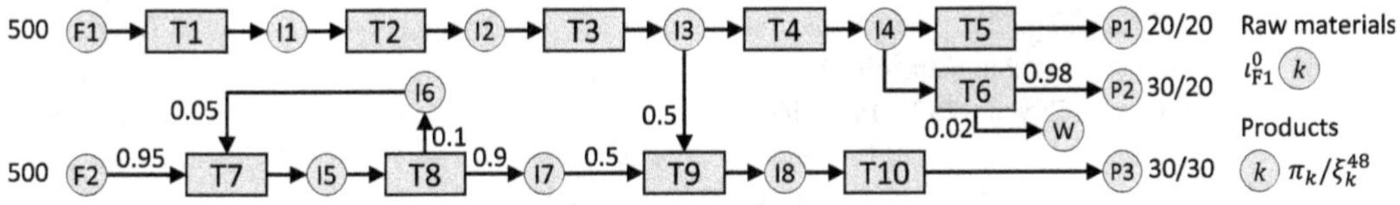

Figure 7.13 STN representation of system considered in Exercise 3 along with material prices, π_k [$/$10^2$ kg]; initial inventories, ι_k^0 [10^2 kg]; demands after 48 h, ξ_k^{48} [10^2 kg]; and conversion coefficients, ρ_{ik}. Material-related data not shown are zero; conversion coefficients not shown are either –1 or 1; task-unit compatibility not shown.

(k) Resolve the profit maximization instance ($\eta = 48$ h, with demand at 24 and 48), using $\delta = 1$ and $\delta = 0.5$. Are the optimal objective function values the same? Why or why not?

(2) Consider the network environment shown in Figure 7.12, with the processing data given in Tables 7.5 through 7.7.

(a) Identify subsets $\mathbf{I}_j$, $\mathbf{J}_i$, $\mathbf{I}_k^+$, $\mathbf{I}_k^-$, $\mathbf{K}_i^+$, and $\mathbf{K}_i^-$.

(b) What does task H represent?

(c) A1 and A2 are additives that are always available. Develop a representation that does not explicitly include them but accounts for their effect on task conversions.

(d) Formulate STN- and RTN-based discrete time models to solve a cost-minimization problem.

(e) Formulate an STN-based continuous time model to minimize cost. Assume that I1 can be temporarily stored in U1. How can this be modeled? How many points/periods are required?

(f) Employ the ideas discussed in Section 7.3 to formulate a continuous time RTN-based model.

(3) Consider the network environment shown in Figure 7.13 along with the processing data given in Tables 7.8 and 7.9. There is unlimited capacity for all materials ($\chi_k^M = \infty$).

(a) Solve the cost minimization problem using an STN-based discrete time model with $\delta = 1$. Is the optimal solution you obtain with this model the best solution for this instance?

(b) What is the largest value of δ that can be used to obtain the best solution? Solve the instance using a model that employs this value.

References

[1] Kondili E, Pantelides CC, Sargent RWH. A General Algorithm for Short-Term Scheduling of Batch-Operations .1. MILP Formulation. *Comput Chem Eng*. 1993;17(2):211–227.

[2] Shah N, Pantelides CC, Sargent RWH. A General Algorithm for Short-Term Scheduling of Batch-Operations .2. Computational Issues. *Comput Chem Eng*. 1993;17(2):229–244.

[3] Pantelides CC, editor Unified Frameworks for Optimal Process Planning and Scheduling. 2nd Conference on Foundations of Computer Aided Process Operations; 1994 1994; Snowmass, CO: CACHE Publications.

[4] BarbosaPovoa APFD, Pantelides CC. Design of Multipurpose Plants Using the Resource-Task Network Unified Framework. *Comput Chem Eng*. 1997;21:S703–S708.

[5] Zentner MG, Pekny JF, Reklaitis GV, Gupta JND. Practical Considerations in Using Model-Based Optimization for the Scheduling and Planning of Batch/Semicontinuous Processes. *Journal of Process Control*. 1994;4(4):259–280.

[6] Subrahmanyam S, Bassett MH, Pekny JF, Reklaitis GV. Issues in Solving Large-Scale Planning, Design and Scheduling Problems in Batch Chemical Plants. *Comput Chem Eng*. 1995;19:S577–S582.

[7] Schilling G, Pantelides CC. A Simple Continuous-Time Process Scheduling Formulation and a Novel Solution Algorithm. *Comput Chem Eng*. 1996;20:S1221–S1226.

[8] Zhang X, Sargent RWH. The Optimal Operation of Mixed Production Facilities – a General Formulation and Some Approaches for the Solution. *Comput Chem Eng*. 1996;20(6–7):897–904.

[9] Zhang XY, Sargent RWH. The Optimal Operation of Mixed Production Facilities – Extensions and Improvements. *Comput Chem Eng*. 1996;20:S1287–S1293.

[10] Mockus L, Reklaitis GV. Continuous Time Representation Approach to Batch and Continuous Process Scheduling. 1. MINLP Formulation. *Ind Eng Chem Res*. 1999;38 (1):197–203.

[11] Mockus L, Reklaitis GV. Continuous Time Representation Approach to Batch and Continuous Process Scheduling. 2. Computational Issues. *Ind Eng Chem Res*. 1999;38 (1):204–210.

[12] Lee KH, Park HI, Lee IB. A Novel Nonuniform Discrete Time Formulation for Short-Term Scheduling of Batch and Continuous Processes. *Ind Eng Chem Res*. 2001;40(22):4902–4911.

[13] Castro P, Barbosa-Povoa APFD, Matos H. An Improved RTN Continuous-Time Formulation for the Short-Term Scheduling of Multipurpose Batch Plants. *Ind Eng Chem Res.* 2001;40(9):2059–2068.
[14] Sundaramoorthy A, Karimi IA. A Simpler Better Slot-Based Continuous-Time Formulation for Short-Term Scheduling in Multipurpose Batch Plants. *Chem Eng Sci.* 2005;60 (10):2679–2702.
[15] Gimenez DM, Henning GP, Maravelias CT. A Novel Network-Based Continuous-Time Representation for Process Scheduling: Part II. *General Framework. Comput Chem Eng.* 2009;33(10):1644–1660.
[16] Gimenez DM, Henning GP, Maravelias CT. A Novel Network-Based Continuous-Time Representation for Process Scheduling: Part I. Main Concepts and Mathematical Formulation. *Comput Chem Eng.* 2009;33(9):1511–1528.
[17] Ierapetritou MG, Floudas CA. Effective Continuous-Time Formulation for Short-Term Scheduling. 1. Multipurpose Batch Processes. *Ind Eng Chem Res.* 1998;37(11):4341–4359.
[18] Ierapetritou MG, Floudas CA. Effective Continuous-Time Formulation for Short-Term Scheduling. 2. Continuous and Semicontinuous Processes. *Ind Eng Chem Res.* 1998;37 (11):4360–4374.
[19] Giannelos NF, Georgiadis MC. A Novel Event-Driven Formulation for Short-Term Scheduling of Multipurpose Continuous Processes. *Ind Eng Chem Res.* 2002;41 (10):2431–2439.
[20] Giannelos NF, Georgiadis MC. A Simple New Continuous-Time Formulation for Short-Term Scheduling of Multipurpose Batch Processes. *Ind Eng Chem Res.* 2002;41 (9):2178–2184.
[21] Janak SL, Lin XX, Floudas CA. Enhanced Continuous-Time Unit-Specific Event-Based Formulation for Short-Term Scheduling of Multipurpose Batch Processes: Resource Constraints and Mixed Storage Policies. *Ind Eng Chem Res.* 2004;43(10):2516–2533.
[22] Janak SL, Floudas CA. Improving Unit-Specific Event Based Continuous-Time Approaches for Batch Processes: Integrality Gap and Task Splitting. *Comput Chem Eng.* 2008;32(4–5):913–955.
[23] Susarla N, Li J, Karimi IA. A Novel Approach to Scheduling Multipurpose Batch Plants Using Unit-Slots. *AIChE J.* 2010;56(7):1859–1879.
[24] Maravelias CT. Mixed-Time Representation for State-Task Network Models. *Ind Eng Chem Res.* 2005;44(24):9129–9145.
[25] Kelly JD, Zyngier D. Unit-Operation Nonlinear Modeling for Planning and Scheduling Applications. *Optimization and Engineering.* 2017;18(1):133–154.
[26] Zyngier D, Kelly JD. Multiproduct Inventory Logistics Modeling in the Process Industries. *Springer Ser Optim A.* 2009;30:61–95.
[27] Sundaramoorthy A, Maravelias CT. A General Framework for Process Scheduling. *AIChE J.* 2011;57(3):695–710.
[28] Velez S, Maravelias CT. Mixed-Integer Programming Model and Tightening Methods for Scheduling in General Chemical Production Environments. *Ind Eng Chem Res.* 2013;52 (9):3407–3423.
[29] Sundaramoorthy A, Maravelias CT. Computational Study of Network-Based Mixed-Integer Programming Approaches for Chemical Production Scheduling. *Ind Eng Chem Res.* 2011;50(9):5023–5040.
[30] Velez S, Maravelias CT. Advances in Mixed-Integer Programming Methods for Chemical Production Scheduling. *Annu Rev Chem Biomol.* 2014;5:97–121.

Part III

Advanced Methods

8 Network Environment: Extensions

In this chapter, we discuss how to model additional processing features that may be present in a chemical facility. To keep the presentation simple, we illustrate how models based on a common discrete grid can be modified to account for these features. Continuous time models can also be extended to account for most of these features, but often lead to more complex and/or nonlinear formulations.

We start, in Section 8.1, with the modeling of material consumption and production during the execution of a batch. In Section 8.2, we discuss the modeling of complex material storage and transfer activities. In Section 8.3, we present how to account for unit and task setups and task families. Finally, in Section 8.4, we present how to model unit deterioration and maintenance tasks.

We will use the sets, subsets, and parameters introduced in the previous chapter to represent a facility using the STN and RTN representations. To simplify the presentation, we will assume that, in the STN representation, each task can be carried out only in one unit (i.e., $|\mathbf{J}_i| = 1$). This is not restrictive because a task that can be carried out in many units (i.e., $|\mathbf{J}_i| > 1$) can be represented as $|\mathbf{J}_i|$ tasks, each one carried out in only one unit without increasing the size of the resulting model. Consequently, variables X_{ijn} and B_{ijn} are replaced by X_{in} and B_{in}. We will also directly use time-related parameters, without overbars, to express them as multiples of time periods; for example, τ_i, rather than $\bar{\tau}_i$, represents the processing time of task i in periods of length δ.

8.1 Material Consumption and Production during Task Execution

A task may require the consumption of some input materials after its start, and may lead to output material production before its end. Using the ideas presented in Chapter 4, to model varying resource consumption during the execution of a task, we introduce parameter ρ_{iks} to represent the conversion coefficient by task i, for material k ($\rho_{iks} < 0$ if $k \in \mathbf{K}_i^-$ and $\rho_{iks} > 0$ if $k \in \mathbf{K}_i^+$), s periods after the start of a batch of i. In the present chapter, we assume that ρ_{iks} are readily available, but we note that they should be calculated based on the actual time the *interbatch* consumption and production occur, and the chosen scheduling horizon step δ. For example, if task T1 requires the addition of solvent M1 in a 1:10 ratio compared to its batchsize, 90 minutes after its start, and the period length is 30 minutes ($\delta = 0.5$), then $\rho_{\mathrm{T1,M2},3} = -0.1$; whereas if $\delta = 1$, then

$\rho_{T1,M2,2} = -0.1$ and $\rho_{T1,M2,3} = 0$. If we were to convert the previously used coefficients, ρ_{ik}, which were used when all consumption/production occurs at the start $(s = 0)$/end $(s = \tau_i)$ of the task, we would have $\rho_{ik,s=0} = \rho_{ik} < 0$ for $k \in \mathbf{K}_i^-$ and $\rho_{ik,s=\tau_i} = \rho_{ik} > 0$ for $k \in \mathbf{K}_i^+$.

The generalization of the STN-based material balance, for $|\mathbf{J}_i| = 1$, is written as follows:

$$I_{k,n+1} = I_{kn} + \sum_{i \in \mathbf{I}_k^+} \sum_{s \in \mathbf{S}_{ik}^+} \rho_{iks} B_{i,n-s} + \sum_{i \in \mathbf{I}_k^-} \sum_{s \in \mathbf{S}_k^-} \rho_{iks} B_{i,n-s} + \xi_{ik} - S_{kn} \leq \chi_k^M, \; k, n, \tag{8.1}$$

where $\mathbf{S}_{ik}^+ = \{s : \rho_{iks} > 0\}$ and $\mathbf{S}_{ik}^- = \{s : \rho_{iks} < 0\}$ are the sets of time points, with respect to the start of task i, where material k is produced and consumed, respectively. The modeling of material consumption and production during a batch and the effect on material balances are further illustrated in Example 8.1.

We note that the approach presented in this section can be used to model a wide range of operations, including semibatch processes, as well as the startup and shutdown phases of continuous processes (see Chapter 9).

Example 8.1 Material Consumption and Production during Task Execution We consider a task T, with processing time $\tau_T = 6$ h, consuming two input materials, A1 and A2, and producing two output materials, C1 and C2. If B is the batchsize, then a batch of T consumes $0.5B$ of A1 at the beginning of the batch, $0.3B$ of A2 1 h after the beginning of the batch, and $0.2B$ of A1 2 h after the beginning of the batch; and produces $0.3B$ of C1 5 h after the beginning of the batch, and $0.5B$ of C2 at the end of the batch. If we use $\delta = 1$, the consumption occurs at $s = 0$ ($0.5B$ of A1), $s = 1$ ($0.3B$ of A2), and $s = 2$ ($0.2B$ of A1); and production occurs at $s = 5$ ($0.3B$ of C1), and $s = 6$ ($0.5B$ of C2). The corresponding subsets $\mathbf{S}_{ik}^-$ and $\mathbf{S}_{ik}^+$ and coefficients ρ_{iks} used to model task T are shown in Figure 8.1A. The inventory evolution, per (8.1), for input A1 and output C1, along with some nonzero terms used for the calculation, are shown in Figure 8.1B.

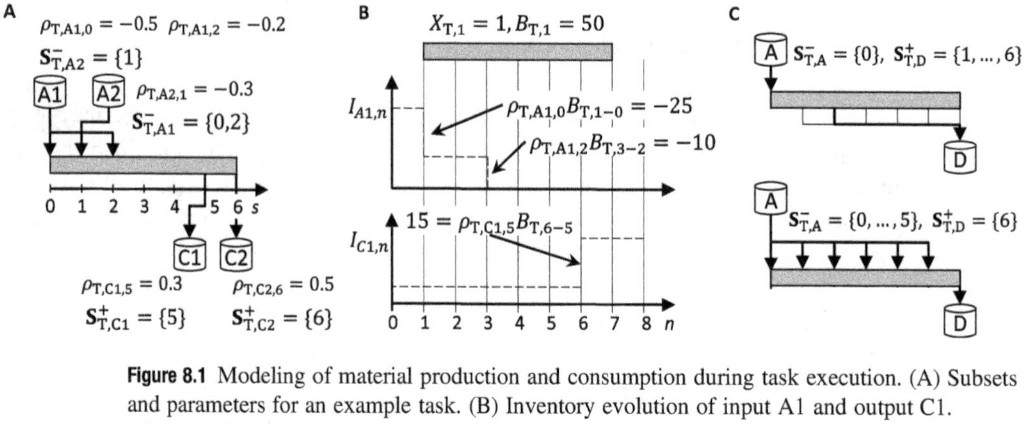

Figure 8.1 Modeling of material production and consumption during task execution. (A) Subsets and parameters for an example task. (B) Inventory evolution of input A1 and output C1. (C) Modeling of semibatch processes.

Finally, Figure 8.1C shows how two semibatch processes, consuming material A and producing material D, can be modeled as tasks with material production and consumption during execution.

8.2 Material Storage and Transfer

We study storage in shared storage vessels, storage in processing units before or after the execution of tasks, and (resource-constrained) transfer tasks. We use the STN-based model and assume no material consumption and production during task execution.

The assumption in the previous chapter was that each material has a dedicated storage vessel, with capacity χ_k^M (STN) or $\chi_r, r \in \mathbf{R}^M$ (RTN); or it cannot be stored, which can be modeled by simply setting $\chi_k^M = 0$ or $\chi_r = 0$. If storage vessels are shared among materials, then, in the STN representation, they are treated explicitly as additional units. The set of units, $\mathbf{J}$, in this case, has two subsets: the set of processing units, $\mathbf{J}^{PU}$, and the set of storage vessels, $\mathbf{J}^{SV}$. Further, we define the subset, $\mathbf{K}_j$, of materials that can be stored in unit $j \in \mathbf{J}^{SV}$. In addition, we use the following subsets:

- $\mathbf{K}^{DV}$: materials stored in dedicated vessels
- $\mathbf{K}^{SV}$: materials stored in shared vessels
- $\mathbf{K}_j^{ST-}/\mathbf{K}_j^{ST+}$: materials that can be stored in $j \in \mathbf{J}^{PU}$ before/after the execution of a batch on j
- $\mathbf{J}_k^{PU-}/\mathbf{J}_k^{PU+}$: processing units where material k can be stored before/after the execution of a batch
- $\mathbf{I}_j^{ST-}/\mathbf{I}_j^{ST+}$: storage tasks executed in processing unit j before/after a batch starts/ends

Figure 8.2 illustrates these subsets using a network with two processing units (U1 and U2), four tasks (T11, T11, T21, and T22), six materials (F1, F2, I1, I2, P1, and P2), four dedicated storage vessels (DF1, DI1, DP1, and DP2), and two shared storage vessels

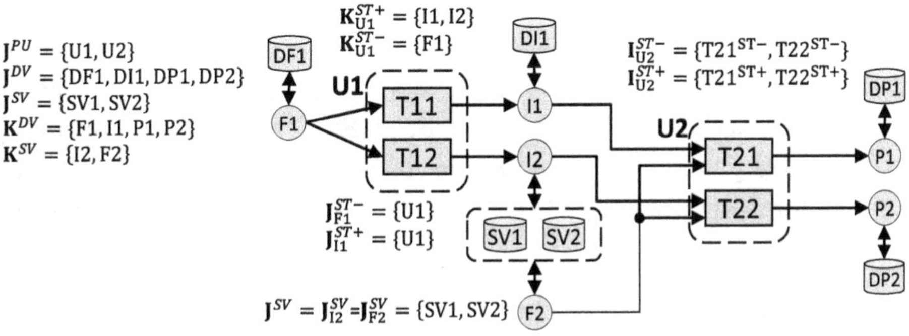

Figure 8.2 Illustration of subsets introduced to model storage in shared vessels and processing units.

(SV1 and SV2). Note that if X is the task that consumes (produces) materials that can be temporarily stored in the processing unit before (after) its execution, then the tasks representing the corresponding storage task is, in general, represented as X^{ST-} (X^{ST+}). Please refer back to Figure 8.2 as you go through Section 8.2.

8.2.1 Storage in Shared Vessels

Variable $X^{ST}_{jkn} \in \{0, 1\}$ is introduced to denote storing of material k in vessel $j \in \mathbf{J}^{SV}$ during period n, and variable $I_{jkn} \in \mathbb{R}_+$ denotes the inventory of material k in vessel j during period n. Only one material can be stored in a vessel at any time :

$$\sum_{k\in\mathbf{K}_j} X^{ST}_{jkn} \leq 1, \;\; j \in \mathbf{J}^{SV}, n \tag{8.2}$$

and if stored, then its inventory cannot exceed the vessel capacity,

$$I_{jkn} \leq \chi^M_k X^{ST}_{jkn}, \;\; j \in \mathbf{J}^{SV}, k \in \mathbf{K}_j, n. \tag{8.3}$$

Note that, in the general case, a material can be stored in multiple vessels, which means that the amount of material consumed or produced by a task may be coming from multiple vessels. If $\mathbf{J}^{SV}_k \subseteq \mathbf{J}^{SV}$ is the set of vessels material k can be stored in, and there is no dedicated storage vessel for k, then the material balance, which replaces (7.6), becomes

$$\sum_{j\in\mathbf{J}^{SV}_k} I_{jk,n+1} = \sum_{j\in\mathbf{J}^{SV}_k} I_{jkn} + \sum_{i\in\mathbf{I}^+_k} \rho_{ik} B_{i,n-\tau_{ij}} + \sum_{i\in\mathbf{I}^-_k} \rho_{ik} B_{in} + \xi_{ik} - S_{kn}, \quad k, n. \tag{8.4}$$

If no materials have dedicated vessels, then the STN-based model consists of (7.4), (7.5), (8.2) through (8.4), and the objective function. Interestingly, (8.2) can be viewed as the counterpart of (7.4), ensuring correct material-vessel assignments; (8.3) can be viewed as the counterpart of (7.5), ensuring that vessel capacity is not exceeded; and (8.4) brings processing units and storage vessels together – the consumption ($\sum_{i\in\mathbf{I}^-_k} \rho_{ik} B_{in}$) and production ($\sum_{i\in\mathbf{I}^+_k} \rho_{ik} B_{i,n-\tau_{ij}}$) terms depend on the task-unit decisions, whereas the storage terms ($\sum_{j\in\mathbf{J}^{SV}_k} I_{k,n+1}$ and $\sum_{j\in\mathbf{J}^{SV}_k} I_{kn}$) depend on the material-vessel decisions.

If some materials, $k \in \mathbf{K}^{DV}$, can be stored only in dedicated vessels and some materials, $k \in \mathbf{K}^{SV}$, can be stored only in shared vessels, with $\mathbf{K}^{DV} \cap \mathbf{K}^{SV} = \emptyset$, then (7.6) is used for $k \in \mathbf{K}^{DV}$ and (8.4) for $k \in \mathbf{K}^{SV}$. If some materials have a dedicated vessel but can also be stored in subset of shared vessels (i.e., $k \in \mathbf{K}^{DV} \cap \mathbf{K}^{SV} = \emptyset$), then the terms $\sum_{j\in\mathbf{J}^{SV}_k} I_{jk,n+1}$ and $\sum_{j\in\mathbf{J}^{SV}_k} I_{jkn}$ in (8.4) are replaced by $(I_{k,n+1} + \sum_{j\in\mathbf{J}^{SV}_k} I_{jk,n+1})$ and $(I_{kn} + \sum_{j\in\mathbf{J}^{SV}_k} I_{jkn})$, respectively, where I_{kn} represents the inventory level in the vessel dedicated to k.

8.2.2 Storage in Processing Units and Material Flows

To model systems where the material(s) produced by a task can be temporarily stored in the processing unit where the task took place, we introduce *storage* tasks $i \in \mathbf{I}^{ST}$ with the associated assignment variables $X_{in}^{ST} \in \{0, 1\}$, inventory variables, $I_{jkn}^{ST+} \in \mathbb{R}_{+}$, and flow variables $F_{jj'kn} \in \mathbb{R}_{+}$ representing transfer of material $k \in \mathbf{K}_i^{+}$ from processing unit j to storage vessel $j' \in \mathbf{J}_k^{SV}$.[1] If the output of processing task i can be stored in j, then we introduce storage task i^{ST+}, with processing time equal to one period, during which materials produced by i can be stored in j. The inventory, I_{jkn}^{ST+}, of material $k \in \mathbf{K}_i^{+}$ in processing unit j during period n is calculated as follows:

$$I_{jk,n+1}^{ST+} = I_{jkn}^{ST+} + \sum\nolimits_{i \in \mathbf{I}_j \cap \mathbf{I}_k^{+}} \rho_{ik} B_{i,n-\tau_{ij}} - \sum\nolimits_{j' \in \mathbf{J}_k^{SV}} F_{jj'kn}, \quad j \in \mathbf{J}^{PU}, k \in \mathbf{K}_j^{ST+}, n, \tag{8.5}$$

where $\mathbf{K}_j^{ST+}$ is the set of materials that can be stored in unit j . Alternatively, (8.5) can be defined for $k \in \mathbf{K}$, $j \in \mathbf{J}_k^{PU+}$, where $\mathbf{J}_k^{PU+}$ is the set of processing units where material k can be stored as a product of batch that finished in j. A unit belongs in $\mathbf{J}_k^{PU+}$ if there is a task producing k which can be carried out in j (i.e., $\exists i \in \mathbf{I}_j \cap \mathbf{I}_k^{+}$).

The inventory of a material that can be stored in $j \in \mathbf{J}^{PU}$, can be nonnegative only if a compatible storage task takes place:[2]

$$I_{jkn}^{ST+} \leq \sum\nolimits_{i \in \mathbf{I}_j \cap \mathbf{I}_k^{+}} \left(\rho_{ik} \beta_{ij}^{MAX}\right) X_{i^{ST+}n}^{ST}, \quad j \in \mathbf{J}^{PU}, k \in K_j^{ST+}, n, \tag{8.6}$$

where $\rho_{ik}\beta_{ij}^{MAX}$ is the maximum amount of material k that can be produced by a batch of task i. Note that, to maintain consistency with variables X_{jkn}^{ST} used for storage in vessels, $X_{i^{ST+}n}^{ST} = 1$ if the storage task is executed during period n, rather than starts at time point n.

The (clique) assignment constraint for processing units is now written as follows:

$$\sum\nolimits_{i \in \mathbf{I}_j} \sum\nolimits_{n' \in \mathbf{N}_{in}^{U}} X_{in'} + \sum\nolimits_{i \in \mathbf{I}_j^{ST}} X_{i,n+1}^{ST} \leq 1, \quad j, n, \tag{8.7}$$

where $\mathbf{I}_j^{ST+}$ is the set of storage tasks executed in processing unit j after a batch is finished, and $\mathbf{N}_{in}^{U} = \{n - \tau_i + 1, \ldots, n\}$. If a processing task is executed during period $n+1$ (i.e., $\sum_{i \in \mathbf{I}_j} \sum_{n' \in \mathbf{N}_{in}^{U}} X_{in'} = 1$), then no storage task can be carried out and thus all inventories should be zero through (8.6).

[1] The processing features described in this subsection can also be viewed as a generalization of the no intermediate storage (NIS) policy. The basic case of NIS, where a task produces one or multiple products that are all removed at the same time, can be readily modeled using a continuous time model. Interestingly, the basic discrete time STN- and RTN-based models, discussed in the previous chapter, cannot account for NIS without significant modifications, even when a task produces a single product. In fact, if no storage vessels are available for a material, the constraints of both STN and RTN discrete time models enforce a zero wait policy. However, as we discuss in this subsection, these models can be extended to account for NIS, including some more complex settings, e.g., multiple material transfers.

[2] Note that during a storage task the inventory is positive. Later, using the idea in the original STN papers, we will model storage through tasks that *consume* materials when they start and *produce* them when they end, so that the inventory during the execution of a storage task is zero.

The basic material balance of the STN-based model, representing inventory in a dedicated storage vessel ((7.6) in the previous chapter), is written as follows if $|\mathbf{J}_i| = 1$:

$$I_{k,n+1} = I_{kn} + \sum_{i \in \mathbf{I}_k^+} \rho_{ik} B_{i,n-\tau_i} + \sum_{i \in \mathbf{I}_k^-} \rho_{ik} B_{in} + \xi_{kn} - S_{kn} \leq \chi_k^M, \quad k, n. \qquad (8.8)$$

If a dedicated storage vessel is available for a material and no shared vessels (i.e., $\mathbf{J}_k^{SV} = \emptyset$), then flow variables $F_{jj'kn}$ in (8.5) can be replaced by F_{jkn}, denoting a flow from processing units to the dedicated vessel for k and, consequently, the term $\sum_{j' \in \mathbf{J}_k^{SV}} F_{jj'kn}$ in (8.5) is replaced by F_{jkn}. Thus, (8.8) is rewritten as follows:

$$I_{k,n+1} = I_{kn} + \sum_{j \in \mathbf{J}_k^{PU}} F_{jkn} + \sum_{i \in \mathbf{I}_k^-} \rho_{ik} B_{in} + \xi_{kn} - S_{kn} \leq \chi_k^M, \quad k, n. \qquad (8.9)$$

Note that if there is no storage in a processing unit, we have $I_{jk,n+1}^{ST+} = I_{jkn}^{ST+} = 0$ which, through (8.5), implies $\sum_{i \in \mathbf{I}_j \cap \mathbf{I}_k^+} \rho_{ik} B_{i,n-\tau_i} = \sum_{j' \in \mathbf{J}_k^{SV}} F_{jj'kn}$, which for the case of dedicated storage becomes $\sum_{i \in \mathbf{I}_j \cap \mathbf{I}_k^+} \rho_{ik} B_{i,n-\tau_i} = F_{jkn}$. Therefore, (8.9) reduces to (8.8) because for the first summation in the RHS of (8.9) we have

$$\sum_{j \in \mathbf{J}_k^{PU}} F_{jkn} = \sum_{j \in \mathbf{J}_k^{PU}} \left[\sum_{i \in \mathbf{I}_j \cap \mathbf{I}_k^+} \rho_{ik} B_{i,n-\tau_i} \right] = \sum_{j \in \mathbf{J}_k^{PU}, i \in \mathbf{I}_j \cap \mathbf{I}_k^+} \rho_{ik} B_{i,n-\tau_i}$$

and the summation over all processing units $\left(j \in \mathbf{J}_k^{PU}\right)$ and all tasks that can be carried out in these units $\left(i \in \mathbf{I}_j\right)$ returns all tasks, so

$$\sum_{j \in \mathbf{J}_k^{PU}} F_{jkn} = \sum_{i \in \mathbf{I} \cap \mathbf{I}_k^+} \rho_{ik} B_{i,n-\tau_i} = \sum_{i \in \mathbf{I}_k^+} \rho_{ik} B_{i,n-\tau_i}.$$

The case where material k can be stored in shared storage vessels $j \in \mathbf{J}_k^{SV}$ will be discussed in Section 8.2.3.

We note the following regarding the modeling of storage in processing units:

(1) The same material can be transferred out of the processing unit at multiple time points.

(2) *Selective* transfer of material out of the processing unit is allowed; for example, $k \in \mathbf{K}_i^+$ can be transferred immediately after a batch ends, while $k' \in \mathbf{K}_i^+$ can be stored in the processing unit.

(3) A storage task assignment where two different storage tasks are carried out in consecutive periods satisfies the aforementioned equations. For example, if two tasks $i, i' \in \mathbf{I}_j$ produce the same material (e.g., $\mathbf{K}_i^+ = \mathbf{K}_{i'}^+ = \{k\}$), then I_{jkn}^{ST+} can be positive if $X_{i^{ST+},n}^{ST} = 1$ or $X_{i'^{ST+},n}^{ST} = 1$. While two different storage tasks should not be allowed to take place in consecutive time periods, the schedules obtained by models employing (8.5) through (8.7) and (8.9) are *implementable,*[3] so we do not add constraints to exclude this assignment.

To remove some of the flexibility described in points (1) and (2), one could define more *restrictive* storage tasks and/or include additional equations. Multiple material transfers,

[3] By *implementable*, we mean that the obtained solution can be executed. In this case, it means that there is a feasible schedule that results in the predicted inventory profiles in the units.

point (1), will be discussed in Section 8.2.4. Selective material transfers, point (2), can be avoided by treating storage tasks as processing tasks with a variable batchsize, $B_{in}^{ST} \in \mathbb{R}_+$; input, $\mathbf{K}_i^-$ and output, $\mathbf{K}_i^+$, materials; and conversion coefficients, ρ_{ik}^{ST}. If i' is the storage task following processing task i (i.e., $i' = i^{ST+}$), then $\mathbf{K}_{i'}^- = \mathbf{K}_{i'}^+ = \mathbf{K}_{i'} = \mathbf{K}_i^+$, and $\rho_{i'k}^{ST} = \rho_{ik}$ for $k \in \mathbf{K}_i^+$. In other words, the storage task consumes and produces the same materials, the outputs of the corresponding processing task, according to the ratio of the conversion (production) coefficients of the processing task. In this case, the material balance for a material stored in a processing unit after the execution of a batch becomes

$$0 = \sum\nolimits_{i \in \mathbf{I}_j \cap \mathbf{I}_k^+} \rho_{ik} B_{i,n-\tau_{ij}} + \sum\nolimits_{i \in \mathbf{I}_{jk}^{ST+}} \rho_{ik}^{ST} (B_{i,n-1} - B_{in}) - \sum\nolimits_{j' \in \mathbf{J}_k^{SV}} F_{jj'kn}, \quad j \in \mathbf{J}_k^{PU}, k \in \mathbf{K}_j^{ST+}, n, \tag{8.10}$$

where $\mathbf{I}_{jk}^{ST+} = \mathbf{I}_j^{ST+} \cap \mathbf{I}_k^{ST}$ is the set of storage tasks that can be carried out in j and involve storage of material k; and rather than defining consumption (< 0) and production (> 0) coefficients, we use only $\rho_{ik}^{ST} > 0$ and subtract B_{in}. Compared to (8.5), in (8.10), we have removed inventory variables or, alternatively, set them to zero; and added the second term in the RHS. Setting inventory variables to zero enforces that materials are stored at the ratio at which they were produced, that is, if there is withdrawal, then the *composition* of the *lumped outgoing flow*[4] is the same as the composition of the mixture in the unit. The second term in the RHS represents the net effect of storage: inventory is *returned* at n if a storage batch started at $n-1$, and is depleted if a storage task starts at n. Note that two consecutive storage batches may have different batchsizes. Finally, since storage tasks are treated as processing tasks, the assignment constraint becomes

$$\sum\nolimits_{i \in \mathbf{I}_j} \sum\nolimits_{n' \in \mathbf{N}_{in}^U} X_{in'} + \sum\nolimits_{i \in \mathbf{I}_j^{ST+}} X_{in}^{ST} \leq 1, \quad j, n. \tag{8.11}$$

8.2.3 Material Storage Extensions

We consider systems where (1) both input and output materials can be temporarily stored in processing units, and (2) materials can be stored in multiple storage vessels.

If storage is also allowed prior to the execution of a task, then we introduce a preexecution storage task, i^{ST-}, along with variables $I_{jkn}^{ST-} \in \mathbb{R}_+$ denoting inventory levels in the units prior to task execution and flows from storage vessels to processing units; and add constraints similar to (8.5) and (8.6):

$$I_{jk,n+1}^{ST-} = I_{jkn}^{ST-} + \sum\nolimits_{j' \in \mathbf{J}_k^{SV}} F_{j'jkn} + \sum\nolimits_{i \in \mathbf{I}_j \cap \mathbf{I}_k^-} \rho_{ik} B_{in}, \quad j \in \mathbf{J}^{PU},\ k \in \mathbf{K}_j^{ST-},\ n \tag{8.12}$$

$$I_{jkn}^{ST-} \leq \sum\nolimits_{i \in \mathbf{I}_j \cap \mathbf{I}_k^+} \left(-\rho_{ik} \beta_{ij}^{MAX}\right) X_{i^{ST-}n}^{ST}, \quad j \in \mathbf{J}^{PU},\ k \in \mathbf{K}_i^{ST-},\ n, \tag{8.13}$$

[4] Material transfer from a processing unit may involve multiple outgoing flows; as we have seen, a unit where a task produces two output materials feeds two storage vessels. We use the term *lumped outgoing flow* to denote the total amount of all materials leaving a processing unit at a given time.

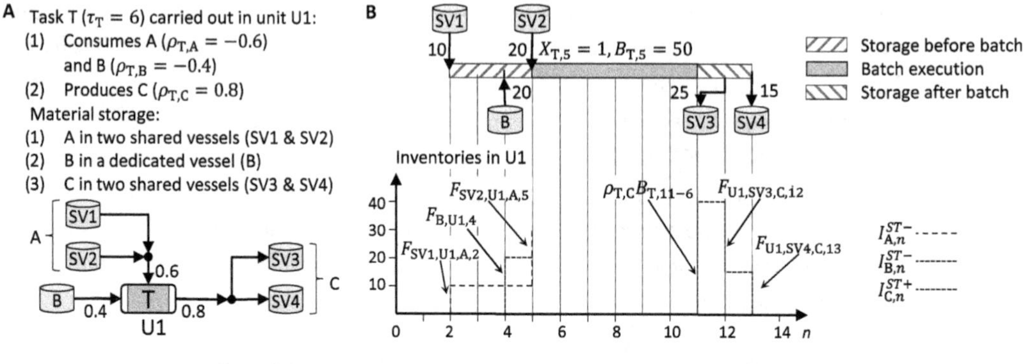

Figure 8.3 Material storage in processing units. (A) Task, data, and representation of the toy example used for illustration. Materials are not shown explicitly (as in STN); they are replaced by the vessels they can be stored in. (B) Depiction of material transfers (top panel) and corresponding inventory levels in the processing unit (bottom panel); some nonzero terms in the material balance are shown.

where $\mathbf{K}_j^{ST-}$ is the set of materials that can be stored in processing unit j prior to a batch execution. The generalization of (8.7) is

$$\sum\nolimits_{i\in\mathbf{I}_j}\sum\nolimits_{n'\in\mathbf{N}_{in}^U} X_{in'} + \sum\nolimits_{i\in\mathbf{I}_j^{ST}} X_{i,n+1}^{ST} \le 1, \quad j,n, \tag{8.14}$$

where $\mathbf{I}_j^{ST} = \mathbf{I}_j^{ST-} \cup \mathbf{I}_j^{ST+}$ is the set of all storage tasks. Figure 8.3 illustrates the modeling of storage in processing units before and after the execution of a task.

For materials stored in a dedicated storage vessel, we also replace the summation of $F_{j'jkn}$ with F_{kjn} in (8.12) (since there is only one storage vessel for k) and modify (8.9) to model material transfer to processing units through flow variables:

$$I_{k,n+1} = I_{kn} + \sum\nolimits_{j\in\mathbf{J}_k^{PU+}} F_{jkn} - \sum\nolimits_{j\in\mathbf{J}_k^{PU-}} F_{kjn} + \xi_{kn} - S_{kn} \le \chi_k^M, \quad k,n, \tag{8.15}$$

where $\mathbf{J}_k^{PU+}$ is the set of processing units where material k can be produced or temporarily stored after the execution of a task; and $\mathbf{J}_k^{PU-}$ is the set of processing units where material k can be consumed or temporarily stored before the execution of a task. Unit j belongs in $\mathbf{J}_k^{PU+}$ and $\mathbf{J}_k^{PU-}$ if there is $i \in \mathbf{I}_j$ with $k \in \mathbf{K}_i^+$ and $\mathbf{K}_i^-$, respectively.

The most general case arises when materials can be stored in multiple storage vessels (shared and dedicated) and processing units, both before and after task execution. To simplify the notation, we treat dedicated storage vessels as shared vessels with $|\mathbf{K}_j| = 1$. Recall that (8.4) accounts for the total transfer of material k (to all vessels where k can be stored) through, essentially, batchsize variables. Here, we replace this *lumped* inventory balance with a vessel-specific balance using unit-to-unit flows:

$$I_{jk,n+1} = I_{jkn} + \sum\nolimits_{j'\in\mathbf{J}_k^{PU+}} F_{j'jkn} - \sum\nolimits_{j'\in\mathbf{J}_k^{PU-}} F_{jj'kn} + F_{kjn}^{s} - S_{kjn} \le \chi_k^M, \quad k, j\in\mathbf{J}_k^{SV}, n, \tag{8.16}$$

where $F^S_{kjn} \in \mathbb{R}$ and $S_{kjn} \in \mathbb{R}_+$ are disaggregated material shipments and sales:

$$\xi_{kn} = \sum\nolimits_{j \in \mathbf{J}_k^{SV}} F^S_{kjn}, \quad k, n \tag{8.17}$$

$$S_{kn} = \sum\nolimits_{j \in \mathbf{J}_k^{SV}} S_{kjn}, \quad k, n. \tag{8.18}$$

Note that direct transfers between processing units are modeled as, essentially, dummy transfers via one or more storage vessels occurring instantaneously. The direct transfer can be computed post optimization. An alternative approach would be to introduce flow variables $F_{jj'kn}$ between processing units, and introduce additional terms in the material balances. For example, the inventory in a processing unit j prior/after the execution of a task will have an additional incoming/outgoing term, $\sum_{j' \in \mathbf{J}^{PU}} F_{j'jkn} / \sum_{j' \in \mathbf{J}^{PU}} F_{jj'kn}$.

Finally, we note that the introduction of flow variables allows us to account for unit connectivity constraints, that is, disallow material transfer between units that are not *connected* with the physical infrastructure that would allow such transfer. This is accomplished by defining flow variables only over the units connected to unit j.

8.2.4 Material Transfer Tasks

To simplify the presentation in this subsection, we assume that only shared storage vessels are available and that materials cannot be stored in processing units, that is, $\mathbf{I}^{ST} = \emptyset$ and $I^{ST-}_{jkn} = I^{ST+}_{jkn} = 0, \forall j \in \mathbf{J}^{PU}, k, n$. Thus, the inventory balance equations for the processing units, (8.5) and (8.12), become

$$\sum\nolimits_{i \in \mathbf{I}_j \cap \mathbf{I}_k^+} \rho_{ik} B_{i,n-\tau_{ij}} = \sum\nolimits_{j' \in \mathbf{J}_k^{SV}} F_{jj'kn}, \quad j \in \mathbf{J}^{PU}, k \in \mathbf{K}_j^{ST}, n \tag{8.19}$$

$$-\sum\nolimits_{i \in \mathbf{I}_j \cap \mathbf{I}_k^-} \rho_{ik} B_{in} = \sum\nolimits_{j' \in \mathbf{J}_k^{SV}} F_{j'jkn}, \quad j \in \mathbf{J}^{PU}, k \in \mathbf{K}_j^{ST}, n. \tag{8.20}$$

The interpretation of (8.19) is that the material produced by a task has to be immediately transferred to, potentially many, shared vessels. Similarly, (8.20) says that a material consumed by a task has to be delivered exactly on time from storage vessels.

So far, we have assumed that material transfers are instantaneous and do not require any shared resources. While these assumptions are reasonable in many systems, there are facilities or subsystems within a facility where the duration of material transfer activities and the associated resource constraints should be accounted for. To do so, we introduce the concept of a transfer task $i \in \mathbf{I}^{TR}$, associated with an origin unit, $j^-(i)$, and destination unit, $j^+(i)$. If $j^-(i) = j$ and $j^+(i) = j'$, then the task is also denoted by $j \to j'$. Note that we adopt material-independent transfer tasks, that is, when the transfer task between units j and j' is active, any compatible material can be transferred. If necessary, material-specific tasks, with associated origin, destination, and material, can be adopted, but this would increase the size of the model.

In terms of transfer duration, we assume that the rate at which material is transferred is such that all outgoing material can be withdrawn from the origin within a period, though the duration of the transfer itself can be longer. In other words, the withdrawal from the origin lasts one period, the charging of the destination unit lasts one period, but removal and addition of material may occur with a delay; it is most often assumed that the delay $(\theta_{jj'})$ is zero or one period. If transfer task i starts at n, then $X_{in}^{TR} = 1$ and the corresponding *batchsize* of the task can be positive:

$$\sum_k F_{jj'kn} \le \beta_j^{MAX} X_{j\to j',n}^{TR}, \quad (j \to j') = i \in \mathbf{I}^{TR}, n \tag{8.21}$$

Note that $F_{jj'kn}$ is positive at the point where the transfer starts; that is, it represents the (beginning of) material transfer and thus inventory decrease at the origin. If transfer task $j \to j'$ has no delay $(\theta_{j\to j'} = 0)$ and starts at n, then the inventory at the origin will decrease at n and the inventory at the destination will increase at $n+1$. Note that while the delay is zero, the processing time of the transfer task is $\tau_{j\to j'} = 1$.

While simultaneous transfer of multiple materials between two units does not violate (8.21), such transfer is infeasible because of the requirement to store only one material in a storage vessels – see (8.2) and (8.3). If j is a storage vessel, then it stores only one material at a given time and thus $F_{jj'kn}$ can be positive for only one k. If j is a processing unit, then j' is a storage vessel that can store only one material, so, again, at most one of the $F_{jj'kn}$ variables in the summation can be positive.

Before we present the inventory balances, we have to *modify* the tasks that require materials that have to be transferred. Consider a task that consumes a material delivered during the period immediately preceding the start of a batch and produces a material transferred during the period immediately following the completion of a batch (see Figure 8.4). Since the processing unit is occupied one period before and one period after the execution of a batch, the processing time should be adjusted, $\tau_i^{INT} = \tau_i + 2$, so no other batch is carried out during transfer operations. We will refer to the new tasks, which include material transfers, as the *integrated* tasks.

If a batch of an integrated task taking place in unit j starts at time point n $(X_{in} = 1, B_{in} > 0)$, then

(1) The transfer of the input materials, assuming $\tau_{j'\to j} = 1$ for all relevant shared vessels j', should start also at n, that is, (8.20) remains unchanged.

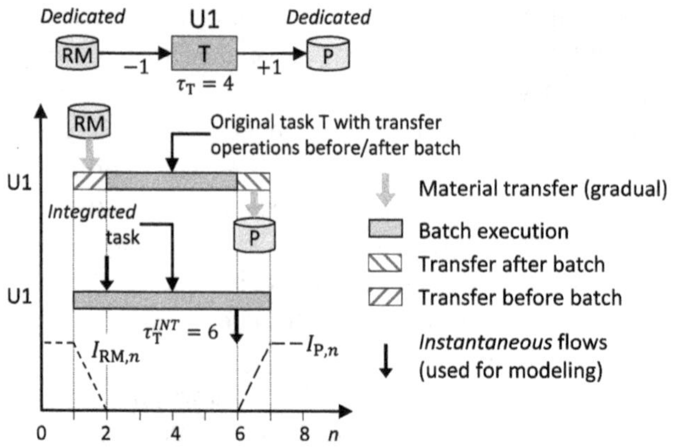

Figure 8.4 Modeling of an integrated task.

(2) The transfer of the output materials should start at $n+\tau_i^{INT}-1$, so it is completed by the end of the batch of the integrated task, which means that (8.19) becomes

$$\sum\nolimits_{i \in \mathbf{I}_j \cap \mathbf{I}_k^+} \rho_{ik} B_{i,n-\tau_i^{INT}+1} = \sum\nolimits_{j' \in \mathbf{J}_k^{SV}} F_{jj'kn}, \quad j \in \mathbf{J}_k^{PU}, k \in \mathbf{K}_j^{ST}, n. \qquad (8.22)$$

The inventory balances in the shared vessels are modified as follows:

$$I_{jk,n+1} = I_{jkn} + \sum\nolimits_{j' \in \mathbf{J}_k^{PU+}} F_{j'jk,n-\tau_{j' \to j}} - \sum\nolimits_{j' \in \mathbf{J}_k^{PU-}} F_{jj'kn} + F_{kjn}^S - S_{kjn} \le \chi_k^M, \quad k, j \in \mathbf{J}_k^{SV}, n \qquad (8.23)$$

The only difference between (8.16) and (8.23) is that the incoming flows (first summation in the RHS) correspond to transfer tasks that started $\tau_{j' \to j}$ periods earlier.

To summarize, a transfer task is activated as follows: the execution of a processing task leads to a $B_{in} > 0$, which through (8.20) and (8.22) leads to positive flow variables, which in turn impact the material balances, (8.23), and activate X_{in}^{TR} variables through (8.21).

Different types of resource constraints can be enforced. If specific shared transfer equipment is necessary, then we add

$$\sum\nolimits_{i \in \mathbf{I}_j^{TR}} X_{in}^{TR} \le 1, \quad j \in \mathbf{J}^{TR}, n, \qquad (8.24)$$

where $\mathbf{J}^{TR}$ is the set of shared transfer equipment (e.g., pipeline or pump) and $\mathbf{I}_j^{TR}$ is the set of transfer tasks that require equipment $j \in \mathbf{J}^{TR}$. For example, it is common to enforce that a storage vessel participates in only one withdrawing or charging activity at any given time. If a transfer task can be carried out using alternative equipment, then the task is decomposed into two tasks, or variables X_{in}^{TR} are indexed by j and the equations are modified, exactly as in the case of processing tasks. Note that if transfer of material k from processing unit j to storage vessel j', via task $j \to j'$ (with $\tau_{j \to j'} = 1$), occurs during period $n+1$, then (1) $X_{j \to j',n}^{TR} = 1$; (2) $F_{jj'kn} > 0$; (3) the inventory in j' will increase at $n+1$; and (4) j' should be assigned to store material k during period $n+1$ (i.e., $X_{j'k,n+1}^{ST} = 1$). Example 8.2 illustrates the modeling of transfer tasks.

Example 8.2 Modeling of Transfer Tasks We consider a network with two tasks (T1 and T2), two units (U1 and U2), and four materials (A, B, C, and D). Task T1, with $\tau_{T1} = 3$, is carried out in unit U1. It consumes materials A $(\rho_{T1,A} = -0.6)$ and B $(\rho_{T1,B} = -0.4)$ and produces material C $(\rho_{T1,C} = 0.8)$, which is stored in a dedicated vessel DC. The transfer from U1 to DC is modeled by transfer task U1→DC with $\tau_{U1 \to DC} = 1$. Task T2, with $\tau_{T2} = 4$, is carried out in unit U2, and it consumes C $(\rho_{T2,C} = -1)$. The transfer task from DC to U2 is DC→U2, with $\tau_{DC \to U2} = 2$. The initial inventory of C is 10. The representation of the network is given in Figure 8.5A. The Gantt chart of a solution, including transfer tasks, along with information regarding amounts transferred, are given in Figure 8.5B. Note that U1→DC $(\tau_{U1 \to DC} = 1)$ requires both U1 and DC during period 4, whereas DC→U2 $(\tau_{DC \to U2} = 2)$ draws from DC during period 8 only, and charges U2 during period 9

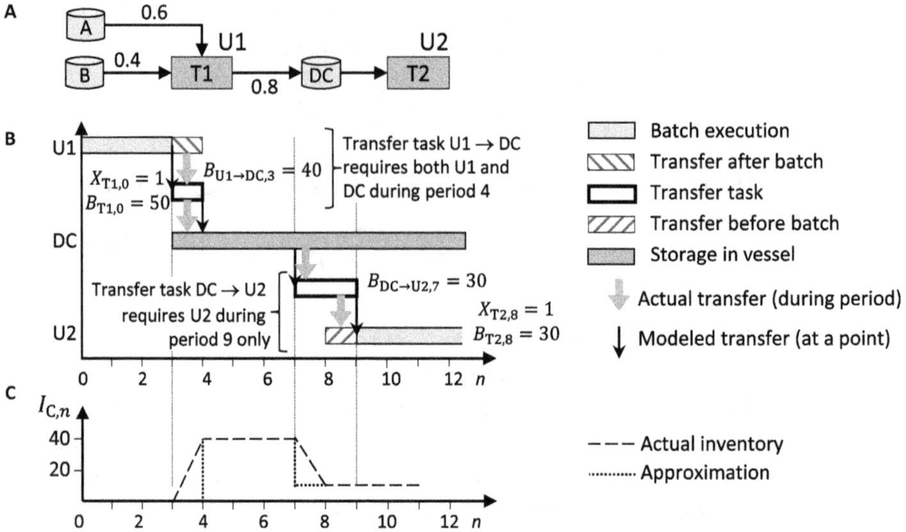

Figure 8.5 Modeling of transfer activities. (A) STN representation of example network. (B) Gantt chart of solution for units U1 and U2, and vessel DC. (C) Actual and calculated inventory profiles of material C.

only. Thus, it requires U2 only during period 9. The evolution of inventory of material C is shown in Figure 8.5C. Note that the actual inventory increases and decreases gradually; for example, material is moved into DC over one period (gray thick arrows). The calculated inventory changes in a stepwise fashion, though, because withdrawals and charges are modeled to occur instantaneously (thin black arrows).

If (a subset of) transfer tasks are subject to an aggregate rate capacity (e.g., the total pumping capacity is bounded), then we enforce

$$\sum_{(j\to j')\in \mathbf{I}_r^{TR}} \sum_k F_{jj'kn} \le \chi_r^{TR}, \quad r \in \mathbf{R}^{TR}, n, \tag{8.25}$$

where $\mathbf{R}^{TR}$ is the set of continuous renewable resources used for material transfers (e.g., pumping capacity), χ_r^{TR} is the capacity of such resource r, and $\mathbf{I}_r^{TR}$ is the subset of transfer tasks requiring this resource.

If the pumping rate is such that material cannot be transferred within a period, then material withdrawal (charging) from (into) the origin (destination) unit lasts more than one period. In this case, the outline of the approach we will have to follow is the following:

(1) Inventory variables I_{jkn}^{ST-} and I_{jkn}^{ST+} are maintained and the corresponding inventory balances in the processing units are written (modified versions of (8.5) and (8.12)).

(2) Positive flow variables activate transfer task binary variables as in (8.21).

(3) Processing tasks are not modified; if a processing task is executed on a unit, the unit cannot engage in a transfer task (either as the origin or the destination), though multiple transfer tasks may occur simultaneously.

(4) Constraints are written to enforce that transfer tasks require both the origin and destination units.

(5) Constraints are written to enforce that incoming transfer batches are followed by a processing task, and the latter are followed by outgoing transfer tasks.

(6) Binaries and flow variables are used to express resource constraints.

8.3 Setups and Task Families

In this section, we discuss a wide range of *auxiliary* activities, which, though not directly leading to material conversion, are necessary for production. In Section 8.3.1, we discuss unit setups, and in Section 8.3.2 task setups. In Section 8.3.3, we discuss task families, a generalization of the concept of product families discussed in Chapter 3 in the context of single-unit problems. We will mostly employ the RTN formulation, which offers a more natural way to model the aforementioned features. To simplify the discussion, we consider problems with dedicated storage vessels only, no material storage in processing units, no material consumption and production during the execution of a task, and instantaneous and unconstrained material transfers. Thus, we use the notation introduced in Chapter 7, without the generalizations presented in Sections 8.1 and 8.2.

8.3.1 Unit Setups

A unit setup occurs the first time a processing unit is used and it is sequence independent. The setup time is σ_j, and the setup cost is γ_j^{SET}. In the RTN formulation, where units are represented as resources, the unit setup cost/time is σ_r/γ_r^{SET}.

In the STN-based model, unit setups can be modeled by introducing $\hat{X}_{jn} \in \{0, 1\}$ denoting the start of the setup of unit j, adding a constraint enforcing that a unit setup is carried out at most once,

$$\sum\nolimits_n \hat{X}_{jn} \leq 1, \quad j, \tag{8.26}$$

and modifying the unit assignment constraint to allow task assignments only after the setup task has been completed:

$$\sum\nolimits_{i \in \mathbf{I}_j} \sum\nolimits_{n' \in \mathbf{N}_{in}^U} X_{in'} \leq \sum\nolimits_{n' \leq n - \sigma_j} \hat{X}_{jn'}, \quad j, n, \tag{8.27}$$

where $\mathbf{N}_{in}^U = \{n - \tau_i + 1, \ldots, n\}$.

Alternatively, a second variable $\bar{X}_{jn} \in \{0, 1\}$ is introduced to indicate that a unit has been setup,

$$\bar{X}_{j,n+1} = \bar{X}_{jn} + \hat{X}_{j,n-\sigma_j}, \quad j, n, \tag{8.28}$$

and the assignment constraint is written as

$$\sum\nolimits_{i \in \mathbf{I}_j} \sum\nolimits_{n' \in \mathbf{N}_{in}^U} X_{in'} \le \bar{X}_{j,n+1}, \quad j, n. \tag{8.29}$$

In the RTN-based model, a unit setup can be modeled by assuming that the capacity of the corresponding unit is zero at the beginning of the scheduling horizon, and introducing a setup task that increases the availability of the unit to 1; in the RTN parlance, a setup *produces* a unit resource. If $i = r^{SET}$ is the *unit setup task* for resource $r \in \mathbf{R}^U$, then the resource balance is expressed as follows,

$$R_{r,n+1} = R_{rn} + X_{i=r^{SET}, n-\sigma_r} + \sum\nolimits_{i \in \mathbf{I}_r} X_{i,n-\tau_i} - \sum\nolimits_{i \in \mathbf{I}_r} X_{in}, \quad r \in \mathbf{R}^U, n, \tag{8.30}$$

with initial availability, replacing R_{rn} when the equation is written for $n = 0$, $R_r^0 = 0$.

8.3.2 Task Setups

A task setup should be performed between the processing of batches of different tasks. The task setup time is σ_i, and the setup cost is γ_i^{SET}. Using the RTN representation, the processing unit is disaggregated into $| \mathbf{I}_r |$ task resources (task modes), each one capable of carrying out a single task, plus one *neutral* resource (neutral mode). Furthermore, one *start-setup* and one *end-setup* task are introduced per processing task to model transition between modes. The former consumes the neutral mode at its start and produces, after σ_i periods, the corresponding task mode. The latter is instantaneous ($\tau_i = 0$) and consumes one of the disaggregated task modes and produces the neutral mode. The *start-setup* and *end-setup* tasks corresponding to processing task i are denoted by i^{SS} and i^{ES}, respectively. If r^N is the neutral mode of unit resource r, then *start-setup* task $\mathrm{T1}^{SS}$ consumes r^N and produces T1, and *end-setup* task $\mathrm{T1}^{ES}$ consumes T1 and produces r^N.

We use r' to denote the newly introduced modes (resources). Let $\mathbf{R}_r^{UM}$ be the subset of task modes of unit resource $r \in \mathbf{R}^U$. If $\sigma_{r'}$ is the setup time of the processing task that can be carried out in task mode r' (i.e., $\sigma_{r'} = \sigma_i$ if $\mathbf{I}_{r'} = \{i\}$), then the resource balance for the task and neutral modes are written, respectively, as follows:

$$R_{r',n+1} = R_{r'n} + X_{i^{SS}, n-\sigma_{r'}} + X_{i,n-\tau_i} - X_{in} - X_{i^{ES},n}, \quad r \in \mathbf{R}^U, r' \in \mathbf{R}_r^{UM}, i = i(r'), n \tag{8.31}$$

$$R_{r',n+1} = R_{r'n} + \sum\nolimits_{i \in \mathbf{I}^{ES}} X_{in} - \sum\nolimits_{i \in \mathbf{I}^{SS}} X_{in}, \quad r' \in \mathbf{R}^{NU}, n, \tag{8.32}$$

where $\mathbf{R}^{UN}$ is the set of all neutral unit resources $\mathbf{R}^{UN} = \cup_{r \in \mathbf{R}^U} \{r^N\}$; $i(r')$ is the task that can be carried out in task more r'; and $\mathbf{I}^{SS}$ and $\mathbf{I}^{ES}$ are the sets of start-setup and end-setup tasks. The modeling of task setups is illustrated in Example 8.3.

Example 8.3 Modeling of Task Setups We consider a unit, U, that carries out four tasks, $\mathbf{I}_U = \{\mathrm{T1}, \mathrm{T2}, \mathrm{T3}, \mathrm{T4}\}$, each one requiring a setup. The processing (τ_i) and setup (σ_i) times are given in Table 8.1. The operation of Unit U can be modeled through its disaggregation into four task modes, for simplicity denoted by T1, T2, T3, T4 with

Table 8.1 Processing and setup times.

Task	τ_i	σ_i
T1	5	2
T2	6	3
T3	10	4
T4	8	1

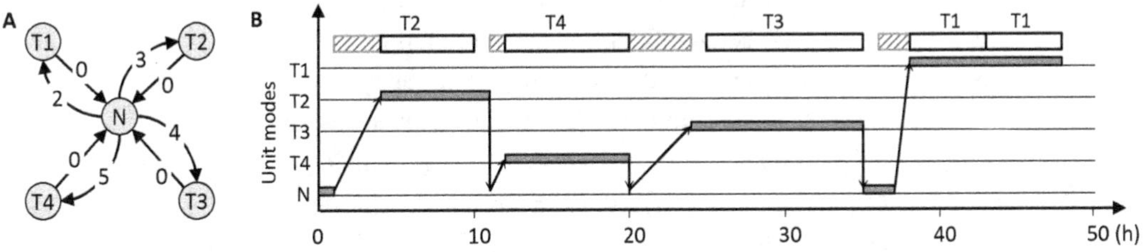

Figure 8.6 Modeling of task setups using task and neutral modes. (A) Modeling of mode consumption/production by startup and shutdown tasks, represented as graph edges. (B) Gantt chart of an example solution.

$\mathbf{I}_{T1} = \{T1\}$, $\mathbf{I}_{T2} = \{T2\}$, $\mathbf{I}_{T3} = \{T3\}$, and $\mathbf{I}_{T4} = \{T4\}$; and a neutral mode N with $\mathbf{I}_N = \emptyset$. If each mode is represented as a graph vertex, then start-setups are represented by N $\rightarrow i$ edges, while end-setups are represented by $i \rightarrow$ N edges, with the traversal time of N $\rightarrow i$ edges being equal to the setup times, while the traversal times of all $i \rightarrow N$ edges being equal to zero (see Figure 8.6A). Figure 8.6B shows a Gantt chart, including both batches and setups, where each mode is represented as a different resource. Note that a setup can appear anywhere between two batches of different tasks; for example, the setup for mode T4 is executed just before a batch of T4, while there is idle time between the setup for T3 and the execution of a batch of T3. Also, idle time can occur while the unit is on a task mode (e.g., after a batch of T2 is finished) or while in the neutral mode (e.g., before the setup to mode T1).

Note that the adoption of the neutral resource means that the modeling of task setups requires the replacement of one resource with $|\mathbf{I}_r| + 1$ resources, and the introduction of $2\ |\mathbf{I}_r|$ additional tasks.

8.3.3 Task Families

In Chapter 3, we discussed the single-unit problem where a set of products are produced in a single unit, each one through a product-specific task, and there is no changeover time and cost associated with a transition between products belonging in the same *product family*, while changeover times, costs, or both are positive for a transition between products (tasks) belonging to different families. We used the term *product* families, rather than *task* families, because each product was produced by a single product-specific task. In the general case, discussed here, a task may be associated with multiple products (e.g., a task producing an intermediate that is then converted to multiple products), and/or a set of tasks associated with a single product may be subject to very different processing restrictions. Thus, it is more appropriate to define *families of tasks* carried out in a unit.

We use the RTN formulation. The set of tasks that can be assigned to unit resource $r \in \mathbf{R}^U$, $\mathbf{I}_r$, is partitioned into subsets $\mathbf{I}_{rf}, f \in \mathbf{F}_r$, where $\mathbf{F}_r$ is the set of task families of resource r ($\mathbf{I}_r = \cup_{f \in \mathbf{F}_r} \mathbf{I}_{rf}$ and $\mathbf{I}_{rf} \cap \mathbf{I}_{rf'} = \emptyset, \forall f, f' \in \mathbf{I}_{rf}$). Pairs of tasks belonging to the same family have no setup or changeover times and costs, while the transition from a task of a family to a task of another family requires a setup time, σ_{rf}, or changeover time, $\sigma_{rff'}$, and incurs a setup cost, γ_{rf}^{SET}, or changeover cost, $\gamma_{rff'}^{CH}$.

Setups. If only setups have to be modeled for $r \in \mathbf{R}^U$ and $\mathbf{F}_r = \{(r,1), (r,2), \ldots, (r, |\mathbf{F}_r|)\}$, then the original unit resource r is replaced with $|\mathbf{F}_r| + 1$ resources, $\mathbf{R}_r = \{(r,1), (r,2), \ldots, (r, |\mathbf{F}_r|), r^N)\}$. The first $|\mathbf{F}_r|$ resources represent family *modes*; when at mode (r,f) the unit can carry out tasks belonging to family f only. The last resource represents the neutral mode; the original unit is in mode r^N during family setups. The next step is to introduce tasks to model setups. For each family mode, we introduce one *start-setup* task, i_{rf}^{SS}, with $\tau_{i_{rf}^{SS}} = \sigma_{rf}$; and one *end-setup* task, i_{rf}^{ES}, with $\tau_{i_{rf}^{ES}} = 0$. The family mode balance is

$$R_{(r,f),n+1} = R_{(r,f)n} + X_{i_{rf}^{SS}, n-\sigma_{rf}} + \sum\nolimits_{i \in \mathbf{I}_{rf}} X_{i,n-\tau_i} - \sum\nolimits_{i \in \mathbf{I}_{rf}} X_{in} - X_{i_{rf}^{ES}, n}, \quad r \in \mathbf{R}^U, \ f \in \mathbf{F}_r, n \tag{8.33}$$

where a disaggregated family mode is represented by pair (r,f). If the sets of start-setup and end-setup tasks for the original resource unit $r \in \mathbf{R}^U$ are $\mathbf{I}_r^{SS}$ and $\mathbf{I}_r^{ES}$, then the resource balance for the neutral mode of r is written as follows:

$$R_{r^N,n+1} = R_{r^N,n} + \sum\nolimits_{i \in \mathbf{I}^{ES}} X_{in} - \sum\nolimits_{i \in \mathbf{I}^{SS}} X_{in}, \quad r \in \mathbf{R}^U, n. \tag{8.34}$$

Changeover Times Only. If there are no changeover or setup costs, then changeover times can, in principle, be modeled by simply calculating task-task changeover times, $\sigma_{jii'}$ ($\sigma_{rii'}$), and employing the STN-based (RTN-based) constraints we have already presented for the modeling of changeover times. Changeover times between tasks in different families are calculated as follows (for the RTN-based model):

$$\sigma_{rii'} = \sigma_{rff'}, \quad r \in \mathbf{R}^U, \quad f \in \mathbf{F}_r, \quad f' \in \mathbf{F}_r, \quad i \in \mathbf{I}_{rf}, \quad i' \in \mathbf{I}_{rf'}. \tag{8.35}$$

Note that since costs do not have to be accounted for, no binary variables denoting changeovers are needed. However, changeovers should be enforced among all tasks carried out in a unit, which may lead to an unnecessarily large formulation, especially when the number of tasks is large but the number of families is small. (Why?) An alternative approach is to define a mode for each task family and introduce changeover tasks that consume a mode and produce another (see the next paragraph).

Changeover Costs. The original unit resource, $r \in \mathbf{R}^U$, is disaggregated into $|\mathbf{F}_r|$ *family* modes (resources) and $|\mathbf{F}_r| \cdot (|\mathbf{F}_r| - 1)$ new *changeover* tasks, $i \in \mathbf{I}_r^{CH}$, consuming a mode and producing another, are introduced. If $i_{rff'}^{CH}$ is the task representing the transition from family mode f to mode f' of resource r, then $\tau_{i_{rff'}^{CH}} = \sigma_{rff'}$. Family mode f is *produced* when a changeover to mode f $(i \in \mathbf{I}_{rf}^{CH+})$ ends or a processing task $(i \in \mathbf{I}_{rf})$ in the family is completed, and it is consumed when a changeover from mode f $(i \in \mathbf{I}_{rf}^{CH-})$ or a processing task in the family $(i \in \mathbf{I}_{rf})$ starts:

$$R_{rf,n+1} = R_{rfn} + \sum\nolimits_{i \in \mathbf{I}_{rf}^{CH+} \cup \mathbf{I}_{rf}} X_{i,n-\tau_i} - \sum\nolimits_{i \in \mathbf{I}_{rf}^{CH-} \cup \mathbf{I}_{rf}} X_{in}, \quad r \in \mathbf{R}^U,\ f \in \mathbf{F}_r, n. \quad (8.36)$$

The idea of family modes is illustrated through Example 8.4. Note that compared to the case of families with setup costs, here we introduce $|\mathbf{F}_r|$, rather than $|\mathbf{F}_r| + 1$, resources, but the number of additional tasks is $|\mathbf{F}_r| \cdot (|\mathbf{F}_r| - 1)$ instead of $2|\mathbf{F}_r|$, which can be significantly larger; for example, if $|\mathbf{F}_r| = 10$, 90 changeover tasks are required instead of 20 setup tasks ($|\mathbf{I}_r^{SS}| = 10$ and $|\mathbf{I}_r^{ES}| = 10$).

Objective Function. The following terms have to be added to the objective function if setup and changeover costs are present:

$$C^{FAM} = \sum\nolimits_n \left(\sum\nolimits_{i \in \mathbf{I}^{SS}} \gamma_i^F X_{in} + \sum\nolimits_{i \in \mathbf{I}^{CH}} \gamma_i^F X_{in} \right), \quad (8.37)$$

where $\mathbf{I}^{SS} = \cup_r \mathbf{I}_r^{SS}$ includes all *start-setup* tasks, $\mathbf{I}^{CH} = \cup_r \mathbf{I}_r^{CH}$ is the set of all family changeover tasks, and the fixed cost γ_i^F of family setup and changeover tasks is

$$\gamma_{i_{rf}^{SS}}^F = \gamma_{rf}^{SET}, \quad r \in \mathbf{R}^U,\ f \in \mathbf{F}_r; \quad \gamma_{i_{rff'}^{CH}}^F = \gamma_{rff'}^{CH}, \quad r \in \mathbf{R}^U,\ f \in \mathbf{F}_r,\ f' \in \mathbf{F}_r. \quad (8.38)$$

Example 8.4 Modeling of Task Families We consider a unit, U, that carries out 14 tasks, grouped into four families, $\mathbf{F}_\mathrm{U} = \{F1, F2, F3, F4\}$. The changeover times between families, along with the tasks belonging in each family and the task processing times, are given in Table 8.2. If each mode is represented as a graph vertex, then changeovers are represented by edges with traversal times equal to changeover times (see Figure 8.7A). Figure 8.7B shows a Gantt chart, along with the utilization of the four family modes. The Gantt chart shows batches of different tasks

Table 8.2 Changeover times ($\sigma_{\mathrm{U},ff'}$), tasks in each family ($I_{\mathrm{U},f}$), and task processing times in parentheses.

	f'				
f	F1	F2	F3	F4	Tasks in each family
F1	–	2	3	4	T11 (2), T12 (3), T13 (3)
F2	3	–	2	1	T21 (2), T22 (1), T23 (3)
F3	3	4	–	2	T31 (2), T32 (3), T33 (1), T34 (4)
F4	2	3	2	–	T41 (3), T42 (2), T43 (1), T44 (4)

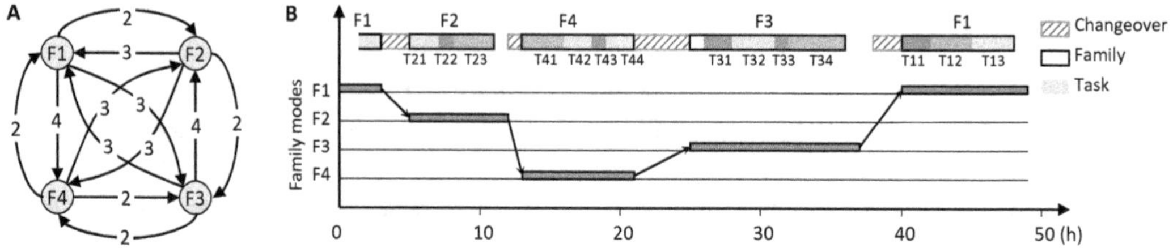

Figure 8.7 Modeling of task families with changeovers times. (A) Modeling of family mode consumption/production by changeover tasks, represented by arcs with changeover times shown along arcs. (B) Gantt chart and corresponding transitions among unit modes (tasks shown as blocks with different colors within a family); gray blocks in the family mode graph represent periods during which the unit is setup to carry out tasks in the corresponding family; they do not represent resource utilization.

within a family, with no idle time in between them, being executed while the unit is in the corresponding mode, along with the changeover times between families. Note, again, that a changeover can occur anywhere between the execution of batches from two different families. Also, while on a mode, the availability of the mode may be 1; for example, the availability of F3 during period 26 is 1, that is, the unit is idle.

8.4 Unit Deterioration and Maintenance

A feature of many (chemical and other) production systems is that the capacity or conversion coefficients of the tasks executed on a unit decrease over time, and thus maintenance is required to *restore* the unit in its *original state.*[5] For example, the activity of a catalyst can be restored after regeneration. There are three cases, in increasing order of complexity:

(1) Maintenance is required to restore the unit to its original state, but the deterioration of the unit does not affect its capacity nor the conversion coefficients of the tasks executed on the unit.

(2) The deterioration of the unit affects its capacity, that is, the execution of batches leads to a decrease in the maximum allowable batchsizes of batches carried out on the unit. The conversion coefficients (STN) or task-resource interaction coefficients (RTN) remain the same.

(3) The deterioration of the unit affects the conversion or task-resource interaction coefficients of the tasks carried out in the unit.

[5] The concept described by the term *state* here is clearly different from the concept of *state* in the STN representation. The use of the term in the present chapter is consistent with its use in Chapter 14, where it describes the *current condition* of the system, that is, its state if it were to be viewed as a dynamic system.

In all three cases, the state of the unit is restored through the execution of a maintenance task, but the three cases will be discussed separately. The maintenance task for unit j is denoted by i_j^M. We assume there is no material consumption and production during the execution of a batch; materials cannot be stored in processing units; there are no family setups and changeovers; and there are no shared utilities. We will illustrate the basic concepts using the STN representation.

8.4.1 No Effect on Capacity and Conversion

In addition to the introduction of maintenance tasks, variable, $H_{jn} \in \mathbb{R}_+$, is introduced to track the *state* (health) of unit j. Whenever a batch of a task $i \in \mathbf{I}_j$ is executed, H_{jn} decreases and a maintenance task can be used to increase its level to its maximum level η_j^{MAX}. The state of the unit, which is modeled essentially as an inventory level, cannot be lower than a threshold level η_j^{MIN}. The deterioration of the unit has task-specific fixed and variable components, modeled through θ_i^F and θ_i^V, respectively; and the maintenance task has processing time $\tau_{i_j^M} = \tau_j^M$. We assume that H_{jn} drops upon the completion of batches and it is restored to η_j^{MAX} upon completion of the maintenance task. If $H_{jn}^+ \in [0, \eta_j^{MAX}]$ represents the improvement in the health of the unit (*recovery*), after τ_j^M periods, due to a maintenance task that started at time point n, then the evolution of the state (health) of the unit is described as follows:

$$H_{j,n+1} = H_{jn} - \sum\nolimits_{i \in \mathbf{I}_j} \left(v_i^F X_{i,n-\tau_i} + v_i^V B_{i,n-\tau_i} \right) + H_{j,n-\tau_j^M}^+, \quad j,n \tag{8.39}$$

with

$$\eta_j^{MIN} \leq H_{jn} \leq \eta_j^{MAX}, \quad j,n. \tag{8.40}$$

Variable H_{jn}^+ can be positive only if the corresponding maintenance task starts at n:

$$H_{jn}^+ \leq \left(\eta_j^{MAX} - \eta_j^{MIN} \right) X_{i_j^M n}, \quad j,n. \tag{8.41}$$

To enforce $H_{jn} = \eta_j^{MAX}$ after a maintenance task is completed, we introduce a dummy maintenance variable $\widehat{H}_{jn}^+ \in \mathbb{R}_+$, satisfying

$$\widehat{H}_{jn}^+ \leq \left(\eta_j^{MAX} - \eta_j^{MIN} \right) \left(1 - X_{i_j^M n} \right), \quad j,n \tag{8.42}$$

and add

$$\eta_j^{MAX} - H_{jn} = H_{jn}^+ + \widehat{H}_{jn}^+, \quad j,n. \tag{8.43}$$

The LHS of (8.43) represents the deterioration of unit j during any period, that is, the *deviation* from η_j^{MAX}. This deviation is equal to either the dummy variable $\widehat{H}_{jn}^+$ or the recovery variable H_{jn}^+ when a maintenance task is carried out. Thus,

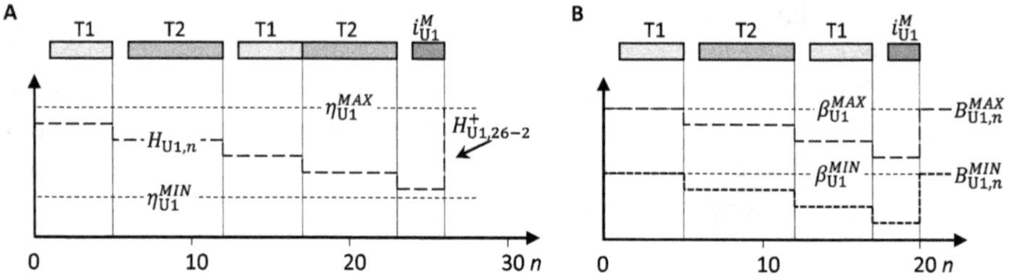

Figure 8.8 Modeling of unit state and minimum/maximum capacity evolution. (A) The state of U1 deteriorates upon completion of batches; it increases to its maximum at $n = 26$ due to the execution of the maintenance task, with $\tau_i = 2$, at $n = 24$. (B) The minimum and maximum capacity of U1 deteriorate upon batch completion; they are back at their maximum values at $n = 20$ upon completion of the maintenance task executed at $n = 18$.

(8.39) and (8.41) through (8.43) ensure that when there is maintenance, H_{jn} returns to η_j^{MAX}.

Finally, no processing tasks can be carried out while a unit undergoes maintenance:

$$\sum\nolimits_{i \in \mathbf{I}_j^G} \sum\nolimits_{n' \in \mathbf{N}_{in}} X_{in'} \leq 1, \quad j, n, \tag{8.44}$$

where $\mathbf{I}_j^G = \mathbf{I}_j \cup \left\{ i_j^M \right\}$ includes the maintenance task. The modeling of the evolution of the unit state is illustrated in Figure 8.8A.

8.4.2 Unit Capacity Reduction

If batch processing leads to capacity decrease, then, similarly to the monitoring of unit state H_{jn} with (8.39) through (8.43), we introduce capacity variable $B_{j,n}^{MAX} \in \mathbb{R}_+$ and capacity recovery variable $B_{j,n}^{MAX+} \in \mathbb{R}_+$, and write

$$B_{j,n+1}^{MAX} = B_{j,n}^{MAX} - \sum\nolimits_{i \in \mathbf{I}_j} \left(v_i^{CF} X_{i,n-\tau_i} + v_i^{CV} B_{i,n-\tau_i} \right) + B_{j,n-\tau_j^M}^{MAX+}, \; j, n, \tag{8.45}$$

where v_i^{CF} / v_i^{CV} represent the fixed and variable capacity decrease due to the execution of a batch of task i.

Variable B_{jn}^{MAX+} can be positive only when there is maintenance,

$$B_{jn}^{MAX+} \leq \beta_j^{MAX} X_{i_j^M n}, \quad j, n, \tag{8.46}$$

and the modeling of capacity recovery is achieved via the introduction of dummy variable $\widehat{B}_{jn}^{MAX+} \in \mathbb{R}_+$ and the addition of the following constraints:

$$\widehat{B}_{jn}^{MAX+} \leq \beta_j^{MAX} \left(1 - X_{i_j^M n} \right), \quad j, n \tag{8.47}$$

$$\beta_j^{MAX} - B_{j,n}^{MAX} = B_{jn}^{MAX+} + \widehat{B}_{jn}^{MAX+}, \quad j, n. \tag{8.48}$$

Finally, batchsizes are constrained by both the variable lower/upper bound constraint,

$$\alpha_{ij}^{MIN}\beta_j^{MIN}X_{in} \le B_{in} \le \alpha_{ij}^{MAX}\beta_j^{MAX}X_{in}, \quad j, i \in \mathbf{I}_j, n \tag{8.49}$$

and

$$B_{in} \le \alpha_{ij}^{MAX}B_{jn}^{MAX}, \quad j, i \in \mathbf{I}_j, n, \tag{8.50}$$

where α_{ij}^{MIN} and α_{ij}^{MAX} are coefficients used to convert unit capacity to task-specific capacity; that is, $\beta_i^{MAX} = \alpha_{ij}^{MAX}\beta_j^{MAX}$.[6]

In the general case, the modeling of unit deterioration with simultaneous capacity decrease is achieved with (8.39) through (8.44), and (8.45) through (8.50). Note that we assume that the *trigger* for maintenance comes from unit state H_{jn}, which cannot drop below η_j^{MIN}. However, if the coefficients ν_i^{CF} and ν_i^{CV} are such that the capacity B_{jn}^{MAX} drops below β_j^{MIN} (which means that no tasks can be carried out due to (8.49)) while H_{jn} is not prohibitively low, then the optimization will schedule a maintenance task, so that the capacity increases and the unit can be utilized again.

The case where the minimum capacity also decreases can be addressed by introducing new variable B_{jn}^{MIN} and either coupling it to $B_{j,n}^{MAX}$ (i.e., $B_{jn}^{MIN} = \alpha B_{jn}^{MAX}$); or explicitly modeling its evolution using equations similar to (8.45) through (8.48) and the corresponding variables representing recovery of the minimum capacity. In both cases, (8.49) and (8.50) are replaced by

$$\alpha_{ij}^{MIN}B_{jn}^{MIN} \le B_{in} \le \alpha_{ij}^{MAX}B_{jn}^{MAX}, \quad j, i \in \mathbf{I}_j, n. \tag{8.51}$$

Figure 8.8B illustrates the concepts underpinning unit capacity deterioration and recovery.

8.4.3 Conversion Reduction

This case arises when the execution of batches leads to lower unit *efficiency*, broadly defined; for example, catalyst deactivation can lead to lower conversion. In the general case, this feature can be addressed, albeit via nonlinear modeling of material production. Specifically, in STN representation, fixed conversion coefficients of outputs ($\rho_{ik} > 0$) are replaced by variable coefficients Q_{ikn}, whose evolution over time is modeled using equations similar to (8.39) through (8.43). The material balance is then written as

$$I_{k,n+1} = I_{kn} + \sum\nolimits_{i \in \mathbf{I}_k^+} Q_{ik,n-\tau_{ij}}B_{i,n-\tau_{ij}} + \sum\nolimits_{i \in \mathbf{I}_k^-} \rho_{ik}B_{in} + \xi_{ik} - S_{kn} \le \chi_k^M, \quad k, n. \tag{8.52}$$

Note that we do not replace consumption coefficients ($\rho_{ik} < 0$), since any change in the efficiency of a task (which can also be viewed as yield) can be represented via ratio

[6] Recall that in the present chapter we assume that each task can be carried out in only one unit, and thus task-specific capacities, β_i^{MIN} and β_i^{MAX}, can be directly used. However, capacity decrease occurs at the unit level, so task capacities should be coupled with the unit capacity. This can be achieved through constant coefficients α_{ij}^{MIN} and α_{ij}^{MAX}. The simplest case arises when $\alpha_{ij}^{MIN} = \alpha_{ij}^{MAX} = 1, \forall i$, which means that unit capacity can be used instead of task capacity. The next case arises when $\alpha_{ij}^{MIN} = \alpha_{ij}^{MAX}, \forall i$ but $\alpha_{ij}^{MAX} \ne \alpha_{i'j}^{MAX}$, which means that each task has different capacity, but the task-specific minimum and maximum capacities have the same proportionality constant. The most general case arises when $\alpha_{ij}^{MIN} \ne \alpha_{ij}^{MAX}$ and $\alpha_{ij}^{MAX} \ne \alpha_{i'j}^{MAX}$.

$Q_{ik}/\rho_{i,\bar{k}(i)}$ of the corresponding production conversion coefficient over the consumption coefficient of a reference material $\bar{k}(i)$ for task i (with $\bar{k}(i) \in \mathbf{K}_i^+$).

In the special case where the batchsize of a task is fixed (i.e., $\beta_i^{MIN} = \beta_i^{MAX} = \beta_i$), then efficiency decrease can be modeled linearly. Production coefficient Q_{ikn}, modeled using equations similar to (8.39) through (8.43), is further disaggregated into Q_{ikn}^A, which is positive when a batch starts at n, and dummy Q_{ikn}^D, which is positive at all other times:

$$Q_{ikn} = Q_{ikn}^A + Q_{ikn}^D, \quad i, k, n \tag{8.53}$$

$$Q_{ikn}^A \leq \rho_{ik}^{MAX} X_{in}, \quad i, k, n \tag{8.54}$$

$$Q_{ikn}^D \leq \rho_{ik}^{MAX}(1 - X_{in}), \quad i, k, n, \tag{8.55}$$

where ρ_{ik}^{MAX} is the nominal (maximum) conversion coefficient. In this case, (8.52) is rewritten as follows:

$$I_{k,n+1} = I_{kn} + \sum\nolimits_{i \in \mathbf{I}_k^+} \beta_i Q_{ik,n-\tau_{ij}}^A + \sum\nolimits_{i \in \mathbf{I}_k^-} \rho_{ik} \beta_i X_{in} + \xi_{ik} - S_{kn} \leq \chi_k^M, \quad k, n. \tag{8.56}$$

Note that the consumption, which remains constant over time, is expressed in terms of binary variable X_{in}, whereas the production term, which varies over time, is expressed using $Q_{ik,n-\tau_{ij}}^A$, which is positive only when $X_{i,n-\tau_{ij}} = 1$. Thus, no batchsize variable is defined, and the bilinear term $Q_{ik,n-\tau_{ij}} B_{i,n-\tau_{ij}}$ in (8.52) is replaced by the linear term $\beta_i Q_{ik,n-\tau_{ij}}^A$.

8.5 Notes and Further Reading

(1) The concept of material consumption and production during the execution of a batch task appeared in the original STN [1] and RTN papers [2].

(2) Material storage in shared vessels was also discussed in Kondili et al. [1]. A more general treatment of material storage in processing units before after the execution of a batch was discussed in Gimenez et al. [3], while resource constrained transfer activities were discussed in Gimenez et al. [4]. Both of these models were based on a common continuous time grid. The presentation here adopts some of the ideas in [3, 4], but has been adjusted for a discrete time model. Storage in units, coupled with modeling of transfers, using a discrete time model are discussed in Velez and Maravelias [5].

(3) Unit deterioration has been discussed in the context of catalyst decay [6] and economic lot sizing [7] under the assumption that the units are continuously used, that is, the schedule does not impact the evolution of the unit state. The case of capacity decrease with constant conversion coefficients is discussed in several papers [8, 9]. Models to account for yield decrease [10], prolonged processing times [11], and increased utility consumption [12] have also been proposed recently. Failures, in addition to performance decay, have also been studied [13].

(4) Wu et al [14] present models for efficiency deterioration and maintenance scheduling in heating ventilation and air conditioning (HVAC) systems for commercial buildings. Their approach, which also includes methods for the solution of the resulting MINLP models, was used as basis for the presentation in Section 8.4.

References

[1] Kondili E, Pantelides CC, Sargent RWH. A General Algorithm for Short-Term Scheduling of Batch-Operations. 1. MILP Formulation. *Comput Chem Eng*. 1993;17(2):211–227.

[2] Pantelides CC, editor. *Unified Frameworks for Optimal Process Planning and Scheduling. 2nd Conference on Foundations of Computer Aided Process Operations*; 1994; Snowmass: CACHE Publications.

[3] Gimenez DM, Henning GP, Maravelias CT. A Novel Network-Based Continuous-Time Representation for Process Scheduling: Part I. Main Concepts and Mathematical Formulation. *Comput Chem Eng*. 2009;33(9):1511–1528.

[4] Gimenez DM, Henning GP, Maravelias CT. A Novel Network-Based Continuous-Time Representation for Process Scheduling: Part II. General framework. *Comput Chem Eng*. 2009;33(10):1644–1660.

[5] Velez S, Maravelias CT. Mixed-Integer Programming Model and Tightening Methods for Scheduling in General Chemical Production Environments. *Ind Eng Chem Res*. 2013;52 (9):3407–3423.

[6] Jain V, Grossmann IE. Cyclic Scheduling of Continuous Parallel-Process Units with Decaying Performance. *AIChE J*. 1998;44(7):1623–1636.

[7] Alle A, Pinto JM, Papageorgiou LG. The Economic Lot Scheduling Problem under Performance Decay. *Ind Eng Chem Res*. 2004;43(20):6463–6475.

[8] Nie Y, Biegler LT, Wassick JM, Villa CM. Extended Discrete-Time Resource Task Network Formulation for the Reactive Scheduling of a Mixed Batch/Continuous Process. *Ind Eng Chem Res*. 2014; 53(44):17112–17123.

[9] Biondi M, Sand G, Harjunkoski I. Optimization of Multipurpose Process Plant Operations: A Multi-Time-Scale Maintenance and Production Scheduling Approach. *Comput Chem Eng*. 2017;99:325–339.

[10] Liu SS, Yahia A, Papageorgiou LG. Optimal Production and Maintenance Planning of Biopharmaceutical Manufacturing under Performance Decay. *Ind Eng Chem Res*. 2014;53 (44):17075–17091.

[11] Aguirre AM, Papageorgiou LG. Medium-Term Optimization-Based Approach for the Integration of Production Planning, Scheduling and Maintenance. *Comput Chem Eng*. 2018;116:191–211.

[12] Xenos DP, Kopanos GM, Cicciotti M, Thornhill NF. Operational Optimization of Networks of Compressors Considering Condition-Based Maintenance. *Comput Chem Eng*. 2016;84:117–131.

[13] Wiebe J, Cecilia I, Misener R. Data-Driven Optimization of Processes with Degrading Equipment. *Ind Eng Chem Res*. 2018;57(50):17177–17191.

[14] Wu Y, Maravelias CT, Wenzel MJ, ElBsat MN, Turney RT. *Predictive Maintenance Scheduling Optimization of Building Heating, Ventilation, and Air Conditioning Systems. Energy and Buildings*, 2021; 110487.

9 Continuous Processes

In this chapter, we discuss models for the scheduling of continuous processes. We focus on models based on a common discrete grid. We use the sets, subsets, and parameters introduced in Chapter 7 to represent a facility using the STN and RTN representations. As in Chapter 8, we assume that, in the STN representation, each task can be carried out in only one unit (i.e., $|\mathbf{J}_i| = 1$), and we directly use time-related parameters, without overbars (e.g., τ_i, rather than $\bar{\tau}_i$, represents the processing time of task i in terms of periods of length δ). Furthermore, to simplify the presentation, we consider problems with the following assumptions: (1) dedicated storage vessels; (2) no storage in processing units; (3) instantaneous and resource unconstrained material transfers; (4) no unit deterioration; and (5) demand can be satisfied (no backlogs).

We start in Section 9.1 with some background and a discussion of the main differences between batch and continuous processing. In Section 9.2, we present the basic, STN-based, model; and we close, in Section 9.3, with numerous extensions including modeling for startups and shutdowns, transitions between tasks, and time delays.

9.1 Preliminaries

9.1.1 Background

Traditionally, continuous processing has been employed for *dedicated* production, that is, a unit operating continuously and carrying out a single task over time. The operational *challenge* then is how to maintain the operation as close to a steady state as possible, given, for example, changes in the quality of input materials and disturbances (which is the subject of process control rather than scheduling). Furthermore, continuous processing is often employed for the production of low-value, high-volume products, which justifies (1) the initial investment to develop a continuous process; and (2) the allocation of large-scale and expensive assets to a single or a small number of products.

On the other hand, (multiproduct) batch processes are typically employed for the production of high-value, low-volume products because (1) the cost to install large-scale equipment, which allows manufacturers to exploit economies of scale, is distributed among multiple products; and (2) it is easier to alternate between different batch tasks on the same unit rather than between different continuous tasks. Naturally, scheduling problems, which deal with the allocation of limited resources to competing tasks, arise in multiproduct facilities, and since the majority of multiproduct facilities are batch facilities, the majority of chemical production scheduling approaches have been developed for batch processes.

Nevertheless, there are multiproduct facilities where at least some of the units operate continuously; for example, polymerization reactors producing polymers of different grades can be viewed as continuous units carrying out different (continuous) tasks. Furthermore, since continuous processing is in general more efficient, there is a trend toward the development of multiproduct continuous processes. Finally, there are systems where continuously operating units producing a single product need to be turned on and off regularly or use different raw materials, a feature giving rise to scheduling problems as well.

In terms of notation, starting in Chapter 7, we use the term *batch* to denote the specific execution of a task, assuming that all tasks in a facility represent batch processes. The corresponding term for tasks representing continuous processes is *run*; that is, a schedule may include multiple runs of the same (continuous) task. Further, in the present chapter, we will use the term *batch* to also describe the type of a process (batch vs. continuous).

9.1.2 Batch versus Continuous Processing

There are two basic differences between batch and continuous processing: (1) the relationship between task execution and material consumption and production; and (2) the duration and *rate* of tasks. Next, we discuss these differences in detail and close with a note on production environments.

Task Execution and Material Consumption/Production. In general, batch tasks consume input materials at the start (of the execution) of a batch and produce output materials upon the completion of batches. Thus, a batch triggers a step change (decrease) in the inventory of inputs when it starts and a step change (increase) in the inventory of outputs when it ends; for example, see inventory decrease of IN1 and increase of IN2 due to a batch of task T2 in Figure 9.1. The same stepwise changes occur also when there is consumption or production during the execution of a task. Continuous tasks, on the other hand, consume input and produce output materials (almost) throughout (the execution of) a run; for example, see decrease (increase) of inventory of RM (IN1) due to a run of task T1 in Figure 9.1. If θ_i is the *time delay*, in terms of time periods of length δ, between consumption and production of materials by task i, then there is no production during the first θ_i time periods, and no consumption during the last θ_i periods of a run; for example, see delay between inventory decrease of IN2 and increase of P during the execution of a run of task T3 in Figure 9.1. To summarize, batch tasks lead to changes at specified time points (*charging* and *withdrawing*), while continuous tasks lead to inventory depletion (accumulation) of inputs (outputs) throughout the execution of a task. This *continuous interaction* between continuous tasks and their input and output materials requires some modeling modifications.

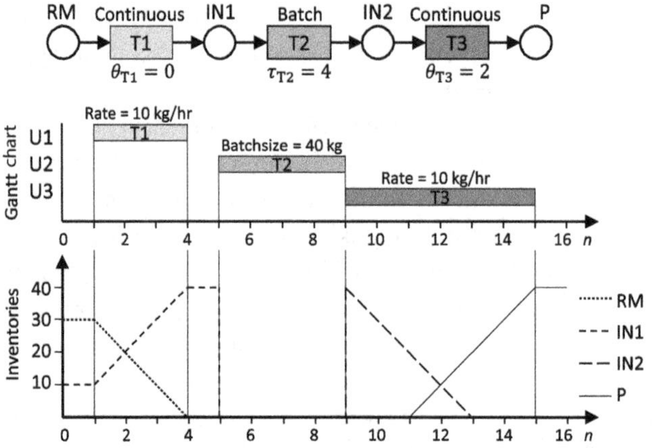

Figure 9.1 Differences in material consumption/production between batch and continuous tasks. Process network shown in the top panel: tasks T1, T2, and T3 are carried out in units U1, U2, and U3, respectively; T1 and T3 (continuous) have fixed processing rates equal to 10 kg/h; task T2 (batch) has fixed batchsize equal to 40 kg; all conversion coefficients are –1 (inputs) and 1 (outputs). An example Gantt chart is shown in the middle panel, and the corresponding inventory profile is shown in the bottom panel.

Duration and Rate of Tasks. Runs of continuous tasks have to be assigned to a unit and sequenced, exactly as batches of batch tasks. However, the decisions regarding the amount of materials processed during a run are different. Specifically, the amount of material processed depends on both the duration and the rate of a run, that is, there are multiple combinations of durations and rates that lead to the same *total processed amount during a run*, which is equivalent to the batchsize of a batch task. Furthermore, the rate of a run, which is an optimization decision similar to the batchsize of a batch, may change during the execution of a run. The duration is also an optimization decision subject, potentially, to lower and upper bounds. Note that even when the processing times of batch tasks are variable (functions of their batchsize), the number and duration of all batches in a schedule are fixed once batching decisions are made. In the case of continuous tasks, schedules including very different numbers of runs can meet the same demand, and even when the number of runs is specified, the corresponding run lengths can be very different. Even when both the number and duration of runs are specified, there are multiple schedules with different amounts processed during each run. The concept is illustrated in Example 9.1. Another interpretation of this difference between batch and continuous tasks is that they have different degrees of freedom. The former have one, the batchsize, which uniquely defines the duration of batches. The latter have two, the rate and length of runs.

Example 9.1 Degrees of Freedom in Continuous Processing We consider a network with two parallel units, U1 and U2, carrying out four continuous tasks with $\mathbf{I}_{U1} = \{T11, T12\}$ and $\mathbf{I}_{U2} = \{T21, T22\}$, converting two raw materials, F1 and F2, into two final products, P1 and P2. The first number in a task name indicates the input

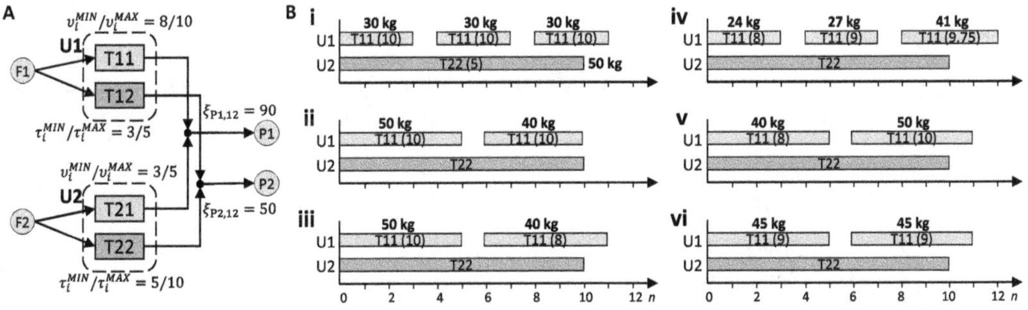

Figure 9.2 Illustration of how the number, processing rate, and duration of runs can be used to generate alternative schedules meeting the same demand. (A) Example process network. (B) Alternative schedules, (i) through (vi), meeting the same demand.

(e.g., T11 and T12 consume F1), and the second number indicates the output product (e.g., T12 and T22 produce P2). All consumption and production coefficients are −1 and +1, respectively. The rate of U1 is bounded in $[8, 10]$ kg/h, while the rate of U2 is bounded in $[4, 5]$ kg/h. The length of runs of both T11 and T12 should be between 3 and 5 h, while the length of runs of T21 and T22 should be between 5 and 10 h. We use $\delta = 1$. There are two orders due after 12 h: one for 90 kg of P1 and one for 50 kg of P2. The STN representation and parameters of the network, along with the product demand parameters, are given in Figure 9.2A.

Figure 9.2B shows six different solutions satisfying the demand for P1 (90) and P2 (50). Note that these solutions are different because of differences in the number, rate, and duration of runs of task T11 only. If decisions for other tasks are considered, the number of alternative solutions is greater than 30. We note the following:

(1) Solutions (i) and (iv) have three T11 batches, whereas all others have two.
(2) Solution (ii) has two runs of different length (5 and 4 h), whereas solutions (iii), (v), and (vi) have two runs of the same duration.
(3) Solutions (iii), (v), and (vi) have the same number and duration of runs, but the amounts processed are different because of different rates.

Production Environments. The continuous multiproduct facilities that have been studied in the literature are relatively simple; they primarily correspond to sequential single- and multi-unit environments. However, in principle, any of the environments discussed in Chapter 1 can be transformed into a continuous facility by maintaining the same structure and simply replacing all batch tasks with continuous ones. Thus, while in the remainder of this section we focus on the modeling of the core entity, a continuous processing unit carrying out continuous tasks, the models we develop are applicable to continuous processing units (and tasks) present in any environment.

9.2 Basic Model

We present a discrete time STN-based model. As in the case of batch processes, the facility is defined in terms of the following sets and subsets:

Indices/sets:

$i \in \mathbf{I}$	Tasks
$j \in \mathbf{J}$	Processing units
$k \in \mathbf{K}$	Materials

Subsets

$\mathbf{I}_k^+/\mathbf{I}_k^-$	Tasks producing/consuming material k
$\mathbf{I}_j$	Tasks that can be executed on unit j
$\mathbf{K}_i^+/\mathbf{K}_i^-$	Materials produced/consumed by task i

Continuous tasks have the following associated parameters:

$\upsilon_i^{MIN}/\upsilon_i^{MAX}$	Minimum/maximum rate [kg/period] of task i
γ_i^F/γ_i^V	Fixed (\$/period)/variable [\$/(kg · period)] cost for carrying out task i
ρ_{ik}	Conversion coefficient of material k produced (>0) or consumed (<0) by task i
$\tau_i^{MIN}/\tau_i^{MAX}$	Minimum/maximum length of a run of task i

Note that rates, costs, and minimum and maximum run lengths are given in terms of period of length δ. If $\widehat{\upsilon}_i^{MIN}/\widehat{\upsilon}_i^{MAX}$ are the actual bounds on rates [kg/h], then we calculate $\upsilon_i^{MIN} = \delta\widehat{\upsilon}_i^{MIN}$ and $\upsilon_i^{MAX} = \delta\widehat{\upsilon}_i^{MAX}$. Similarly, if the per-hour costs are $\widehat{\gamma}_i^F$ [\$/h] and $\widehat{\gamma}_i^V$ [\$/(kg·h)], we calculate $\widehat{\gamma}_i^F$ and $\widehat{\gamma}_i^V$ by multiplying $\widehat{\gamma}_i^F$ and $\widehat{\gamma}_i^V$, respectively, by δ.

In addition, we are given the following:

χ_k^M	Capacity of storage vessel dedicated to material k
π_k	Price of material k
ξ_{kn}	Net shipment of material k at time point n

The four key ideas that allow us to readily handle continuous tasks, with processing delay $\theta_i = 0$, using the already introduced modeling concepts are the following:

(1) A continuous task is treated as a batch task with processing time equal to one period ($\tau_i = 1$), consuming/producing input/output materials at its start/end.

(2) Given that the duration of a batch is one period, the minimum/maximum per period rate constraints are equivalent to minimum/maximum batchsize constraints.

(3) A *run* of a continuous task is represented via *multiple* consecutive one-period batches.

(4) Constraints are added to enforce minimum and maximum run lengths.

Note that the processing delay ($\theta_i = 0$) and the duration of the batch ($\tau_i = 1$) are different. The generalization for $\theta_i > 0$ will be discussed in Section 9.3.3. The representation of a run of a continuous task via multiple one-period batches is further discussed in Example 9.2.

Based on these ideas, the STN-based model for the scheduling of continuous processes consists of the three main constraints presented in Chapter 7 plus constraints regarding the length of a run. In addition to $X_{in} \in \{0, 1\}$, which denotes the start of an one-period batch of task i, we introduce the following two binary variables (see Example 9.2):

- $Y^S_{in} \in \{0, 1\} = 1$ if a run of task i starts at time point n.
- $Y^E_{in} \in \{0, 1\} = 1$ if a run of task i ends at time point n.

The assignment constraint is

$$\sum\nolimits_{i \in \mathbf{I}_j} X_{in} \leq 1, \quad j, n, \tag{9.1}$$

where the LHS does not include a summation over $n' \in \mathbf{N}^U_{in} = \{n - \tau_i + 1, \ldots, n\}$ because $\tau_i = 1, \forall i \in \mathbf{I}_j$. Variables Y^S_{in} and Y^E_{in} should satisfy

$$Y^S_{in} = X_{in} - X_{i,n-1} + Y^E_{in}, \quad i, n, \tag{9.2}$$

which can also be written as

$$X_{in} = Y^S_{in} + X_{i,n-1} - Y^E_{in}, \quad i, n. \tag{9.3}$$

Note that (9.3) is similar to a constraint we have already discussed. (Which is that constraint? What are the differences between (9.3) and that constraint?)

The material balance becomes

$$I_{k,n+1} = I_{kn} + \sum\nolimits_{i \in \mathbf{I}^+_k} \rho_{ik} B_{i,n-1} + \sum\nolimits_{i \in \mathbf{I}^-_k} \rho_{ik} B_{in} + \xi_{ik} - S_{kn} \leq \chi^M_k, \quad k, n, \tag{9.4}$$

where $B_{i,n-1}$ is the batchsize of the one-period batch that starts at point $n-1$ and represents the processing rate of task i during period n. We assume that if the task is executed during period n, all material consumption occurs at time point $n-1$ and all production at time point n, which leads to a reasonable stepwise approximation of the linearly changing inventory level (see Example 9.2). Again, note that in the first summation in the RHS of (9.4) we use $B_{ij,n-1}$ instead of $B_{ij,n-\tau_i}$ since $\tau_i = 1, \forall i$.

The batchsize constraints, which now represent rate constraints, become

$$\upsilon^{MIN}_i X_{in} \leq B_{in} \leq \upsilon^{MAX}_i X_{in}, \quad i, n. \tag{9.5}$$

Finally, constraints to enforce the minimum and maximum run lengths are added. A run of a task should last at least τ^{MIN}_i periods:

$$X_{in} \geq \sum\nolimits_{n' \in \mathbf{N}^{MIN}_{in}} Y^S_{in'}, \quad i, n, \tag{9.6}$$

where $\mathbf{N}^{MIN}_{in} = \{n - \tau^{MIN}_i + 1, \ldots, n\}$. If a run of task i starts at any $n' \in \mathbf{N}^{MIN}_{in}$, then the run should continue through at least point n (i.e., $X_{in} = 1$). Note that $\left|\mathbf{N}^{MIN}_{in}\right| = \tau^{MIN}_i$ and there can be only one start of a run over any τ_i^{MIN} points, which means that the RHS will always be less than or equal to 1.

A straightforward, though not *tight*, constraint to enforce that a run of a task cannot last more than τ_i^{MAX} periods is

$$\sum_{n' \in \mathbf{N}_{in}^{MAX}} X_{in'} \le \tau_i^{MAX} \quad i, n, \tag{9.7}$$

where $\mathbf{N}_{in}^{MAX} = \{n - \tau_i^{MAX}, \ldots, n\}$. Note that $|\mathbf{N}_{in}^{MAX}| = \tau_i^{MAX} + 1$, so (9.7) enforces that a run can be active for at most τ_i^{MAX} periods over any $\tau_i^{MAX} + 1$ consecutive periods. (Can you develop a tighter constraint enforcing the maximum run length?)

Equations (9.1) through (9.7) constitute the basic model for the scheduling of continuous processes. The objective function is

$$\max \sum_{k,n} \pi_k S_{kn} - \sum_{i,n} \left(\gamma_i^F X_{in} + \gamma_i^V B_{in}\right). \tag{9.8}$$

Startup costs, expressed using Y_{in}^S, as well as other types of costs (e.g., utilities), can be readily considered.

Example 9.2 Runs of Continuous Tasks We illustrate how a run of a continuous task can be modeled through consecutive one-period ($\tau_i = 1$) batches of a batch task. We use $\delta = 1$. We consider a single continuous task, T1, consuming RM and producing P, with $\rho_{\mathrm{T1,RM}} = -1$, $\rho_{\mathrm{T1,P}} = 1$, and $\upsilon_{\mathrm{T1}}^{MIN} = \upsilon_{\mathrm{T1}}^{MAX} = 10$ kg/h (see Figure 9.3A). Figure 9.3B shows how a four-period run, starting at $n = 1$ and ending at $n = 5$, is represented as four consecutive one-period batches. Note that four binary variables $X_{\mathrm{T1},n}$, for $n \in \{1, 2, 3, 4\}$, are equal to 1, and the four corresponding B_{in} are positive. Also, note that $X_{\mathrm{T1},5} = 0$ and $Y_{\mathrm{T1},5}^E = 1$. Figure 9.3C shows the difference between the actual inventory evolution, assuming $\theta_{\mathrm{T1}} = 0$, and the inventory profile calculated, through (9.4), using the one-period batch approximation. Note that the approximation is exact at all time points.

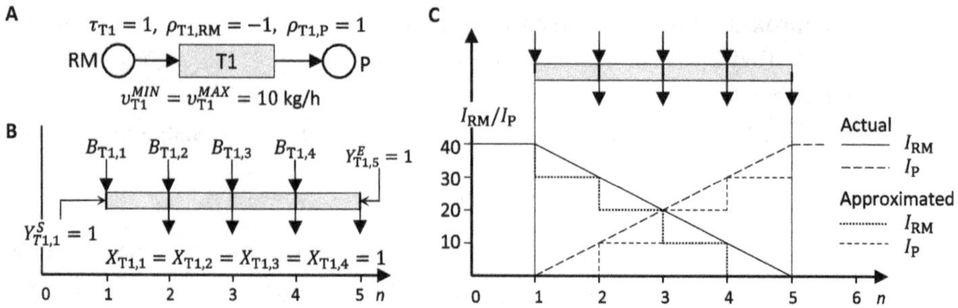

Figure 9.3 Modeling of a run of a continuous task. (A) Continuous task used for illustration and associated parameters (δ = 1 h). (B) Representation of a four-period run as four consecutive one-period batches. (C) Actual and approximated inventory evolution.

9.3 Extensions

A wide range of additional processing features can be handled using the modeling approaches we have discussed for batch processes. Shared utility constraints, including time-varying capacity and cost, can be modeled using variables X_{in} and B_{in} as in batch processes (see Chapters 4 and 7). Materials cannot be stored in processing units before and after the execution of a run (why?), but storage in shared vessels and material transfers can be modeled using the ideas presented in Section 8.2. Unit and task setups, including task families, can be modeled adopting the ideas presented in Section 8.3; and unit deterioration and maintenance can be readily modeled using the approaches introduced in Section 8.4. In addition to the aforementioned ones, there are three features that are specific to continuous processes. They are discussed in more detail next.

9.3.1 Startups and Shutdowns

In the previous subsection, we assumed that the processing rate is constrained uniformly across the periods a run is executed – see (9.5).[1] Continuous processes, however, require some time to reach a *steady state,*[2] and if this *startup* time is significant when compared to the length δ of the time period, then the maximum cumulative production (based on average rate)[3] during the first period of a run can be significantly lower. If the startup phase is *well defined,*[4] then its effect on the average production during the first period can be calculated prior to optimization. If ${}^{SU}\upsilon_i^{MAX} < \upsilon_i^{MAX}$ is the maximum production rate during the startup (first) period of a run, then we add

$$B_{in} \leq {}^{SU}\upsilon_i^{MAX} Y_{in}^S, \quad i, n. \tag{9.9}$$

Since B_{in} should satisfy both (9.5) and (9.9), the latter becomes binding.

[1] Batch tasks may also have startup and shutdown phases; for example, a cell growth phase precedes the *production* phase in a batch bioreactor. However, the *trajectory* of actual material consumption and production rates during a batch task is irrelevant if all input (output) materials are modeled as being consumed (produced) at the beginning (end) of a batch. On the contrary, material consumption and production during continuous tasks are modeled explicitly, so significant deviations from their nominal values during, for example, startup and shutdown should be taken into account.

[2] Strictly speaking, we do not actually require that a steady state is achieved and maintained throughout the execution of a run – recall that the processing rate during a run is allowed to vary, subject to (9.3). Here, we use the term *steady state* to denote the set of states at which the continuous task runs *normally*, that is, it consumes and produces materials according to its conversion coefficients ρ_{ik}, and its production is between υ_i^{MIN} and υ_i^{MAX}.

[3] We use the term *cumulative production*, during a period of length δ, represented by B_{in}, to differentiate it from the instantaneous rate that can change during a period. The cumulative production can be viewed as the production if the rate was constant and equal to the average instantaneous rate over the period.

[4] By well defined, we mean that the distance between the trajectory of the instantaneous lower and upper bound on processing rate during the first period is small. By integrating (from 0 to δ) the two functions defining these trajectories, we can obtain production bounds during startup, and the same can be done for shutdown. If the startup and shutdown phases can be significantly tuned to meet, say, certain constraints and objectives, then this leads to an integrated scheduling-control problem that will be studied in Chapter 14.

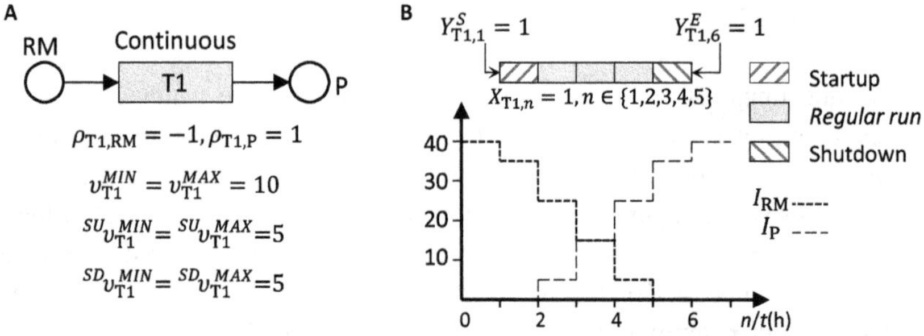

Figure 9.4 Modeling of one-period startup and shutdown phases. (A) Parameters of the task used for illustration. (B) Activation of startup, shutdown, and batch execution binaries for a run of task T1 and corresponding inventory evolution for input and output materials; note that consumption (at $n = 1$) and production (at $n = 2$) corresponding to the startup phase are reduced (5 instead of 10), and the same can be observed during the shutdown phase.

However, if ${}^{SU}v_i^{MAX} < v_i^{MIN}$, then keeping both (9.5) and (9.9) leads to infeasibility. Furthermore, a different lower bound on the processing rate during the first period, ${}^{SU}v_i^{MIN} < v_i^{MIN}$, may have to be enforced. To address these cases, a more general way to enforce different processing rate bounds during the startup (first) period is to replace (9.5) with

$$v_i^{MIN} X_{in} - \left(v_i^{MIN} - {}^{SU}v_i^{MIN}\right) Y_{in}^S \le B_{in} \le v_i^{MAX} X_{in} - \left(v_i^{MAX} - {}^{SU}v_i^{MAX}\right) Y_{in}^S, \quad i, n. \tag{9.10}$$

Note that when $X_{in} = 1$ and $Y_{in}^S = 0$, (9.10) reduces to (9.5). When $X_{in} = 1$ and $Y_{in}^S = 1$, it enforces ${}^{SU}v_i^{MIN} Y_{in}^S \le B_{in} \le {}^{SU}v_i^{MAX} Y_{in}^S$.

Similarly, during the last period of a run, the process has to be *shut down,* which may also lead to lower processing rate. If ${}^{SD}v_i^{MIN} / {}^{SD}v_i^{MAX}$ are the lower/upper bounds on production during the shutdown period, then the processing rate is constrained as follows:

$${}^{SU}v_i^{MIN} Y_{i,n+1}^E \le B_{in} \le {}^{SD}v_i^{MAX} Y_{i,n+1}^E, \quad i, n. \tag{9.11}$$

Note that, from (9.2), if $Y_{i,n+1}^E = 1$, then $X_{in} = 1$ $(B_{in} > 0)$ but $X_{i,n+1} = 0$, which means that the constraint should be enforced for B_{in}.

Finally, if ${}^{SU}\bar{v}_i^{MIN} = v_i^{MIN} - {}^{SU}v_i^{MIN}$, ${}^{SU}\bar{v}_i^{MAX} = v_i^{MAX} - {}^{SU}v_i^{MAX}$, ${}^{SD}\bar{v}_i^{MIN} = v_i^{MIN} - {}^{SD}v_i^{MIN}$, and ${}^{SD}\bar{v}_i^{MAX} = v_i^{MAX} - {}^{SD}v_i^{MAX}$, then all processing rate constraints can be lumped into and enforced by[5]

$$v_i^{MIN} X_{in} - \bar{v}_i^{SUMIN} Y_{in}^S - \bar{v}_i^{SDMIN} Y_{i,n+1}^E \le B_{in} \le v_i^{MAX} X_{in} - \bar{v}_i^{SUMAX} Y_{in}^S - \bar{v}_i^{SDMAX} Y_{i,n+1}^E, \quad i, n, \tag{9.12}$$

where we assume $\tau_i^{MIN} > 1$, and thus if $Y_{in}^S = 1$, then $Y_{i,n+1}^E = 0$, and if $Y_{i,n+1}^E = 1$, then $Y_{i,n}^S = 0$. Figure 9.4 illustrates how reduced production during startup and shutdown phases are modeled.

[5] Parameters with an overbar represent, essentially, deviations from the nominal bounds.

9.3.2 Transitions between Steady States

In many cases, the transition between runs (of different continuous tasks) may not involve a shutdown of the preceding and a startup of the succeeding run, but rather a transition between runs.[6] For example, if the distillation of two different crude oils is represented through two different tasks, then the distillation column (i.e., the unit in the STN framework) will continue to operate during the transition between the two crude oils. Thus, we consider two types of *transitions*: (1) the *direct* one, which is the transition between two steady states and (2) the *indirect* one, which involves the shutdown of the preceding run (to idle), and then a startup of the succeeding run (from idle).

If the second type is not allowed, then we adopt an RTN-based approach with additional constraints to avoid idle time. Specifically, if a unit resource can carry out tasks in $\mathbf{I}_r = \{T1, \ldots, TR\}$, then it is modeled as $|\mathbf{I}_r|$ *task modes,* which, for simplicity, are also denoted by i. Thus, for each original unit resource $r \in \mathbf{R}^U$, we introduce task modes $\mathbf{R}_r^{UM}$, and we use index i to denote both a task and the corresponding task mode. Further, we introduce a direct transition (changeover) between all pairs of tasks, denoted for simplicity by $i \rightarrow i'$. If the transition between modes lasts $\sigma_{i' \rightarrow i}$ periods, then the task mode balance over period n is then written as

$$R_{ri,n+1} = \sum_{i'} X_{i' \rightarrow i, n-\sigma_{i' \rightarrow i}} + X_{i,n-1} - X_{in} - \sum_{i'} X_{i \rightarrow i',n} = 0, \quad r \in \mathbf{R}^U, i \in \mathbf{R}_r^{UM} = \mathbf{I}_r, \; n. \tag{9.13}$$

Note that a task mode is produced upon completion of a transition and consumed at the point a transition starts, which implies that no mode is available during a transition. Thus, by setting $R_{rin} = 0$ for all modes and periods, we enforce that the unit is either undergoing a transition or executing a task ($X_{i,n-1} = X_{in} = 1$).

If both types of transitions are allowed, then for each original resource unit, in addition to the task modes and direct transition tasks, we also use an *idle* mode represented by r'; and introduce the corresponding transitions (modeled as setups) from a task mode to the idle mode ($i \rightarrow r'$) and from the idle mode to a task mode ($r' \rightarrow i$). Assuming that the startup and shutdown phases last one period, the balance of the task and idle modes are described by (9.14) and (9.15), respectively

$$R_{ri,n+1} = \sum_{i'} X_{i' \rightarrow i, n-\sigma_{i' \rightarrow i}} + X_{r' \rightarrow i,n} + X_{i,n-1} - X_{in} - \sum_{i'} X_{i \rightarrow i',n} - X_{i \rightarrow r',n} = 0, \; r, i \in \mathbf{R}_r^{UM}, \; n \tag{9.14}$$

$$R_{r',n+1} = R_{r',n} + \sum_i X_{i \rightarrow r',n-1} - \sum_i X_{r' \rightarrow i,n}, \; r' \in \mathbf{R}^U, \; n. \tag{9.15}$$

[6] Interestingly, the modeling of transitions is accomplished using the ideas already used to model startups and changeovers of batch tasks in the previous chapter (Section 8.3).

By setting $R_{rin}=0, \forall i \in \mathbf{R}_r^{UM}, n$ we enforce that no idle time can occur while the unit is in a task mode. Note that $X_{r' \to i,n}$ is essentially equivalent to Y_{in}^S, so $(X_{r' \to i,n} - X_{in})$ is needed to describe the effect on resource availability during the first period of a run; that is, if a run starts from idle at n, then $X_{r' \to i,n} = X_{in} = 1$, which means that the mode is not available during period $n+1$ $(R_{i,n+1}=0)$. If the run is active during both periods n and $n+1$, then $X_{i,n-1} = X_{in} = 1$ and $R_{i,n+1}=0$. On the other hand, $X_{r' \to i,n}$ is equivalent to $Y_{i,n+1}^E$: if $X_{r' \to i,n}=1$, then the run ends at n.

9.3.3 Time Delays

So far, we have assumed that for tasks with zero conversion delay $(\theta_i = 0)$ the consumption of input materials leads to the production of output materials one time period (δ hours) later, which means that consumption (production) stops (starts) one period before (after) a run ends (starts).[7] However, some processes may exhibit slower conversion (larger residence time), which means that output material production will start more than one period after the start of a run, and input material consumption will end more than one period before the run ends, while the unit remains active from the moment material consumption starts through the last period production occurs. A run of such a task cannot be modeled as a set of consecutive batches with processing time equal to the delay of the task. (Why?)

This case is handled through a generalization of startup and shutdown tasks, if a run is modeled to start (end) with a startup (shutdown) task; and a generalization of the transition tasks if a run starts and ends with a transition. We will present the model for startups and shutdowns. As we have seen, we model a unit resource via $|\mathbf{I}_r| + 1$ modes. We employ startup $(r' \to i)$ and shutdown $(i \to r')$ tasks, which, in addition to enabling the transition from/to the idle mode, also enable the modeling of continuous task delays via material consumption and production during task execution (see discussion in Section 8.1). If θ_i is the delay of task i, and s represents the number of periods after the start of a batch, then

(1) Startup task $r' \to i$ has duration $\tau_{r' \to i} = \theta_i$, it consumes input materials at $s = 0, 1, \ldots, \tau_{r' \to i} - 1$, and produces no materials.

(2) Shutdown task $r' \to i$ has duration $\tau_{r' \to i} = \theta_i$, consumes no materials, and produces materials at $s = 1, 2, \ldots, \tau_{r' \to i}$.

Recall that continuous tasks with conversion delay $\theta_i = 0$ were modeled using *regular* batch tasks with duration $\tau_i = 1$. Here, tasks with $\theta_i > 0$ are modeled using regular batch tasks with $\tau_i = 1$, again, but, in addition, startup and shutdown tasks with

[7] While one might think that the employment of a one-period batch can also be viewed as modeling the continuous task as a time delay system with delay constant equal to δ, it is important to note that the conversion delay of the underlying continuous task is different; it is equal to zero. Can you see why?

$\tau_{r' \to i} = \theta_i = \tau_{i \to r'}$. The modeling of continuous tasks with delays is further explained in Example 9.3.

A single run of the original task i starts with (a batch of) task $r' \to i$, followed by one-period *batches* (as in the $\tau_i = 1$ case), the last of which is followed by task $i \to r'$. The unit is in no mode during the startup and shutdown, while the regular one-period batches have to be executed consecutively, with no idle time in between. The equations describing the transitions between modes, assuming that there are no transitions between steady states, are

$$R_{ri,n+1} = X_{r' \to i,n-\theta_i} + X_{i,n-1} - X_{in} - X_{i \to r',n} = 0, \qquad r \in \mathbf{R}^U, i \in \mathbf{I}_r = \mathbf{R}_r^{UM}, \ n \tag{9.16}$$

$$R_{r',n+1} = R_{r',n} + \sum_i X_{i \to r',n-\theta_i} - \sum_i X_{r' \to i,n}, \quad r' \in \mathbf{R}^U, n. \tag{9.17}$$

Minimum and maximum run length constraints can still be enforced using binaries X_{in}. A run of task i can only be performed if the startup and shutdown tasks, lasting $2\theta_i$ periods, are carried out. Thus, if τ_i^{MIN} is the original minimum length, then (9.6) is rewritten using $\left(\tau_i^{MIN} - 2\theta_i\right)$ instead of τ_i^{MIN}. Similar adjustment can be made to constraint the maximum length of a run.

Finally, the material balance is written using (8.1) from the previous chapter,[8]

$$I_{k,n+1} = I_{kn} + \sum_{i \in \mathbf{I}_k^+} \sum_{s \in \mathbf{S}_{ik}^+} \rho_{iks} B_{i,n-s} + \sum_{i \in \mathbf{I}_k^-} \sum_{s \in \mathbf{S}_{ik}^-} \rho_{iks} B_{i,n-s} + \xi_{ik} - S_{kn} \le \chi_k^M, \ k, n, \tag{9.18}$$

where both regular one-period and startup/shutdown tasks are considered. If the consumption (production) coefficients during the startup (shutdown) phases are equal to the nominal conversion coefficients, then we simply use the following:

(1) For the startup task, $\rho_{r' \to i,ks} = \rho_{ik} < 0, \ k \in \mathbf{K}_i^-, \ s \in \{0, 1, \ldots, \tau_i - 1\}$; $\rho_{r' \to i,ks} = 0, k \in \mathbf{K}_i^+$.

(2) For the shutdown task, $\rho_{i \to r',ks} = \rho_{ik} > 0, \ k \in \mathbf{K}_i^+, \ s \in \{1, 2, \ldots, \tau_i\}$; $\rho_{r' \to i,ks} = 0, k \in \mathbf{K}_i^-$.

Example 9.3 Continuous Tasks with Delays To illustrate the modeling of processing delays, we consider a continuous task, T1, consuming RM and producing P, with $\rho_{\text{T1,RM}} = -1$, $\rho_{\text{T1,P}} = 1$, and $\theta_{\text{T1}} = 1$ (see Figure 9.5A). We introduce two new tasks: (1) $r' \to i$ (startup, T1-SU) with $\tau_{r' \to i} = \tau_{\text{T1-SU}} = 1$; and (2) $i \to r'$ (shutdown, T1-SD) with $\tau_{i \to r'} = \tau_{\text{T1-SD}} = 1$. Figure 9.5B shows the Gantt chart for a run that starts at $n = 1$ and ends at $n = 6$ and the corresponding inventory profiles

[8] Note how, for this model, we use a combination of ideas from RTN (transition modeling) and STN (material balances).

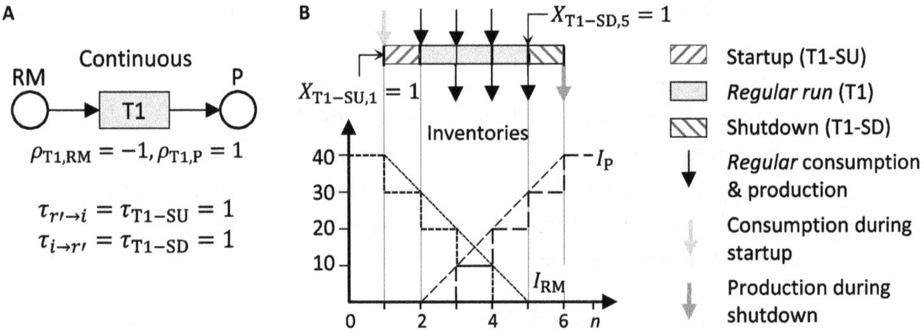

Figure 9.5 Modeling of delays using setup and shutdown tasks. (A) Original continuous task and corresponding startup and shutdown tasks. (B) Illustrative Gantt chart and corresponding inventory profile of a run of T1.

of RM and P. Note that $r' \rightarrow i/i \rightarrow r'$ does not produce (consume) P (RM), leading to a delay in inventory decrease (RM) and increase (P).[9]

9.3.4 General Startups and Shutdowns with Time Delays

In addition to enabling time delay modeling, the introduction of startup and shutdown tasks allows us to model startup and shutdown phases during which the consumption and production rates are positive but deviate from the nominal values. For example, we can readily model gradual increase of material consumption and production during startup and gradual decrease during shutdown. In fact, any *transition profile* can be modeled through an *evolution* of conversion coefficients during the startup and shutdown tasks. The concepts are illustrated in Example 9.4. Time-varying utility engagement during the startup and shutdown can be modeled similarly using the ideas presented in Section 4.7.2.

Example 9.4 Modeling of General Startups and Shutdowns As in Example 9.3, we consider a continuous task, T1, consuming RM and producing P, but with varying consumption and production rates during its execution. Specifically:

(1) The regular (steady-state) rate is $v_{T1}^{MIN} = v_{T1}^{MAX} = v_{T1} = 10$, with conversion coefficients $\rho_{T1,RM} = -1$ and $\rho_{T1,P} = 1$.
(2) During the startup phase, which lasts two hours and is modeled as task T1-SU, the processing rate is reduced $\left(v_{T1-SU}^{MIN} = v_{T1-SU}^{MAX} = v_{T1-SU} = 5\right)$, leading to reduced consumption of RM.

[9] Note that the first production occurs two periods after the start of the run but then production occurs at every subsequent point. Similarly, the last consumption occurs two periods before the end of the run. In the first paragraph of this subsection, it was stated that a task with delay cannot be modeled as a set of consecutive batches with processing time equal to the delay of the task. Can you now see why?

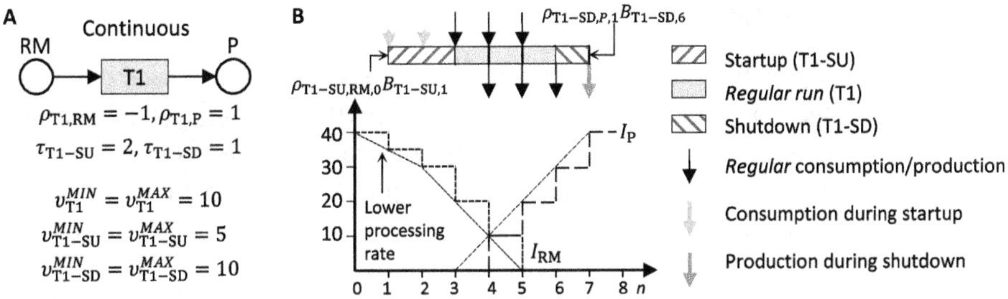

Figure 9.6 Modeling of general delays and startup and shutdown operations. (A) Continuous task used for illustration. (B) Gantt chart showing how startup, regular, and shutdown tasks are used to represent a six-period run; corresponding inventory evolution of input and output materials are also shown.

(3) During the shutdown phase, which lasts one hour and is modeled as task T1-SD, the processing rate is the same as the nominal rate $\left(\upsilon_{T1-SD}^{MIN} = \upsilon_{T1-SD}^{MAX} = \upsilon_{T1-SD} = 10\right)$, so the production of P is at its nominal value, but there is no RM consumption.

The three subtasks and associated parameters used to model T1 are shown in Figure 9.6A.

Figure 9.6B shows the Gantt chart for a six-hour run, starting at $n = 1$ and the inventory evolution of RM and P from $n = 0$ to $n = 8$. We note the following:

(1) The run is represented by an execution of (a batch of) the startup task, T1-SU, followed by three batches of the regular task, T1, followed by (a batch of) the shutdown task, T1-SD.
(2) The reduced consumption of RM during startup (first two periods) can be modeled via reduced parameters and $\rho_{T1-SU,RM,s}$ parameters.
(3) Note that the slope of inventory decrease of RM during the first two periods is less steep than the slope during regular production.
(4) There is no P production during the first two periods and no RM consumption during the last period, which can be viewed as having two delays: $\theta_{T1-SU} = 2$ and $\theta_{T1-SD} = 1$.
(5) The total consumption of RM and production of P are equal, that is, they satisfy the nominal conversion coefficients ($\rho_{T1,RM} = -1$ and $\rho_{T1,P} = 1$), but this is not necessary. Processes where, for example, some output should be discarded (and thus not considered in the material balance) during the startup phase can easily be modeled using appropriate conversion coefficients for the startup subtask.

In general, the introduction of the startup and shutdown tasks with potentially unequal processing times coupled with the flexibility offered by the selection of conversion coefficients ρ_{iks} for these tasks allows us to model a wide range of operational features and constraints.

9.3.5 General Transitions

Finally, multiperiod direct changeover tasks, $i \rightarrow i'$, with consumption and production during their execution (i.e., conversion coefficients $\rho_{i \rightarrow i', ks}$) can be used to model conversion delays during direct transitions between tasks (*steady states*) as well as, in general, nonstandard direct transition profiles. The main ideas are illustrated in Example 9.5.

Example 9.5 Ideas for Modeling of Direct Transitions We consider a facility with two continuous tasks, T1 and T2, converting raw material RM to products P1 and P2, respectively, as shown in Figure 9.7A. Both tasks are carried out in unit U and have fixed and equal production rates resulting in $\upsilon_{T1} = \upsilon_{T2} = 10$; the conversion coefficients given in Figure 9.7A; and zero processing delays.[10] During the transition between runs of the two tasks, which is modeled with task TRN and lasts three periods, RM continuous to be consumed, at lower rate; and some reduced amounts of P1 and P2 are also produced. The *irregular* material consumption and production during the transition phase is modeled using $\upsilon_{TRN} = 10$ and different conversion coefficients $(\rho_{TRN,ks})$: $\rho_{TRN,RM,s} = -0.5, s \in \{0, 1, 2\}$ $\rho_{TRN,P1,1} = 0.5$, and $\rho_{TRN,P2,3} = 0.5$. Note that the same material production and consumption can be modeled using lower capacity (e.g., $\upsilon_{TRNS} = 5$) and higher conversion coefficients (e.g., $\rho_{TRNS,RM,s} = -1, s \in \{0, 1, 2\}$). The Gantt chart of an example solution along with the resulting inventory profiles of all materials are shown in Figure 9.7B.

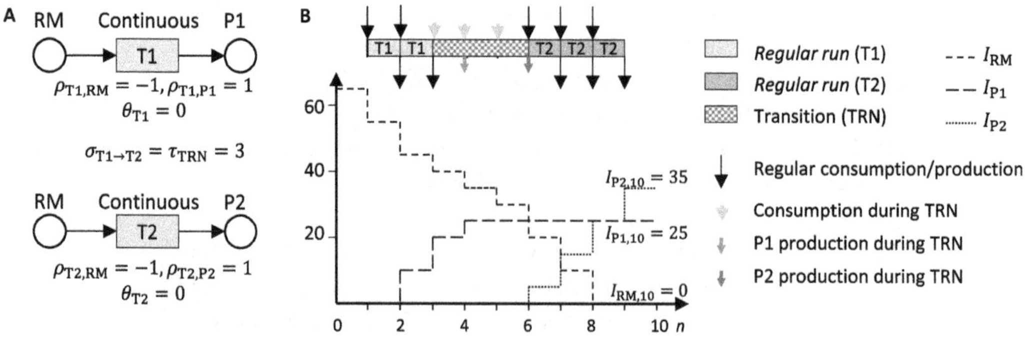

Figure 9.7 Modeling of direct transitions between tasks. (A) Tasks and processing parameters used for illustration. (B) Example Gantt chart and corresponding inventory profiles.

---

[10] This could mean that no startup and shutdown tasks are needed, though this is not necessary; see, for instance, the task discussed in Example 9.4. In any case, the existence or not of startup and shutdown phases is not critical for this example and thus omitted.

9.4 Notes and Further Reading

(1) The presentation in the present chapter, including the major modeling concepts, is based on Wu and Maravelias [1].

(2) The basic modeling of continuous processes was addressed in the original STN [2] and RTN [3] papers. Nie et al. proposed an extended RTN model, based on discrete modeling of time, that can handle systems with both batch and continuous processes [4].

(3) The majority of approaches for continuous process scheduling adopt continuous modeling of time. Zhang and Sargent [5] and Castro et al. [6] developed RTN-based models for scheduling of both batch and continuous processes. Ierapetritou and Floudas [7] and Lee et al. [8] proposed models using STN representation.

9.5 Exercises

(1) Consider the process, and associated parameters, shown in Figure 9.2A, where the units for time, capacity, and costs are h, kg, and \$, respectively. All materials have unlimited capacity and the fixed processing costs are $\gamma_i^F = \$5, \forall i$. Formulate a discrete time model ($\delta = 1$ h) and find the schedule with the minimum cost.

(2) Consider the process used for the previous problem. In addition, tasks are subject to one-hour startups and shutdowns with the processing rate bounds given in Table 9.1. Formulate a discrete time model ($\delta = 1$ h) and find the schedule with the minimum cost.

(3) Consider a single-unit process carrying out two continuous tasks TA and TB, producing two products, A and B, from the same raw material, with unlimited storage capacity for all materials. All consumption/production coefficients are –1/+1. The maximum processing rate of TA and TB are 2.5 and 2 kg/h, respectively; the inventory cost for products is \$1/(kg·h); and the fixed processing cost for both tasks is \$10/h. Formulate discrete time RTN-based models ($\delta = 1$ h) to solve the following cost minimization instances:

(a) Direct transition is considered: during the transition, which lasts 3 h, 1 kg of raw material is consumed per hour; during the A $\rightarrow$ B (B $\rightarrow$ A) transition, 1 kg of B (A) is produced during the third hour; the transition cost is \$25. Ten kilograms of A and 10 kg of B should be shipped at $t = 12$ h.

Table 9.1 Processing rate lower and upper bounds [kg/h] during startup and shutdown; all tasks carried out in a unit have the same bounds.

	$SU\upsilon_i^{MIN}$	$SU\upsilon_i^{MAX}$	$SD\upsilon_i^{MIN}$	$SD\upsilon_i^{MAX}$
U1	6	9	7	9
U2	1	4	2	4

(b) Indirect transitions are considered: all transitions last 1 h; 10 kg of A and 10 kg of B should be shipped at $t = 11$ h.

(c) Time delay is considered: the startup and shutdown phases last two hours, and the processing rate of both tasks during these two phases are 0.5 kg/h. Ten kilograms of A, and 10 kg of B should be shipped at $t = 15$ h.

References

[1] Wu Y, Maravelias CT. A General Framework and Optimization Models for the Scheduling of Continuous Chemical Processes. Submitted for publication.

[2] Kondili E, Pantelides CC, Sargent RWH. A General Algorithm for Short-Term Scheduling of Batch-Operations. 1. MILP Formulation. *Comput Chem Eng*. 1993;17(2):211–227.

[3] Pantelides CC, editor, *Unified Frameworks for Optimal Process Planning and Scheduling. 2nd Conference on Foundations of Computer Aided Process Operations*. 1994; Snowmass: CACHE Publications.

[4] Nie Y, Biegler LT, Wassick JM, Villa CM. Extended Discrete-Time Resource Task Network Formulation for the Reactive Scheduling of a Mixed Batch/Continuous Process. *Ind Eng Chem Res*. 2014; 53(44): 17112–17123.

[5] Zhang X, Sargent RWH. The Optimal Operation of Mixed Production Facilities – a General Formulation and Some Approaches for the Solution. *Comput Chem Eng*. 1996;20(6–7):897–904.

[6] Castro PM, Barbosa-Povoa AP, Matos HA, Novais AQ. Simple Continuous-Time Formulation for Short-Term Scheduling of Batch and Continuous Processes. *Ind Eng Chem Res*. 2004;43(1):105–118.

[7] Ierapetritou MG, Floudas CA. Effective Continuous-Time Formulation for Short-Term Scheduling. 2. Continuous and Semicontinuous Processes. *Ind Eng Chem Res*. 1998;37 (11):4360–4374.

[8] Lee KH, Park HI, Lee IB. A Novel Nonuniform Discrete Time Formulation for Short-Term Scheduling of Batch and Continuous Processes. *Ind Eng Chem Res*. 2001;40(22):4902–4911.

10 Periodic Scheduling

In Chapter 1, we discussed how the market environment within which a company operates affects the scheduling of its manufacturing facilities primarily because the volume and variability of product demand determine the regularity and frequency in which scheduling is performed. For example, the production of products with irregular and/or small demand is scheduled as needed (make-to-order), which typically means that a short-term schedule should be generated frequently based on actual orders; this is the class of problems we have studied so far in this book. However, the production of high-volume products with relatively constant demand or high-volume intermediates can be based on demand forecasts rather than specific orders. If the forecasts do not exhibit significant fluctuations over time, one approach is to generate a schedule that can be repeated periodically, often referred to as *production wheel*, so as to maintain a certain level of stock (make-to-stock). The process of generating such schedules, which is the topic of the present chapter, is termed *periodic scheduling*.[1]

In Section 10.1, we use the single-unit environment to motivate the need for periodic scheduling, define relevant notation, and present some preliminary concepts and models. In Section 10.2, we use the single-stage environment to discuss some additional concepts and one formulation. Finally, in Section 10.3, we present periodic scheduling in network environments. We consider processes without any of the processing features discussed in Chapter 8 (e.g., shared storage vessels, storage in processing units, transfer tasks, etc.). If units are not specified, demand-related parameters are in kilograms (kg) and time-related parameters are in hours (h). If not specified, we use $\delta = 1$. Also, throughout the present chapter, all time-related parameters will be without overbars.

10.1 Single-Unit Environment

The problem statement is given in Section 10.1.1, followed by a discussion of the basic considerations and trade-offs in Section 10.1.2, and the formal introduction of the notation of periodic scheduling in Section 10.1.3. A basic STN-based model, which requires an iterative solution procedure, is presented in Section 10.1.4, while a model

[1] The term *cyclic scheduling* has also been used widely to describe this activity. We prefer to use the term *periodic* because the output is a *periodic solution* rather than a *cycle*, with the latter being a special case of the former. The distinction between the two will be further discussed in Section 10.2.4.

that yields the optimal solution directly is presented in Section 10.1.5. We close with some remarks in Section 10.1.6.

10.1.1 Problem Statement

We consider a single unit that can carry out multiple tasks each one of which leads to the production of a final product. Using the STN framework, we are given the following:

- We have processing unit U and a set of tasks, $i \in \mathbf{I} = \mathbf{I}_{\mathrm{U}}$.
- All tasks consume the same raw material, RM (i.e., $\mathbf{K}_i^- = \{\mathrm{RM}\}, \forall i$), with $\rho_{i,\mathrm{RM}} = -1, \forall i$.
- Each task produces product $k^+(i)$ which, for simplicity, is also denoted by index i; that is, $\mathbf{K}_i^+ = \{i\}$.
- The minimum and maximum batchsize of task i are β_i^{MIN} and β_i^{MAX}, respectively.
- The processing time of task i is τ_i and its setup time and cost are σ_i and γ_i^{SET}, respectively.
- The availability and storage capacity for RM are unlimited; products are stored in dedicated vessels with capacity χ_i^M.
- The per unit time demand for product i is ξ_i, that is, $-\xi_i$ units of product should be shipped *continuously* per unit of time.
- The inventory cost for product i is ζ_j.

The goal is to find the least per unit time cost[2] [\$/h] schedule that can be repeated indefinitely to meet the demand. The cost includes a setup component and an inventory component. Note that the length of this periodic solution is unknown. In fact, one of the major differences between short-term and periodic scheduling is that the scheduling horizon, that is, the period of the periodic solution, is a variable.

The presence of task setup times impacts feasibility while the presence of inventory and task setup costs leads to an interesting trade-off that makes this seemingly simple problem nontrivial. Significant setup times may lead to systems where the transitions among products are infrequent in order to minimize total setup time and thus meet demand. If the setup costs are negligible while inventory costs are substantial, then the optimal solution is to alternate among tasks as fast as possible to keep inventories low. If the setup costs are significant, however, then alternating among products fast is suboptimal because it leads to high setup costs.

10.1.2 Preliminaries and Motivation

To illustrate the concepts, we consider an example with three products, $\mathbf{I} = \{\mathrm{A, B, C}\}$, with the processing times, setup time and costs, fixed batchsizes (i.e., $\beta_i^{MIN} = \beta_i^{MAX}$),

[2] There are two types of costs in periodic scheduling: the total cost over a periodic solution (in \$) and the corresponding cost divided by the period of the solution (in \$/h). Henceforth, we will refer to the former as *total cost* and the latter as simply *cost*, not *per unit time cost*. We will use the same convention for other solution features; for example, total setup time (h) and, simply, setup time (h/h).

Table 10.1 Data for motivating example.

Task/Product	τ_i (h)	σ_i (h)	γ_i^{SET} (\$)	β_i^{MAX} (kg)	ζ_i (\$$\cdot$kg$^{-1}\cdot$h^{-1})	$-\xi_i$ (kg$\cdot$h^{-1})
A	3	2	40	10	1	1
B	4	1	50	10	1	0.5
C	5	3	60	8	1	0.4

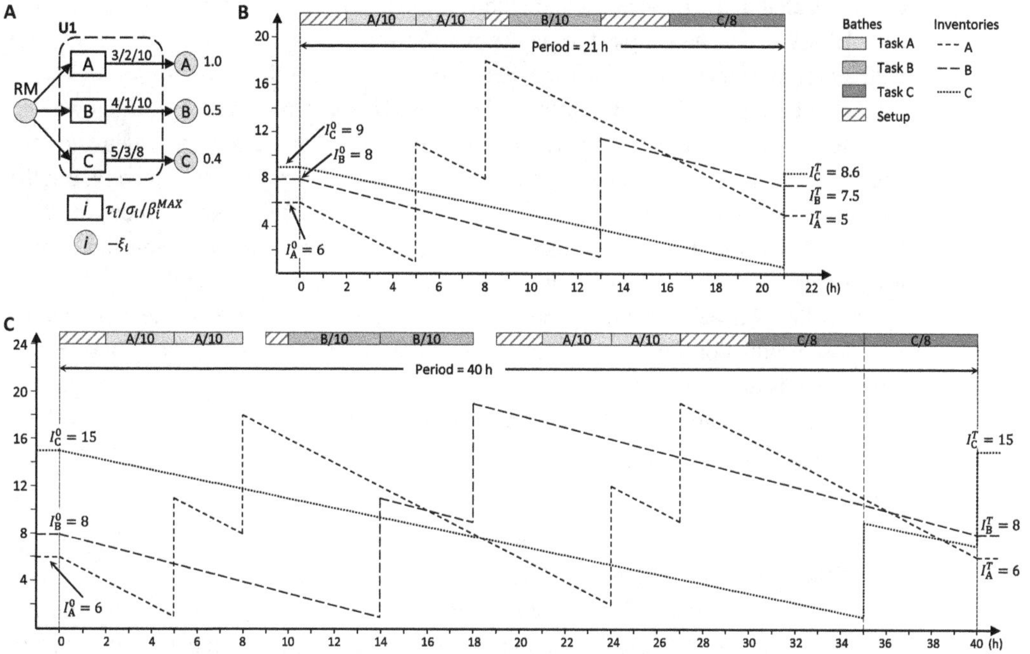

Figure 10.1 STN representation (A), and Gantt chart and inventory profiles of Campaign 1 (B) and Campaign 2 (C) of the motivating example. The system shown in A is used in all solutions shown in Figures 10.2 and 10.3.

inventory costs, and demands given in Table 10.1 (see also Figure 10.1A). All consumption/production conversion coefficients are –1/+1.

Our first goal is to find a sequence[3] of batches of the three tasks that meet the demand and thus can be repeated indefinitely. An example of a batch sequence is [A → A → A → B → C → C], where it is understood that the last batch of C will be followed by a batch of A; that is, the sequence of batch executions in two periodic schedules is [A → A → A → B → C → C] →[A → A → A → B → C → C] →. The time it takes to execute all batches, including setup times and idle time, is the *period* of the solution, denoted by T. Note that the sequence can start at any batch, that is, sequences

[3] In the previous chapters, we used the term *sequence* (e.g., in sequence-based models) to refer to a pairwise relationship. In the present chapter, we use it to refer to a sequence (of execution) of multiple batches.

[A → A → A → B → C → C] and [C → C→ A → A → A → B] can lead to the same solution. A batch *campaign* is defined by a batch sequence and the corresponding timing (start time) of batches. For a campaign to be feasible, the inventory of each product at the end of the horizon, which in this case is equal to the period of the solution, should be equal to the inventory of the same product at the beginning of the horizon. We will use I_k^0 and I_k^T to represent the initial and final, respectively, inventory of material k.

Going back to the motivating example, we note that if N_i is the number of batches of task i repeated in a sequence, then the ratio $N_A : N_B : N_C$ in any feasible solution should be 2:1:1. (Why?) The simplest sequence we can study is, clearly, [A → A → B → C] with a period, if there is no idle time, equal to 21 h $(= \sigma_A + 2\tau_A + \sigma_B + \tau_B + \sigma_C + \tau_C)$ and batch timing {2, 5, 9, 16} as shown in Figure 10.1B. This campaign, which we will refer to as Campaign 1, is infeasible because production cannot meet demand: the demand for products A, B, and C over 21 h is 21, 10.5, and 8.4 kg, respectively, but the maximum amount that can be produced is 20, 10, and 8 kg, respectively. In Figure 10.1B, this demand–production imbalance is addressed by having the ending inventories $\left(I_i^T\right)$ being lower than the starting inventories $\left(I_i^0\right)$, and the difference between the two being equal to the shortfall of production; for example, for product A, whose production shortfall is 1 (21 – 20), the difference between starting and ending inventories is also 1 (6 – 5).

An alternative campaign (Campaign 2), with sequence [A → A → B → B → A → A → C→ C] and timing {2, 5, 11, 14, 21, 24, 30, 35}, is shown in Figure 10.1C. We note the following:

(1) The ratio of batches in Campaign 2 is 4:2:2 = 2:1:1, the same as in Campaign 1.

(2) The period of Campaign 2 is 40 h, and Campaign 2 is feasible: the total production (40 kg of A, 20 kg of B, and 16 kg of C) is equal to the total demand over the same period, and the starting inventories are equal to the ending inventories for all products (please repeat all calculations yourself).

(3) There are multiple different sequences with the same number of batches, for example, [A → A → A → A → B → B → C→ C] leading to Campaign 3, and [A → A → B → C→ C → A → A → B] leading to Campaign 4.

(4) The minimum time sequence [A → A → B → B → A → A → C→ C] can be executed in is 38 h, if there are no idle times, but a campaign with period equal to 38 h would be infeasible. (Why?)

(5) The minimum period of an infeasible solution with $N_{\rm A} = 4$, $N_{\rm B} = 2$ and $N_{\rm C} = 2$ is 36 h. (What batch sequence results in a campaign with $T = 36$? Why is it infeasible?)

(6) The total inventory costs, over the 40 h solution, for products A, B, and C are \$400, \$400, and \$320, respectively. (You should be able to calculate these numbers based on the information given in Figure 10.1C). The total setup cost is (2·40 + 50 + 60 =) \$190.

The comparison between Campaigns 1 and 2 reveals one of the two fundamental considerations in periodic scheduling: the impact of the batch sequence, through setups,

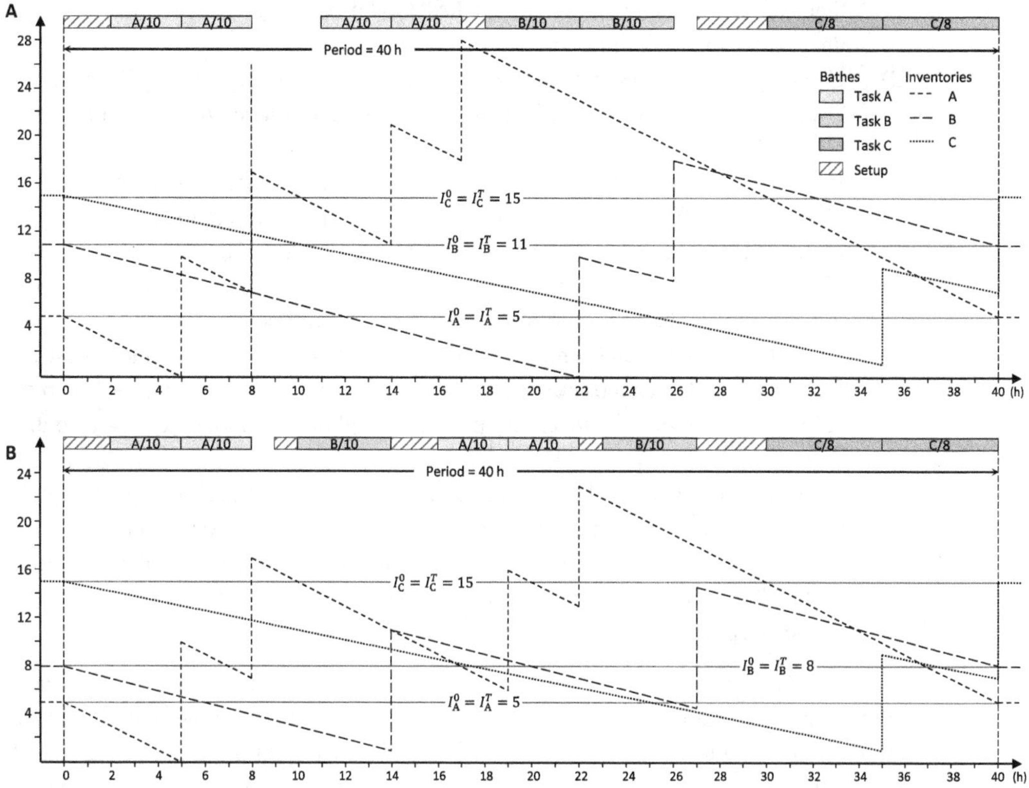

Figure 10.2 Gantt chart and inventory profiles of Campaign 3 (A) and Campaign 4 (B) of the motivating example.

to the production capacity. Compared to Campaign 1, Campaign 2 is feasible because less time is devoted on task setups and thus production can be higher: the setup time in Campaign 1 is 6 h of setups over 21 h, or 0.2857, while the setup time in Campaign 2 is 8/40 = 0.2. The lower setup time allows increased production: the production in Campaign 1, which has no idle time, is 1.81 kg/h (= 38 kg/21 h), while the production in Campaign 2 increases to 1.9 kg/h (76 kg/40 h), despite the presence of idle times.

To study the second consideration, the trade-off between setup and inventory costs, we compare Campaign 2 against Campaign 3 with sequence [A → A → A → A → B → B → C→ C] (Figure 10.2A) and Campaign 4 based on sequence [A → A → B → A → A → B → C→ C][4] (Figure 10.2B). We observe the following:

(1) Campaign 3 has the lowest total setup cost (40 + 50 + 60 = \$150) and lowest setup time (6/40 = 0.15) among all campaigns with $T = 40$. Thus, all sequences

[4] In item (3) of the preceding numbered list, we mentioned that Campaign 4 has the following batch sequence: [A → A → B → C→ C → A → A → B]. Why are we now using a different sequence? Are the two sequences the same?

that lead to Campaign 3 have the smallest total processing + setup time (36 h) and the highest total idle time (4 h).

(2) Campaign 4 has the highest setup time (9/40 = 0.225) among Campaigns 2 through 4, thus the lowest idle time (1 h) and highest setup cost (2·40 + 2·50 + 60 = \$240).

(3) Even when the period and sequence are given, which means that setup cost is given, there is flexibility in the start time of batches; for example, the third and fourth batches of A in Campaign 3 could start earlier. The timing of batches affects inventory costs. (What would be the impact on inventory cost for A, if the third and fourth batches start at $t = 8$ and $t = 11$, respectively?)

(4) Given the sequence and timing of batches in Campaign 3, the inventory cost for products A and B cannot be reduced (why?), but the inventory cost for C can be reduced by lowering the inventory profile throughout the horizon by 1 kg, so that $I_{\mathrm{C}}^0 = I_{\mathrm{C}}^T = 14$ and $I_{\mathrm{C},35} = 0$.

(5) In general, even when the period, batch sequence, and batch timing are given, there are multiple inventory profiles satisfying the periodic constraint $\left(I_i^0 = I_i^T\right)$. As explained next, all these profiles can be obtained by *shifting* a basic inventory profile. The minimum inventory cost can be obtained when the inventory profile becomes zero at at least one time point. (Why?)

Let $c \in \mathbf{C}$ be a campaign (i.e., given batch sequence and timing), and $s \in \mathbf{S}_c$ be a solution that corresponds to c; there are infinite such solutions with the only difference among them being the inventory levels. If $I_{in}^{c,s}$ is the inventory of product i during period n in solution s, assuming we use a discrete time model, then we have the following:

$$I_{in}^{c,s} = \hat{I}_{in}^{c} + \hat{I}_{i}^{s}, \quad c, s \in \mathbf{S}_c, i, n, \tag{10.1}$$

where $\hat{I}_{in}^{c}$ gives a *basic* profile, unique for each c; and $\hat{I}_{i}^{s}$ is a *shifting* parameter, which is product specific but constant over time. The $I_{in}^{c,s}$ profile is uniquely defined by $\hat{I}_{i}^{s}$ or, alternatively, by the starting inventory, $I_i^{0,s}$, in solution s. Though the basic solution could be chosen arbitrarily, if we require that all profiles can be obtained from $\hat{I}_{in}^{c}$ by adding $\hat{I}_{i}^{s} \geq 0$, then $\hat{I}_{in}^{c}$ should at some point become equal to zero. (Can you see why?)[5]

The concept is illustrated in Figure 10.3 using the sequence used for Campaign 2 (C2) but with different batch timing from the one shown in Figure 10.1B. Note that the inventories of A and C in the basic solution ($\hat{I}_{\mathrm{A},n}^{c}$ and $\hat{I}_{\mathrm{C},n}^{c}$) become zero at time 5 and 35, respectively. The profiles in solution 1, $I_{\mathrm{A},n}^{c,s}$ and $I_{\mathrm{C},n}^{c,s}$, are obtained by adding $\hat{I}_{\mathrm{A}}^{1} = 6$ and $\hat{I}_{\mathrm{C}}^{1} = 4$ to the basic profiles. Based on this analysis, the total inventory cost for each of products A, B, and C in the solution shown in Figure 10.1C can be decreased by \$40. (Can you see why?)

[5] There are infinite inventory profiles that are feasible for a given period, batch sequence, and timing. Among all of them, the minimum cost, for the deterministic problem, is obtained by the (unique) profile in which $I_{in}^{c,s} = 0$ for some n. In problems under demand uncertainty, low inventories may lead to stockouts and thus increased cost, due to backlog or unmet demand penalties. In that context, the concept of *safety stock* is introduced leading to optimal solutions with positive minimum inventory levels.

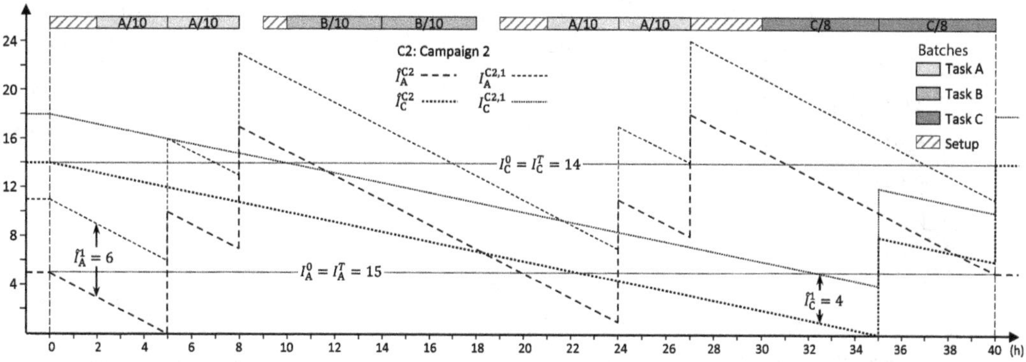

Figure 10.3 Different inventory profiles corresponding to the same batch sequence and timing; index n is dropped from inventory variables for simplicity.

We close this section with a note regarding the available *degrees of freedom* in periodic scheduling. Compared to short-term scheduling, where product demand can follow any pattern, periodic scheduling appears to have limited flexibility. However, even for the simple one-unit problem we considered in this subsection, we observed that finding a periodic solution includes the following decisions: (1) length of periodic solution (period), (2) batch sequence, (3) batch timing, and (4) inventory profiles.

10.1.3 Notation

In this subsection, we more formally define a number of terms that have been introduced so far. The terms are illustrated in Figure 10.4 *Batch sequence* is the sequence that corresponds to the batches that are carried out in a solution, chronologically ordered. *Batch timing* is the sequence of start times, in increasing order, of the scheduled batches. A *campaign* is defined by a batch sequence and a batch timing. A *periodic solution*, or simply *solution*, is defined by a campaign (i.e., a batch sequence and timing), the corresponding batchsizes, and the inventory profiles of all *relevant* materials.[6] We use the term *periodic solution* to describe the solution that is repeated indefinitely. We use the term *solution execution* to describe an instance of this solution. For example, in Figure 10.4 two executions are shown, one between 0 and 16 h, and one between 16 and 32 h.

We note the following:

(1) Two different batch sequences can lead to the same solution.
(2) Two different batch timings, corresponding to the same batch sequence, may also lead to the same solution. (Can you see why?)

[6] As we will see in the subsequent subsections, we often make assumptions that allow us to not consider, for example, inventory profiles of raw materials.

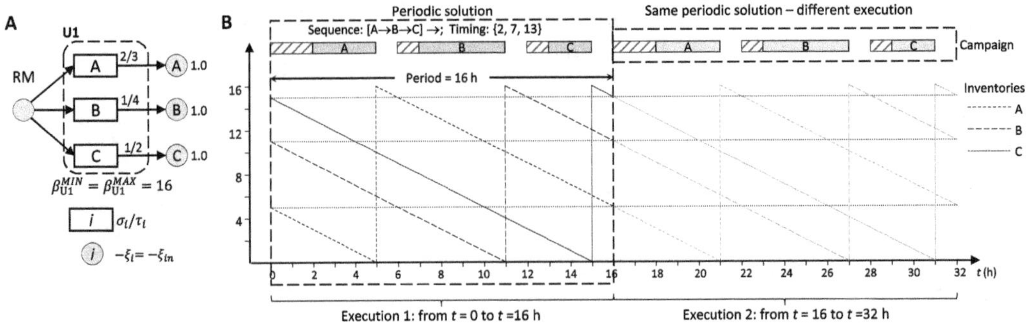

Figure 10.4 Illustration of batch sequence, batch timing, campaign, periodic solution, period of solution, and solution execution. (A) Example used for illustration. (B) Gantt chart and inventory profile over two executions.

(3) The same campaign can lead to different solutions because of differences in (a) batchsizes and (b) inventory profiles.

If we seek to find the optimal periodic solution for a given period, then, as we will see in the next subsection, we can solve a problem with a horizon equal to the given period. In this case, *scheduling horizon* and *period* (of periodic solution) are the same. In the general case, however, the scheduling horizon of the model we consider and the period of the solution we find can be different.

10.1.4 Basic Discrete Time Model

A consequence of the period of the optimal solution being unknown is that the number of batches in that solution is also unknown, which in turn means that there is no efficient way to determine how many periods (slots)[7] would be needed in a continuous time model to represent the optimal solution. If z_t^* is the optimal solution obtained using t time periods, then obtaining a series of solutions with $z_{t'}^* \geq z_t^*$ for, say, $t' \in \{t+1, t+2, t+3\}$ does not mean that a significantly better solution does not exist if, say, $t+5$ periods are used. In fact, it is rather common to see no improvement as the number of periods increases, and then a significant decrease.[8] Employing a discrete time approach requires an iterative procedure as well, because the period of the solution is unknown, but this procedure is well defined. If δ is the length of the period of the

[7] Note that in the present chapter we use the term *period* to describe two concepts: (1) an old one, the period of the time grid, that is, the time interval between two time points; and (2) a new one, the period of a periodic solution. No disambiguation is necessary, though, because the meaning of the term will be clear based on the context in which it is used.

[8] To illustrate, consider the example in Section 10.1.2 but with the setup time for product C being 2 ($\sigma_C = 2$) instead of 3 h. Batch sequence [A → A → B → C] leads to a feasible campaign with batch starting times equal to 2, 5, 9, and 15 h; the only difference compared against the starting times shown in Figure 10.1A is that the batch of C starts at 15 instead of 16. This feasible solution would be obtained using four time periods. No different solution can be obtained using five, six, and seven time periods. Using eight periods, however, allows us to find all the solutions with eight batches discussed in Section 10.1.2.

discrete time grid, and a lower bound, T^{LO}, and an upper bound, T^{UP}, on the length of the periodic solution can be obtained from instance-specific knowledge and constraints,[9] then we have to solve $\left(\lfloor T^{UP}/\delta \rfloor - \lceil T^{LO}/\delta \rceil + 1\right)$ problems with $\lceil T^{LO}/\delta \rceil, \lceil T^{LO}/\delta \rceil + 1, \cdots, \lfloor T^{UP}/\delta \rfloor - 1, \lfloor T^{UP}/\delta \rfloor$ time periods. Furthermore, inventory costs can be trivially modeled using linear terms in discrete time models.[10] Therefore, in this subsection we present a discrete time model based on the RTN representation.

Task setups are modeled using the ideas presented in Chapter 8. The unit is disaggregated into $|\mathbf{I}|$ task resources (task modes) and one *neutral* resource (neutral mode). One *start-setup* and one *end-setup* task are introduced per task to model transition between modes. The former consumes the neutral mode at its start and produces the corresponding task mode after σ_i periods. The latter is instantaneous ($\tau_i = 0$), consumes one of the task modes, and produces the neutral mode. The *start-setup* and *end-setup* tasks corresponding to processing task i are denoted by i^{SS} and i^{ES}, respectively.

If $N \in \left\{\lceil T^{LO}/\delta \rceil, \ldots, \lfloor T^{UP}/\delta \rfloor\right\}$ is the number of grid periods we would like to consider, then we define a time grid with points $n \in \mathbf{N} = \{0, 1, \ldots, N\}$, and we employ the following time indexed variables:

- $X_{in} \in \{0, 1\}$: $= 1$ if a batch of task i starts at time point n.
- $R_{in} \in \{0, 1\}$: $= 1$ if the unit is idle in mode i during period n.
- $R_n^N \in \{0, 1\}$: $= 1$ if the unit is in the neutral mode during period n.
- $B_{in} \in \mathbb{R}_+$: batchsize of batch of task i that starts at time point n.
- $I_{in} \in \mathbb{R}_+$: inventory of product i during period n.

The resource balance for the task modes is written as follows:

$$R_{i,n+1} = R_{in} + X_{i^{SS},n-\sigma_i} + X_{i,n-\tau_i} - X_{in} - X_{i^{ES},n}, \quad i, n. \tag{10.2}$$

The balance for the neutral mode is

$$R_{n+1}^N = R_n^N + \sum\nolimits_{i \in \mathbf{I}^{ES}} X_{in} - \sum\nolimits_{i \in \mathbf{I}^{SS}} X_{in}, \quad n, \tag{10.3}$$

where $\mathbf{I}^{SS}$ and $\mathbf{I}^{ES}$ are the sets of start-setup and end-setup tasks. Note that for a batch to end within the horizon,[11] it should start at $n \in \{0, 1, \ldots, N - \tau_i\}$, and that no batch can start on a unit after $N - \min_{i \in \mathbf{I}_j}\{\tau_i\}$. Thus, we set $X_{in} = 0, n > N - \tau_i$.

[9] We may be able to calculate the minimum time it would take to produce (minimum amounts of) all products; for instance, in the example in Section 10.1.2, we were able to easily calculate $T^{LO} = 21$ h. Also, an upper bound can be obtained from the maximum acceptable length of a periodic solution; a long periodic solution results in either complex batch sequences or large inventories.

[10] There exist simple linear reformulations for inventory cost calculations in continuous time models if certain assumptions hold; for example, one setup per product in a periodic solution. However, in the general case, the inventory cost calculations in continuous time models lead to nonlinear (bilinear) terms.

[11] In the single-unit problem, all feasible solutions can be converted to equivalent solutions in which all batches start and end within the horizon, that is, there are no batches starting at $n \in \{N - \tau_i + 1, \ldots N - 1\}$. Can you see why?

Since the same solution is repeated, the setup status and the decisions to start a setup or batch at the beginning of the next solution execution (i.e., at $n = |\mathbf{N}|$) should be the same as the ones at the beginning of the current one (at $n = 0$):

$$X_{i,|\mathbf{N}|} = X_{i,0}, \quad i; R_{i,|\mathbf{N}|} = R_{i,0}, \quad i; R^N_{|\mathbf{N}|} = R^N_0. \tag{10.4}$$

We will refer to the equalities enforced by (10.4) as *periodic constraints* or *boundary continuity*.

The batchsize constraints are

$$\beta_i^{MIN} X_{in} \le B_{in} \le \beta_i^{MAX} X_{in}, \quad i, n. \tag{10.5}$$

The inventory balance for products is written as follows:

$$I_{i,n+1} = I_{in} + \rho_i B_{i,n-\tau_i} + \xi_i \delta \le \chi_k^M, \quad i, n. \tag{10.6}$$

Recall that the demand for product i is $-\xi_i$ [kg·h^{-1}], so the shipments per period are $\xi_{in} = \xi_i \delta$ [(kg·h^{-1})·(h·period^{-1})] = [kg·period^{-1}]. We assume that the demand during period n is all shipped at time point n, so the inventory during period n is constant. Also, unlike the assumption made in short-term scheduling, where the starting inventory is known (i.e., variable I_{in} is replaced by parameter I_i^0 when the material balance is written for $n = 0$), here we enforce that the (variable) starting and ending inventories should be equal:

$$I_{i,|\mathbf{N}|} = I_{i,0}, \quad i \tag{10.7}$$

or, alternatively, (10.6) for $n = 1$ is written as follows:

$$I_{i,1} = I_{i,|\mathbf{N}|} + \rho_i B_{i,|\mathbf{N}|-\tau_i} + \xi_i \delta \le \chi_k^M, \quad i \tag{10.8}$$

Finally, the objective function [\$·h^{-1}] is:

$$\min \frac{1}{|\mathbf{N}|\ \delta} \sum\nolimits_{i,n} \left(\gamma_i^{SET} X_{iss_n} + \zeta_i \delta I_{in} \right) \tag{10.9}$$

If z_N^* is the optimal objective function value of the model employing $N \in \{\lceil T^{LO}/\delta \rceil\}, \ldots, \{\lfloor T^{UP}/\delta \rfloor\}$ time periods, then the optimal objective function value for the original problem is $\min_N \{z_N^*\}$.

10.1.5 Advanced Discrete Time Model

If it is desirable to solve a single optimization model to determine the optimal periodic solution, then we define a time grid with points $n \in \mathbf{N} = \{0, 1, \ldots, \lfloor T^{UP}/\delta \rfloor\}$ and introduce variable $T \in [T^{LO}, T^{UP}]$ to denote the period of the optimal solution. Further, we introduce $Z_n \in \{0, 1\}$ to represent the consecutive time periods of the grid that will be chosen to be *active* in the periodic solution; that is, $Z_n = 1$ if $n\delta \le T$ (see Figure 10.5). Conceptually, the constraints presented in the previous subsection should be enforced during active periods $(n\text{: } n\delta \le T)$ and relaxed during the inactive ones.

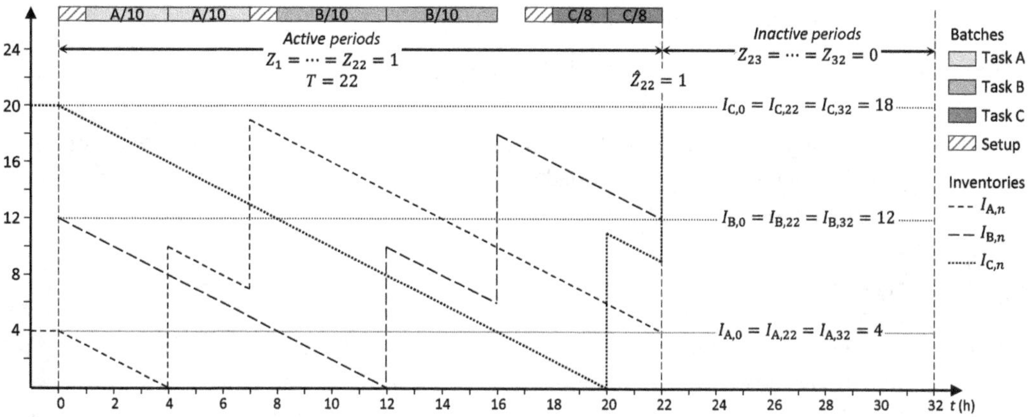

Figure 10.5 Illustration of grid period activation through Z_n variables; scheduling horizon contains 36 periods; periodic solution has 22 (activated) periods.

The setup status of the unit is described by (10.2) and (10.3). To enforce that no batch or setup is carried out during inactive periods, we add

$$\sum\nolimits_{i\in\mathbf{I}}\sum\nolimits_{n'\in\mathbf{N}_{in}^{U}}X_{in'}+\sum\nolimits_{i\in\mathbf{I}^{SS}}\sum\nolimits_{n'\in\mathbf{N}_{in}^{US}}X_{in'}\leq Z_{n+1},\quad n, \tag{10.10}$$

where $\mathbf{N}_{in}^{U} = \{n-\tau_i+1,\ldots,n\}$ and $\mathbf{N}_{in}^{US} = \{n-\sigma_i+1,\ldots,n\}$.

Equation (10.4) is used to enforce periodic constraints. Batch capacity constraints are enforced using (10.5).

The inventory balance is modified as follows:

$$I_{i,n+1} = I_{in} + \rho_i B_{i,n-\tau_i} + \xi_i \delta Z_n \leq \chi_k^M,\quad i,n, \tag{10.11}$$

subject to (10.7). Note that during inactive periods the inventory levels do not change because (1) there is no material production, because $X_{i,n-\tau_i} = 0$ from (10.10), and then $B_{i,n-\tau_i} = 0$ from (10.5); and (2) there are no shipments, because $Z_n = 0$ and thus $\xi_i \delta Z_n = 0$.

To enforce that active periods are consecutive, we can add

$$Z_{n+1} \leq Z_n,\quad n. \tag{10.12}$$

Alternatively, we can introduce $\hat{Z}_n$ to denote the last active period or the time point at which the active solution ends (i.e., $\hat{Z}_n = 1$ if $Z_n = 1$ and $Z_{n+1} = 0$), add

$$\hat{Z}_n = Z_n - Z_{n+1},\quad n \tag{10.13}$$

and also fix $Z_n = 1$ and $\hat{Z}_n = 0$ for $n \leq T^{LO}/\delta$.

Finally, the length of the periodic solution can be calculated by either of the following two equations:

$$T = \sum\nolimits_n \delta Z_n \tag{10.14}$$

$$T = \sum\nolimits_n n\delta\hat{Z}_n, \tag{10.15}$$

and the objective function becomes[12]

$$\min \frac{1}{T}\sum\nolimits_{i,n}\left(\gamma_i^{SET}X_{i^{SS}n} + \zeta_i\delta I_{in}\right), \tag{10.16}$$

which is nonlinear. The model presented in this section allows us to identify the optimal periodic solution directly, provided that there is a good estimate of T^{UP}, but it requires the solution of an MINLP, instead of multiple MILPs.

10.1.6 Remarks

Changeovers. If sequence-dependent changeovers, instead of setups, are present, production capacity and the trade-off between setup and inventory costs remain the two key considerations, but there is an added layer of complexity. First, production capacity depends not only on the number (of setups) but also the type of changeovers. For example, in the case of setups, if there is a single setup per product, the total setup time and thus production capacity does not depend on the batch sequence. In the presence of changeovers, though, the batch sequence determines the total changeover time and thus production capacity. Second, the trade-off between inventory and changeover costs becomes more nuanced. For example, even in the case of a single changeover per product, a given batch sequence may lead to high changeover costs, which would lead one to believe that a longer periodic solution would be better, whereas another sequence would result in low changeover costs compared to inventory costs. Despite the presence of more complex trade-offs, the treatment of sequence-dependent changeovers from a modeling standpoint is straightforward. For example, in the basic model presented in Section 10.1.4, (10.2) and (10.3) would have to be replaced by the equations we have already presented in previous chapters to model changeovers. Finally, if changeover costs are present, the term $\sum_{i,n}\gamma_i^{SET}X_{i^{SS}n}$ in the objective function is replaced by the corresponding term.

Profit Maximization. If, in addition to meeting some minimum demand, extra production can be sold at price π_i, then the basic model (Section 10.1.4) is modified in two ways. First, sales variables $S_{in} \in \mathbb{R}_+$ are introduced and the material balance is written as follows:

$$I_{i,n+1} = I_{in} + \rho_i B_{i,n-\tau_i} + \xi_i\delta - S_{in} \leq \chi_k^M, \quad i,n. \tag{10.17}$$

Note that if sales can occur at any point, then the introduction of S_{in} in (10.17) will lead to zero inventories through the entire horizon to minimize inventory costs. If sales are allowed only at prespecified points $n \in \mathbf{N}^S$, then we fix $S_{in} = 0$, $n \in \mathbf{N}\backslash\mathbf{N}^S$ and inventories can be positive.

Second, the objective function is modified as follows:

$$\max \frac{1}{|\mathbf{N}|\,\delta}\sum\nolimits_{i,n}\left(\pi_i S_{in} - \gamma_i^{SET}X_{i^{SS}n} - \zeta_i\delta I_{in}\right). \tag{10.18}$$

[12] Note that, to avoid using two domains for index i, we use $\sum_{i,n}\gamma_i^{SET}X_{i^{SS}n}$ instead of $\sum_{i\in\mathbf{I}^{SS},n}\gamma_i^{SET}X_{in}$.

Continuous Processing. Continuous processing can be addressed using the methods presented in Chapter 9: (1) a continuous task is treated as a batch task with processing time equal to one period ($\tau_i = 1$); (2) an one-period batch consumes/produces input/output materials at its start/end; (3) a *run* is represented via *multiple* consecutive one-period batches; and (4) constraints are added to enforce minimum and maximum run lengths. The additional features discussed in Chapter 9 (time delays, transitions, etc.) can be readily modeled.

Solution Method. In Section 10.1.4, we described a simplistic iterative solution procedure, for the fixed-period model, where the horizon/period is increased iteratively from $N = \lceil T^{LO}/\delta \rceil$ to $N = \lfloor T^{UP}/\delta \rfloor$. Interestingly, the nonlinear model in Section 10.1.5 can be solved using an iterative procedure, known as Dinkelbach's algorithm, that employs a linear objective function (see Section 10.4 for further discussion and references). The advantage of using the model presented in Section 10.1.5 coupled with Dinkelbach's algorithm is that in the general case fewer than ($\lfloor T^{UP}/\delta \rfloor - \lceil T^{LO}/\delta \rceil + 1$) iterations will be necessary.

10.2 Single-Stage Environment

The problem statement for single-stage environments is given in Section 10.2.1. In Section 10.2.2, we present a discrete time model, and, based on this model, in Section 10.2.3, we present how shipments at specified times can be addressed. In Sections 10.2.4 and 10.2.5, we discuss under what circumstances some common assumptions/simplifications can be made and how they impact the corresponding solutions. Finally, in Sections 10.2.6 and 10.2.7 we overview some basic ideas for the formulation of continuous time models (for batch processes) and continuous processes, respectively.

10.2.1 Problem Statement

We consider a single-stage (parallel units) facility producing multiple products from possibly multiple raw materials. Using the STN framework, we are given the following:

- Processing units $j \in \mathbf{J}$.
- Materials $k \in \mathbf{K}$, including raw materials $k \in \mathbf{K}^{RM}$ and products $k \in \mathbf{K}^{P}$: all materials have dedicated storage vessels with capacity χ_k^M; the output of batches producing the same product can be mixed in the corresponding storage vessel; the inventory cost of product k is ζ_k; and the *continuous* demand for product k is ξ_k [kg/h].
- Tasks $i \in \mathbf{I}$: for simplicity, we assume that each task can be carried out in only one unit (i.e., $|\mathbf{J}_i| = 1$); $\mathbf{I}_j$ is the subset of tasks that can be carried out in unit j; a task consumes a raw material, $k^-(i)$, and produces one final product, $k^+(j)$, according to conversion coefficients ρ_{ik}; the processing time, setup time, and setup cost are τ_i, σ_i, and γ_i^{SET}, respectively; the minimum and maximum batch-sizes are β_i^{MIN} and β_i^{MAX}, respectively.

The goal is to find the least cost ($/h) periodic solution that meets product demand. As in Section 10.1, the cost includes a setup and an inventory component. In the periodic solution, a product may be produced by multiple tasks and in multiple units; and a unit can produce multiple products and/or the same product through multiple batches of the same task. Also, all batches start and finish within the periodic solution, that is, there are no batches *crossing over* the boundary of the solution.[13]

10.2.2 Basic Model

We present the extension of the model, used in an iterative procedure, presented in Section 10.1.4. The counterpart of the model presented in Section 10.1.5 can be developed using the same ideas: define a horizon equal to T^{UP}, introduce variables $Z_n \in \{0,1\}$, and ensure that the schedule during the *active* horizon (i.e., periods with $Z_n = 1$) satisfies all the constraints of a periodic solution.

We introduce the following variables:

- $X_{in} \in \{0,1\}$: =1 if a batch of task i starts at time point n.
- $R_{ijn} \in \{0,1\}$: = 1 if unit j is idle in mode i during period n.
- $R^N_{jn} \in \{0,1\}$: = 1 if unit j is the neutral mode during period n.
- $B_{in} \in \mathbb{R}_+$: batchsize of batch of task i that starts at time point n.
- $I_{kn} \in \mathbb{R}_+$: inventory of material k during period n.

Note that, to simplify the presentation, variables R_{ijn} are indexed by j, although specifying i would suffice (since $|\mathbf{J}_i| = 1$). The other difference compared to the model in Section 10.1.4 is that here we use variables I_{kn}, for both raw materials and products, instead of variables I_{in}, for products only.

The resource balance for the task modes is written as follows:

$$R_{ij,n+1} = R_{ijn} + X_{i^{SS},n-\sigma_i} + X_{i,n-\tau_i} - X_{in} - X_{i^{ES},n}, \quad j, i \in \mathbf{I}_j, n. \tag{10.19}$$

The balance for the neutral mode is

$$R^N_{j,n+1} = R^N_{j,n} + \sum\nolimits_{i \in \mathbf{I}_j^{ES}} X_{in} - \sum\nolimits_{i \in \mathbf{I}_j^{SS}} X_{in}, \quad j, n, \tag{10.20}$$

where $\mathbf{I}_j^{SS}$ and $\mathbf{I}_j^{ES}$ are the sets of start-setup and end-setup tasks in unit j.

Periodic constraints are enforced using equations similar to (10.4), and the batchsize capacity constraints are

$$\beta_i^{MIN} X_{in} \le B_{in} \le \beta_i^{MAX} X_{in}, \quad j, i, n. \tag{10.21}$$

[13] As we will see in Section 10.3, enforcing that all batches finish within the periodic solution (referred to as no batch cross over) may cut off the optimal solution or even lead to infeasibility. Thus, in network environments, discussed in Section 10.3, this requirement will be relaxed. In single-stage environments, however, a solution very *similar* to the optimal one, which would be found if this restriction was relaxed, can almost always be identified. Thus, the no cross over restriction is enforced in Section 10.2.

The inventory balance for raw materials is

$$I_{k,n+1} = I_{kn} + \sum\nolimits_{i \in \mathbf{I}_k^-} \rho_{ik} B_{in} \leq \chi_k^M, \quad k \in \mathbf{K}^{RM}, n, \tag{10.22}$$

where we assume that there are no deliveries, but we start from a sufficiently large initial inventory $I_{k,n=0} = I_k^0 \leq \chi_k^M$. In addition, we do not enforce the periodic inventory constraint $(I_{k,|\mathbf{N}|} = I_{k,0})$; that is, we do not consider the (periodic) replenishment of raw materials and how it affects the schedule. This assumption will be relaxed in Section 10.3.

The inventory balance for final products is

$$I_{k,n+1} = I_{kn} + \sum\nolimits_{i \in \mathbf{I}_k^+} \rho_{ik} B_{i,n-\tau_i} + \xi_k \delta \leq \chi_k^M, \quad k \in \mathbf{K}^P, n \tag{10.23}$$

subject to $I_{k,|\mathbf{N}|} = I_{k,0}$, $k \in \mathbf{K}^P$.

Finally, the objective function [$\$\cdot h^{-1}$] is

$$\min \frac{1}{|\mathbf{N}|\,\delta} \left\{ \sum\nolimits_{j, i \in \mathbf{I}_j} \gamma_i^{SET} \sum\nolimits_n X_{i^{SS} n} + \sum\nolimits_{k,n} \zeta_k \delta I_{kn} \right\}. \tag{10.24}$$

Fixed $\left(\sum_{i,n} \gamma_i^F X_{in}\right)$ and variable $\left(\sum_{i,n} \gamma_i^V B_{in}\right)$ production cost terms can be added, with the latter incorporating raw material costs, in the case where different raw materials, with different prices, can be used for the production of the same product.

10.2.3 Shipments at Specified Times

Two major assumptions we have employed so far are that (1) the supply of raw materials is unconstrained, and (2) products are continuously shipped. Both of these assumptions can be relaxed, allowing us to more accurately model realistic situations, by introducing materials shipments at specified points. Since we seek periodic solutions, it is reasonable to consider *periodic* shipments, though the periods of these shipments need not be the same across materials, nor are they equal to the period of the solution.

Let $\hat{T}_k$ be the period (h) of shipments of material k, with, in general, $\hat{T}_k \neq \hat{T}_{k'}$; and $\hat{\xi}_k$ the corresponding net amount (kg) shipped. Further, without loss of generality, we assume that the delivery of one raw material, say k^*, occurs at times $0, \hat{T}_{k^*}, \; 2\hat{T}_{k^*}, \ldots,$ which then allows us to define a *lag* between shipments in terms of parameter $\hat{\omega}_k$: the shipments of material $k \neq k^*$ occur at $\hat{\omega}_k, \hat{\omega}_k + \hat{T}_k, \hat{\omega}_k + 2\hat{T}_k, \ldots$. In other words, we *anchor* the horizon with respect to the shipments of k^*. Importantly, introducing this point of reference does not cut off any solution, as discussed in Example 10.1. We assume that all $\hat{T}_k$ and $\hat{\omega}_k$ are multiples of δ.

The next step is to introduce the shipment information, originally represented in terms of parameters $\hat{T}_k$, $\hat{\xi}_k$, and $\hat{\omega}_k$, into the inventory balances. The key idea is that the period of the solution should be an integer multiple of the least common multiple (LCM) of all shipment periods $(\hat{T}_k)$. If $N_k = \left(T/\hat{T}_k\right) \in \mathbb{N}$ is the number of shipments of material k during the solution, then the shipments of k occur at points

$\hat{\omega}_k/\delta, (\hat{\omega}_k + \hat{T}_k)/\delta, \ldots, (\hat{\omega}_k + (N_k - 1)\hat{T}_k)/\delta$, and the shipment parameters, ξ_{kn}, entering the material balances can be calculated as follows:

$$\xi_{kn} = \begin{cases} \hat{\xi}_k & \text{if } n\delta = \hat{\omega}_k + m\hat{T}_k, m \in \{0, 1, .., N_k - 1\}, \\ 0 & \text{otherwise} \end{cases} \quad k, n \tag{10.25}$$

Once parameters ξ_{kn} are calculated, material balances become

$$I_{k,n+1} = I_{kn} + \sum\nolimits_{i \in \mathbf{I}_k^-} \rho_{ik} B_{in} + \xi_{kn} \leq \chi_k^M, \quad k \in \mathbf{K}^{RM}, n \tag{10.26}$$

$$I_{k,n+1} = I_{kn} + \sum\nolimits_{i \in \mathbf{I}_k^+} \rho_{ik} B_{i,n-\tau_i} + \xi_{kn} \leq \chi_k^M, \quad k \in \mathbf{K}^P, n \tag{10.27}$$

subject to $I_{k,|\mathbf{N}|} = I_{k,0}, k \in \mathbf{K}^{RM} \cup \mathbf{K}^P$. The setup and task-unit assignment constraints, (10.19) and (10.20), batchsize constraint (10.21), as well as the objective function (10.24), remain the same.

Example 10.1 Modeling of Shipments at Specified Times We consider an instance with shipments at specified points for two raw materials, A and B, and two products, C and D. Each delivery of A, which is used as the reference material, is 20 kg, and there is one delivery every two days, starting at time 0. Thus, if $\delta = 1$ h is adopted, the parameters describing the deliveries of A are $\hat{T}_A = 24$ h, $\hat{\xi}_A = 20$ kg, and $\hat{\omega}_A = 0$. Using these parameters in (10.25) we obtain $\xi_{A,0} = \xi_{A,48} = 20$. Figure 10.6 shows the raw data, in terms of days; the secondary parameters $\hat{T}_k$, $\hat{\xi}_k$, and $\hat{\omega}_k$; and model parameters, ξ_{kn}, for the remaining materials. Note that the LCM of all periods is 48 h and the chosen horizon/period is 96 h.

A campaign with periodic shipments is shown in Figure 10.7. The batch sequence and timing are the same as in Figure 10.2B (Campaign 4) but the inventory profiles are

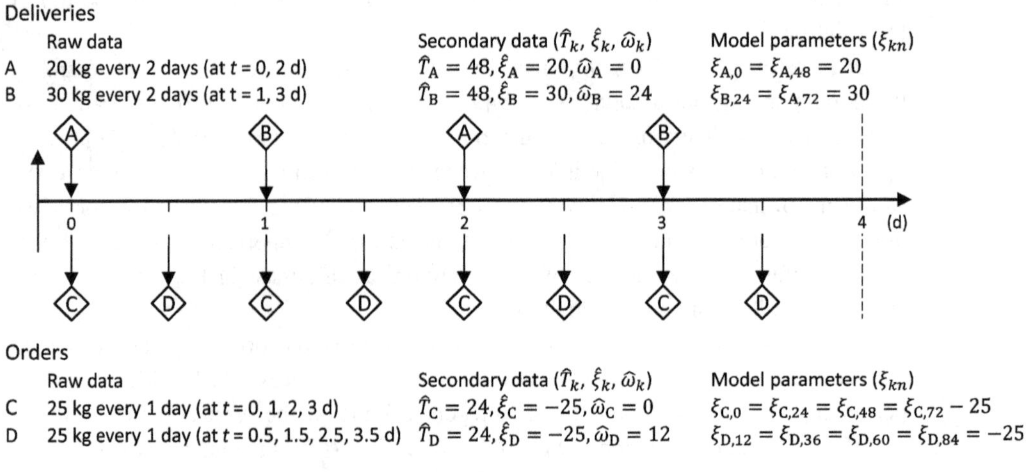

Figure 10.6 Calculation of parameters ξ_{kn} from period, shipment amount, and phase lag (given in days) of periodic shipments; parameters are calculated for $\delta = 1$ h.

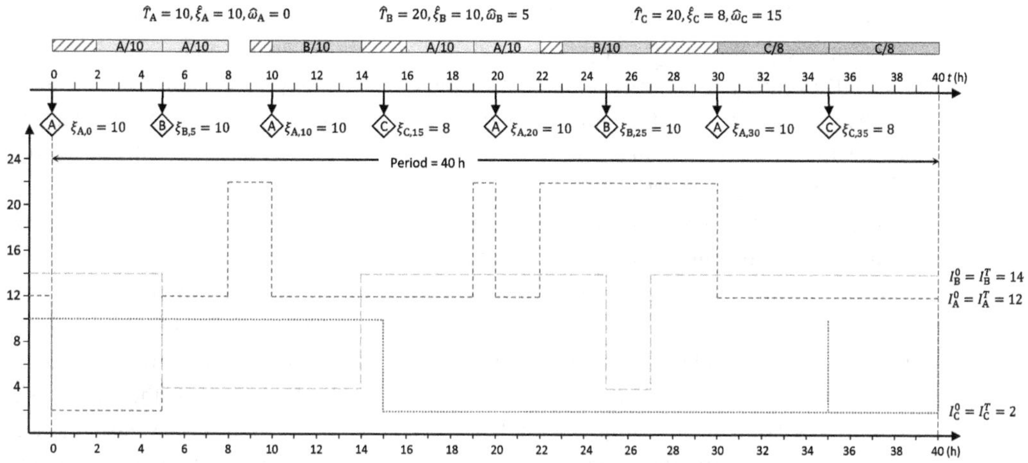

Figure 10.7 Periodic solution with product shipments at specified time points; parameters describing period shipments are shown above blocks of the Gantt chart; corresponding ξ_{kn} parameters, used in material balances, are shown along the timeline below the Gantt chart.

fundamentally different: they increase when a batch is finished (as in Figure 10.2B) but drop suddenly when there is an order due (rather than continuously as in Figure 10.2B), and remain constant if no batch is finished nor an order is due. The starting and ending levels are the same. Note that the total demand is the same as in Figure 10.2B; for example, the total shipments of A are 4·10 = 40 kg or 40 h = 1 kg/h (same as in Table 10.1). Compared to Figure 10.2B, the inventory cost for product A is higher. (Can you see why? Is this always the case when demand is due at specific points? Can you think of a situation where continuous consumption would lead to higher inventory cost?)

10.2.4 Simplifying Assumptions and Solution Features

Two common assumptions in single-stage periodic scheduling are that (1) each product is produced in only one unit and (2) all batches producing a product are carried out consecutively in a unit. The first assumption may be valid in systems where the number of products is significantly higher than the number of units and, importantly, the time needed to carry out the batches to meet the demand is balanced across products.[14] However, evaluating whether the required production times are balanced is nontrivial

[14] Note that the total processing times necessary to meet demand being balanced is different from the demands across products being balanced. For example, consider two products, A and B, produced in two units, U1 and U2, with demand $\xi_A = \xi_B = 1$, and $\beta_i^{MIN} = \beta_i^{MAX} = \beta_i$ for all tasks. Product A is produced by T1A $\in \mathbf{I}_{U1}$ and T2A $\in \mathbf{I}_{U2}$ with $\beta_{T1A} = \beta_{T2A} = 10$ and $\tau_{T1A} = \tau_{T2A} = 4$; and Product B is produced by T1B $\in \mathbf{I}_{U1}$ and T2B $\in \mathbf{I}_{U2}$ with $\beta_{T1B} = \beta_{T2B} = 5$ and $\tau_{T1B} = \tau_{T2B} = 5$. Meeting the demand for products A and B over say 20 h (20 kg) requires two and four batches for A and B, respectively, which translates into 8 and 32 h, respectively, of processing. Thus, though the demands are equal, the processing time required to meet these demands are rather different, and, as a result, product B will have to be produced in both units.

because it depends on unit capacities and task processing times, that is, the periodic solution itself. Note that if a product can be produced in many units, then the final product inventory, which is *continuously* depleted, has multiple incoming flows, which means that the units producing this product are coupled through the inventory balance. In the model presented in Section 10.2.2, this coupling is represented by the term $\sum_{i \in \mathbf{I}_k^+} \rho_{ik} B_{i,n-\tau_i}$ in (10.23).

The second assumption is valid in systems where (1) the demand can be met only by solutions with the minimum number of setups; or (2) setup costs are significantly higher than inventory costs. In the general case, however, as we discussed in Section 10.1.2, the optimal solution may include multiple setups for the same task because of the setup versus inventory trade-off. Large differences in demand across products may also lead to solutions with long periods and, consequently, multiple setups for the same task within a period. Finally, binding final product storage constraints may also induce multiple setups for the same product during a periodic solution to maintain low inventory.

The solution features discussed in the previous two paragraphs are shown in Figure 10.8 through a three-unit process producing five products. Unit U1 is the only unit capable of producing product A, so it cannot be used exclusively for high-demand product B, which is then produced in both units U1 and U2. Product C is also produced in unit U2, while low-demand products D and E are produced in unit U3, but due to low storage capacity, they have to be produced using two setups each.

We close with an interesting observation regarding problem classification with an important notation implication. The batch sequence in solutions satisfying the two aforementioned assumptions can be viewed as a cycle, in a graph theoretic sense (see definition of cycle in a graph in Section 2.1.3). Specifically, if the batch sequence is viewed as a *closed walk* (again, see Section 2.1.3 for a definition) in a graph with products (not batches) represented as vertices, then the sequence is also a cycle; for example, sequence [A → A → C → B → B] can be viewed as cycle A → C → B → A. If the second assumption is relaxed, then sequence [A → C → B → A → B] is feasible

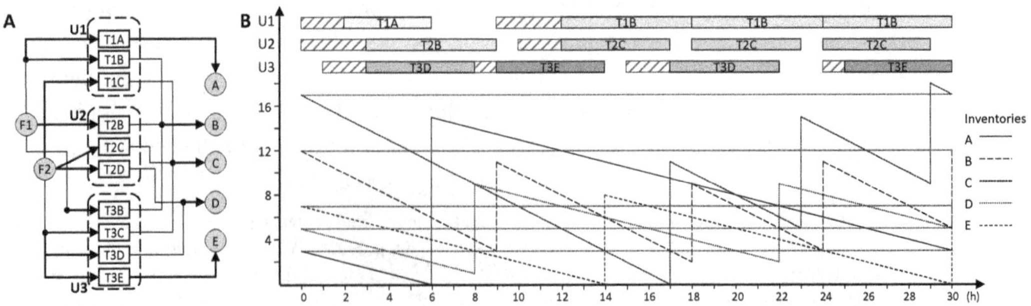

Figure 10.8 Illustration of solution features: a product produced in multiple units (product B in units U1 and U2) and two setups for the same product in the same unit (products D and E in unit U3). (A) Process network used for illustration. (B) Gantt chart and inventory profiles of the periodic solution.

but corresponds to a closed walk in the graph (A → C → B → A → B → B), not a cycle. This explains why in footnote 1 in the introduction of the present chapter we said that we will seek a periodic solution rather than a cycle, with the former being more general. In graph theoretic terms, we seek a closed walk rather than a cycle; all cycles are closed walks, but there are closed walks that are not cycles.

10.2.5 Unit-Specific Solutions

If products with very different demands and/or frequency of periodic shipments are produced in the same facility, then it is reasonable to have units operating under different periods. To illustrate, consider a facility, shown in Figure 10.9A, consisting of two units (U1 and U2); producing seven products (A, B, …, G), with demands $\xi_A = \xi_B = 0.5$ and $\xi_C = \ldots = \xi_G = 0.2$ (kg/h); and the following parameters for all i: $\tau_i = 4$, $\sigma_i = 1$, and $\beta_i^{MIN} = \beta_i^{MAX} = \beta_i = 5$. Further, all tasks consume raw material RM (not shown) with coefficient $\rho_{i,RM} = -1, \forall i$, and all production coefficients are $\rho_{i,k^+(i)} = \rho_i = 1, \forall i$. A single batch of a task producing products A or B, is sufficient to meet demand for 10 h, so one periodic solution for one unit is based on sequence [A → B] → with timing {1,6}. This campaign will be the optimal if inventory costs are high compared to setup costs. A single batch of a task producing products C, D, E, F, and G is sufficient to meet demand for 25 h, so it is reasonable to have one unit dedicated to producing these five products, with sequence [C → D → E → F → G] → and timing {1, 6, 11, 16, 21} (see Figure 10.9B). Note that the solutions for unit U1 based on the [A → B] → sequence have periods that are multiples of 10, whereas the

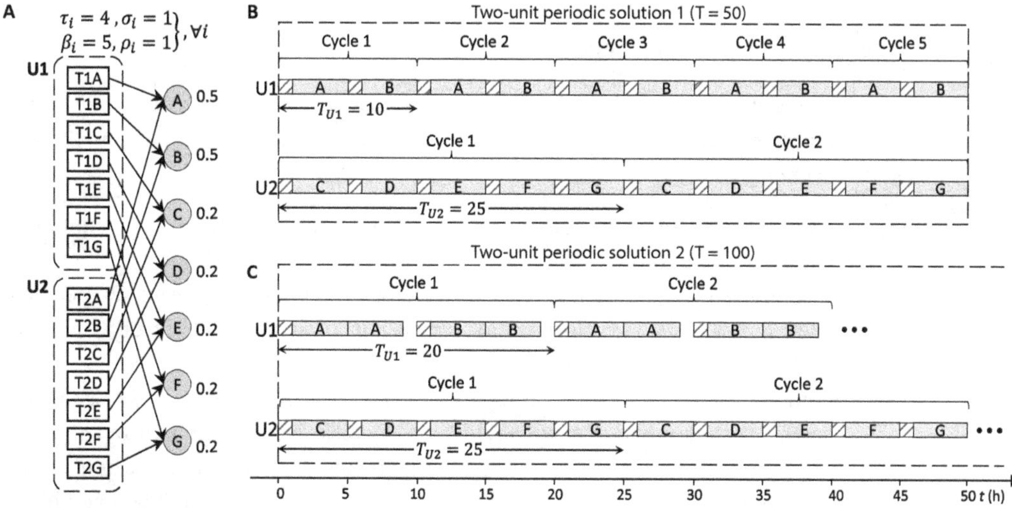

Figure 10.9 Unit-specific periodic solutions. (A) Network used for illustration; demand is given next to product nodes. (B) Solution where the period of units U1 and U2 is 10 and 25 h, respectively; the two-unit periodic solution has period 50 h. (C) Alternative solution, preferred if setup costs for A and B are large compared to inventory costs. (The term *cycle* is used to represent the partial batch sequence within a unit-specific periodic solution.)

solutions for unit U2 have periods that are multiples of 25. Thus, the shortest single-stage periodic solution, including both units, will have $T = 50$ h (see Figure 10.9B). An alternative solution, preferred if the setup costs for A and B are large compared to the corresponding inventory costs, is shown in Figure 10.9C. Note that the period of the solution for unit U1 is now 20 h, allowing for one setup per two batches of product A or B, thereby lowering the relative contribution of setup cost. The solution for U2 remains the same, with $T_{U2} = 25$, which means that the two-unit solution has period $T = 100$.

Allowing units to operate under different periods can lead to a simpler problem, especially if there are no raw material–related coupling restrictions (e.g., tasks producing different products in different units consume the same raw material that is delivered periodically). We distinguish three cases, all with no raw material restrictions:

(1) Each product is produced in one unit and products can be preassigned to units: the single-stage problem can be decomposed into $|\mathbf{J}|$ single-unit problems. This decomposition is valid even when final products are due periodically.

(2) Each product is produced in one unit, but the assignment of products to units is unknown: a single-stage problem needs to be solved, but tightening constraints, expressing the one-unit-per-product restriction, can be generated.

(3) Each product is produced in one unit, and there is only one setup per product in each cycle (but the assignment of products to units is unknown): the problem cannot be decomposed into single-unit problems, but a series of major simplifications can be made.

10.2.6 Continuous Time Models: Basics

Based on the discussion in the last three subsections, we note the following regarding the formulation of continuous time models for the general single-stage batch production environment:

(1) A single, common across units, time grid is necessary to account for the production of the same product in multiple units.

(2) A common grid is also necessary if there are (periodic) shipments at specified times; in this case, a synchronization between the unit(s) producing a product and the shipments is necessary.

(3) Similarly, a common grid is necessary to model the consumption of the same raw material, potentially delivered periodically, by tasks carried out in different units.

(4) To model task setups, the RTN-inspired approach using task modes can be used.

(5) If inventory costs cannot be modeled linearly, then the objective function will be nonlinear, even for fixed period T.

(6) If variable period T is considered, then solutions with different periods can be identified using the same time grid $n \in \mathbf{N}$ by simply using variable horizon $\eta = \mathrm{T}_{|\mathbf{N}|}$.

10.2.7 Continuous Processing: Basics

Periodic solutions of continuous facilities may exhibit the same features we discussed in the previous subsections: (1) a product is produced in more than one unit; (2) multiple setups for the same task (product) in the same unit may be necessary; and (3) multiple runs of the same task may be executed while the unit is set up for that task due to, for example, maximum run length constraints (see discussion in Chapter 9). If the maximum run length constraint is relaxed and storage constraints are not binding, then we can assume that each setup is followed by a single run, which implies that no time points are needed to map the beginning (and end) of multiple runs between two setups. This assumption, which is common in the literature, leads to the formulation of continuous time models employing fewer time points.

The major difference between batch and continuous processing, therefore, as discussed in Chapter 9, is the evolution of inventory. If the demand is continuous, then continuous processing implies that both production and consumption (shipments) of products occur *continuously*, which leads to linear inventory changes only (see Example 10.2). During time intervals where there is no production, the slope of the inventory profile is equal to the demand (rate), ξ_k; while when there is production, the slope is equal to the production rate minus demand. In discrete time models, the production of material k during period n is $\sum_{i\in\mathbf{I}_k^+}\rho_{ik}B_{i,n-1}$, so (10.23) is replaced with

$$I_{k,n+1} = I_{kn} + \sum_{i\in\mathbf{I}_k^+}\rho_{ik}B_{i,n-1} + \xi_k\delta \leq \chi_k^M, \quad k\in\mathbf{K}^P, n \tag{10.28}$$

Note that when there is no production, the inventory decreases by $-\xi_k\delta$ (i.e., at rate ξ_k), and when there is production the inventory changes by $\sum_{i\in\mathbf{I}_k^+}\rho_{ik}B_{i,n-1} + \xi_k\delta$.

Example 10.2 Periodic Scheduling of Continuous Processes: Inventory Profiles We consider a continuous process, consisting of two parallel units (U1 and U2), producing four final products $\left(\mathbf{K}^P = \{\mathrm{A,B,C,D}\}\right)$, from two raw materials $\left(\mathbf{K}^{RM} = \{\mathrm{F1,F2}\}\right)$, through six tasks (T1A, T1B, T1C, T2B, T2C, and T2D). All tasks have (fixed) processing rates (i.e., $\upsilon_i\delta = \beta_i = \beta_i^{MIN} = \beta_i^{MAX}, \forall i$), and it is assumed that there is no delay between consumption and production. Figure 10.10A gives the STN representation of the facility, along with the corresponding (fixed) processing rates and demands. The maximum run length for T1B is $\tau_{\mathrm{T1B}}^{MAX} = 8$. The Gantt chart and the corresponding inventory profiles of the products in a periodic solution are shown in Figure 10.10B. Note that the actual profiles, as opposed to the stepwise ones calculated by the discrete time model, are shown. Since a run of task T1A cannot last more than 8 h, two runs are executed. (Why is this necessary?) We also observe the following:

(1) When there is no production, the rate of inventory decrease is equal to ξ_k (please verify).
(2) The rate of production by a run is equal to β_i because the processing rate is fixed and all consumption/conversion coefficients are –1/+1.
(3) At any time, a product is produced by at most one run of a task.

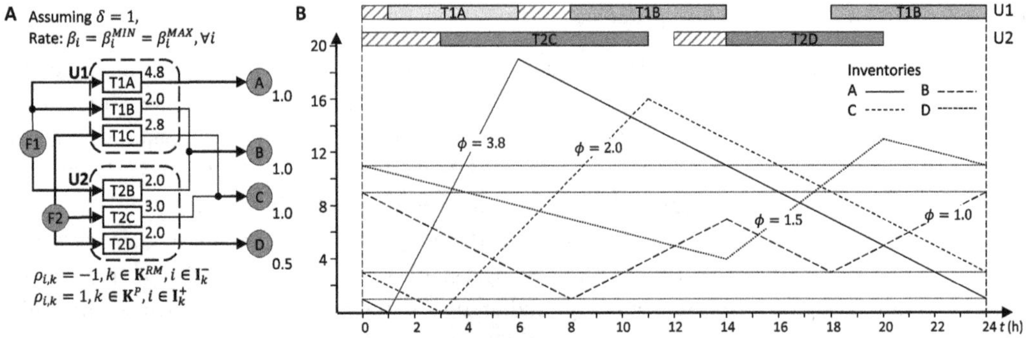

Figure 10.10 Inventory evolution in periodic solution of continuous single-stage process. (A) Example facility used for illustration; processing rates (next to task rectangles) and demands (next to material circles) are given. (B) Gantt chart of periodic solution and corresponding inventory profiles; rate of change (slope) during selected intervals is also given.

Based on (2) and (3), we conclude that when a run of $i \in \mathbf{I}_k^+$ is producing k, then the rate of inventory change (slope) is $(\beta_i - \xi_k)$. Please verify using the given parameters and inventory profiles in Figure 10.10B.

The modeling of inventory evolution using a continuous time formulation requires the introduction of some new ideas. Assuming that all starts and ends or runs are mapped onto time points, the two key concepts are the following. First, inventory levels have to be calculated at time points only, because a slope change can only occur at a time point, which implies that if the inventory satisfies no negativity and storage capacity constraints at time points, then it satisfies these constraints everywhere. Second, the *slope* of change between time points $n-1$ and n (i.e., period n) has to be calculated based on (the processing rates of) the tasks that are running during period n; this can be accomplished by introducing binary variables to denote the start and end of a run, a binary variable to denote the processing task during a period, and then a nonnegative variable for the corresponding rate. Conceptually, the idea is similar to the use of start, end, and processing binary variables used in Section 7.3.1 for the modeling of batch tasks.

10.3 Network Environment

Periodic scheduling in network environments requires to account for batch *cross over*, that is, batches starting in one execution of the periodic solution and ending in the next. The problem statement and introduction of batch cross over is given in Section 10.3.1, and an STN-based discrete time model is presented in Section 10.3.2. We assume that raw materials are available as needed, and thus not modeled, and that final products are continuously shipped.

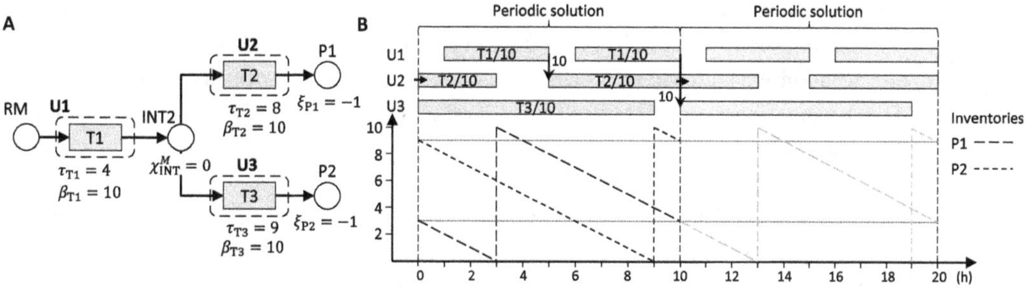

Figure 10.11 Example of a periodic solution with batch cross over. (A) STN-representation and process parameters of instance used for illustration. (B) Periodic solution with T = 10 h; batch cross over is represented by small horizontal arrows; intermediate material transfer is represented by vertical arrows; inventory of RM is not shown for simplicity.

10.3.1 Problem Statement

The facility is described in terms of the sets and parameters introduced in Chapter 7, assuming that each task can be carried out in only one unit. For simplicity, we assume that there are no task setups.[15] In short, the following are given:

- Tasks, $i \in \mathbf{I}$, with processing time τ_i, minimum/maximum batchsize $\beta_i^{MIN}/\beta_i^{MAX}$, and conversion coefficients ρ_{ik}.
- Units $j \in \mathbf{J}$; $\mathbf{I}_j$ is the subset of tasks that can be executed on unit j.
- Materials $k \in \mathbf{K}$, stored in dedicated vessels with capacity χ_k^M, and demand rate for products $\left(k \in \mathbf{K}^P\right)$ ξ_k; $\mathbf{I}_k^+/\mathbf{I}_k^-$ is the set of tasks producing/consuming material k; $\mathbf{K}_i^+/\mathbf{K}_i^-$ is the set of materials produced/consumed by task i.

The goal is to find the least cost-periodic solution that meets product demand.

To illustrate the concept of batch cross over, consider the network shown in Figure 10.11A, along with the associated parameters, based on a discretization of one hour ($\delta = 1$). In addition, all consumption/production coefficients are –1/+1. From the *continuous* demand ($\xi_{P1} = \xi_{P2} = -1$), we calculate $\xi_{P1,n} = \xi_{P2,n} = -1$, and note that to meet the demand one batch of T2 and T3 have to be executed every 10 h. The key feature of the network is that there is no storage for intermediate INT $\left(\chi_{INT}^M = 0\right)$, which means that all batches of T2 and T3 should run immediately upon completion of T1 batches. This, in turn, implies that in all feasible solutions there will be a lag between the start of a batch of T2 and the start of a batch of T3. (Why?) The lag will be between 4 and 6 h. (Can you see why?) Thus, there is no periodic solution in which all batches that start at $n \in \mathbf{N}_T = \{1, 2, \ldots, T\}$ can finish in $n' \in \mathbf{N}_T$. In other words, there is no solution in which all batches can start and finish within the same execution of the

[15] In Section 10.1.2, we showed that the trade-off between inventory and setup costs plays a key role in periodic scheduling. In this section, we do not consider setups because finding periodic solutions is relevant even in the absence of setups, and there are other features that require the introduction of new concepts. Also, from a modeling viewpoint, the existence of setups can be readily addressed, as has already been shown.

solution. However, if we allow batches that start at $n \in \mathbf{N}_T$ to finish at $n' > T$, then we can obtain a periodic solution with batch cross over, that is, a batch that starts during one solution execution (i.e., at $n = T - \nu$, with $\nu < \tau_i$) *ends* in the next one (i.e., at $n' = \tau_i - \nu$). Figure 10.11B shows a solution with period $T = 10$ h, in which a batch of T3 starts at $n = 5 = \mathrm{T} - \nu = 10 - 5 = 5 > 2 = 10 - 8 = T - \tau_{\mathrm{T3}}$, and ends at $n' = 3 = 8 - 5 = \tau_{\mathrm{T3}} - \nu$. A second execution, over interval $[10, 20]$, is shown to illustrate how the batch of T3 that ends, in reality, at $n' = 13$ can be viewed as a batch continuing at the beginning of the solution ending at $n' = 13$. Note that inventories at the beginning and end of the solution are the same despite batch cross over.

10.3.2 Model

We employ the following variables:

- $X_{in} \in \{0, 1\}$: = 1 if a batch of task i starts at time point n.
- $B_{in} \in \mathbb{R}_+$: batchsize of batch of task i that starts at time point n.
- $I_{kn} \in \mathbb{R}_+$: inventory of material k during period n.

As in previously presented STN-based models, the model consists of task-unit assignment, batchsize capacity, and inventory balance constraints. There are two differences:

(1) A unit may be busy during period n because it carries out a batch that started at $n' > n$; for example, in Figure 10.11B, during $n = 2$ unit U2 is executing a batch of T2 that started at $n' = 5$.

(2) Inventory levels at n may be affected by batches that start at $n' > n$; for example, in Figure 10.11B, the inventory of P2 at $n = 3$ increases because a batch that started at $n' = 5$ ends.

To account for the former, we replace $\mathbf{N}_{in} = \{n - \tau_i + 1, \ldots, n\}$, which is used in the standard task-unit assignment constraint,[16] with $\mathbf{N}_{in}^{CO}$, defined as follows:

$$\mathbf{N}_{in}^{CO} = \begin{cases} \mathbf{N}_{in} & n \geq \tau_i - 1 \\ \{|\mathbf{N}| - \tau_i + n + 1, |\mathbf{N}| - \tau_i + n + 2, \ldots, 0, 1, \ldots, n - 1, n\} & n < \tau_i - 1 \end{cases}, \; i. \tag{10.29}$$

The concept is illustrated in Figure 10.12 using $\delta = 1$, a single task T1 with $\tau_{\mathrm{T1}} = 5$, and a periodic solution with $T = 10$. When $n < \tau_i - 1$, the task-unit assignment constraint should include decisions at time points at the end of of the previous execution of the solution. However, these decisions are identical to the decisions made at the end of the current execution. If we use $n < 0$ to represent points before the execution of interest, then the decisions made, for example, at $n = -2$ are the same as the decisions made at $n = 8$. Thus, the task-unit assignment constraint becomes

$$\sum_{i \in \mathbf{I}_j} \sum_{n' \in \mathbf{N}_{in}^{CO}} X_{in'} \leq 1, \quad j, n. \tag{10.30}$$

[16] In (10.3), (10.11), and (10.21) we use $\mathbf{N}_{in}^{U}$ instead of $\mathbf{N}_{in}$, to differentiate between processing tasks and setups, but the definitions of $\mathbf{N}_{in}^{U}$ and $\mathbf{N}_{in}$ are identical.

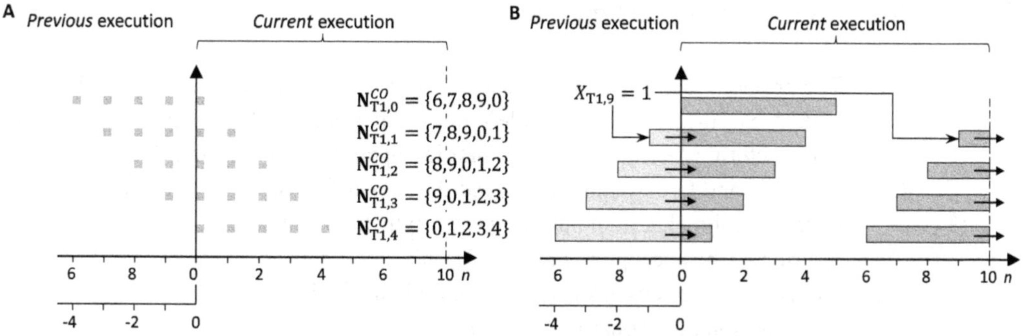

Figure 10.12 Modeling of batch cross over using for task T1 with $\tau_{T1} = 5$, and solution with $T = 10$. (A) Definition of $\mathbf{N}^{CO}_{in}$ for $n \leq \tau_i - 1$ through (10.29). (B) Representation of batches starting before the beginning of the current solution as batches starting in the current solution and finishing beyond the end of the current horizon.

Note that the decisions at $n = 10$ are the same as the decisions made at $n = 0$, so $X_{i,10}$ are not defined, or $n' = 10 \notin \mathbf{N}^{CO}_{in}$ for any i, n. This means that no equations are necessary to enforce periodic constraints. Further, the batchsize capacity constraint, (10.21), remains the same.

To address the latter, we use a similar concept: if production from batches that started before the current solution starts should be taken into account, which is necessary at $n < max_{i\in\mathbf{I}^+_k}\{\tau_i\}$ (can you see why?), then we modify the material balance equation to include such batches, equivalently, as batches that start in the current execution. This is accomplished by replacing $B_{i,n-\tau_i}$ in the traditional STN-based material balance,

$$I_{k,n+1} = I_{kn} + \sum\nolimits_{i\in\mathbf{I}^+_k}\rho_{ik}B_{i,n-\tau_i} + \sum\nolimits_{i\in\mathbf{I}^-_k}\rho_{ik}B_{in} + \xi_k\delta \leq \chi^M_k, \quad k,n \tag{10.31}$$

with $B_{i,n-m(i,n)}$:

$$I_{k,n+1} = I_{kn} + \sum\nolimits_{i\in\mathbf{I}^+_k}\rho_{ik}B_{i,n-m(i,n)} + \sum\nolimits_{i\in\mathbf{I}^-_k}\rho_{ik}B_{in} + \xi_k\delta \leq \chi^M_k, \quad k,n, \tag{10.32}$$

where $m(i,n)$ is an operator defined as

$$m(i,n) = \begin{cases} \tau_i, & n \geq \tau_i \\ |\mathbf{N}| - \tau_i + n + 1, & n < \tau_i \end{cases}, \quad i. \tag{10.33}$$

The basic model consists of (10.30), (10.21), and (10.32), and the objective function

$$\min \frac{1}{T}\sum\nolimits_{i,n}\zeta_i\delta I_{in}. \tag{10.34}$$

A number of problem features can be readily handled: (1) setups and changeovers; (2) periodic material shipments, as discussed in Section 10.2.3; (3) utility constraints, including time-varying capacity and cost; (4) additional sales (in a max Profit problem); and (5) complex storage and material transfer constraints. In general, the modeling of all processing features we have discussed in this book, using discrete time models, can be readily used in the fixed-horizon models for periodic scheduling without any modifications.

10.4 Notes and Further Reading

(1) The presentation in the current chapter is based on Wu and Maravelias [1].

(2) Many of the concepts underpinning periodic scheduling in chemical manufacturing facilities appeared first in papers by Wellons and Reklaitis [2–4].

(3) Sahinidis and Grossmann considered the scheduling of continuous parallel units under the following assumptions: (1) a product is produced in only one unit (to be determined by the optimization); (2) only one changeover is required for each product; (3) the length of a run is unlimited [5]. Under these assumptions, the resulting solution consists of cycles. They developed a slot-based continuous time model that handles products with given demand as well as products whose production can be adjusted to maximize profit. The model identifies unit-specific solutions with, potentially, different periods. Many elements of the discussion in Section 10.1.2, although for the single-unit problem, are taken from [5].

(4) An extension to multistage continuous processes, with one unit per stage, was proposed by Pinto and Grossmann [6]. A continuous time slot-based model was developed based on the following assumptions: (1) only one changeover is required for each product within a cycle and (2) each product must be processed in the same sequence at each stage.

(5) Models based on continuous modeling of time accounting for variable processing times have also been proposed [7–9].

(6) The first approach to periodic scheduling in network environments was developed by Shah et al. [10], based on the STN-based discrete time model of Shah et al. [11]. They discussed the concept of continuity across cycles, what we termed as batch cross over, and developed the *wrap-around time operator* to address it. The introduction of set $\mathbf{N}_{in}^{CO}$ and operator $m(i,n)$ in Section 10.3.2 are, essentially, alternative ways to achieve the same goal. They considered fixed horizon (period) problems, described a solution procedure, and discussed how to address solution degeneracy.

(7) The general algorithm to solve fractional programming problems appeared in Dinkelbach [12]. It was used in the context of cyclic scheduling first from Pochet and Warichet [13] and then You et al. [14].

(8) Period scheduling has also been used for multipurpose plant design [8] as well as integrated scheduling and control [15, 16].

10.5 Exercises

(1) Consider a single-unit process producing two products, A and B, both with a continuous demand of 0.5 kg/h. Product A (B) are produced by batch tasks TA (TB) with processing time 3 (5) h. The unit capacity is 15 kg, inventory cost for both products is \$1/(kg·h), and the fixed processing cost for both tasks is \$10. Find the best periodic solution given that $T^{UP} = 30$ h.

(2) Consider a single-stage batch process with three units (U1, U2, and U3) producing five products (A, B, C, D, and E). The demand for products A, B, and D is 0.5 kg/h, while the demand for C and E is 1 kg/h. Task TA produces product A, task TB produces B, and so on. All tasks can be executed in all units, and the processing times, in h, are $\tau_{TA,j} = 3$, $\tau_{TB,j} = 5$, $\tau_{TC,j} = 6$, $\tau_{TD,j} = 5$, and $\tau_{TE,j} = 5$. The unit capacities are 15 kg. Inventory cost for both products is \$1/(kg·h), and the fixed processing cost for both tasks is \$10. Find the best periodic solution for the following instances, using $T^{UP} = 30$ h:

   (a) Each product has to be produced in one unit; there is unlimited storage capacity for all products.

   (b) Each product can be produced in more than one unit; there is unlimited storage capacity for all products.

   (c) Each product can be produced in more than one unit; there is unlimited storage capacity for all products except E, for which the maximum storage capacity is 8 kg.

(3) Consider the three-unit, five-product system introduced in Exercise 2. The total demand for all products is the same but must be met every three hours. For example, $\xi_{A,0} = \xi_{A,3} = \xi_{A,6} \cdots = -1.5$ kg, and $\xi_{C,0} = \xi_{C,3} = \xi_{C,6} \cdots = -3$ kg. Find the minimum cost periodic solution given $T^{UP} = 30$ h.

(4) Consider the process shown in Figure 10.13 with the unit information given in Table 10.2. All inventory costs are \$0.5/(kg·h), and the price of products P1 and P2 is \$10/kg. Find the maximum profit periodic schedule assuming $T^{UP} = 30$ h. The wrap-around operator should be included in your model.

Table 10.2 Unit capacities [kg], β_j^{MAX}; and unit-task suitability, $\mathbf{J}_i$.

	HT	R1	R2	DC
β_j^{MAX}	100	80	50	200
$\mathbf{J}_i$	H	R1, R2, R3	R1, R2, R3	S

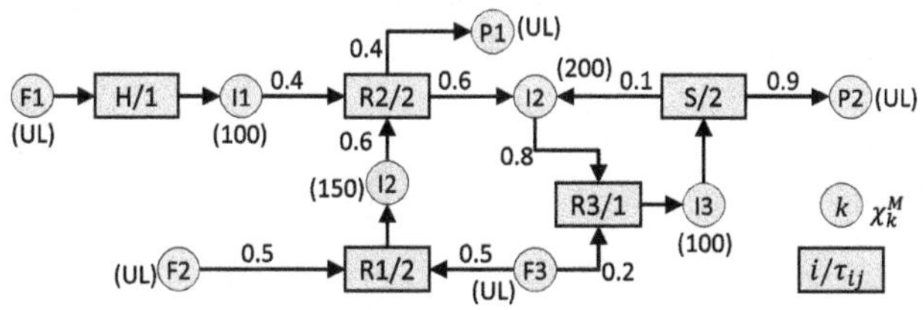

Figure 10.13 STN representation of facility for Exercise (4); conversion coefficients given along corresponding arcs; coefficients not given are equal to –1 or +1; storage capacities, in kg, are given in parentheses next to corresponding materials (UL: unlimited storage); processing times are given inside tasks.

References

[1] Wu Y, Maravelias CT. A General Model for Periodic Chemical Production Scheduling. *Ind Eng Chem Res*. 2020; 59 (6): 2505–2515.

[2] Wellons MC, Reklaitis GV. Optimal Schedule Generation for a Single-Product Production Line. 1. Problem Formulation. *Comput Chem Eng*. 1989;13(1–2):201–212.

[3] Wellons MC, Reklaitis GV. Optimal Schedule Generation for a Single-Product Production Line. 2. Identification of Dominant Unique Path Sequences. *Comput Chem Eng*. 1989;13(1–2):213–227.

[4] Wellons MC, Reklaitis GV. Scheduling of Multipurpose Batch Chemical-Plants. 2. Multiple-Product Campaign Formation and Production Planning. *Ind Eng Chem Res*. 1991;30(4):688–705.

[5] Sahinidis NV, Grossmann IE. Minlp Model for Cyclic Multiproduct Scheduling on Continuous Parallel Lines. *Comput Chem Eng*. 1991;15(2):85–103.

[6] Pinto JM, Grossmann IE. Optimal Cyclic Scheduling of Multistage Continuous Multiproduct Plants. *Comput Chem Eng*. 1994;18(9):797–816.

[7] Schilling G, Pantelides CC. Optimal Periodic Scheduling of Multipurpose Plants in the Continuous Time Domain. *Comput Chem Eng*. 1997;21:S1191–S1196.

[8] Castro PM, Barbosa-Povoa AP, Novais AQ. Simultaneous Design and Scheduling of Multipurpose Plants Using Resource Task Network Based Continuous-Time Formulations. *Ind Eng Chem Res*. 2005;44(2):343–357.

[9] Wu D, Ierapetritou M. Cyclic Short-Term Scheduling of Multiproduct Batch Plants Using Continuous-Time Representation. *Comput Chem Eng*. 2004;28(11):2271–2286.

[10] Shah N, Pantelides CC, Sargent RWH. Optimal Periodic Scheduling of Multipurpose Batch Plants. *Ann. Oper. Res*. 1993;42(1):193–228.

[11] Shah N, Pantelides CC, Sargent RWH. A General Algorithm for Short-Term Scheduling of Batch-Operations. 2. Computational Issues. *Comput Chem Eng*. 1993;17(2):229–244.

[12] Dinkelbach W. On Nonlinear Fractional Programming. *Manage Sci*. 1967;13(7):492–498.

[13] Pochet Y, Warichet F. A Tighter Continuous Time Formulation for the Cyclic Scheduling of a Mixed Plant. *Comput Chem Eng*. 2008;32(11):2723–2744.

[14] You FQ, Castro PM, Grossmann IE. Dinkelbach's Algorithm as an Efficient Method to Solve a Class of MINLP Models for Large-Scale Cyclic Scheduling Problems. *Comput Chem Eng*. 2009;33(11):1879–1889.

[15] Flores-Tlacuahuac A, Grossmann IE. Simultaneous Cyclic Scheduling and Control of a Multiproduct CSTR. *Ind Eng Chem Res*. 2006;45(20):6698–6712.

[16] Moniz S, Barbosa-Povoa AP, de Sousa JP. Simultaneous Regular and Non-regular Production Scheduling of Multipurpose Batch Plants: A Real Chemical-Pharmaceutical Case Study. *Comput Chem Eng*. 2014;67:83–102.

11 Multiperiod Blending

In the problems we have considered so far, we have either chosen to ignore the actual consumption and production of materials (majority of models in sequential environments) or assumed that materials are consumed and produced at fixed proportions (models for network environments), given, for example, by parameters ρ_{ik} in the STN representation. There are problems, however, where the proportions in which input materials are consumed can vary provided that some product specifications are satisfied. In fact, the input materials that can be used to produce a given product are not fixed, that is, various materials can be chosen to be consumed. We will refer to this problem as *multiperiod blending* or simply *blending*. Blending is a fundamentally different problem from the ones discussed thus far because it leads to nonlinear models regardless of the employed time representation. Specifically, the requirement to meet certain property specifications necessitates the inclusion of nonlinear equations for monitoring these properties.

While *blending* arises in numerous industrial sectors, it is particularly important in oil refining. Specifically, it arises in (1) the operation planning of the crude oil distillation units (CDUs), where different crude oils (crudes) are blended before they are sent to the distillation unit and (2) blending of refinery streams to produce final products. We term the first problem, where feed blending is followed by processing of the blend, as *process* blending, and in oil refining, specifically, as *crude distillation* blending. The second problem is termed *final product* or simply *product* blending. We first discuss product blending because it is simpler than process blending and has more characteristics in common with problems in sectors other than oil refining.

In Section 11.1, we introduce some preliminary concepts and a formal problem statement for product blending. In Section 11.2, we present two alternative formulations for product blending, and in Section 11.3 we present two approximate linear reformulations. We close, in Section 11.4, with a discussion of models for process blending.

We focus on the equations necessary to account for the key new features of blending problems: (1) the selection of input materials and their blending in variable proportions and (2) the requirement to satisfy given property specifications. Thus, we present all models employing a common discrete time grid but note that the basic modeling ideas can be readily applied in models employing continuous time grids. Finally, we do not consider utility constraints or other processing features, except for some material handling restrictions (see Section 11.2.3).

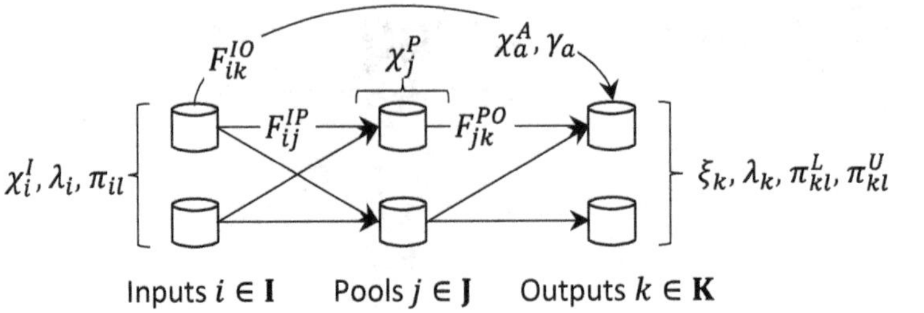

Figure 11.1 Representation of pooling problem: structure, parameters, and flow variables.

11.1 Preliminaries

In Section 11.1.1, we introduce *pooling*, a problem that serves as a building block for blending, and then, in Section 11.1.2, we present two formulations for pooling. We close, in Section 11.1.3, with the problem statement for final product blending.

11.1.1 Pooling

We are given the following (see Figure 11.1):

- A set of inputs (raw materials), $i \in \mathbf{I}$, with availability χ_i^I and unit cost λ_i.[1]
- A set of *pools* (units), $j \in \mathbf{J}$, with capacity χ_j^P.
- A set of outputs (products), $k \in \mathbf{K}$, with demand ξ_k and price λ_k.
- A set of properties $l \in \mathbf{L}$[2]; the value (level) of property l of input i is denoted by π_{il}; each product has a lower and/or upper bound specification for (a subset of) properties, denoted by π_{kl}^L and π_{kl}^U, respectively.
- A set of arcs (connections) between inputs and pools, $\mathbf{A}^{IP} \subseteq \mathbf{I} \times \mathbf{J}$; inputs and outputs, $\mathbf{A}^{IO} \subseteq \mathbf{I} \times \mathbf{K}$; and pools and outputs, $\mathbf{A}^{PO} \subseteq \mathbf{J} \times \mathbf{K}$.
- Each arc $a \in \mathbf{A} = \mathbf{A}^{IP} \cup \mathbf{A}^{IO} \cup \mathbf{A}^{PO}$ has unit cost γ_a and capacity χ_a^A.

The goal is to find the flows $F_a \in \mathbb{R}_+$ along $a \in \mathbf{A}$ that minimize the total cost, subject to minimum demand constraints; or maximize profit (revenue from sales minus cost).

We note the following (see Figure 11.1):

(1) An arc a can also be represented by its tail and head, as $(i,j)/(j,k)/(i,k)$ or $i \to j/j \to k/i \to k$.

(2) Depending on the arc, flows F_a can be across input $\to$ pool (F_{ij}^{IP}), pool $\to$ output (F_{jk}^{PO}), and input $\to$ output (F_{ik}^{IO}) arcs.

[1] In the previous chapters, we used index i to denote tasks and index k to denote materials. In the present chapter, there are no well-defined tasks, consuming/producing materials at fixed proportions. Furthermore, inputs and outputs are fundamentally different in that the former have fixed property levels, whereas the latter have only property specifications. Thus, we choose to use different indices to describe them: i for inputs and k for outputs. We continue to use j for equipment units.

[2] Index l has been used far to denote batches of orders (batching problem in Chapter 4) and utilities (STN-based models in Chapter 7). These features are not present here, so we use l to denote properties.

(3) For consistency, all parameters that lead to constraints on flows are represented using Greek letter χ: input availability (χ_i^I), arc capacity (χ_a^A), and pool capacity (χ_j^P).

11.1.2 Pooling Formulations

There are two basic formulations for the pooling problem: (1) the p-formulation, also known as the *flow* model or the *concentration* model and (2) the q-formulation, also known as the *proportional* model. Here, we present the flow model, where the decisions are the flows and property levels, $C_{jl} \in \mathbb{R}_+$, in the pools.

The objective function is to maximize profit, that is, revenue from sales minus raw material and *flow* costs:

$$\max \sum_{i,k} \lambda_k F_{ik}^{IO} + \sum_{j,k} \lambda_k F_{jk}^{PO} - \sum_{i,j} \lambda_i F_{ik}^{IP} - \sum_{i,k} \lambda_i F_{ik}^{IO} - \sum_{a} \gamma_a F_a, \tag{11.1}$$

where flow costs can be used to represent transportation as well as processing cost at the pools.

The input availability, pool capacity, and output demand constraints are written as follows:

$$\sum_j F_{ij}^{IP} + \sum_k F_{ik}^{IO} \leq \chi_i^I, \quad i \tag{11.2}$$

$$\sum_i F_{ij}^{IP} \leq \chi_j^P, \quad j \tag{11.3}$$

$$\sum_i F_{ik}^{IO} + \sum_j F_{jk}^{PO} \geq \xi_k, \quad k, \tag{11.4}$$

where flows are bounded, $F_a \in \left[0, \chi_a^A\right]$.

The material (flow) balance at each pool is

$$\sum_i F_{ij}^{IP} = \sum_k F_{jk}^{PO}, \quad j, \tag{11.5}$$

where we assume that volume (flow) is preserved under blending.

The property levels (concentrations[3]) at pools are governed by the following equation,

$$\sum_i \pi_{il} F_{ij}^{IP} = C_{jl} \sum_k F_{jk}^{PO}, \quad j, l, \tag{11.6}$$

which can be viewed as property balances: the *flow of a property into a pool* (LHS) should be equal to the outflow (RHS).[4]

[3] The term *property* is used to describe elements of set **L**; e.g., density, boiling point, and sulfur concentration. The term *property level* is used for the numerical value of the property; e.g., 0.98 gr/cm^3 (density), 95^{0}C (boiling point), 0.05 gr/lt (concentration). We will be using the terms *property level* and *concentration* interchangeably, although not all properties are related to concentrations.

[4] The rationale for treating the LHS and RHS of (11.6) as *property flows* stems from the treatment of property levels as concentrations. For example, if C_{jl} is in gr/lt and F_{jk}^{PO} is in lt/h (assuming $\delta = 1$ h.), then the units of (11.6) are gr/h. We henceforth liberally use the term *property flow* for all properties.

Finally, the property specifications for the outputs are expressed as follows:

$$\pi_{kl}^{L}\left(\sum_{i}F_{ik}^{IO}+\sum_{j}F_{jk}^{PO}\right)\leq\sum_{i}\pi_{il}F_{ik}^{IO}+\sum_{j}C_{jl}F_{jk}^{PO}\leq\left(\sum_{i}F_{ik}^{IO}+\sum_{j}F_{jk}^{PO}\right)\pi_{kl}^{U},\ k,l,\tag{11.7}$$

where the total *property flow* is bounded to be within the products between the total flow toward an output $\left(\sum_i F_{ik}^{IO}+\sum_j F_{jk}^{PO}\right)$ and the minimum/maximum property specifications.

We note the following:

(1) The nonlinearity of the model comes from the bilinear terms in (11.6) and (11.17).
(2) Pooling is a *static* (one-period) problem.
(3) The total inflow to a pool should be equal to the total outflow, as enforced by (11.5).
(4) A pool property is a linear combination of the properties of the inputs, with the weights being the input flows, that is, the *blending rules* are linear.
(5) Similarly, output properties are linear combinations of the properties of all incoming flows.
(6) The properties of the streams toward an output need not satisfy property specifications; it is the mixture (blend) of these flows that should satisfy them.
(7) We do not consider pool–pool arcs, though this variant has been also studied in the literature.

11.1.3 Product Blending

The blending problem is essentially the extension of the pooling problem into a multiperiod setting, where input availability may be affected by deliveries and output demand can vary across periods. Also, mixing occurs in blenders, that is, we replace the term *pool* with the term *blender*. A key difference between the two problems is that in blending the properties of the streams from a blender toward an output should satisfy all output specifications. We will first consider the final product blending problem employing a common discrete time grid with points $n \in \mathbf{N}$.

Given are the following (see Figure 11.2)[5]:

(1) A set of inputs, $i \in \mathbf{I}$, with capacity χ_i^I, unit cost λ_i, deliveries ξ_{in}, and property levels π_{il}.
(2) A set of blenders, $j \in \mathbf{J}$, with capacity χ_j^B.
(3) A set of outputs, $k \in \mathbf{K}$, with demand $-\xi_{kn}$, price λ_k, property specifications π_{kl}^L and π_{kl}^U, and capacity χ_k^O.
(4) Arcs $a \in \mathbf{A}$ connecting the various nodes of the network. Inputs are connected to (a subset of) blenders only; blenders have incoming arcs from inputs and

[5] Strictly positive initial inventory and concentration levels can be handled using the methods described in previous chapters.

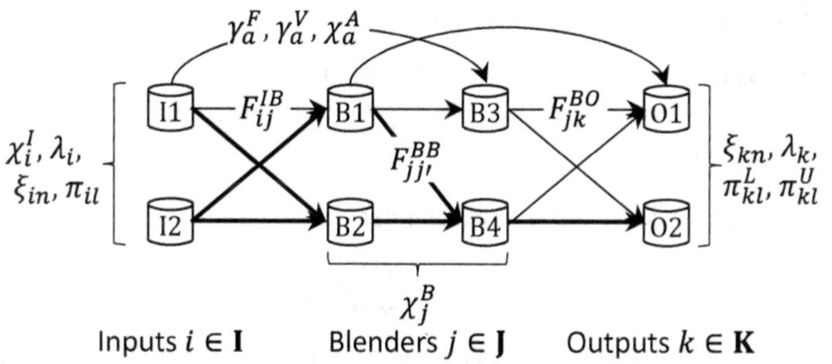

Figure 11.2 Representation of blending: example network and associated parameters; thick lines represent positive flows, and thin lines represent possible connections (existing arcs).

blenders and outgoing arcs to other blenders and outputs; and outputs have incoming arcs from blenders only.

(5) Each arc has fixed cost, γ_a^F, variable cost, γ_a^V, and capacity χ_a^A.

(6) Operating logic (material handling) constraints; for example, material cannot be simultaneously fed and withdrawn from the same blender.

The goal is to meet the demand at the minimum total cost or maximize profit. Note that the fixed charges for flows and the modeling of material handling constraints necessitates the introduction of binary variables for the *activation* of arcs, thereby leading to mixed-integer nonlinear programming models.

11.2 Product Blending: Nonlinear Models

We discuss two models: the so-called *concentration* model, in Section 11.2.1, and the so-called *source-based* model in Section 11.2.2. We focus on the modeling of material balances and constraints to monitor property levels at the blenders, and constraints to enforce output property specification satisfaction. Example constraints to enforce operating logic restrictions are presented in Section 11.2.3.

11.2.1 Concentration-Based Model

We introduce the following time-indexed, nonnegative, continuous variables:

- C_{jln} : Value (concentration) of property l of the blend in blender j during time period n
- F_{ijn}^{IB}: Flow from input i to blender j at time point n
- $F_{jj'n}^{BB}$: Flow from blender j to blender j' at time point n
- F_{jkn}^{BO}: Flow from blender j to output k at time point n
- I_{in}^{I} : Inventory of input i during time period n

- I^B_{jn}: Inventory in blender j during time period n
- I^O_{kn} : Inventory of output k during time period n
- V_{kn}: Sales of output k at time point n in addition to given demand (additional sales)

The material balances for the inputs, blenders, and outputs are expressed as follows:

$$I^I_{i,n+1} = I^I_{in} + \xi_{in} - \sum_j F^{IB}_{ijn} \le \chi^I_i, \quad i, n \tag{11.8}$$

$$I^B_{j,n+1} = I^B_{jn} + \sum_i F^{IB}_{ijn} + \sum_{j'} F^{BB}_{j'jn} - \sum_{j'} F^{BB}_{jj'n} - \sum_k F^{BO}_{jkn} \le \chi^B_j, \quad j, n \tag{11.9}$$

$$I^O_{k,n+1} = I^O_{kn} + \sum_j F^{BO}_{jkn} + \xi_{kn} - V_{kn} \le \chi^O_k, \quad k, n. \tag{11.10}$$

Note that initial and final inventories need not be zero, which means that the total inflow (over the horizon) should not be equal to the total outflow.

The property balances in the blenders are

$$I^B_{j,n+1} C_{jl,n+1} = I^B_{jn} C_{jln} + \sum_i \pi_{il} F^{IB}_{ijn} + \sum_{j'} C_{j'ln} F^{BB}_{j'jn} - \sum_{j'} C_{jln} F^{BB}_{jj'n} - \sum_k C_{jln} F^{BO}_{jkn}, \quad j, l, n \tag{11.11}$$

where the term $I^B_{jn} C_{jln}$ can be viewed as the *amount* (or inventory) of property l in the blender during period n.[6]

The constraints enforcing that the material transferred from a blender to an output should satisfy the output's property specifications can be expressed using big-M type constraints:

$$\pi^L_{kl} - M\left(1 - X^{BO}_{jkn}\right) \le C_{jln} \le \pi^U_{kl} + M\left(1 - X^{BO}_{jkn}\right), \quad j, k, l, n. \tag{11.12}$$

Note that the property levels in a blender vary over time and thus during some periods may, for example, violate specifications of outputs whose demand is, at different time points, satisfied from the same blender. Thus, specification satisfaction should be enforced only when there is positive flow between a blender and an output (i.e., $X^{BO}_{jkn} = 1$).

The calculation of concentrations is illustrated in Figure 11.3, which shows an example network along with the associated parameters and the flows corresponding to a four-period solution. The blender starts with an inventory of 10 and concentration of 80, with only input I1. At $n = 1$, 10 units of I2 are sent to the blender increasing its inventory to 20 and its concentration to 85. (Can you calculate it?) At $n = 2$, 10 units are removed to meet the demand for O1, lowering the inventory level while keeping the concentration unchanged. At $n = 3$, flow from I3 raises the inventory to 20 and the

[6] Equation (11.11) is the counterpart of (11.6). If we enforce that the inflow should be equal to the outflow (by, for example setting, $I^B_{j,n+1} C_{jl,n+1} = I^B_{jn} C_{jln}$), then we obtain a property flow equality similar to the one in (11.6). In that respect, *property flows* (e.g., term $C_{j'ln} F^{BB}_{j'jn}$) modify the *property amount* (inventory) in the blender. Again, the motivation to view $I^B_{jn} C_{jln}$ as property inventory comes from the treatment of C_{jln} as concentration.

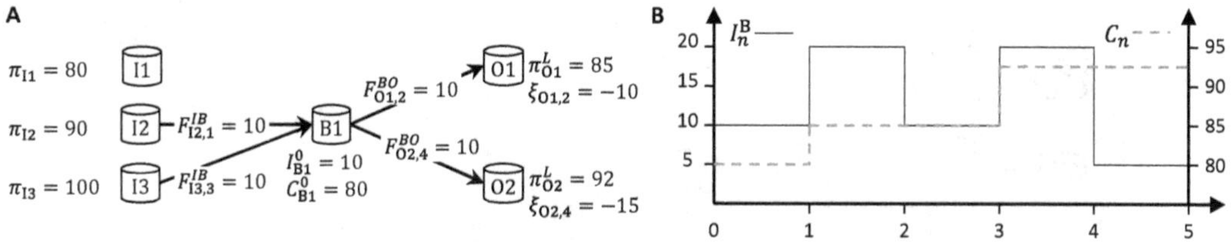

Figure 11.3 Concentration profile calculation. (A) Example system and associated parameters: input property values, blender initial inventory and concentration, output property specifications (lower bounding), and output orders; flows are shown along corresponding arcs. (B) Inventory (left axis) and concentration (right axis) profile resulting from the flows given in panel (A).

concentration to 92.5. (Can you calculate it?) Finally, at $n = 4$, 15 units are removed to meet the order for O2 lowering the inventory to 5.

If the given demand is satisfied, the general objective function includes revenue from additional sales, and input and operating (fixed and proportional) costs:

$$\max \sum\nolimits_{k,n} \lambda_k V_{kn} - \sum\nolimits_{i,j,n} \lambda_k F^{IB}_{ijn} - \sum\nolimits_{a \in \mathbf{A},n} \left(\gamma^F_a X_{an} + \gamma^V_a F_{an}\right), \tag{11.13}$$

where $\mathbf{A}$ represents all existing arcs, and X_{an} and F_{an} represent the corresponding activation variables and flows (i.e., F^{IB}_{ijn}, $F^{BB}_{jj'n}$, and F^{BO}_{jkn}) at time point n. We also assume that input costs occur upon the utilization of the input.

The concentration-based model, henceforth referred to as $\mathbb{M}^{CB}$, consists of (11.8) through (11.13).

11.2.2 Source-Based Model

To simplify the presentation in this subsection, we assume that there are no arcs between pairs of blenders, that is, the only incoming/outgoing arcs of blenders are from/to inputs/outputs. Thus, (11.9) becomes

$$I^B_{j,n+1} = I^B_{jn} + \sum\nolimits_i F^{IB}_{ijn} - \sum\nolimits_k F^{BO}_{jkn} \le \chi^B_j, \quad j, n. \tag{11.14}$$

We further assume that initial blender and output inventories are zero.

The model is based on the disaggregation of blender inventories into input-specific components; and the disaggregation of flows from the blenders to outputs into input-specific (source) components:

- $I^{IB}_{ijn} \in \mathbb{R}_+$: Inventory of input i during period n in blender j
- $F^{IBO}_{ijkn} \in \mathbb{R}_+$: Flow of input i at point n allocated to output k via blender j (i.e., flow of i from blender j to output k at point n)

The disaggregated material balance, for input i in blender j, during period n is

$$I^{IB}_{ij,n+1} = I^{IB}_{ijn} + F^{IB}_{ijn} - \sum\nolimits_k F^{IBO}_{ijkn}, \qquad i, j, n, \tag{11.15}$$

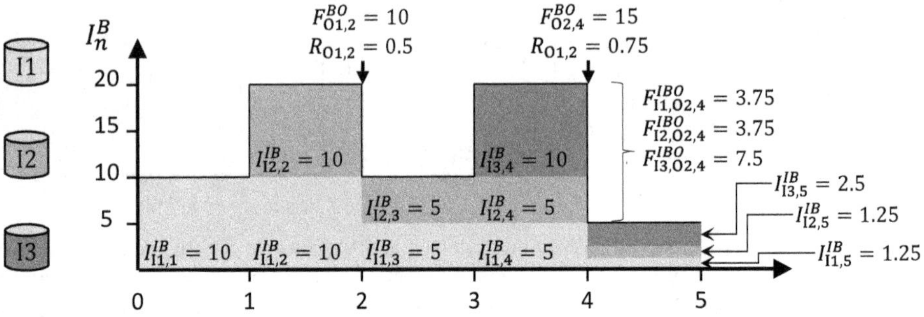

Figure 11.4 Illustration of calculation of disaggregated variables (input-specific blender inventories and input-specific blender-output flows) and split fractions for the blender in the example shown in Figure 11.3 (index B1 not shown). Shades of gray are used for three inputs shown at the left of the graph.

where the disaggregated inventories satisfy

$$I_{jn}^B = \sum_i I_{ijn}^{IB}, \quad j, n \tag{11.16}$$

and the disaggregated flows satisfy

$$F_{jkn}^{BO} = \sum_i F_{ijkn}^{IBO}, \quad j, k, n. \tag{11.17}$$

Next, we enforce that the proportion of an input in the blend flowing from a blender to an output should be the same as the proportion of the same input inside the blender. This is accomplished via the introduction of the *split fraction* variables, $R_{jkn} \in [0, 1]$, which represent the fraction of inventory in blender j that is sent to output k at time point n. The composition equality between the blend in the blender and the blend in the stream is accomplished by the following nonlinear equations:

$$F_{jkn}^{BO} = R_{jkn} I_{jn}^B, \quad j, k, n \tag{11.18}$$

$$F_{ijkn}^{IBO} = R_{jkn} I_{ijn}^{IB}, \quad i, j, k, n. \tag{11.19}$$

An example calculation of variables I_{ijn}^{IB}, F_{ijkn}^{IBO}, and R_{jkn} is shown in Figure 11.4 based on the network and solution given in Figure 11.3. The blender starts with only I1, that is, $I_{\text{I1,B1,1}}^{IB} = 10$. The flow of I2 to B1 at $n = 1$ ($F_{\text{I2,B1,1}}^{IB} = 10$) leads to (the addition of) $I_{\text{I2,B1,2}}^{IB} = 10$. At $n = 2$, half the inventory is used to meet the order of O1 (i.e., $R_{\text{B1,O1,2}} = 0.5$), resulting in both input-specific inventories to be halved ($I_{\text{I1,B1,3}}^{IB} = I_{\text{I2,B1,3}}^{IB} = 5$). The inflow of I3 to B1 at $n = 3$ ($F_{\text{I3,B1,3}}^{IB} = 10$) leads to $I_{\text{I3,B1,4}}^{IB} = 10$. At $n = 4$, 75% of the inventory ($R_{\text{B1,O2,4}} = 0.75$) is used to meet the order of output O2, resulting in all input-specific inventories to be reduced by 75%.

Output property specifications can now be enforced using linear equations:

$$\pi_{kl}^L F_{jkn}^{BO} \le \sum_i \pi_{il} F_{ijkn}^{IBO} \le \pi_{kl}^U F_{jkn}^{BO}, \quad j, k, l, n. \tag{11.20}$$

The source-based model, henceforth referred to as $\mathbb{M}^{SB}$, consists of (11.8), (11.10), (11.14) through (11.20), and objective function (11.13).

11.2.3 Remarks and Extensions

Material Handling Constraints. Binary variables X_{an} can be used to enforce a wide range of operating restrictions. For example, the restriction that a blender can receive material from only one input at a given time, for the source-based model where there are no $j \rightarrow j'$ arcs, can be enforced by

$$\sum_i X_{ijn} \leq 1, \quad j,n \tag{11.21}$$

$$F_{ijn}^{IB} \leq \chi_{ij}^{A} X_{ijn}, \quad i,j,n. \tag{11.22}$$

Output Inventory. If the output has positive capacity, then it is essentially another blender that is dedicated to a single output. Notably, the properties of the blend in it may vary over time, but they will always satisfy the property specifications. (Can you see why?) If the output is viewed as a node at which the exact demand should be shipped (it has zero holding capacity), then we set $I_{k,n+1}^{O} = I_{kn}^{O} = 0$ in (11.10) and obtain (recalling that $\xi_{kn} < 0$ for outputs)

$$\sum_j F_{jkn}^{BO} = -\xi_{kn} + V_{kn}, \quad k,n. \tag{11.23}$$

Blender-Blender Arcs in Source-Based Model. Model $\mathbb{M}^{SB}$ can be extended to account for blender–blender arcs. Specifically, disaggregated variables $F_{ijj'n}^{IBB} \in \mathbb{R}_+$ are introduced to keep track of the input amount in a *second-layer* blender, and the second-layer blender material and property balances; (11.14) and (11.15) are modified to account for flows to/from blenders, and split fractions are considered for blender–blender flows.

Nonlinearities. If linear blending rules are applied, then the only nonlinear terms in both models are bilinear terms, which appear in the property balance, (11.11), in $\mathbb{M}^{CB}$; and the definition of the split fractions, (11.18) and (11.19), in $\mathbb{M}^{SB}$. If nonlinear mixing rules are used, then (11.11) in $\mathbb{M}^{CB}$ should be replaced by an equation describing the chosen rules (which can be different for different properties). In $\mathbb{M}^{SB}$, (11.18) and (11.19) remain the same, but (11.20) should be replaced by equations describing the chosen rules, which means that (11.20) will also become nonlinear.

Continuous Time Models. The ideas described so far can be readily implemented to models based on continuous time grids. As we have seen, the materials balances of grid-based discrete and continuous time models are very similar, while the same approach can be followed to monitor properties. Thus, if costs based on profiles over time (e.g., inventory and backlog) need not be considered, the same types of nonlinearities will appear in the two types of models. Since solving the resulting MINLP models for blending requires global optimization solvers, whose performance depends more heavily than the performance of MILP solvers on model size, models based on continuous time grids may be better suited for some blending instances.

11.3 Product Blending: Linear Approximate Models

The MINLP formulations presented in the previous section can lead to large models (even if the number of inputs, blenders, and outputs is small) due to the typically large number of properties that need to be considered and the multiperiod nature of the problem. One approach to overcome this challenge is to develop linear approximations or relaxations of the nonlinear equations. Using the concentration-based model as basis, in this section we present one linear model that leads to an approximation whose feasible region is a subset of the feasible region of $\mathbb{M}^{CB}$ (Section 11.3.1) and another linear model that is a relaxation of $\mathbb{M}^{CB}$ (Section 11.3.2). To simplify the presentation, we assume that there are no blender–blender arcs, so the blender material balance is given by (11.14).

11.3.1 Discretization-Based Model

The basic idea is that concentrations in the blenders are allowed to assume values only from a predefined discrete set, that is, $C_{jln} \in \{\psi_{l,1}, \psi_{l,2}, \ldots, \psi_{l,|\mathbf{M}_l|}\}$, $\forall l$, with $\psi_{l,1} < \psi_{l,2} < \ldots, < \psi_{l,|\mathbf{M}_l|}$, where $\mathbf{M}_l$ is the index set for the discrete values property l can assume, and the cardinality of $\mathbf{M}_l$ can vary across properties. We introduce binary (SOS1 with respect to index m) variables Z_{jlmn} to model the selection of the discrete value of property l in blender j during period n:

$$\sum\nolimits_{m \in \mathbf{M}_l} Z_{jlmn} = 1, \quad j, l, n. \tag{11.24}$$

The concentration can then be calculated as follows:

$$C_{jln} = \sum\nolimits_{m \in \mathbf{M}_l} \psi_{lm} Z_{jlmn}, \quad j, l, n. \tag{11.25}$$

Figure 11.5 shows how the discretization of C_{jln} leads to a reduced feasible space. The original feasible space $(C_{jln} \in [\min_i\{\pi_{il}\}, \max_i\{\pi_{il}\}])$ is replaced by $C_{jln} \in \{\psi_{l,1}, \psi_{l,2}, \ldots, \psi_{l,|\mathbf{M}_l|}\}$. It also shows how a given property *amount* (i.e., fixed $C_{jln}I^B_{jn}$) can be achieved by a finite number of inventory levels (which can be precalculated for given $C_{jln}I^B_{jn}$).

The goal, using this discretization, is to linearize the property balance, (11.11), which, when there are no blender–blender flows, is rewritten as follows:

$$C_{jl,n+1}I^B_{j,n+1} = C_{jln}I^B_{jn} + \sum\nolimits_i \pi_{il}F^{IB}_{ijn} - \sum\nolimits_k C_{jln}F^{BO}_{jkn} \quad j, l, n. \tag{11.26}$$

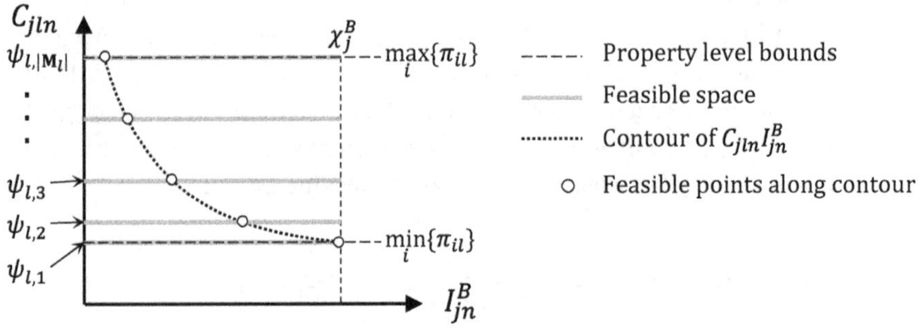

Figure 11.5 Feasible space of model based on property level discretization.

We introduce new variables to replace the bilinear terms:

- $W^B_{jln} \in \mathbb{R}_+ := C_{jln} I^B_{jn}$, *amount* of property l in blender j during period n
- $W^{BO}_{jkln} \in \mathbb{R}_+ := C_{jln} F^{BO}_{jkn}$, *flow* of property l from blender j to output k at time point n

Thus, the property balance becomes

$$W^B_{jl,n+1} = W^B_{jln} + \sum_i \pi_{il} F^{IB}_{ijn} - \sum_k W^{BO}_{jkln}, \quad j, l, n. \tag{11.27}$$

The last step is to add constraints to enforce, linearly, $W^B_{jln} = C_{jln} I^B_{jn}$ and $W^{BO}_{jkln} = C_{jln} F^{BO}_{jkn}$. The inventory in the blender is disaggregated based on the active discrete property value

$$I^B_{jn} = \sum_{m \in \mathbf{M}_l} I^B_{jlmn}, \quad j, l, n \tag{11.28}$$

subject to

$$I^B_{jlmn} \le \chi^B_j Z_{jlmn}, \quad j, l, m, n. \tag{11.29}$$

Blender-output flows are disaggregated in a similar manner:

$$F^{BO}_{jkn} = \sum_{m \in \mathbf{M}_l} F^{BO}_{jklmn}, \quad j, k, l, n \tag{11.30}$$

$$F^{BO}_{jklmn} \le \chi^A_{jk} Z_{jlmn}, \quad j, k, l, m, n. \tag{11.31}$$

Finally, the disaggregated inventories and blender-output flows are coupled with the corresponding property amounts in the blender (W^B_{jln}) and property flows (W^{BO}_{jkln}) as follows:

$$W^B_{jln} = \sum_{m \in \mathbf{M}_l} \psi_{lm} I^B_{jlmn}, \quad j, l, n \tag{11.32}$$

$$W^{BO}_{jkln} = \sum_{m \in \mathbf{M}_l} \psi_{lm} F^{BO}_{jklmn}, \quad j, k, l, n. \tag{11.33}$$

The discretized concentration-based linear model, henceforth referred to as $\mathbb{M}^{CB}_D$, consists of material balances, (11.8), (11.10), and (11.14); the discretization of concentrations, (11.24) and (11.25); the linearized property balance, (11.27); the output property specification constraints, (11.12); the equations for the disaggregation of I^B_{jlmn} and F^{BO}_{jklmn}, (11.28) through (11.31); the calculation of W^B_{jln} and W^{BO}_{jklmn}, (11.32) and (11.33); and the objective function, (11.13). Figure 11.6 shows how the concentration discretization is used to calculate the various variables and ultimately leads to a linear property balance.

Figure 11.7 shows the values of the variables in $\mathbb{M}^{CB}_D$ for the example shown in Figure 11.3 (indices l and $j = $ B1 are omitted for brevity). The concentration in the

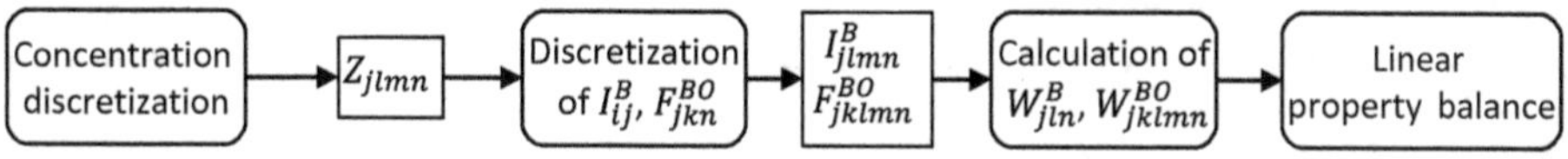

Figure 11.6 Calculation of new variables based on property level discretization.

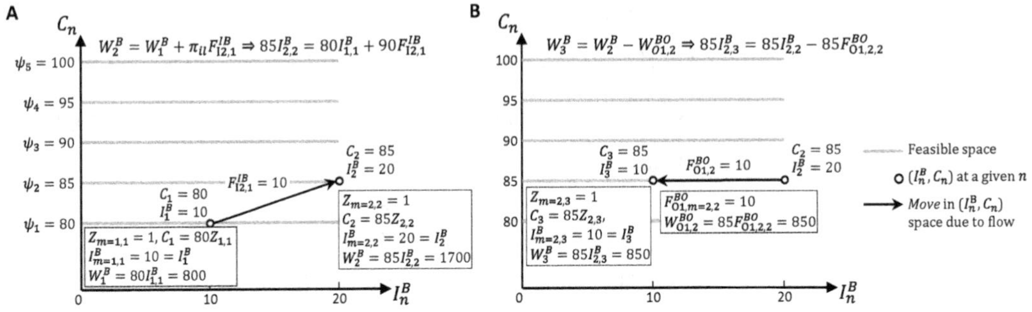

Figure 11.7 Examples of disaggregated variable calculations using the example shown in Figure 11.3; disaggregated variable calculations and values are shown in dashed boxes; indices l and j are omitted for simplicity; overall property balance is shown at the top of each graph. (A) Variables used to describe property balance between periods 1 and 2. (B) Variables used to describe property balance between periods 2 and 3.

blender is discretized into five values, $C_n \in \{80, 85, 90, 95, 100\}$. During the first period, we have $C_1 = 80$ and $I_1^B = 10$, which is represented by $Z_{m=1,1} = 1$ and $I_{m=1,1}^B = 10$ to yield a total property amount of 800. The incoming flow from I2 at $n = 1$ ($F_{I2,1}^{IB} = 10$) adds 10·90 = 900 units of property, for a total property inventory in the blender during period 2 equal to 1,700 (= 800 + 900). The disaggregated variables for the blender inventory during period 2 are $Z_{m=2,2} = 1$ and $I_{m=2,2}^B = 20$. The *move*, in the (I_n^B, C_n) space, from $n = 1$ to $n = 2$, along with the corresponding variables and the property balance, is shown in Figure 11.7A. The corresponding move from $n = 2$ to $n = 3$, due to the outflow to output O1, is shown in Figure 11.7B. The outflow at $n = 2$ ($F_{O1,2}^{BO} = 10$), which is represented by $F_{O1,m=2,2}^{BO} = 10$, lowers the inventory to 10; and removes 850 property units, bringing the property amount in the blender down to 850 (= 1,700 – 850). The new (I_n^B, C_n) point, at $n = 3$, is represented by $Z_{m=2,3} = 1$ and $I_{m=2,3}^B = 10$.

11.3.2 Discretization-Relaxation-Based Model

The feasible space of $\mathbb{M}_D^{CB}$ is a subset of the feasible space of the original problem, so the optimal solution may be cut off, or, if the discretization is too coarse, the feasible space can be empty. An alternative approach is to generate a linear model whose feasible space is a superset of the original problem, that is, a relaxation of $\mathbb{M}^{CB}$. In this subsection, we discuss a relaxation that is, specifically, generated as a relaxation of $\mathbb{M}_D^{CB}$, and is henceforth referred to as $\mathbb{M}_{D-R}^{CB}$.

Model $\mathbb{M}_{D-R}^{CB}$ consists of the same material balances as $\mathbb{M}_D^{CB}$: input, (11.8); blender, (11.14); and output, (11.10).

The relaxation of the discretized concentration is achieved through the introduction of variables $Z_{jlmn}^- \in [0, 1]$ and $Z_{jlmn}^+ \in [0, 1]$, which should satisfy

$$Z_{jlmn}^0 + Z_{jlmn}^+ = Z_{jlmn}, \quad j, l, m \in \mathbf{M}_l, n \tag{11.34}$$

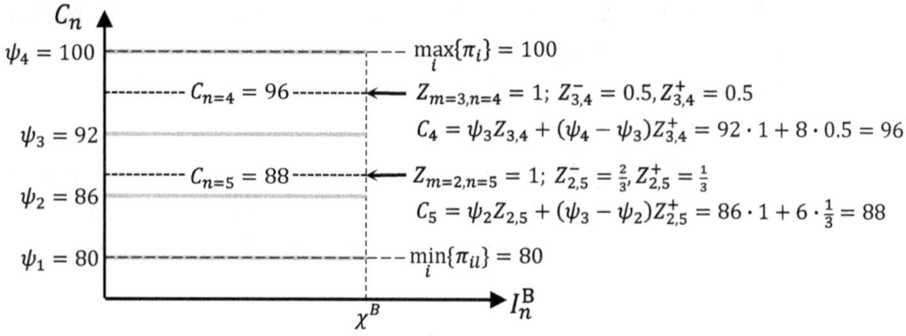

Figure 11.8 Modeling of continuous concentration in blenders (indices j and l are omitted).

while Z_{jlmn} (SOS1 variables with respect to m) satisfies

$$\sum_{m \in \mathbf{M}_l^*} Z_{jlmn} = 1, \quad j, l, n, \tag{11.35}$$

where the summation in the LHS is over $\mathbf{M}_l^* = \{1, 2, \ldots, |\mathbf{M}_l| - 1\}$, rather than $\mathbf{M}_l$ as in (11.24). The concentration, which now assumes continuous values, can be expressed as follows (see Figure 11.8):

$$C_{jln} = \sum_{m \in \mathbf{M}_l^*} \psi_{lm} Z_{jlmn} + \sum_{m \in \mathbf{M}_l^*} (\psi_{l,m+1} - \psi_{lm}) Z^+_{jlmn}, \quad j, l, n. \tag{11.36}$$

Conceptually, variables Z_{jlmn} determine the interval $[\psi_{lm}, \psi_{l,m+1}]$ in which C_{jln} lies, and variables Z^-_{jlmn} and Z^+_{jlmn}, which can be viewed as weights of endpoints ψ_{lm} and $\psi_{l,m+1}$, respectively, determine the exact value of C_{jln} within $[\psi_{lm}, \psi_{l,m+1}]$. Alternatively, the first term in the RHS of (11.36) corresponds to a *base* concentration, based on the active discrete value, and the second term corresponds to a *residual* concentration, based on the exact value within the chosen interval.

Figure 11.8 illustrates how variables Z_{jlmn}, Z^-_{jlmn}, and Z^+_{jlmn} are used to model continuous property levels. Specifically, it shows how two levels (88 and 96), which do not coincide with the discretized values (80, 86, 92, and 100), are modeled. Note that levels in the last interval, [92, 100], are modeled using the variables with $l = 3$. Also, intervals may have unequal size.

The property balance in the relaxed model is the same as in $\mathbb{M}_D^{CB}$,[7]

$$W^B_{jl,n+1} = W^B_{jln} + \sum_i \pi_{il} F^{lB}_{ijn} - \sum_k W^{BO}_{jkln}, \quad k, l, n, \tag{11.37}$$

but the terms in this equation should be *corrected* to account for the *residual* concentration within the selected interval $[\psi_{lm}, \psi_{l,m+1}]$.

[7] We repeat (11.27) here so the reader can easily refer to it, since it is the modeling of the terms of this equation that is further discussed in this subsection.

The variables describing the actual blend inventories (I^B_{jn}) and blend flows (F^{BO}_{jkn}) are used, exactly as in $\mathbb{M}^{CB}_D$, to keep track of material balances. Furthermore, as in $\mathbb{M}^{CB}_D$, they are disaggregated into I^B_{jlmn} and F^{BO}_{jklmn}, through (11.28) and (11.29) and (11.30) and (11.31), respectively, to model property flows. However, I^B_{jlmn} and F^{BO}_{jklmn} account for the *base flow* that would correspond to the discrete concentration. To model the *residual* property flow, due to the residual concentration within interval $[\psi_{lm}, \psi_{l,m+1}]$, we introduce and calculate two *residual* disaggregated inventory (I^-_{jlmn} and I^+_{jlmn}) and flow (F^{BO-}_{jklmn} and F^{BO+}_{jklmn}) variables as follows:

$$I^B_{jlmn} = I^{B-}_{jlmn} + I^{B+}_{jlmn}, \quad j, l, m \in \mathbf{M}^*_l, n \tag{11.38}$$

$$F^{BO}_{jklmn} = F^{BO-}_{jklmn} + F^{BO+}_{jklmn}, \quad j, k, l, m \in \mathbf{M}^*_l, n. \tag{11.39}$$

Note that the new disaggregated variables are the counterparts of Z^-_{jlmn} and Z^+_{jlmn}.

The last step is to calculate property flows accounting for both the *base* values, as in (11.32) and (11.33), and the residual values:

$$W^B_{jln} = \sum\nolimits_{m \in \mathbf{M}^*_l} \psi_{lm} I^B_{jlmn} + \sum\nolimits_{m \in \mathbf{M}^*_l} (\psi_{l,m+1} - \psi_{lm}) I^{B+}_{jlmn}, \quad j, l, n \tag{11.40}$$

$$W^{BO}_{jklmn} = \sum\nolimits_{m \in \mathbf{M}^*_l} \psi_{lm} F^{BO}_{jklmn} + \sum\nolimits_{m \in \mathbf{M}^*_l} (\psi_{l,m+1} - \psi_{lm}) F^{BO+}_{jlmn}, \quad j, k, l, n. \tag{11.41}$$

Note that the calculations in (11.40) and (11.41) follow exactly the logic of the calculation in (11.36). Further, the concentration, from (11.36), is used in the property specification constraint, (11.12); and variables W^B_{jln} and W^{BO}_{jklmn}, from (11.40) and (11.41), respectively, are used in the property balance, (11.37).

Figure 11.9 shows the modeling of the concentrations shown in Figure 11.3, during periods 1 and 2, when, unlike Figure 11.7A, the concentrations are not equal to any ψ_{lm}. Values of $\mathbb{M}^{CB}$ variables are shown next to (I^B_n,C_n) points; the corresponding values of disaggregated variables of $\mathbb{M}^{CB}_{D-R}$ are shown in dashed boxes; the calculations of property flows are shown above each point; and the overall property balance is given at the top of the figure. Note that Figure 11.9 shows a feasible solution. However, the same concentrations can be represented by different solutions.[8]

Finally, note that if the distance between all pairs of discretization points is the same (i.e.,$\psi_{l,m+1} - \psi_{lm} = \Delta\psi_l, \forall m \in \mathbf{M}^*_l$), then only one pair of residual variables can be defined for concentration selection (Z^-_{jln} and Z^+_{jln}), property inventory (I^{B-}_{jln} and I^{B+}_{jln}), and property flow (F^{BO-}_{jkln} and F^{BO+}_{jkln}); and (11.34), (11.38), and (11.39) can be written, respectively, as follows:

$$Z^-_{jln} + Z^+_{jln} = 1, \quad j, l, n \tag{11.42}$$

[8] The variable values in Figure 11.9 are based on *proportional allocation* of the original variables to the disaggregated ones. For example, $C_1 = 80$ lies in the middle of interval $[\psi_1, \psi_2] = [78, 82]$, so the allocation is shown to be even: $Z^-_{1,1} = Z^+_{1,1} = 0.5$ and $I^{B-}_{1,1} = I^{B+}_{1,1} = 5$. However, there is no constraint enforcing this proportional allocation. For example, $I^{B-}_{1,1} = 9, I^{B+}_{1,1} = 1$ would also be feasible.

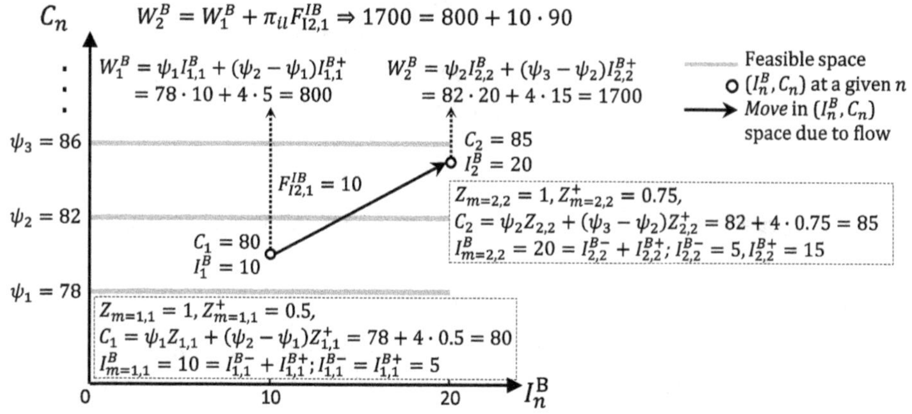

Figure 11.9 Modeling of continuous concentrations using model $\mathbb{M}_{D-R}^{CB}$; indices l and j omitted for simplicity.

$$I_{jn}^{B} = I_{jln}^{B-} + I_{jln}^{B+}, \quad j, l, n \tag{11.43}$$

$$F_{jkn}^{BO} = F_{jkln}^{BO-} + F_{jkln}^{BO+}, \quad j, k, l, n, \tag{11.44}$$

which leads to significant reduction in the number of variable and constraints.

11.4 Process Blending

The problem we addressed in the previous two sections concerns outputs that are obtained directly through the blending of inputs and, importantly, each unit (blender) at a given period contains and therefore can send only one blend. A common industrial problem, however, involves the intermediate processing of inputs to become (different) outputs simultaneously: inputs are blended to form a stream that is fed to a processing unit that produces, simultaneously, multiple outgoing streams to meet demand for different outputs. This is the process blending problem, mentioned in the introduction of the present chapter. The problem statement is presented in Section 11.4.1, a basic model is discussed in Section 11.4.2, and an illustrative example is presented in Section 11.4.3. Finally, some extensions are discussed in Section 11.4.4.

11.4.1 Problem Statement

Since the most important application area of this problem is oil refining, we use the terminology from this sector. In the general case, there are multiple inputs, blended in *feeding* blenders, which then feed multiple distillation columns, which should, collectively, meet demand for multiple outputs. To keep the notation and model simple, we will consider the problem with multiple inputs and outputs but a single blender and distillation column. We will also ignore operating logic constraints.

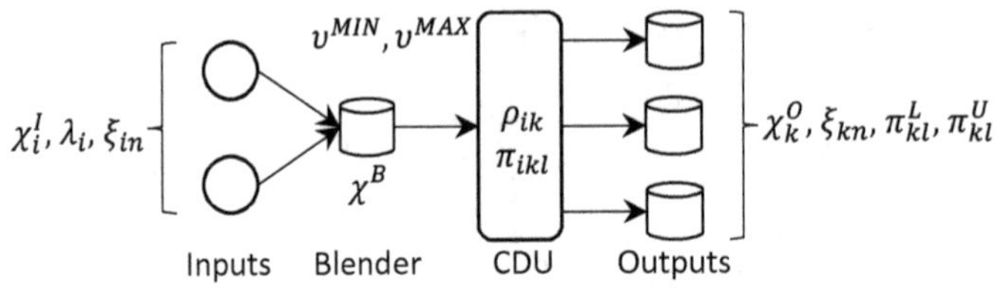

Figure 11.10 Graphical representation of crude oil distillation unit blending.

The following are given (see Figure 11.10):

- Inputs (crudes), $i \in \mathbf{I}$, with capacity χ_i^I, unit cost λ_i, and deliveries ξ_{in}
- A blender with capacity χ^B
- Outputs, $k \in \mathbf{K}$, with demand ξ_{kn}, price λ_k, property specifications π_{kl}^L and π_{kl}^U, and capacity χ_k^O
- A crude distillation unit (CDU) converting inputs to outputs, according to an *operating model*, with a minimum (v^{MIN}) and maximum (v^{MAX}) production rate
- Arcs $a \in \mathbf{A}$ connecting the various nodes of the network (see details in Section 11.1.3)

The goal is to minimize total cost for fixed demand.

We consider two simple CDU operating models, in both of which the feed to the CDU is treated as a linear combination of the constituent inputs:

(1) Model 1: the CDU converts each input to outputs at fixed conversion coefficients, ρ_{ik}; the so produced outputs satisfy property specifications.

(2) Model 2: the CDU converts each input to outputs at fixed conversion coefficients, ρ_{ik}, and property levels, π_{ikl}; an output from a given input may not satisfy property specifications.

We will use the first oversimplified model to introduce the basic concepts, and then the second, more realistic, model. Note that the property levels of inputs are of no interest and thus are not tracked in the blender. Instead, it is the conversion to outputs and the corresponding input–output-specific property levels (in the second CDU model) that are of importance.

11.4.2 Basic Model

The variables used in the previous section to represent input, blender, and output inventories (I_{in}^I, I_{jn}^B, I_{kn}^O) and the corresponding flows $\left(F_{ijn}^{IB}\right)$ are also used here, with the simplification that the index j is omitted, because there is only one blender. Furthermore, blender-CDU $\left(F_n^{BC} \in \mathbb{R}_+\right)$ and CDU-output $\left(F_{kn}^{CO} \in \mathbb{R}_+\right)$ flows are introduced.

The inventory of the inputs and the blend in the blender are described as follows:

$$I_{i,n+1}^I = I_{in}^I + \xi_{in} - F_{in}^{IB} \leq \chi_i^I, \quad i,n \tag{11.45}$$

$$I_{n+1}^B = I_n^B + \sum_i F_{iin}^{IB} - F_n^{BC} \leq \chi^B, \quad n \tag{11.46}$$

The composition of the mixture in the blender determines the blend sent to the CDU and thus the overall conversion. This composition can be modeled by employing input-specific blender inventories $\left(I_{in}^{B}\right)$ and blender-CDU $\left(F_{ikn}^{BC}\right)$ flows:

$$I_{i,n+1}^{B} = I_{in}^{B} + F_{in}^{IB} - F_{in}^{IBC}, \qquad i,n \tag{11.47}$$

which should satisfy splitting equations, similar the ones in Section 11.2.2,[9] using blender-to-column split ratios, $R_n^B \in [0,1]$:

$$F_n^{BC} = R_n^B\left(I_n^B + \sum\nolimits_i F_{in}^{IB}\right), \qquad n \tag{11.48}$$

$$F_{in}^{IBC} = R_n^B\left(I_{in}^B + F_{in}^{IB}\right), \qquad i,n \tag{11.49}$$

where $\sum_i F_{in}^{IB}$ and F_{in}^{IB} are included in (11.48) and (11.49), respectively, because a blender can receive and send material at the same time (recall that we do not consider any operating logic constraints).[10]

The total flow into the CDU is subject to capacity bounds

$$v^{MIN} \leq F_n^{BC} \leq v^{MAX}, \qquad n, \tag{11.50}$$

which, if necessary, can also be expressed as variable bounds, using a binary CDU operating variable.

The total output production, F_{kn}^{CO}, for both CDU models, is calculated by

$$F_{kn}^{CO} = \sum\nolimits_i \rho_{ik} F_{in}^{IBC}, \qquad k,n \tag{11.51}$$

and used in the output inventory balance (recall $\xi_{kn} < 0$):

$$I_{k,n+1}^{O} = I_{kn}^{O} + F_{kn}^{CO} + \xi_{kn} \leq \chi_k^O, \qquad k,n. \tag{11.52}$$

If the first CDU model is used, then output streams automatically satisfy specifications and no constraints for specification satisfaction are needed. Thus, the model consists of (11.45) through (11.52) and objective function (11.53):

$$\min \sum\nolimits_{i,n} \lambda_i F_{in}^{IB} + \sum\nolimits_{a\in\mathbf{A},n} \left(\gamma_a^F X_{an} + \gamma_a^V F_{an}\right). \tag{11.53}$$

If the second CDU model is used, then each output inventory becomes, essentially, a blender with variable property level profiles over time, as in the final product blending problem, but with only one possible outgoing arc and thus a unique set of property

[9] The equality $I_n^B = \sum_i I_{in}^B$ and the corresponding equality for F_n^{BC} are automatically satisfied (can you see why?), so they need not be added.

[10] When material cannot be simultaneously fed and withdrawn from a blender (see problem statement in Section 11.1.3), then if there is outflow at point n, the split fraction is based on the inventory during period n, I_n^B or I_{in}^B, through (11.18) and (11.19). If there is inflow at the time material is withdrawn, then the split fraction should also account for this inflow, which is assumed to occur just before the outflow (but both, still at the same time point). Strictly speaking, the capacity constraint should then be $I_n^B + \sum_i F_{in}^{IB} \leq \chi^B$, but we consider the approximation in (11.46).

specifications.[11] As we have seen in Section 11.2 that property specification constraints can be enforced in two ways. The first is to introduce concentration variables (in this case, C_{kln}), express property balances, and enforce that whenever there is flow to meet output demand, concentrations are within acceptable ranges.

The second approach, the details of which we present here, relies on the monitoring of the output property levels through the introduction of input-specific CDU-output flows $\left(F_{ikn}^{\mathrm{ICO}}\right)$ and output inventories $\left(I_{ikn}^{\mathrm{O}}\right)$:

$$F_{ikn}^{ICO} = \rho_{ik} F_{in}^{IBC}, \qquad i, k, n \tag{11.54}$$

$$I_{ik,n+1}^{O} = I_{ikn}^{O} + F_{ikn}^{ICO} - F_{ikn}^{\Xi}, \qquad i, k, n, \tag{11.55}$$

where F_{ikn}^{Ξ} represents the contribution of input i toward the satisfaction of demand for output k at time n, so it should satisfy

$$\sum_i F_{ikn}^{\Xi} = -\xi_{kn}, \qquad k, n. \tag{11.56}$$

Finally, split ratios, $R_{kn}^{O} \in [0, 1]$, are introduced for the output-to-demand flow

$$-\xi_{kn} = R_{kn}^{O}\left(I_{kn}^{O} + \sum_i F_{ikn}^{ICO}\right), \qquad k, n \tag{11.57}$$

$$F_{ikn}^{\Xi} = R_{kn}^{O}\left(I_{ikn}^{O} + F_{ikn}^{ICO}\right), \qquad i, k, n \tag{11.58}$$

and the property specification constraints are enforced as follows:

$$-\pi_{kl}^{L}\xi_{kn} \le \sum_i \pi_{ikl} F_{ikn}^{\Xi} \le -\pi_{kl}^{U}\xi_{kn}, \qquad k, l, n. \tag{11.59}$$

If we use the second CDU model, the overall blending model consists of (11.45) through (11.52), (11.54) through (11.59), and objective function (11.53).

11.4.3 Illustrative Example

We consider a two-input (I1 and I2), three-output (O1, O2, and O3), one-property network (same as the one shown in Figure 11.10) over four periods. We consider CDU Model 2. The corresponding ρ_{ik} and π_{ik} parameters along with the input starting inventories $\left(\iota_i^0\right)$ and deliveries (ξ_{in}) are given in Table 11.1, where index l is dropped because we consider only one property. The output property specifications (lower bounds) and output orders are given in Table 11.2. Note that O1 produced from I1 $(\pi_{\mathrm{I1,O2}} = 0.88)$ violates the property specification $\left(\pi_{\mathrm{O1}}^{L} = 0.9\right)$, and the same is true for O3 produced from I2 $\left(\pi_{\mathrm{I2,O3}} = 0.65 < 0.7 = \pi_{\mathrm{O3}}^{L}\right)$. At $n = 0$, there are 15 units of I1, and the blender has 10 units of I1. The feed rate to the CDU should be between 15 and 20. To reduce the number of periods studied, it is assumed that material can immediately be transferred through the blender to the CDU and that the CDU has zero residence time.

[11] Treating outputs as blenders and allowing temporary violation of the specifications, when there are no order satisfaction outflows, means that, as in the pooling problem, an inflow to an output may violate specifications but, unlike the pooling problem, there are no constraints for the total inflow (sum of incoming flows) at a time point to satisfy the specifications.

Table 11.1 Data for illustrative example (ι_i^0: initial input inventory); index l dropped from π_{ikl}.

	ρ_{ik}			π_{ik}				ξ_{in}	
Input	O1	O2	O3	O1	O2	O3	ι_i^0	$n=1$	$n=2$
I1	0.3	0.4	0.3	0.88	0.80	0.75	15		
I2	0.3	0.5	0.2	0.95	0.80	0.65		25	

Table 11.2 Data for illustrative example; index l dropped from π_{kl}^L.

		$-\xi_{kn}$	
Output	π_k^L	$n=1$	$n=2$
O1	0.9		12
O2	0.8	8	8
O3	0.7	4	6.4

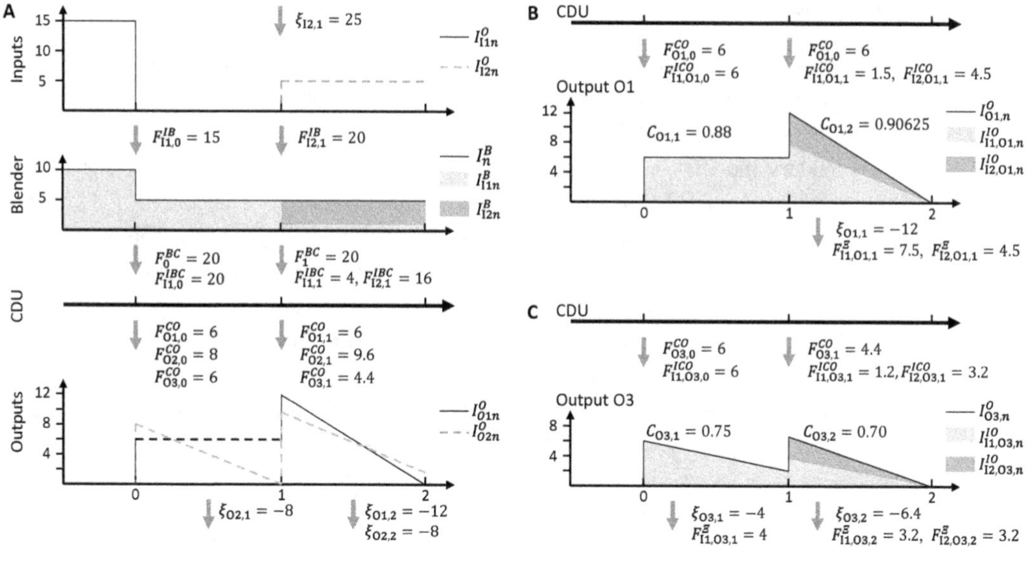

Figure 11.11 Illustration of CDU blending model (to facilitate the illustration, output demand is assumed to be continuous, so output inventories decrease linearly). (A) Feasible solution to the illustrative example: inventory profiles of inputs, blender mixture, and outputs (except O3) are shown in the three panels; input-specific blender inventories shown in middle panel; flows shown as thick gray arrows (not scaled). (B) and (C). Solution details for output O1 (B) and O3 (C): input-specific (i) CDU-output flows, (ii) output inventories, and (iii) demand flows; and concentrations.

Figure 11.11A shows the inventory profiles and major flows in a feasible solution to the illustrative example. Note that at $n = 0$ only I1 is fed to the CDU, so the output flows are based on the conversion coefficients of I1. At $n = 1$, both I1 and I2 are fed into the CDU, so the output flows are linear combinations of the I1 and I2 flows.

(Can you calculate $F^{CO}_{k,n=1}$ from $F^{ICO}_{ik,n=1}$?) Also, note that the blender inventory during $n = 2$ has two positive input-specific inventories because there is I1 left from the previous period and new I2 flow (at $n = 1$).

Figure 11.11B shows the details regarding the production and inventory of output O1 (i.e., input-specific CDU-output flows, inventory profiles, and order satisfaction flows) and concentrations. The same variables are shown in Figure 11.11C for output O3. Both figures show that the processing of two inputs by the CDU at $n = 1$ leads to positive F^{ICO}_{ikn} at $n = 1$, I^{IO}_{ikn} during $n = 2$, and F^{Ξ}_{ikn} during $n = 2$, for both I1 and I2. Note that the inventory of O1 during $n = 1$ does not meet the specification $(C_{O1,1} = 0.88)$, but the O1 flow at $n = 1$ raises the concentration above the specification $(C_{O1,2} = 0.90625)$, and thus the order during $n = 2$ can be met. (Can you calculate the concentration of the stream coming from the CDU at $n = 1$?) Also, note that the flow of O3 at $n = 1$ lowers the concentration from $C_{O3,1} = 0.75$ to $C_{O3,2} = 0.70385$, just above the specification. (Can you calculate the concentration of the O3 stream coming from the CDU at $n = 1$?)

11.4.4 Extensions

We overview modifications and extensions that are necessary when modeling systems with multiple blenders and CDUs and more complex CDU models.

Multiblender, Multi-CDU Systems. The major changes needed to account for multiple blenders and CDUs are (1) the addition of new nodes, arcs, and the corresponding flow variables; and (2) the modification of the inventory balances. If $j \in \mathbf{J}^B$ and $j' \in \mathbf{J}^{CDU}$ are used to denote blenders and CDUs, respectively, then the following flow variable changes are made: F^{IB}_{in} are replaced by F^{IB}_{ijn}, F^{BC}_{n} are replaced by $F^{BC}_{jj'n}$, and F^{CO}_{kn} are replaced by $F^{CO}_{j'kn}$. Input-specific variables are also modified, to I^B_{ij} and $F^{IBC}_{ij'n}$. The overall concept and the modified inventory balances are shown in Figure 11.12, where CDUs have unit-specific conversion coefficients, $\rho_{j'ik}$.

Complex CDU Models. CDU Models 1 and 2 assume that conversion coefficients are fixed and consequently the properties of the outputs, for a given input, are also fixed. However, in practice a CDU can be operated in different ways so the conversion coefficients, and as a result the properties, can vary subject to relationships that couple

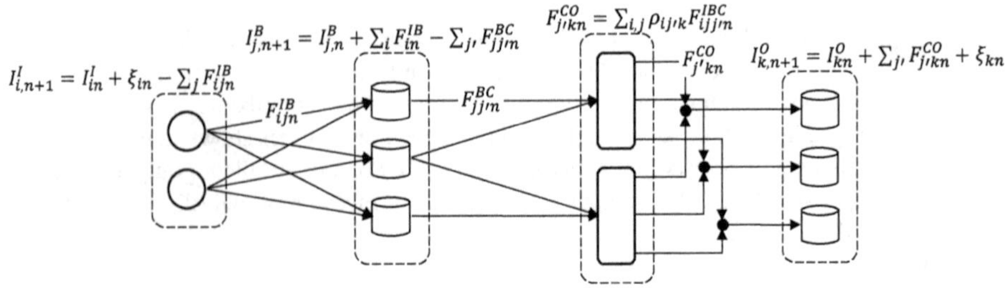

Figure 11.12 Multiblender, multi-CDU final product blending problem: representation and inventory balances.

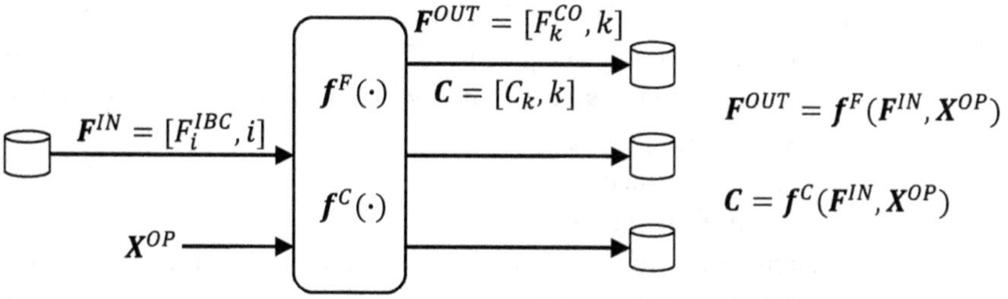

Figure 11.13 General modeling of CDU operation (bold letters used for variable vectors, index n omitted from flow and concentration variables).

conversions and properties.[12] Furthermore, the simplified models presented in the previous subsection assume that *inputs blend linearly*, that is, the output flows can be viewed as linear combinations of the flows that would be obtained if the column feed was a single input, resulting in (11.51). However, in the general case, this linear relationship is not valid, or sufficiently accurate, so a more complex model may be necessary. Regardless of its specifics, a CDU model can be viewed as, essentially, a set of equations that yield (1) the output flows ($\boldsymbol{F}^{OUT}$) as a function, $\boldsymbol{f}^F(\cdot)$, of the input flows $\boldsymbol{F}^{IN}$ and, potentially, some additional operational (degrees of freedom) variables ($\boldsymbol{X}^{OP}$); and (2) the properties of the output streams ($\boldsymbol{C}$) as a function, $\boldsymbol{f}^C(\cdot)$, of the input flows and the operational variables. Figure 11.13 illustrates the idea. Alternatively, the model can be viewed as a set of equations coupling input flows, output flows, output properties, and operational variables, that is, it is a vector function $\boldsymbol{f}(\boldsymbol{F}^{IN}, \boldsymbol{X}^{OP}, \boldsymbol{F}^{OUT}, \boldsymbol{C}) = 0$. The second approach is especially useful when no operational variables are used explicitly.

11.5 Notes and Further Reading

(1) The pooling problem was introduced in Haverly [1]. The presentation in Section 11.1 of the pooling problem is primarily based on Rajagopalan and Sahinidis [2].

(2) While the concepts underpinning the two models for the blending problem (concentration-based and source-based) in Section 11.2 have appeared in many papers, the discussion in the present chapter borrows elements primarily from Kolodziej et al. [3] and Lotero et al. [4].

(3) The discussion of the approximate linear models in Section 11.3 is based on Kolodziej et al. [5]. Gupte et al. [6] also discuss the discretization of bilinear terms.

[12] To give an example, consider the operation of a CDU with the goal to maximize the top (gas) cut. To achieve this, some heavier components must be included in the cut, which means that the cut, on average, will become heavier. Thus, properties such as volatility will be impacted.

(4) Some important early contributions in the area of poling came from the process systems engineering community [7–9]. Computational-focused contributions, including methods based on piecewise linear approximations, appeared in the 2000s [10–13].

(5) The interested reader can find a number of models and methods for crude oil distillation blending [14, 15].

(6) A number of different approaches are available for product blending [16–19].

(7) Practical aspects of blending, including its importance in refinery operations are discussed in a series of papers by Kelly and coauthors [20–22].

(8) An excellent source for general material on global optimization is the book by Tawarmalani and Sahinidis [23]. Chapter 9, in particular, addresses the pooling problem. Appendix A of the book provides GAMS codes for many pooling problems from the literature.

11.6 Exercises

(1) Consider a pooling system with three inputs (I1, I2, and I3), one pool, two outputs (O1 and O2), and one property. Inputs I1 and I2 are connected to the pool but not connected to outputs; I3 is connected to both outputs but not connected to the pool. The pool, with a capacity of 300 mg, is connected to both outputs. Other parameters are given in Table 11.3. Initial input inventories are equal to 200 Mg. Solve a profit maximization pooling problem.

(2) Consider a pooling system with five inputs (I1–I5), two pools (J1 and J2), four outputs (O1–O4), and four properties. All inputs are connected to all pools, and all pools are connected to all outputs. The capacity of each pool is 75 Mg. Other parameters are given in Tables 11.4 and 11.5. Solve a profit maximization pooling problem.

(3) Consider a system with three inputs (I1, I2, and I3), all with initial inventory of 75 Mg; one blender, with capacity of 25 Mg; two outputs (O1 and O2), and one property. Each input is connected to the blender, and the blender is connected to all outputs. All other parameters are given in Table 11.3. Material cannot be

Table 11.3 Input- and output-related data: input capacity (χ_i^I), price (λ_i), and property level (π_i); and output demand at the end of the horizon (ξ_k), price (λ_k), and upper bounding property specification (π_k^U).

	I1	I2	I3	O1	O2
χ_i^I [Mg]	200	200	200	-	-
λ_i [$10^3/Mg]	6	16	10	-	-
π_i	3	1	2	-	-
ξ_k [Mg]	-	-	-	100	100
λ_k [$10^3/Mg]	-	-	-	9	15
π_k^U	-	-	-	2.5	1.5

Table 11.4 Input- and output-related data: input capacity, χ_i^I [Mg], and price, λ_i [$\$10^3$/Mg]; and output demand at the end of the horizon, ξ_k [Mg], and price, λ_k [$\$10^3$/Mg].

	I1	I2	I3	I4	I5	O1	O2	O3	O4
χ_i^I/λ_i	75/7	75/3	75/2	75/10	75/5				
ξ_k/λ_k						0/16	0/25	0/15	0/10

Table 11.5 Input property level (π_i) and output (upper bounding) property specification (π_k^U).

	I1	I2	I3	I4	I5	O1	O2	O3	O4
L1	1	4	4	3	1	3	4	1.5	3
L2	6	1	5.5	3	2.7	3	2.5	5.5	4
L3	4	3	3	3	4	3.25	3.5	3.9	4
L4	0.5	2	0.9	1	1.6	0.75	1.5	0.8	1.8

simultaneously fed/withdrawn to/from blenders. Solve a profit maximization blending problem over two periods using the following models:

(a) Concentration-based model.

(b) Source-based model.

(c) Discretized concentration-based model: discretize variable C_{jln} into 100 discrete points uniformly distributed in $[\min_i\{\pi_{il}\}, \max_i\{\pi_{il}\}]$ for each j and n.

(d) Discretization-relaxation-based model: discretize variable C_{jln} into 100 discrete points uniformly distributed in $[\min_i\{\pi_{il}\}, \max_i\{\pi_{il}\}]$ for each j and n.

(4) Consider a system with five inputs (I1–I5), all with initial availability of 75 Mg; two blenders (J1 and J2), both with capacity of 25 Mg; four outputs (O1–O4); and four properties. Each input is connected to all blenders, and each blender is connected to all outputs. The remaining parameters are given in Tables 11.4 and 11.5. No simultaneous feeding to and withdrawing from blenders is allowed. Solve the profit maximization blending problem over a horizon of four periods using the following models.

(a) Concentration-based model.

(b) Source-based model.

(c) Discretized concentration-based model: discretize variable C_{jln} into 100 discrete points uniformly distributed in $[\min_i\{\pi_{il}\}, \max_i\{\pi_{il}\}]$ for each j and n.

(d) Discretization-relaxation-based model: discretize variable C_{jln} into 100 discrete points uniformly distributed in $[\min_i\{\pi_{il}\}, \max_i\{\pi_{il}\}]$ for each j and n.

(5) Consider the system shown in Figure 11.10 with parameters ρ_{ik}, ι_i^O and ξ_{in} given in Table 11.1; parameters ξ_{kn} given in Table 11.2; parameters λ_i and λ_k given in

Table 11.6 Input (λ_i) and output (λ_k) prices [$\$10^3$/Mg].

	I1	I2	O1	O2	O3
λ_i	7	3			
λ_k			16	25	15

Table 11.6; and $v^{MIN} = 20$ Mg/h and $v^{MAX} = 25$ Mg/h. Solve a profit maximization process blending problem using Model 1 from Section 11.4.1 over two periods.

(6) Consider the system from Exercise 5. Solve a profit maximization process blending problem using Model 2 from Section 11.4.1 over two periods. Parameters π_{ik} and π_k^L can be found in Tables 11.1 and 11.2, respectively.

References

[1] Haverly CA. Studies of the Behavior of Recursion for the Pooling Problem. *SIGMAP Bull.* 1978(25):19–28.

[2] Rajagopalan S, Sahinidis NV. The Pooling Problem. In *Advances and Trends in Optimization with Engineering Applications*, eds. Terlaky T, Anjos MF, Ahmed, S. Philadelphia: Society for Industrial and Applied Mathematics; Mathematical Optimization Society, 2017; pp. 207–217.

[3] Kolodziej S, Castro PM, Grossmann IE. Global Optimization of Bilinear Programs with a Multiparametric Disaggregation Technique. *J Global Optim.* 2013;57(4):1039–1063.

[4] Lotero I, Trespalacios F, Grossmann IE, Papageorgiou DJ, Cheon MS. An MILP-MINLP Decomposition Method for the Global Optimization of a Source Based Model of the Multiperiod Blending Problem. *Comput Chem Eng.* 2016;87:13–35.

[5] Kolodziej SP, Grossmann IE, Furman KC, Sawaya NW. A Discretization-Based Approach for the Optimization of the Multiperiod Blend Scheduling Problem. *Comput Chem Eng.* 2013;53:122–142.

[6] Gupte A, Ahmed S, Seok Cheon M, Dey S. Solving Mixed Integer Bilinear Problems Using MILP Formulations. *SIAM Journal on Optimization.* 2013; 23(2): 721–744.

[7] Floudas CA, Visweswaran V. A Global Optimization Algorithm (GOP) for Certain Classes of Nonconvex NLPs – I. Theory. *Comput Chem Eng.* 1990;14(12):1397–1417.

[8] Visweswaran V, Floudas CA. A Global Optimization Algorithm (GOP) for Certain Classes of Nonconvex NLPs – II. Application of Theory and Test Problems. *Comput Chem Eng.* 1990;14(12):1419–1434.

[9] Adhya N, Tawarmalani M, Sahinidis NV. A Lagrangian Approach to the Pooling Problem. *Ind Eng Chem Res.* 1999;38(5):1956–1972.

[10] Gounaris CE, Misener R, Floudas CA. Computational Comparison of Piecewise – Linear Relaxations for Pooling Problems. *Ind Eng Chem Res.* 2009;48(12):5742–5766.

[11] Misener R, Thompson JP, Floudas CA. APOGEE: Global Optimization of Standard, Generalized, and Extended Pooling Problems via Linear and Logarithmic Partitioning Schemes. *Comput Chem Eng.* 2011;35(5):876–892.

[12] Ceccon F, Kouyialis G, Misener R. Using Functional Programming to Recognize Named Structure in an Optimization Problem: Application to Pooling. *AIChE J.* 2016;62 (9):3085–3095.

[13] Baltean-Lugojan R, Misener R. Piecewise Parametric Structure in the Pooling Problem: From Sparse Strongly-Polynomial Solutions to NP-Hardness. *J Global Optim.* 2018;71 (4):655–690.

[14] Castro PM, Grossmann IE. Global Optimal Scheduling of Crude Oil Blending Operations with RTN Continuous-Time and Multiparametric Disaggregation. *Ind Eng Chem Res.* 2014;53(39):15127–15145.

[15] Kelly JD, Menezes BC, Grossmann IE. Distillation Blending and Cutpoint Temperature Optimization Using Monotonic Interpolation. *Ind Eng Chem Res.* 2014;53 (39):15146–15156.

[16] Li J, Karimi IA, Srinivasan R. Recipe Determination and Scheduling of Gasoline Blending Operations. *AIChE J.* 2010;56(2):441–465.

[17] Castro PM. New MINLP Formulation for the Multiperiod Pooling Problem. *AIChE J.* 2015;61(11):3728–3738.

[18] Castillo PAC, Castro PM, Mahalec V. Global Optimization of Nonlinear Blend-Scheduling Problems. *Engineering.* 2017;3(2):188–201.

[19] Mendez CA, Grossmann IE, Harjunkoski I, Kabore P. A Simultaneous Optimization Approach for Off-Line Blending and Scheduling of Oil-Refinery Operations. *Comput Chem Eng.* 2006;30(4):614–634.

[20] Kelly JD, Mann JL. Crude Oil Blend Scheduling Optimization: An Application with Multimillion Dollar Benefits. Part 2. The Ability to Schedule the Crude Oil Blendshop More Effectively Provides Substantial Downstream Benefits. *Hydrocarbon Processing.* 2003;82(7):72–79.

[21] Kelly JD, Mann JL. Crude Oil Blend Scheduling Optimization: An Application with Multimillion Dollar Benefits. Part 1. The Ability to Schedule the Crude Oil Blendshop More Effectively Provides Substantial Downstream Benefits. *Hydrocarbon Processing.* 2003;82(6):72–79.

[22] Kelly JD. Logistics: The Missing Link in Blend Scheduling Optimization. *Hydrocarbon Processing.* 2006;85(6):45–51.

[23] Tawarmalani M, Sahinidis NV. *Convexification and Global Optimization in Continuous and Mixed-Integer Nonlinear Programming : Theory, Algorithms, Software, and Applications.* Dordrecht and Boston: Kluwer Academic Publishers; 2002. xxv, 475 p. p.

Part IV

Special Topics

12 Solution Methods: Sequential Environments

In Part IV (Chapters 12 through 15), we discuss some special topics. Specifically, we discuss solution methods (Chapters 12 and 13), real-time scheduling (Chapter 14), and the integration of scheduling with planning (Chapter 15). The overarching approach is to introduce some basic concepts and outline different approaches that can be applied to a range of problems, rather than give implementation specifics. The interested reader can find the details of the discussed methods in the corresponding journal papers given at the end of each chapter. The goal of the present chapter, as well as Chapter 13, is to illustrate how problem features can be exploited to develop more efficient models and/or specialized algorithms.

We start, in the present chapter, with solution methods for problems in sequential environments. Specifically, we discuss four methods: (1) a decomposition approach, in Section 12.1; (2) preprocessing algorithms and tightening constraints, in Section 12.2; (3) a reformulation and tightening constraints based on time windows, in Section 12.3; and (4) a two-step algorithm, combining the advantages of discrete and continuous time models, in Section 12.4. While all presented methods can be applied to a wide range of problems, we present them for a subset of problems for the shake of brevity. Also, all methods can be applied to problems under different processing features, but to keep the presentation simple, we discuss problems with no shared utilities and no storage constraints. For simplicity and without loss of generality, when discussing models based on a discrete time grid, we assume $\delta = 1$ and that all time-related data are integers, so $\tau_{ij} - \bar{\tau}_{ij}$, $\rho_i = \bar{\rho}_i$, and so on; and thus we use τ_{ij}, ρ_i, ε_i (without overbars).

12.1 Decomposition Methods

We start, in Section 12.1.1, with some preliminary concepts on decomposition, and then apply them to cost minimization problems in single- (Section 12.1.2) and multistage Section 12.1.3) environments. Then, in Section 12.1.4, we discuss how the approach can be modified to address makespan minimization problems, and we close, in Section 12.1.5, with a number of remarks and extensions.

12.1.1 Preliminaries

In Chapters 3 through 6 we studied problems in sequential environments, where, if the batching problem is solved, a fixed set of batches have to undergo processing, without splitting of mixing, through a set of stages. If there are no constraints for shared general resources, then sequence-based models can be employed, as they are intuitive, accurate, and computationally effective for small- and medium-size instances. However, these models become expensive when the number of batches and/or the number of units and stages increases. One of the reasons for the poor performance of these models is that they include big-M-type constraints, which tend to lead to weak relaxations.

To address large-scale instances, many researchers have proposed decomposition approaches where the original problem is decomposed into two subproblems, which are significantly easier to solve, though multiple iterations may be necessary to prove optimality. The basic idea is to keep the *easy* constraints, which can be used to make a subset of decisions, into one subproblem; and then, with these decisions fixed, make the remaining decisions using a second subproblem. For example, one approach is to decompose the original problem into a high-level batch-unit assignment subproblem, and a lower-level sequencing subproblem, which, interestingly, can be solved effectively by methods other than mathematical programming. The two subproblems may have to be solved multiple times because an assignment determined at the high-level subproblem may lead to an infeasible lower-level problem. To avoid finding the same assignment decisions in subsequent iterations, the high-level subproblem has to be modified; this is typically achieved through the addition of *integer cuts* (see discussion in Section 2.2.1).

Conceptually, the original scheduling problem $(\mathbb{P})$:

$$\begin{aligned} &\min f\left(X_{ij}, S_{ij}\right) \\ &st \;\; g\left(X_{ij}\right) = 0 \qquad (\mathbb{P}) \\ &\qquad h\left(X_{ij}, S_{ij}, U\right) = 0 \end{aligned} \tag{12.1}$$

is decomposed into an assignment subproblem $(\mathbb{P}^A)$:

$$\begin{aligned} &\min f^A\left(X_{ij}\right) \\ &st \;\; g^A\left(X_{ij}\right) = 0 \quad \left(\mathbb{P}^A\right) \\ &\qquad g^{IC}\left(X_{ij}\right) = 0 \end{aligned} \tag{12.2}$$

and a sequencing subproblem $(\mathbb{P}^S)$:

$$\begin{aligned} &\min f\left(\bar{X}_{ij}^n, S_{ij}\right) \\ & \qquad\qquad\qquad\qquad\qquad\qquad\qquad\qquad \left(\mathbb{P}^S\right), \\ &st \;\; h^S\left(\bar{X}_{ij}^n, S_{ij}, U^S\right) = h_n^S\left(S_{ij}, U^S\right) = 0 \end{aligned} \tag{12.3}$$

where $X_{ij} \in \{0, 1\}$ denotes the assignment of batch i to unit j, S_{ij} is the start time of batch i on unit j, and U and U^S represent other auxiliary variables used in models $(\mathbb{P})$ and $(\mathbb{P}^S)$, respectively.

The objective function can depend on assignment (e.g., processing cost minimization), timing (e.g., tardiness minimization), or both assignment and timing decisions. In the general case, a different objective function, $f^A(X_{ij})$, is used in $(\mathbb{P}^A)$ because variables S_{ij} are removed. The objective function of $(\mathbb{P}^S)$ is the same as $(\mathbb{P})$, though some terms may be fixed in the former, because $(\mathbb{P}^S)$ is formulated for given X_{ij}. The original model consists of assignment constraints, $g(X_{ij}) = 0$, and sequencing constraints, $h(X_{ij}, S_{ij}, U) = 0$[1]. Subproblem $(\mathbb{P}^A)$ is obtained by keeping the assignment constraints, potentially modified, removing sequencing constraints and adding integer cuts, $g^{IC}(X_{ij}) = 0$;[2] it contains variables X_{ij} only. At iteration m, $(\mathbb{P}^A)$ yields solution $X_{ij} = \bar{X}_{ij}^m$. Subproblem $(\mathbb{P}^S)$ is, essentially, equivalent to $(\mathbb{P})$ but with fixed assignment decisions, $X_{ij} = \bar{X}_{ij}^m$, which means that assignment constraints can be removed and (some of) the sequencing constraints can be simplified and/or become tighter. Also, note that some of the original auxiliary variables U may be fixed or removed, so, in general, the auxiliary variables used in $(\mathbb{P}^S)$, U^S, are different.

The most popular approach is to decompose the problem into a high-level MIP subproblem and a low-level constraint programming (CP) subproblem, though other modeling and solution paradigms, with complementary strengths, can also be integrated.

12.1.2 Single-Stage Environment: Cost Minimization

We consider the single-stage problem with fixed batching decisions (see Chapter 4): we are given a set, $\mathbf{I}$, of batches and a set, $\mathbf{J}$, of units. Each batch $i \in \mathbf{I}$ has a release, ρ_i, and (hard) due, ε_i, time, and has to be carried out in exactly one *compatible* unit $j \in \mathbf{J}_i \subseteq \mathbf{J}$. The set of batches that can be processed on unit j is $\mathbf{I}_j$. The processing time of batch i on unit j is τ_{ij} and the processing cost γ_{ij}^P.

The global-sequence-based model, $(\mathbb{P})$[3], for this problem consists of assignment constraints

$$\sum\nolimits_{j \in \mathbf{J}_i} X_{ij} = 1, \quad i; \tag{12.4}$$

constraints activating global sequencing binary variables ($Y_{ii'j} = 1$ if batch i preceeds i' on unit j),

$$Y_{ii'j} + Y_{i'ij} \geq X_{ij} + X_{i'j} - 1, \quad i, i' > i, j \in \mathbf{J}_i \cap \mathbf{J}_{i'}; \tag{12.5}$$

constraints for the disaggregation of batch start times, S_i, into unit-specific start times, S_{ij},

$$S_i = \sum\nolimits_j S_{ij}, \quad i; \tag{12.6}$$

$$S_{ij} \leq MX_{ij}, \quad i, j; \tag{12.7}$$

[1] The distinction between assignment and sequencing constraints is not unique because, depending on the problem and employed model, there are constraints containing, for example, both X_{ij} and S_{ij} variables or both X_{ij} and a subset of auxiliary variables U. In the present chapter, we limit our discussion to assignment constraints that contain variables X_{ij} only. All remaining constraints are classified as sequencing ones.

[2] Recall that inequalities can always be rewritten as equalities using nonnegative slack variables.

[3] In Section 12.1.1, we used $(\mathbb{P})$, $(\mathbb{P}^A)$, and $(\mathbb{P}^S)$ to describe *problems*. Henceforth, to simplify the presentation, we use them to denote both problems and specific models to address the corresponding problems; for example, in this subsection we use $(\mathbb{P})$ to refer to a global sequence model for cost minimization in a single-stage environment.

the disjunctive constraint,

$$S_{ij} + \tau_{ij} \le S_{i'j} + M\left(1 - Y_{ii'j}\right), \quad i, i', j \in \mathbf{J}_i \cap \mathbf{J}_{i'}; \tag{12.8}$$

and release and due time constraints:

$$S_i \ge \rho_i, \quad i \tag{12.9}$$

$$S_i + \sum\nolimits_j \tau_{ij} X_{ij} \le \varepsilon_i, \quad i. \tag{12.10}$$

The objective function for cost minimization is

$$\min \sum\nolimits_{i,j} \gamma_{ij}^P X_{ij}. \tag{12.11}$$

We note the following regarding model $(\mathbb{P})$:

(1) Its objective function depends only on assignment decisions.
(2) The *auxiliary* variables include sequencing binary variables $Y_{ii'j}$.
(3) Equation (12.4) is the only constraint containing only variables X_{ij}.

Following the ideas presented in the previous subsection, we decompose this model into an assignment subproblem, $(\mathbb{P}^A)$, consisting of objective function (12.11), assignment constraint (12.4), integer cuts, and the following two constraints, which can be added to exclude some clearly infeasible assignments:

$$\sum\nolimits_{i \in \mathbf{I}_j} \tau_{ij} X_{ij} \le \max_{i \in \mathbf{I}_j} \{\varepsilon_i\} - \min_{i \in \mathbf{I}_j} \{\rho_i\}, \quad j \tag{12.12}$$

$$\rho_i + \sum\nolimits_{i \in \mathbf{J}_i} \tau_{ij} X_{ij} \le \varepsilon_i, \quad i. \tag{12.13}$$

The former ensures that the sum of processing times of the batches assigned to a unit does not exceed the available processing time; and the latter enforces that a batch ends before its due time. Solving $(\mathbb{P}^A)$ yields an assignment solution, and a lower bound, z_m, at iteration m, on the optimal objective function value of the original problem. (Can you see why?) Importantly, if the assignment decisions are given, the sequencing subproblem is decomposed further into $|\mathbf{J}|$ independent single-unit subproblems.

Specifically, if $\mathbf{I}_{mj}$ is the set of batches assigned to unit j in the solution of $\left(\mathbb{P}^A\right)$ at iteration m, henceforth referred to as subproblem $\left(\mathbb{P}_m^A\right)$, then the sequencing subproblem is equivalent to $|\mathbf{J}|$ independent single-unit feasibility problems, $(\mathbb{P}_{mj}^S)$ each one of which can be written as follows:

$$Y_{ii'}^j + Y_{i'i}^j \ge 1, \quad i \in \mathbf{I}_{mj}, i' \in \mathbf{I}_{mj}, i' > i \tag{12.14}$$

$$S_i^j + \tau_{ij} \le S_{i'}^j + M^j\left(1 - Y_{i'i}^j\right), \quad i \in \mathbf{I}_{mj}, i' \in \mathbf{I}_{mj} \tag{12.15}$$

$$\rho_i \le S_i^j, \quad S_i^j + \tau_i^j \le \varepsilon_i, \quad i \in \mathbf{I}_{mj}, \tag{12.16}$$

where we use index j, as superscript, to differentiate the subproblems. Note that (12.14) is obtained from (12.5) by fixing the corresponding assignment binary variables to 1; S_i dissagregation is not needed because the only positive S_{ij} can be identified, and replaced

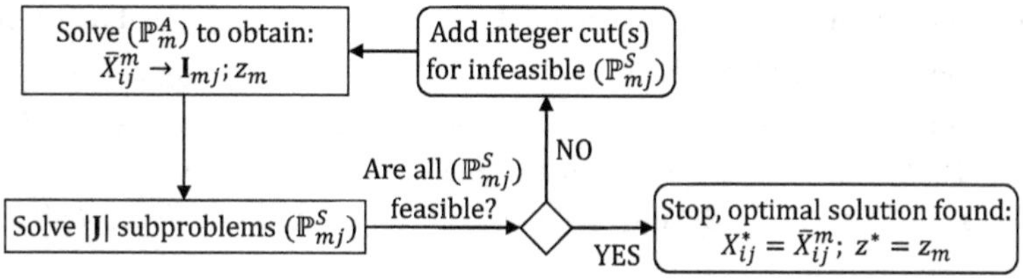

Figure 12.1 Flow chart of decomposition algorithm for the single-stage cost minimization problem.

by S^j_i, based on $\mathbf{I}_{mj}$; and a tighter big-M parameter, M^j, can be used in (12.15), compared to (12.5), based on the subset of batches assigned to unit j.

Subproblems $(\mathbb{P}^S_{mj})$ are feasibility problems because if a subproblem is feasible, then the cost of the assignment is the same as the cost calculated by the assignment subproblem $(\sum_{i \in \mathbf{I}_{mj}} \gamma^P_{ij})$; if it is infeasible, then an integer cut has to be added to the assignment subproblem:

$$\sum_{i \in \hat{\mathbf{I}}^n_j} X_{ij} \le |\mathbf{I}_{mj}| - 1, \quad j \in \mathbf{J}^{INF}_m, \tag{12.17}$$

where $\mathbf{J}^{INF}_m$ is the set of units whose corresponding sequencing subproblem in iteration m was found to be infeasible. Note that (12.17) excludes not only the current assignment, but also all assignments where additional batches are assigned to j, that is, it is not a *no-good* integer cut (see (2.10) and (2.11) in Section 2.2.1). In other words, if $\hat{\mathbf{I}}_j = \{i : X_{ij} = 1\}$ represents a general assignment, then (12.17) excludes all $\hat{\mathbf{I}}_j$ that are supersets of $\mathbf{I}_{mj}$. (Can you see why this is a valid constraint?)

If all subproblems are feasible, then the current assignment is feasible, and the iterative algorithm terminates. If one or more sequencing subproblems are infeasible, then new integer cuts are added to the assignment subproblem, and one more iteration is carried out. The algorithm can also terminate if $(\mathbb{P}^A_m)$ becomes infeasible, which would mean that $(\mathbb{P})$ is infeasible. (Can you see why?) A flowchart of the iterative decomposition algorithm is given in Figure 12.1.

While solving the sequencing feasibility subproblems using MIP can be fast, other solution methods are better suited for detecting infeasibility. For example, constraint programming (CP) has often been employed to solve the sequencing subproblems (see discussion in Section 12.5). The resulting algorithm in this case is a MIP/CP hybrid decomposition algorithm.

12.1.3 Multistage Environment: Cost Minimization

We consider cost minimization with fixed batching decisions. The facility consists of processing stages $k \in \mathbf{K}$, and each stage has units $j \in \mathbf{J}_k$ with $\cup_k \mathbf{J}_k = \mathbf{J}$ and $\mathbf{J}_k \cap \mathbf{J}_{k'} = \emptyset$ for all k, k'. We are given a set, $\mathbf{I}$, of batches that have to be processed on exactly one unit in each stage, and the processing of a batch in stage $k+1$ can start only after its processing in stage k is completed (precedence constraint). The subset of units in stage k suitable for processing batch i is $\mathbf{J}_{ik}$; and the set of batches that can be carried out in unit j is $\mathbf{I}_j$. Each batch $i \in \mathbf{I}$ has a release time, ρ_i, and (hard) due time, ε_i. The processing time of batch i on unit j is τ_{ij} and the processing cost is γ_{ij}^P.

In addition to X_{ij} and S_{ij}, we introduce unit-specific sequencing variables $Y_{ii'j} \in \{0,1\}$ and stage-specific start times $S_{ik} \in \mathbb{R}_+$. The global-sequence-based model (see Chapter 5 for details) includes constraints enforcing batch-unit assignments,

$$\sum\nolimits_{j \in \mathbf{J}_{ik}} X_{ij} = 1, \qquad i,k \tag{12.18}$$

activating global sequencing binary variables,

$$Y_{ii'j} + Y_{i'ij} \geq X_{ij} + X_{i'j} - 1, \qquad i, i' > i, k, j \in \mathbf{J}_{ik} \cap \mathbf{J}_{i'k}, \tag{12.19}$$

which then are used to enforce the precedence and disjunctive constraints,

$$S_{ik} + \sum\nolimits_{j \in \mathbf{J}_{ik}} \tau_{ij} X_{ij} \leq S_{i,k+1}, \qquad i, k < |\mathbf{K}| \tag{12.20}$$

$$S_{ij} + \tau_{ij} \leq S_{i'j} + M\left(1 - Y_{ii'j}\right), \quad i, i', k, j \in \mathbf{J}_{ik} \cap \mathbf{J}_{i'k}, \tag{12.21}$$

where stage-specific start time variables S_{ik} are disaggregated into unit-specific variables S_{ij}

$$S_{ik} = \sum\nolimits_{j \in \mathbf{J}_{ik}} S_{ij}, \qquad i,k \tag{12.22}$$

$$S_{ij} \leq MX_{ij}, \qquad i, k, j \in \mathbf{J}_{ik} \tag{12.23}$$

and release and due time constraints

$$S_{i,1} \geq \rho_i, \quad i; \quad S_{i,|\mathbf{K}|} + \sum\nolimits_{j \in \mathbf{J}_{i,|\mathbf{K}|}} \tau_{ij} X_{ij} \leq \varepsilon_i, \quad i. \tag{12.24}$$

The minimization of processing cost objective function is

$$\min \sum\nolimits_{i,j} \gamma_{ij}^P X_{ij}. \tag{12.25}$$

Following the ideas presented in the previous subsection, we can formulate an assignment subproblem consisting of assignment constraint (12.18), a constraint defining an operating window for each unit,

$$\sum\nolimits_{i \in \mathbf{I}_j} \tau_{ij} X_{ij} \leq \varepsilon_j - \rho_j, \qquad k, j \in \mathbf{J}_k, \tag{12.26}$$

where

$$\rho_j = \min_{i \in \mathbf{I}_j} \left\{ \rho_i + \sum\nolimits_{k' < k} \min_{j \in \mathbf{J}_{ik'}} \tau_{ij} \right\}, \qquad k, j \in \mathbf{J}_k \tag{12.27}$$

$$\varepsilon_j = \max_{i \in \mathbf{I}_j} \left\{ \varepsilon_i - \sum\nolimits_{k' > k} \min_{j \in \mathbf{J}_{ik'}} \tau_{ij} \right\}, \qquad k, j \in \mathbf{J}_k, \tag{12.28}$$

a constraint based on release and due times,

$$\rho_i + \sum\nolimits_k \sum\nolimits_{j \in \mathbf{J}_{ik}} \tau_{ij} X_{ij} \leq \varepsilon_i, \quad i, \tag{12.29}$$

and the integer cuts added at each iteration.

As in the single-stage environment, the sequencing subproblem $(\mathbb{P}^S)$ is a feasibility problem: the first feasible assignment is also the optimal one, provided that both subproblems are solved to optimality.

Based on the batch-unit assignment, $\mathbf{I}_{mj}$, in iteration m, subproblem $(\mathbb{P}^S)$ can be formulated starting from the original full-space MIP model: (1) many variables are fixed to zero, (2) constraints trivially satisfied are removed, and (3) many constraints are simplified. For example, the following variables can be fixed, $X_{ij} = 0, S_{ij} = 0, i, j : i \notin \mathbf{I}_{mj}$; (12.18) can be removed; and (12.19) can be replaced by

$$Y_{ii'j} + Y_{i'ij} \geq 1, \quad i \in \mathbf{I}_{mj}, i' \in \mathbf{I}_{mj}, i' > i, k, j \in \mathbf{J}_k, \tag{12.30}$$

while (12.20) remains the same. (What other modifications should be made to $(\mathbb{P}^S)$? Can you formulate the MIP-based sequencing subproblem?)

However, given a task-unit assignment, the sequencing subproblem in multistage environments is equivalent to the widely studied job-shop scheduling problem.[4] Hence, any algorithm used for job-shop scheduling can be used for the solution of the sequencing subproblem. For example, the Shifting Bottleneck Procedure (SBP), one of the most effective algorithms for job-shop problems, can be used (see discussion in Section 12.5).

Note that both the sequencing among different batches assigned to the same unit and the precedence constraints for the same batch between consecutive stages should be accounted for. This implies that, first, $(\mathbb{P}^S)$ cannot be decomposed into independent subproblems, and, second, $(\mathbb{P}^S)$ can be infeasible due to: (1) the assignment of batches to a single unit, (2) the precedence constraint for a single batch, and (3) a combination of (1) and (2). Interestingly, one advantage of SBP is that it provides information that can be used for the generation of strong integer cuts. Specifically, in the first stage of the SBP it is checked whether a feasible schedule can be obtained given the assignments at each unit independently; that is, SBP detects infeasibilities that are due to the assignments on a single unit.

Regardless of the approach adopted to model and solve the sequencing subproblem, if all single-unit assignments are feasible but the overall assignment is infeasible, a single no-good integer cut is added. If the current assignment is infeasible due to κ single-unit assignments, κ strong single-unit integer cuts are added to the cut pool of the assignment subproblem.

12.1.4 Makespan Minimization

If the objective function is makespan minimization, then the basic idea of the algorithm remains the same, but there are two critical differences:

(1) The makespan cannot be calculated from assignment decisions only, especially in multistage problems, which means that the objective function of $(\mathbb{P}^A)$ will be an estimate of the actual makespan based on the assignments determined in $(\mathbb{P}^A)$.

[4] Can you see why? What is the relationship between job-shop and flexible job-shop discussed in Chapter 1?

To keep $(\mathbb{P}^A)$ as a relaxation of $(\mathbb{P})$, an underestimation of the makespan should be used, therefore providing a lower bound on the optimal makespan.

(2) While a given assignment may lead to an infeasible subproblem,[5] the sequencing subproblem $(\mathbb{P}^S)$ is, in the general case, an optimization problem because there are multiple schedules with different objective function values for a given assignment. The optimal objective function value of $(\mathbb{P}^S_m)$ will be equal to or larger than the optimal objective function value of $(\mathbb{P}^A_m)$.

The optimal objective function value of $(\mathbb{P}^A_m)$, if solved to optimality, monotonically increases with m, while, interestingly, the optimal objective function value of $(\mathbb{P}^S_m)$ does not monotonically change with m. (Can you see why?) Also, $(\mathbb{P}^S_{m'})$ may be feasible although $(\mathbb{P}^S_m)$, with $m < m'$, was infeasible.

In terms of modeling, the key difference is that the best-known makespan value can be used in lieu or in conjunction with due times to calculate time windows. To illustrate, we consider the single-stage problem. The full-space MIP model $(\mathbb{P})$ consists of (12.4), (12.10), and

$$MS \geq S_i + \sum\nolimits_j \tau_{ij} X_{ij}, \quad i \tag{12.31}$$

$$\min MS. \tag{12.32}$$

For the assignment subproblem, an underestimator of MS, in terms of assignment variables only, is

$$MS \geq \min{}_i\{\rho_i\} + \sum\nolimits_i \tau_{ij} X_{ij}, \quad j. \tag{12.33}$$

Subproblem $(\mathbb{P}^A)$ consists of (12.4), integer cuts, and the following two constraints, replacing (12.12) and (12.13):

$$\sum\nolimits_{i \in \mathbf{I}_j} \tau_{ij} X_{ij} \leq \min\left\{MS^{UB}, \max{}_{i \in \mathbf{I}_j}\{\varepsilon_i\}\right\} - \min{}_{i \in \mathbf{I}_j}\{\rho_i\}, \quad j \tag{12.34}$$

$$\rho_i + \sum\nolimits_{j \in \mathbf{J}_i} \tau_{ij} X_{ij} \leq \min\left\{MS^{UB}, \varepsilon_i\right\}, \quad i, \tag{12.35}$$

where, $z^{UB} = MS^{UB}$ is the current best feasible solution. Note that, compared to (12.12) and (12.13), the RHS of (12.34) and (12.35) are modified to restrict the search to solutions that are at least as good as the current best.

Subproblem $(\mathbb{P}^S_m)$ consists of (12.31), (12.32), and (12.14) through (12.16). As in the cost minimization case, it can be decomposed to $|\mathbf{J}|$ independent single-unit subproblems, $(\mathbb{P}^S_{mj})$, each one of which, if feasible, yields a unit-specific makespan MS_{mj}. If all $(\mathbb{P}^S_{mj})$ are feasible, then a feasible solution is found with $MS_m = z^{UB}_m = \max{}_j\left\{MS_{mj}\right\}$, providing an upper bound on the optimal objective function of the original problem. The iterative algorithm is terminated when the two bounds converge. Note that the algorithm may terminate if $(\mathbb{P}^A_m)$ and $(\mathbb{P}^S_m)$ yield the same

[5] In the absence of due times, all assignments lead to a feasible sequencing subproblem. However, if there are due times, then a given assignment may be infeasible.

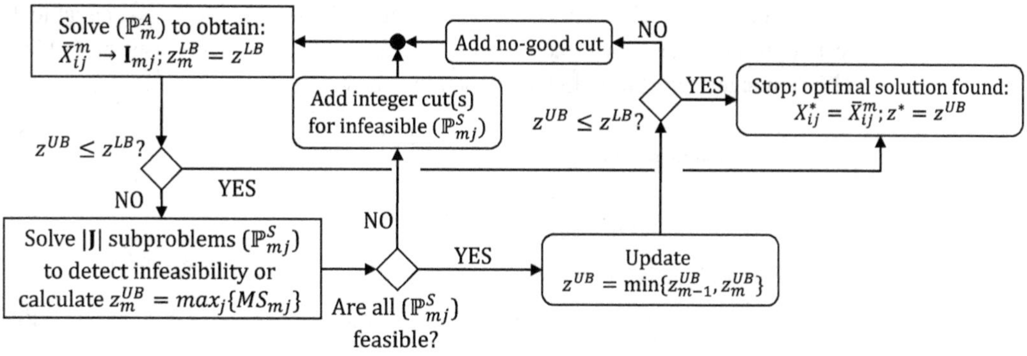

Figure 12.2 Flow chart of decomposition algorithm for single-stage makespan minimization problem.

objective function value, or if the objective function of $(\mathbb{P}_{m'}^A)$ becomes larger than the current upper bound z^{UB} (found by $\mathbb{P}_m^S$ with $m < m'$). The flowchart of the iterative decomposition algorithm is shown in Figure 12.2.

12.1.5 Remarks and Extensions

Computational Performance. The performance of the type of decomposition algorithm presented in this section depends on (1) how fast the two subproblems are solved, and (2) the number of iterations needed to yield the optimal solution and prove optimality. One way to reduce the number of iterations is by reducing the number of assignments that are feasible for $(\mathbb{P}^A)$ but infeasible for $(\mathbb{P})$. Another way is to generate integer cuts during the search that cut off many infeasible or suboptimal assignments. Subproblem $(\mathbb{P}^A)$ can be tightened through problem-specific preprocessing that aims to identify infeasible assignments before the iterative algorithm starts and generates integer cuts that are added to the cut pool of $(\mathbb{P}^A)$, as will be discussed in the next section. In that respect, methods that allow us to identify (combinations of) constraints of $(\mathbb{P}^S)$ that are violated can be exploited for the generation of stronger cuts. Similarly, methods that allow us to identify subsets of assignment decisions that lead to an infeasible $(\mathbb{P}^S)$ enable the generation of integer cuts with fewer binary variables, which exclude more assignment solutions and therefore lead to the exploration of new assignments faster.[6]

Batching Decisions. The algorithm can be readily applied to solve problems in which batching decisions have to be made, through a modification of $(\mathbb{P}^A)$. To illustrate, consider a single-stage problem with units of unequal capacity, β_j, and products, $i \in \mathbf{I}$ with demand ξ_i (see Section 4.5 for details). If $l \in \mathbf{L}_i = \{1, 2, \ldots, \lambda_i^{MAX}\}$ is the set of potential batches for product i, then (12.4) is replaced by

$$\sum\nolimits_{l \in \mathbf{L}_i} \beta_i Z_{il} \geq \xi_i, \quad i \tag{12.36}$$

$$\sum\nolimits_{j \in \mathbf{J}_i} X_{ilj} = Z_{il}, \quad i, l \in \mathbf{L}_i, \tag{12.37}$$

[6] Can you see why, in general, cuts with (positive coefficients for) fewer binary variables X_{ij} are stronger?

where $Z_{il} \in \{0, 1\}$ is equal to 1 if batch (i, l) is selected; and $X_{ilj} \in \{0, 1\}$ is equal to 1 if (selected) batch (i, l) is assigned to unit j. Subproblem $(\mathbb{P}^A)$ consists of (12.36) and (12.37), integer cuts, the following tightening constraints

$$\sum_{i \in \mathbf{I}_j} \sum_{l \in \mathbf{L}_i} \tau_{ij} X_{ilj} \leq \max_{i \in \mathbf{I}_j} \{\varepsilon_i\} - \min_{i \in \mathbf{I}_j} \{\rho_i\}, \ \ j \tag{12.38}$$

$$\rho_i + \sum_{j \in \mathbf{J}_i} \tau_{ij} X_{ilj} \leq \varepsilon_i, \qquad i, l \in \mathbf{L}_i \tag{12.39}$$

and the following objective function:

$$\min \sum_{i,j} \gamma_{ij}^P \sum_{l \in \mathbf{L}_i} X_{ilj}. \tag{12.40}$$

Subproblem $(\mathbb{P}^S)$ remains the same.

Profit Maximization. The algorithm can also be applied to address profit maximization problems. To illustrate, we consider the prize collection problem in a single-stage environment (see Section 3.5.1 for the single-unit variant), where our goal is to maximize the benefit from carrying out a subset of available batches. If π_{ij} is the benefit from carrying out batch i in unit j, then the objective function of $(\mathbb{P}^A)$ is

$$\max \sum_{i,j} \pi_{ij} X_{ij} \tag{12.41}$$

and the assignment of batches is constrained by

$$\sum_{j \in \mathbf{J}_i} X_{ij} \leq 1 \tag{12.42}$$

instead of (12.4). Subproblem $(\mathbb{P}^A)$ consists of (12.41), (12.42) (instead of (12.4)), (12.12) and (12.13), and integer cuts. Subproblem $(\mathbb{P}^S)$ can be decomposed into $|\mathbf{J}|$ independent sequencing subproblems $(\mathbb{P}_j^S)$. One approach is to treat $(\mathbb{P}_j^S)$ as feasibility subproblems, identical to the ones discussed in Section 12.1.2. In this case, if all $(\mathbb{P}_j^S)$ are feasible, then the current assignment is feasible with an objective function value equal to the one calculated in $(\mathbb{P}^A)$ and the algorithm terminates; otherwise, integer cuts are added and one more iteration is executed. An alternative approach is to treat each $(\mathbb{P}_j^S)$ as a single-unit prize collecting (optimization) problem with the set of potential batches coming from the current solution of $(\mathbb{P}^A)$. The generation of effective integer cuts in this case is more challenging.

Other Processing Features. The algorithm can be applied to problems with additional features and constraints, though the complexity of both the assignment and sequencing subproblems increases significantly. For example, if variable batchsizes are considered in $(\mathbb{P})$ but not accounted for in $(\mathbb{P}^A)$, then $(\mathbb{P}^S)$ is typically an optimization problem, even in the case of cost minimization. Similarly, in multistage environments with storage constraints, an assignment determined in $(\mathbb{P}^A)$ may be infeasible for $(\mathbb{P}^S)$ due to storage constraints, which makes the generation of strong integer cuts challenging.

Bound Convergence. The lower (upper) bound provided by $(\mathbb{P}^A)$ monotonically increases (decreases) for a minimization (maximization) problem. Subproblem $(\mathbb{P}^A)$ becomes infeasible only if there are no more feasible assignments; and since $(\mathbb{P}^A)$ is a

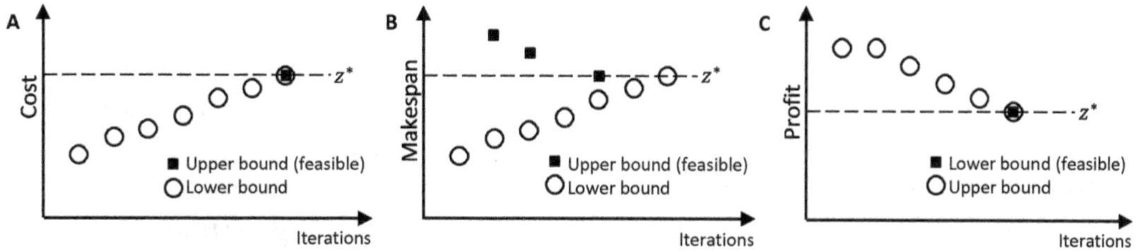

Figure 12.3 Evolution of lower and upper bounds for single-stage cost minimization (A), makespan minimization (B) and profit maximization (C) problems.

relaxation of $(\mathbb{P})$, the algorithm terminates if $(\mathbb{P}^A)$ is infeasible. While the evolution of the best possible bound, provided by $(\mathbb{P}^A)$, is simple and the basic termination criterion of the algorithm (convergence of two bounds) remains the same, the outcome of the solution of the sequencing (or more generally, the low level problem) and the evolution of the bound obtained by it depend on the objective function. Here, we summarize the main observations for single-stage problems:

(1) In cost minimization problems, the objective function value of $(\mathbb{P}^S)$, if feasible, is the same as the objective function value of $(\mathbb{P}^A)$, and the algorithm terminates when the first feasible assignment is found. Thus, $(\mathbb{P}^S)$ can be treated as a feasibility problem.

(2) In makespan minimization problems, the objective function value of $(\mathbb{P}^S)$, which is an optimization problem, is, in general, larger than the objective function value provided by the corresponding solution of $(\mathbb{P}^A)$. Thus, the algorithm terminates when the two bounds converge.

(3) In the profit maximization problem with feasibility subproblems $(\mathbb{P}^S)$, the first feasible $(\mathbb{P}^S)$ corresponds to the optimal solution to $(\mathbb{P})$ and the algorithm terminates.

The evolution of the bounds for the preceding three cases is illustrated in Figure 12.3.

Branch-and-Bound Implementation. The two subproblems can also be used in a branch-and-cut algorithm, in which the assignment subproblem is used as a relaxation of the original problem. When a solution to $(\mathbb{P}^A)$ is obtained, $(\mathbb{P}^S)$ is called to check feasibility. If $(\mathbb{P}^S)$ is feasible, then the current node yields a feasible solution and an upper bound to the original (minimization) problem (P). If $(\mathbb{P}^S)$ is infeasible, then a global[7] integer cut is added to the cut pool of the relaxed problem, $(\mathbb{P}^A)$, the node is pruned, and the branch-and-cut search is continued. A node is pruned if (1) the LP-relaxation of $(\mathbb{P}^A)$ is infeasible, (2) the bound of a node is worse than the current best bound, and (3) $(\mathbb{P}^S)$ is called and found to be infeasible (see Figure 12.4). The

[7] In the context of a branch-and-bound search, a cut is *global* if is applicable to all nodes. A cut is *local* if it is applicable to the descendant nodes of the node where the cut was added (i.e., the nodes in the subtree *originating* at that node).

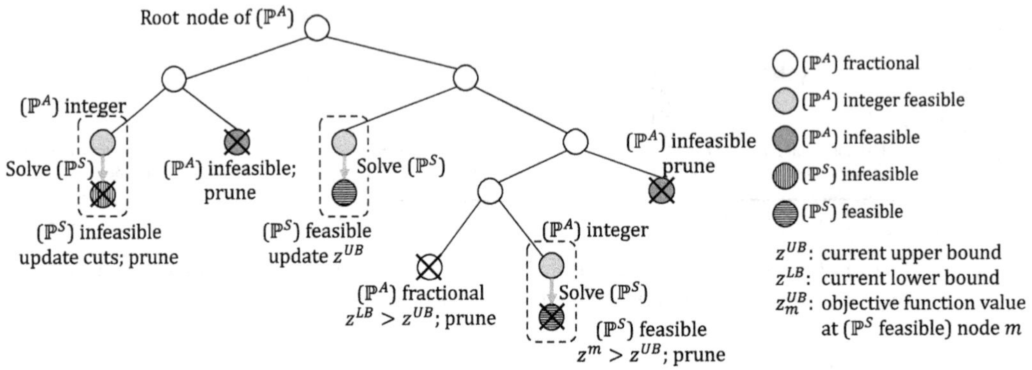

Figure 12.4 Schematic representation of branch-and-bound search employing the assignment-sequencing decomposition.

advantage of the branch-and-cut implementation is that $(\mathbb{P}^A)$ is not solved from scratch every time an integer solution is found. Instead, the search tree is maintained and the solution time needed to obtain the next integer solution of $(\mathbb{P}^A)$ is reduced. The disadvantage of this implementation is that in order to preserve the structure of the incidence matrix of $(\mathbb{P}^A)$ and add cuts during the search, (some) model reductions carried out during MIP presolve may have to be deactivated, resulting in a more expensive assignment subproblem.

12.2 Tightening and Preprocessing

While commercial MIP solvers employ highly efficient preprocessing and tightening methods, domain-specific knowledge can often be exploited to carry out simplifications and add tightening constraints that would not be added by MIP solvers. Accordingly, we present methods based on problem-specific features. We start with the generation of tightening constraints in single-stage (Section 12.2.1) and multistage (Section 12.2.2) environments. These constraints can be added to a MIP model that employs assignment binaries X_{ij} or a model in which X_{ij} can be calculated (see, for example, $X_{ij} = \sum_t X_{ijt}$ in Section 4.3). Interestingly, the same constraints can also be used to strengthen the assignment subproblem in the decomposition algorithm discussed in the previous section. Section 12.2.3 discusses some ideas on how to use release and due times to define windows and then, using these windows, fix sequencing binary variables in sequence-based models.

12.2.1 Tightening Based on Batch-Unit Assignments: Single-Stage

We illustrate how time windows can be exploited to generate tightening constraints through Example 12.1.

Example 12.1 Tightening Knapsack (and Cover) Constraints Consider a single-stage system with two parallel units $(j \in \mathbf{J} = \{U1, U2\})$, and three batches $(i \in \mathbf{I} = \{A, B, C\})$ with the associated processing data given in Table 12.1. We further consider two sets of release and hard due times. Note that the processing time of all batches in both units is 2 h, but the processing cost in unit U1 is lower. Thus, the LP relaxation of a sequence-based model for cost minimization would have assignments of more batches to U1.

Using the first set of release and due times (ρ_i^1 and ε_i^1), we calculate that all batches assigned to U1 cannot start earlier than $\min_i\{\rho_i\} = \min\{0,1,2\} = 0$ and should be finished by $\max_i\{\varepsilon_i\} = \max\{6,3,4\} = 6$, which means that any assignment should satisfy:

$$2X_{A,U1} + 2X_{B,U1} + 2X_{C,U1} \leq \max\{6,3,4\} - \min\{0,1,2\} = 6. \tag{12.43}$$

Equation (12.43) does not cut off any solution, which means that the $X_{A,U1} = X_{B,U1} = X_{C,U1} = 1$ solution would be admissible by, for example, the assignment subproblem of the decomposition algorithm presented in the previous subsection. Obviously, this assignment is infeasible because batch B has to start at or after $t = 1$, which means that there is not enough time for batch A to be performed before B, and there is not enough time for batches A and C to be performed after B. Note that while the same solution may be infeasible for the LP relaxation of a MIP model presented in Chapter 3, solutions where $X_{A,U1} + X_{A,U2} + X_{A,U3} = 2.5$ can be feasible (depending on the selection of the big-M parameters) for the LP relaxation. Thus, the next question is: Can we systematically develop constraints that would enforce $X_{A,U1} + X_{B,U1} + X_{C,U1} \leq 2$?

Interestingly, applying the logic we used to write (12.43) to a subset of batches, namely batches B and C, leads to

$$2X_{B,U1} + 2X_{C,U1} \leq \max\{3,4\} - \min\{1,2\} = 3, \tag{12.44}$$

which cuts off $X_{A,U2} = X_{A,U3} = 1$. In fact, (12.44) is a knapsack constraint (see discussion in Section 2.3.1) and (if we ignore index j which does not change), set $\{B, C\}$ is a cover, which means we can write

$$X_{B,U1} + X_{C,U1} \leq 1. \tag{12.45}$$

Table 12.1 Process data for Example 12.1: processing times, τ_{ij}; processing costs, γ_{ij}^P; release, ρ_i^m, and due, ε_i^m, times ($m \in \{1,2\}$: instance).

i	$\tau_{i,U1}$	$\tau_{i,U2}$	$\gamma_{i,U1}^P$	$\gamma_{i,U2}^P$	ρ_i^1	ε_i^1	ρ_i^2	ε_i^2
A	2	2	2	5	0	6	0	4
B	2	2	3	6	1	3	1	3
C	2	2	3	7	2	4	3	6

Since $X_{\mathrm{A,U1}} \leq 1$, (12.45) also implies

$$X_{\mathrm{A,U1}} + X_{\mathrm{B,U1}} + X_{\mathrm{C,U1}} \leq 2,$$

which is the desired inequality.

Example 12.1 shows that time windows for a subset of batches can be used to develop tightening (knapsack) constraints. The general form of the constraint (type I) we seek to develop is

$$\sum\nolimits_{i \in \mathbf{I}_{jl}^{l}} \tau_{ij} X_{ij} \leq \max_{i \in \mathbf{I}_{jl}^{l}} \{\varepsilon_i\} - \min_{i \in \mathbf{I}_{jl}^{l}} \{\rho_i\}, \quad j, l, \qquad (12.46)$$

where $\mathbf{I}_{jl}^{l}$ are subsets of $\mathbf{I}_j$ which are used as the domain for the summation in the LHS of (12.46). We use index l to denote these subsets. The pseudocode for a straightforward algorithm to generate such constraints is as follows:

Algorithm 1
forall $j \in \mathbf{J}$
  initialize: $l = 0$
  forall $i \in \mathbf{I}_j, i' \in \mathbf{I}_j : \rho_i \leq \rho_{i'}, \varepsilon_i \leq \varepsilon_{i'}$
    $\mathbf{S} = \{i'' : \rho_i \leq \rho_{i''}, \varepsilon_{i''} \leq \varepsilon_{i'}\}$
    if $\sum_{i'' \in \mathbf{S}} \tau_{i''j} > \varepsilon_{i'} - \rho_i$ then
      $l = l + 1$
      $\mathbf{I}_{jl}^{l} = \mathbf{S}$
      add (12.46) for current (j, l)

Although a more efficient algorithm, that exploits the ordering of tasks by ascending release time and descending due time, can be developed, the computational time required to run Algorithm 1 is negligible (less than five seconds even for large-scale problems).

Figure 12.5 illustrates how the consideration of subset of batches leads to tighter formulations. Panel A shows that the time window based on all batches does not

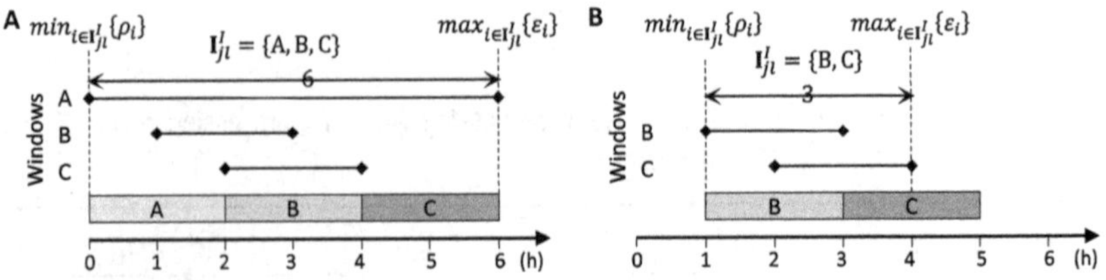

Figure 12.5 Tightening based on subsets of batches (Type I, Algorithm 1). (A) Window based on (release and due time data of) batches A, B, and C: RHS of constraint is 6, and thus no assignment is excluded. (B) Window based on batches B and C: RHS is 3, forbidding the simultaneous assignment of both B and C to U1.

exclude any assignment to U1, whereas the window based on subset $\{B, C\}$ leads to a knapsack constraint that, essentially, disallows the assignment of both B and C to U1.

While the consideration of time windows based on subsets of batches can lead to tighter knapsack constraints, there are cases where these constraints can be further strengthened. The idea is illustrated using Example 12.1 with the second set of data.

Example 12.1 (continued) Consider the two-unit, three-batch example introduced earlier but with the second set of release and due times (ρ_i^2 and ε_i^2) given in Table 12.1. The knapsack constraints for all $\mathbf{I}_{jl}^I$ with $|\mathbf{I}_{jl}^I| > 1$ are the following:[8]

$$\begin{aligned}
&\mathbf{I}_{j,1}^I = \{\text{A,B,C}\}: && 2X_\text{A} + 2X_\text{B} + 2X_\text{C} \le \max\{4,3,6\} - \min\{0,1,3\} = 6\\
&\mathbf{I}_{j,2}^I = \{\text{A,B}\}: && 2X_\text{A} + 2X_\text{B} \le \max\{4,3\} - \min\{0,1\} = 4\\
&\mathbf{I}_{j,3}^I = \{\text{A,C}\}: && 2X_\text{A} + 2X_\text{C} \le \max\{4,6\} - \min\{0,3\} = 6\\
&\mathbf{I}_{j,4}^I = \{\text{B,C}\}: && 2X_\text{B} + 2X_\text{C} \le \max\{3,6\} - \min\{1,3\} = 5,
\end{aligned}$$

where index $j = \text{U1}$ is dropped for simplicity. All constraints are satisfied if all binaries are equal to 1, which means that none of these assignments can be excluded by the constraints generated by Algorithm 1. However, in any feasible assignment batch B must start at $t = 1$ and finish at $t = 3$, which means that batch A cannot be assigned in the same unit as batch B because the due time of A is 4. This assignment could have been excluded if the RHS of the constraint generated for subset $\mathbf{I}_{j,2}^I = \{\text{A, B}\}$ is adjusted to account for the fact that both the maximum due time and the minimum release time for this subset correspond to batch A, so 1 h has to be subtracted from the available time:

$$\mathbf{I}_{j,2}^I = \{\text{A, B}\}: \qquad 2X_\text{A} + 2X_\text{B} \le \max\{4,3\} - \min\{0,1\} - 1 = 3.$$

The preceding example shows that even when batch subsets are considered, the knapsack constraints can be strengthened by reducing the RHS constant. This is achieved through a modification of (12.46), whose general form (type II) is as follows:

$$\sum\nolimits_{i \in \mathbf{I}_{jl}^{II}} \tau_{ij} X_{ij} \le \varepsilon_{i^*} - \rho_{i^*} - \min\left\{\varepsilon_{i^*} - \max_{i \in \mathbf{I}_{jl}^{II} \setminus \{i^*\}}\{\varepsilon_i\},\ \min_{i \in \mathbf{I}_{jl}^{II} \setminus \{i^*\}}\{\rho_i\} - \rho_{i^*}\right\}, \quad j, l, \tag{12.47}$$

where i^* is the batch in subset $\mathbf{I}_{jl}^{II}$ with both the smallest release and largest due time. The concept is illustrated in Figure 12.6, using the system discussed in Example 12.1. The pseudocode for the routine that generates the adjusted knapsack constraints is as follows:

[8] The constraints are not generated using Algorithm 1; thus, the value of index l in subsets $\mathbf{I}_{jl}^I$ is different from the value that would have been assigned if the algorithm was used to generate these subsets.

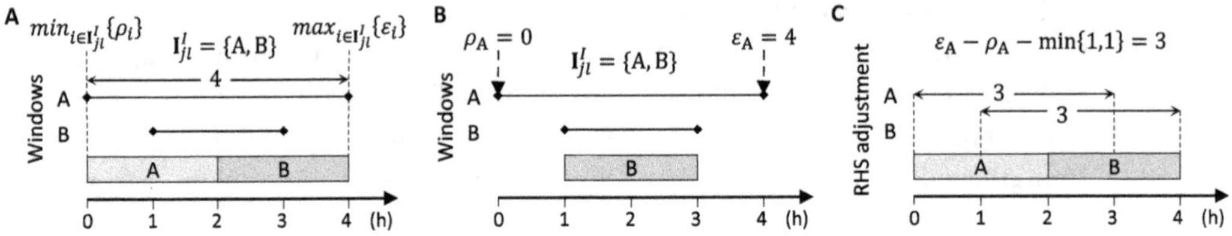

Figure 12.6 Tightening based on subset of batches and strengthening of RHS of knapsack constraints (type II, Algorithm 2). (A) Window based on subset {A, B}: RHS is 4, no assignment is excluded. (B) Both window boundaries are based on batch A release and due times, so RHS should be decreased. (C) Adjusted RHS calculation.

Algorithm 2

forall $j \in \mathbf{J}$
  initialize: $l = 0$
  forall $i \in \mathbf{I}_j$
    $\mathbf{S} = \{i' : \rho_i \leq \rho_{i'}, \varepsilon_{i'} \leq \varepsilon_i\}$
    $A = \min_{i' \in \mathbf{S} \setminus \{i\}} \{\rho_{i'}\} - \rho_i; B = \delta_i - \max_{i' \in \mathbf{S} \setminus \{i\}} \{\delta_{i'}\}$
    $C = \min\{A, B\}$
    if $\sum_{i' \in \mathbf{S}} \tau_{ij} > \varepsilon_i - \rho_i - C$ then
      $l = l + 1$
      $\mathbf{I}_{jl}^{II} = \mathbf{S}$
      add (12.47) for current (j, l)

Once knapsack constraints (12.46) and (12.47) are added, strong cover cuts can also be generated, though commercial MIP solvers can automatically generate such cuts efficiently. Furthermore, MIP preprocessing removes redundant knapsack constraints. Thus, it suffices to add (12.46) and (12.47). The only cover cuts that can be added are for pairs of batches that are not examined by Algorithms 1 and 2 (type III):

$$X_{ij} + X_{i'j} \leq 1, \quad j, l, i \in \mathbf{I}_{jl}^{III}, i' \in \mathbf{I}_{jl}^{III}, \tag{12.48}$$

where $\mathbf{I}_{jl}^{III}$ are pairs of batches that cannot be assigned to the same unit. The pseudocode for generating (12.48) is as follows:

Algorithm 3

forall $j \in \mathbf{J}$
  initialize: $l = 0$
  forall $i \in \mathbf{I}_j,\ i' \in \mathbf{I}_j, i' > i$
    if $\tau_{ij} + \tau_{i'j} > \max\{\varepsilon_i - \rho_{i'}, \varepsilon_{i'} - \rho_i\}$ then
      $l = l + 1$
      $\mathbf{I}_{jl}^{III} = \{i, i'\}$
      add (12.48) for current (j, l)

We close with a note regarding the application of (12.46) through (12.48) to the assignment subproblem ($\mathbb{P}^A$) of the decomposition algorithm discussed in the previous section. When added to a sequence-based model for problem ($\mathbb{P}$) (e.g., the model that consists of (12.4) through (12.11)), these constraints are tightening, that is, they tighten its LP-relaxation but do not alter the feasible space of the MIP model. However, when applied to ($\mathbb{P}^A$), which is a relaxation of ($\mathbb{P}$), they cut off integer solutions of ($\mathbb{P}^A$) (which would later be identified as infeasible by ($\mathbb{P}^S$)). Thus, the addition of (12.46) through (12.48) to ($\mathbb{P}^A$) leads to a model for a different problem, denoted by ($\mathbb{P}^{A*}$), which has a different feasible space, with fewer feasible integer points, but is still a relaxation of ($\mathbb{P}$). Thus, the addition to (12.46) through (12.48) does not necessarily reduce the computational time for the solution of the assignment subproblem, but is likely to reduce the number of iterations.

12.2.2 Tightening Based on Batch-Unit Assignments: Multistage

The algorithms presented in the previous subsection can be extended to multistage problems. First, we can calculate the earliest start time, ρ_{ik}, and latest finish time, ε_{ik}, of a batch in a stage:

$$\rho_{ik} = \rho_i + \sum\nolimits_{k'<k} \min_{j\in\mathbf{J}_{ik'}} \{\tau_{ij}\}, \qquad i,k \tag{12.49}$$

$$\varepsilon_{ik} = \varepsilon_i - \sum\nolimits_{k'>k} \min_{j\in\mathbf{J}_{ik'}} \{\tau_{ij}\}, \qquad i,k, \tag{12.50}$$

where the time horizon η can be used instead of ε_i if there are no due times. Parameters ρ_{ik} and ε_{ik} define the window within which a batch should be processed in a stage, and can be used to calculate parameters ω^I_{kjl} and ω^{II}_{kjl} and sets $\mathbf{I}^I_{kjl}$, $\mathbf{I}^{II}_{kjl}$, and $\mathbf{I}^{III}_{kjl}$ that allow us to express the counterparts of the tightening knapsack and cover inequalities presented in the previous subsection:

$$\sum\nolimits_{i''\in\mathbf{I}^I_{kjl}} \tau_{i''j} X_{i''j} \leq \omega^I_{kjl}, \quad k, j\in\mathbf{J}_k, l \tag{12.51}$$

$$\sum\nolimits_{i\in\mathbf{I}^{II}_{kjl}} \tau_{ij} X_{ij} \leq \omega^{II}_{kjl} \quad k, j\in\mathbf{J}_k, l \tag{12.52}$$

$$X_{ij} + X_{i'j} \leq 1, \qquad k, j\in\mathbf{J}_k, l, i\in\mathbf{I}^{III}_{kjl}, i'\in\mathbf{I}^{III}_{kjl}. \tag{12.53}$$

Note that parameters ω^I_{kjl} and ω^{II}_{kjl} represent, conceptually, the same windows, extended to the multistage problem, used in the RHS of (12.46) and (12.47), respectively. Algorithms 1 to 3 can be modified as follows to obtain parameters ω^I_{kjl} and ω^{II}_{kjl} and sets $\mathbf{I}^I_{kjl}$, $\mathbf{I}^{II}_{kjl}$ and $\mathbf{I}^{III}_{kjl}$:

Algorithm 4

forall $k \in \mathbf{K}$
  forall $j \in \mathbf{J}_k$
  initialize: $l = 0$
   forall $i \in \mathbf{I}_j, i' \in \mathbf{I}_j : \rho_{ik} \leq \rho_{i'k}, \varepsilon_{ik} \leq \varepsilon_{i'k}$
    $\mathbf{S} = \{i'' : \rho_{ik} \leq \rho_{i''k}, \varepsilon_{i''k} \leq \varepsilon_{i'k}\}$
    if $\sum_{i'' \in \mathbf{S}} \tau_{i''j} > \varepsilon_{i'k} - \rho_{ik}$ then
     $l = l + 1$
     $\mathbf{I}^{I}_{kjl} = \mathbf{S}$, $\omega^{I}_{kjl} = \varepsilon_{i'k} - \rho_{ik}$
     add (12.51) for current (k, j, l)

Algorithm 5

forall $k \in \mathbf{K}$
  forall $j \in \mathbf{J}_k$
  initialize: $l = 0$
   for all $i \in \mathbf{I}_j$
    $\mathbf{S} = \{i' : \rho_{ik} \leq \rho_{i'k}, \varepsilon_{i'k} \leq \varepsilon_{ik}\}$
    $A = \min_{i' \in \mathbf{S} \setminus \{i\}} \{\rho_{i'k}\} - \rho_{ik}; B = \delta_{ik} - \max_{i' \in \mathbf{S} \setminus \{i\}} \{\delta_{i'k}\}$
    $C = \min\{A, B\}$
    if $\sum_{i' \in \mathbf{S}} \tau_{i'j} > \varepsilon_{ik} - \rho_{ik} - C$ then
     $l = l + 1$
     $\mathbf{I}^{II}_{kjl} = \mathbf{S}$; $\omega^{II}_{kjl} = \varepsilon_{ik} - \rho_{ik} - C$
     add (12.52) for current (k, j, l)

Algorithm 6

forall $k \in \mathbf{K}$
  forall $j \in \mathbf{J}_k$
  initialize: $l = 0$
   for all $i \in \mathbf{I}_j, i' \in \mathbf{I}_j, i' > i$
    if $\tau_{ij} + \tau_{i'j} > \max\{\varepsilon_{ik} - \rho_{i'k}, \varepsilon_{i'k} - \rho_{ik}\}$ then
     $l = l + 1$
     $\mathbf{I}^{III}_{kjl} = \{i, i'\}$
     add (12.53) for current (k, j, l)

12.2.3 Fixing Sequencing Binary Variables: Multistage

In addition to the earliest start time (ρ_{ik}) and latest finish time (ε_{ik}), release and due times can also be used to calculate the latest start time, ρ^{U}_{ik}, and earliest finish time, ε^{L}_{ik}, of batch i in stage k:

$$\rho_{ik}^{U} = \varepsilon_i - \sum\nolimits_{k' \geq k} \min_{j \in \mathbf{J}_{ik'}} \{\tau_{ij}\}, \qquad i, k \tag{12.54}$$

$$\varepsilon_{ik}^{L} = \rho_i + \sum\nolimits_{k' \leq k} \min_{j \in \mathbf{J}_{ik'}} \{\tau_{ij}\}, \qquad i, k, \tag{12.55}$$

thereby defining, along with ρ_{ik} and ε_{ik}, windows for the execution of a batch in each stage. Specifically, if $\rho_{ik} = \rho_{ik}^L$ and $\varepsilon_{ik} = \varepsilon_{ik}^U$, then the start time, S_{ik}, and finish time, E_{ik}, are bounded as follows:

$$S_{ik} \in \left[\rho_{ik}^L, \rho_{ik}^U\right], \qquad E_{ik} \in \left[\varepsilon_{ik}^L, \varepsilon_{ik}^U\right], \qquad i,\ k. \tag{12.56}$$

Also, note that the following equations are true:

$$\varepsilon_{ik}^{L} = \rho_{ik}^{L} + \min_{j \in \mathbf{J}_{ik}} \{\tau_{ij}\}, \qquad i, k \tag{12.57}$$

$$\varepsilon_{ik}^{U} = \rho_{ik}^{U} + \min_{j \in \mathbf{J}_{ik}} \{\tau_{ij}\}, \qquad i, k, \tag{12.58}$$

which means that only two of the four parameters are necessary.

Clearly, some sequencing binary variables can be fixed to zero:

$$Y_{ii'j} = 0, \qquad k, j \in \mathbf{J}_k, i \in \mathbf{I}_j, i' \in \mathbf{I}_j : \varepsilon_{ik}^L > \rho_{i'k}^U. \tag{12.59}$$

In addition, we can calculate an upper bound, $\rho_{i \to i', j}^U$ on the start time of batch i in unit j assuming that it is followed by batch i':[9]

$$\rho_{i \to i', j}^{U} = \min\left\{\varepsilon_{ik}^U - \tau_{ij}, \varepsilon_{i'k}^U - \tau_{i'j} - \tau_{ij}\right\}, \qquad i, i', k, j \in \mathbf{J}_k. \tag{12.60}$$

Obviously, i can precede i' in stage k only if its earliest start time in k, ρ_{ik}^L, is smaller than this upper bound, which allows us to define a set of feasible predecessors of i' in stage k:

$$\mathbf{I}_{i'k}^{P} = \left\{ i \in \mathbf{I} : \rho_{ik}^L \leq \max_{j \in \mathbf{J}_{ik}} \left\{\rho_{i \to i', j}^U\right\}\right\}, \qquad i', k. \tag{12.61}$$

Some sequencing binary variables can then be fixed as follows:

$$Y_{ii'j} = 0, \qquad k, j \in \mathbf{J}_k, i', i \notin \mathbf{I}_{i'k}^P. \tag{12.62}$$

12.3 A Reformulation and Tightening Based on Variable Time Windows

While using the *fixed* time windows, $\left[\rho_{ik}^L, \rho_{ik}^U\right]$ and $\left[\varepsilon_{ik}^L, \varepsilon_{ik}^U\right]$, can lead to tighter models, as discussed in the previous section, additional strengthening can be achieved by recognizing that multiple batches are processed in each unit. To keep the discussion general, we define the tail of a batch at a stage, $T_{ik} \in \mathbb{R}_+$, as the time elapsed between

[9] Can you see why both τ_{ij} and $\tau_{i'j}$ are subtracted from $\varepsilon_{i'k}^U$ in the argument of the min function in (12.60)?

the end of processing of the batch at that stage, $E_{ik} \in \mathbb{R}_+$, and the end of the scheduling horizon:

$$T_{ik} = \eta - E_{ik}, \quad i, k. \tag{12.63}$$

The shortest tail, σ_{ik}, is a lower bound on T_{ik}.[10] Note that the following relationship holds:

$$\sigma_{ik} = \eta - \varepsilon_{ik}^{U}, \quad i, k. \tag{12.64}$$

To motivate the reformulation, we present Example 12.2.

Example 12.2 Variable Time Windows Consider a two-stage system with two parallel units in each stage ($\mathbf{J}_1 = \{\text{U1}, \text{U2}\}$, $\mathbf{J}_2 = \{\text{U3}, \text{U4}\}$) and four batches, with the associated data given in Table 12.2. Note that all batches have release time $\rho_i = 2$ and due time $\varepsilon_i = 10$. The earliest start time of all batches in stage 1 is 2, and for stage 2, we calculate, using (12.49), $\rho_{\text{A},2} = \rho_{\text{C},2} = 4$, and $\rho_{\text{B},2} = \rho_{\text{D},2} = 5$. Also, note that no sequencing decisions can be fixed using the methods presented in Section 12.2.3.

In terms of batch start times in stage 1, we observe that, at most, two batches could start exactly at their release time, that is, at least two batches would have to start after 4 h or later. Specifically, as shown in Figure 12.7A, batches B and D could start at the earliest (at $t = 4$) because the shortest processing times in stage 1 are for batches A and C ($\tau_{ij} = 2, i \in \{\text{A}, \text{C}\}, j \in \{\text{U1}, \text{U2}\}$). Thus, we observe that the start times in any feasible solution satisfy

$$\sum_i S_{i,1} \geq 2 + 2 + 4 + 4 = 12 > 8 = \sum_i \rho_{i,2}.$$

Similarly, in stage 2, at least two batches would have to start after their earliest start time. For example, in the schedule shown in Figure 12.7A, batches B and D start at $t = 7 > 5 = \rho_{\text{B},2} = \rho_{\text{D},2}$. Note that if batches B and D were executed at $t = 2$ in stage 1, then batches A and C would also start at $t = 7$ (see Figure 12.7B). Thus, in any feasible solution, the following inequality is satisfied:

$$\sum_i S_{i,k=2} \geq 4 + 4 + 7 + 7 = 22 > 18 = \sum_i \rho_{i,2}$$

Let's now consider an earliness minimization problem that, equivalently, can be viewed as a tail minimization problem in stage 2 (i.e., $\min \sum_i T_{i,2}$). We observe that at most two batches can finish exactly at their due times, and, at best, two other batches will finish at $t = 8$, which means that the optimal solution would be $z^* = \min \sum_i T_{i,2} = 4$, and that any feasible solution satisfies the following (see Figure 12.7C):

$$\sum_i T_{i,2} \geq 0 + 0 + 2 + 2 = 4 > 0 = \sum_i (\eta - \varepsilon_i) = \sum_i (\eta - \varepsilon_{i,2}^{U}) = \sum_i \sigma_i.$$

[10] In the present chapter, we do not consider changeover times ($\sigma_{ii'j}$) nor models based on continuous time grids (employing variables T_{ik}), so we use σ_{ik} and T_{ik} to denote the shortest tail (parameter) and tail (variable), respectively, of batch i in stage k.

Table 12.2 Example data (processing times, τ_{ij}, release times, ρ_i, and due times, ε_i) and operating window parameters ($\rho^L_{i,2}$, $\varepsilon^U_{i,1}$).

	τ_{ij}					
$j=$	U1, U2	U3, U4	ρ_i	ε_i	$\rho^L_{i,2}$	$\varepsilon^U_{i,1}$
A	2	3	2	10	4	7
B	3	2	2	10	5	8
C	2	3	2	10	4	7
D	3	2	2	10	5	8

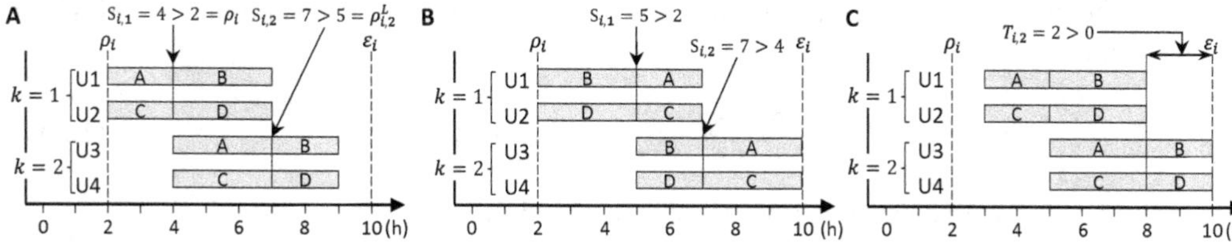

Figure 12.7 Illustration of relationships among start times and tails in feasible solutions. (A) Start times (of batches B and D) larger than the earliest start times in example solution. (B) Start times (of batches A and C) larger than the earliest start times in alternative solution. (C) Tails (of batches A and C) larger than the shortest tails in optimal solution (earliness minimization).

Finally, for stage 1, based on the latest finish times, we can calculate the following shortest tails: $\sigma_{A,1} = \sigma_{C,1} = 10 - 7 = 3$ and $\sigma_{B,1} = \sigma_{D,1} = 10 - 8 = 2$. Again, at most two batches can finish exactly at their latest finish time, which means that in any feasible solution the sum of the tails will be greater than the sum of the shortest tails. For example, in Figure 12.7C, batches B and D finish at their latest finish time, $\varepsilon^U_{i,1} = 8$, while A and C finish before their latest finish time, $\varepsilon^U_{i,1} = 7$. In general, any feasible solution should satisfy

$$\sum_i T_{i,1} \geq 2 + 2 + 5 + 5 = 14 > 10 = \sum_i (\eta - \varepsilon^U_{i,1}) = \sum_i \sigma_{i,1}.$$

Having shown that constraints using the summation of either start times or tails (or, equivalently, end times) can cut off solutions that would be feasible in the LP-relaxation of the MIP model, it is then natural to ask: How can we systematically generate constraints, in terms of variables S_{ik} and E_{ik}, similar to the ones generated for the simple instance studied in Example 12.2? Before we address this question, we note that the so generated constraints can be applied to all the models discussed in Chapter 5, provided that start and end times can be expressed in terms of the variables used in the chosen model.

Accordingly, to show how tightening can be applied to a model that is not sequence based, we consider the discrete time model based on aggregated start and finish batch times that employs the following variables (see Chapter 5 for details):

- $X_{ijn} \in \{0, 1\}$: represents the start of processing of batch i in unit j at time point n.
- $U_{ikn} \in \{0, 1\}$: denotes the *availability* of batch i to undergo processing in stage $k+1$ at time point n; it is defined for $k \in \{0, 1, \ldots |\mathbf{K}|\}$ with $U_{i,k=0,n}$ being parameters.

For cost minimization, the model consists of assignment constraints:

$$\sum\nolimits_{j \in \mathbf{J}_{ik}} \sum\nolimits_{n} X_{jn} = 1, \qquad i, k, \tag{12.65}$$

the clique constraint ensuring that only one batch is processed on a unit at any given time,

$$\sum\nolimits_{i} \sum\nolimits_{n'=n-\tau_{ij}+1}^{n'=n} X_{ijn'} \leq 1, \qquad j, n, \tag{12.66}$$

and the batch balance constraint (see Section 5.4):

$$U_{ik,n+1} = U_{ikn} + \sum\nolimits_{j \in \mathbf{J}_{ik}} X_{j,n-\tau_{ij}} - \sum\nolimits_{j \in \mathbf{J}_{i,k+1}} X_{ijn}, \qquad i, k \in \{0, 1, \ldots, |\mathbf{K}|-1\}, n. \tag{12.67}$$

satisfying, for every batch i, the following:

$$U_{i,k,0} = 0, \;\; k \geq 1; \quad U_{i,0,n} = 0, \;\; n < \rho_i; \quad U_{i,1,\rho_i} = 1; \quad U_{i,|\mathbf{K}|,\varepsilon_i+1} = 1 \tag{12.68}$$

The objective function is

$$\min \sum\nolimits_{i,j} \gamma_{ij}^{P} \sum\nolimits_{n} X_{ijn}. \tag{12.69}$$

Furthermore, we have already seen

$$S_{ik} = \sum\nolimits_{j \in \mathbf{J}_k} \sum\nolimits_{n} n X_{jn}, \;\; i, k. \tag{12.70}$$

Similarly, tail variables can be calculated as follows:

$$T_{ik} = \eta - \sum\nolimits_{j \in \mathbf{J}_k} \sum\nolimits_{n} (n + \tau_{ij}) X_{ijn}, \;\; i, \;\; k. \tag{12.71}$$

Interestingly, we can use S_{ik} and T_{ik} to explicitly enforce precedence constraints (replacing (12.67) and thus removing W_{ikn} variables):

$$S_{i,k+1} \geq S_{ik} + \sum\nolimits_{j \in \mathbf{J}_k} \sum\nolimits_{n} \tau_{ij} X_{ijn}, \qquad i, k \tag{12.72}$$

$$T_{ik} \geq T_{i,k+1} + \sum\nolimits_{j \in \mathbf{J}_{k+1}} \sum\nolimits_{n} \tau_{ij} X_{ijn}, \qquad i, k. \tag{12.73}$$

Our goal, as motivated in Example 12.2, is to calculate lower bounds on the sum of start times and tails at a given stage:

$$\sum_i S_{ik} \geq \hat{\rho}_k, \quad k \tag{12.74}$$

$$\sum_i T_{ik} \geq \hat{\sigma}_k, \quad k. \tag{12.75}$$

In simple cases, bounds $\hat{\rho}_k$ and $\hat{\sigma}_k$ can be estimated using simple induction. For instance, in Example 12.2 we could calculate $\hat{\rho}_1 = 12(= 2+4+2+4), \hat{\rho}_2 = 22(= 4+7+4+7)$, $\hat{\sigma}_1 = 14(= 5+2+5+2)$, and $\hat{\sigma}_2 = 4(= 2+0+2+0)$. In general, $\hat{\rho}_k$ and $\hat{\sigma}_k$ can be calculated by solving $2|\mathbf{K}|$ minimization problems, $(\mathbb{P}_k^S)$ and $(\mathbb{P}_k^T)$:

$$\begin{aligned} \hat{\rho}_k &= \min \sum_i S_{ik} \\ st \quad & \boldsymbol{x}^X \in \mathbf{X}^A \qquad k, \; (\mathbb{P}_k^S) \\ & (\boldsymbol{x}^X, \boldsymbol{x}^S) \in \mathbf{X}^S \end{aligned} \tag{12.76}$$

$$\begin{aligned} \hat{\sigma}_k &= \min \sum_i T_{ik} \\ st \quad & \boldsymbol{x}^X \in \mathbf{X}^A \qquad k, \; (\mathbb{P}_k^T) \\ & (\boldsymbol{x}^X, \boldsymbol{x}^T) \in \mathbf{X}^T, \end{aligned} \tag{12.77}$$

where $\boldsymbol{x}^X$, $\boldsymbol{x}^S$, and $\boldsymbol{x}^T$ are the vectors of X_{ijn}, S_{ik}, and T_{ik} variables, respectively; $\mathbf{X}^A$ is the feasible region defined by (12.65), (12.66), and $X_{ijn} \in \{0,1\}$; $\mathbf{X}^S$ is the feasible region defined by (12.70), (12.72), and bounds on S_{ik} $\left(S_{ik} \in \left[\rho_{ik}^L, \rho_{ik}^U\right]\right)$; and $\mathbf{X}^T$ is the feasible region defined by (12.71), (12.73), and bounds on T_{ik}. Note that neither $(\mathbb{P}_k^S)$ nor $(\mathbb{P}_k^T)$ include constraints to enforce precedence relations because they are now enforced by (12.72) or (12.73).

While problems $(\mathbb{P}_k^S)$ and $(\mathbb{P}_k^T)$ can be computationally expensive to solve to optimality, reasonable values for $\hat{\rho}_k$ and $\hat{\sigma}_k$ can be obtained fast, even at the root node. These lower bounds can then be used in (12.74) and (12.75), respectively, which can be added to any model. Finally, we note that (12.71), which appears in $(\mathbb{P}_k^S)$, is based on η. Therefore, in makespan minimization problems, where η should be replaced by *MS*, we use an overestimation of *MS*, and relax the due dates so that no time is unnecessarily added to the summation of tails due to the presence of early due dates, that is, we set $\varepsilon_i = \eta$.

12.4 Discrete-Continuous Algorithm

One potential disadvantage of discrete time models is the low solution accuracy when a coarse time discretization is used for computational reasons. An approach to address this shortcoming is to use a coarse discrete time grid to obtain an approximate solution fast, and then *refine* this solution using a continuous time model. This approach, which will be further discussed in Chapter 13, allows the solution of large-scale instances fast without loss of accuracy. The key idea is to solve the problem in two steps:

(1) Obtain assignment and sequencing decisions using a discrete time model.
(2) Refine the solution with a (simplified) continuous time model: fix the assignment and sequencing decisions obtained in step 1, and solve for the remaining (timing) decisions.

Any discrete time model can be used in step 1, while in step 2 it is preferable to use a sequence-based, as opposed to grid-based, continuous time model because the former can be more readily reduced to a simple and computationally cheap linear programming (LP) model for the remaining decisions. Here, we use the discrete time model presented in Section 12.3, consisting of (12.65) through (12.68); and the sequence-based model presented in Section 12.1.3, consisting of (12.18) through (12.24).

The solution from step 1 is used to identify assignment decisions for the continuous time model,

$$X_{ij}^1 = \sum\nolimits_n X_{ijn}, \qquad i,j, \tag{12.78}$$

and calculate stage start times,

$$S_{ik}^1 = \sum\nolimits_{j \in \mathbf{J}_{ik}} \sum\nolimits_n n X_{ijn}, \qquad i,j, \tag{12.79}$$

which are used to detect the sequencing decisions in the solution of step 1.

Specifically, sequence $i \to i'$ on unit j is active if (1) batches i and i' are assigned to unit j, that is, $X_{ij}^1 = X_{i'j}^1 = 1$, as calculated in (12.78); and $S_{i'k}^1 \geq S_{ik}^1 + \sum_{j \in \mathbf{J}_{ik}} \tau_{ij} X_{ij}^1$, with $S_{i'k}^1$ and S_{ik}^1 calculated in (12.79). If these two conditions are satisfied, then the precedence constraint should be satisfied in the second step solution;[11] the information is recorded as $Y_{ii'j}^1 = 1$. Note that S_{ik}^1 are only used to identify and fix precedence relationships, that is, they are not fixed in the model in step 2.

If X_{ij}^1 and $Y_{ii'j}^1$ are the assignment and sequencing decisions inferred from the solution in step 1, obtained as described in the previous paragraph, then the continuous time model can be simplified in several ways. First, (12.18) is trivially satisfied and thus removed. Second, all sequencing decisions are fixed, that is, turned into known precedence constraints, and thus (12.19) is removed. Third, (12.20) is simplified as follows (note that $\sum_{j \in \mathbf{J}_{ik}} \tau_{ij} X_{ij}^1$ is a parameter):

$$S_{ik} + \sum\nolimits_{j \in \mathbf{J}_{ik}} \tau_{ij} X_{ij}^1 \leq S_{i,k+1} \quad i,k < |\mathbf{K}|. \tag{12.80}$$

Fourth, (12.21) is written only if $Y_{ii'j}^1 = 1$ and, in this case, is simplified as follows:

$$S_{ij} + \tau_{ij} \leq S_{i'j}, \qquad i,i',j : Y_{ii'j}^1 = 1. \tag{12.81}$$

Fifth, we can set $S_{ij} = 0$ if $X_{ij}^1 = 0$; bound $S_{ij} \in \left[\rho_i, \varepsilon_i - \sum_{j \in \mathbf{J}_{i,|\mathbf{K}|}} \tau_{ij} X_{ij}^1\right]$ if $X_{ij}^1 = 1$; and remove (12.22) and (12.23). Finally, we can replace (12.24) with

$$S_{ij} \geq \rho_i, \qquad i,j \in \mathbf{J}_{i,1}\text{: } X_{ij}^1 = 1 \tag{12.82}$$

$$S_{ij} + \tau_{ij} \leq \varepsilon_i \qquad i,j \in \mathbf{J}_{i,|\mathbf{K}|}\text{: } X_{ij}^1 = 1. \tag{12.83}$$

[11] We use the term *precedence constraint* because in the second step this will be a constraint that will have to be satisfied. It will not be a (sequencing) decision available to the optimizer.

Note that the only variables of the simplified sequence-based model, consisting of (12.80) through (12.83), are S_{ij}, which means that the model in the second step is an LP, which can be solved in seconds. Thus, a schedule with continuous timing of events and an improved objective function value can be found effectively, though we note that this method is not guaranteed to yield the best solution.

12.5 Notes and Further Reading

(1) The presentation in Section 12.1, as well as the discussion of the tightening constraints and algorithms in Sections 12.2.1 and 12.2.2, are based, primarily, on the discussion in Maravelias [1].

(2) Extensions of the algorithms presented in Section 12.2.2 to multistage problems with batching, assignment, and sequencing decisions have been proposed [2, 3], including extensions to address multistage problems with batching decisions and variable processing times (functions of batchsizes) and changeover times [3].

(3) The preprocessing methods for fixing sequencing binary variables discussed in Section 12.2.3 are adjusted from Castro et al. [4] and Prasad and Maravelias [2].

(4) The discussion of the reformulation and tightening in Section 12.3 and the two-step algorithm in Section 12.4 is based on Merchan et al. [5].

(5) Several researchers have proposed hybrid schemes, where multiple solution methods are combined, for a wide range of optimization problems. In general, hybrid methods can be very effective, but their development requires advanced domain knowledge. Also, the effort to modify them or extend them to account, for example, for additional processing features can be significant.

(6) Hybrid approaches that combine MIP and CP techniques have been particularly popular due to the expressiveness of CP methods and the effectiveness of local search CP algorithms [6–10]. In these approaches, the original problem is typically decomposed into a high-level MIP subproblem and a low-level CP subproblem, though other integration approaches have been adopted.

(7) Hybrid MIP/CP algorithms have been proposed for problems in both single-stage [11–14] and multistage [15, 16] environments.

(8) The integration of a MIP assignment subproblem with problem-specific sequencing algorithms was discussed in Maravelias [1], where (1) the single-unit subproblems, arising in the single-stage lower-level subproblem, are solved using the one-unit algorithm described in Balas et al. [17]; and (2) the jobshop subproblem, which is the sequencing subproblem in multistage environments, was solved using the shifting bottleneck procedure [17, 18].

(9) Castro et al. proposed an algorithm, for a special case of a multistage problems, that integrates mathematical programming and discrete-event simulation model [19].

(10) You, Wassick, and coworkers explored the integration of MIP and agent-based methods [20, 21].

(11) Iterative MIP-based decomposition methods which gradually construct a solution have also been shown to be computationally effective [22–24].

12.6 Exercises

(1) The goal of the first two exercises is to help you understand how the decomposition algorithms described in Section 12.1 work – what are their advantages and disadvantages, and what are the challenges in developing them. We start with the single-stage problem.

(a) Retrieve (of generate anew) your favorite full-space model you developed for single-stage problems in Chapter 4 for cost minimization.
(b) Formulate the two MIP models for the subproblems described in Section 12.1.2.
(c) Code a method that allows you to automatically generate cuts, based on the results of $\mathbb{P}^S$, that can be added to($\mathbb{P}^A$).
(d) Generate test instances based on random data generation as described in Maravelias (section 7.1, table 9) [1].
(e) Solve the instances using the full-space model and the decomposition algorithm. What do you observe? How does the perfor mance of the methods change with instance size?
(f) Repeat (d) and (e) using tighter or looser processing windows.[12] What do you observe?

(2) Consider multistage problems.
(a) Retrieve (of generate anew) your favorite two full-space models for cost minimization in multistage problems from Chapter 5.
(b) Formulate the MIP model, described in Section 12.3.1, for ($\mathbb{P}^A$) .
(c) Formulate a new MIP model for ($\mathbb{P}^S$) (not described in Section 12.3.1).
(d) What information can you obtain from an infeasible ($\mathbb{P}^S$)? What (subset of) assignments can be identified as leading to infeasibilities? Develop code to automatically detect such assignments and generate the corresponding cuts that can be added to the MIP for ($\mathbb{P}^A$).
(e) Generate test instances based on random data generation as described in Maravelias (section 7.2, table 13) [1].
(f) Solve the instances using the full-space model and the decomposition algorithm. What do you observe? How does the performance of the methods change with instance size?

[12] How can you generate such instances? Can you do it by changing the values of only one parameter (vector)? Is there only one choice of such a vector?

(3) Consider the preprocessing and tightening methods for single-stage problems.
   (a) Code the three algorithms described in Section 12.2.1.
   (b) Apply them to the first instance you generated in Exercise 1(d).
   (c) Add the generated cuts to $(\mathbb{P}^A)$ and solve the instance using the decomposition algorithm.
   (d) Repeat (b) and (c) for larger instances. What do you observe? How does the number of iterations change? How does the time for the solution of $(\mathbb{P}^A)$ change?
   (e) Are your findings consistent with the finding in section 7.1 of [1]? Why (or why not)?

(4) Consider the preprocessing and tightening methods for multistage problems.
   (a) Code the three algorithms described in Section 12.2.2.
   (b) Apply them to the first instance you generated in Exercise 2(e).
   (c) Add the generated cuts to $(\mathbb{P}^A)$ and solve the instance using the decomposition algorithm.
   (d) Repeat parts (b) and (c) for larger instances. What do you observe? How does the number of iterations change? How does the time for the solution of $(\mathbb{P}^A)$ change?
   (e) Are your findings consistent with the finding in section 7.2 in [1]? Why (or why not)?

(5) We next consider the reformulation and tightening methods described in Section 12.3.
   (a) Formulate the MIP models for problems $(\mathbb{P}_k^S)$ and $(\mathbb{P}_k^T)$.
   (b) Select your favorite two discrete time models for multistage problems. Formulate the *tightened* versions of these models using (12.74) and (12.75).
   (c) Use the models developed in part (a) and the tightened models developed in part (b) to solve Example 1 (section 6.1.1) from Merchan et al. [5].
   (d) Repeat (c) for Example 2 (section 6.1.2) from Merchan et al. [5].

(6) We next consider the reformulation and tightening methods described in Section 12.4.
   (a) Generate code that can automatically detect and record the assignment and sequencing decisions made using a discrete time model for multistage problems.
   (b) Formulate the continuous time model for solution refinement (step 2).
   (c) Generate code that automatically generates the continuous time model given a solution of the discrete time model.
   (d) Apply the discrete-continuous algorithm to solve instance P1 in Merchan et al. [5] (originally from Harjunkoski and Grossmann [15]).
   (e) Repeat part (d) for instances P2-P3, then P4, and then P5-P6 from Merchan et al. [5].

References

[1] Maravelias CT. A Decomposition Framework for the Scheduling of Single- and Multi-stage Processes. *Comput Chem Eng*. 2006;30(3):407–420.

[2] Prasad P, Maravelias CT. Batch Selection, Assignment and Sequencing in Multi-stage Multi-product Processes. *Comput Chem Eng*. 2008;32(6):1106–1119.

[3] Sundaramoorthy A, Maravelias CT. Simultaneous Batching and Scheduling in Multistage Multiproduct Processes. *Ind Eng Chem Res*. 2008;47(5):1546–1555.

[4] Castro PM, Grossmann IE, Novais AQ. Two New Continuous-Time Models for the Scheduling of Multistage Batch Plants with Sequence Dependent Changeovers. *Ind Eng Chem Res*. 2006;45(18):6210–6226.

[5] Merchan AF, Lee H, Maravelias CT. Discrete-Time Mixed-Integer Programming Models and Solution Methods for Production Scheduling in Multistage Facilities. *Comput Chem Eng*. 2016;94:387–410.

[6] Althaus E, Bockmayr A, Elf M, Junger M, Kasper T, Mehlhorn K. SCIL – Symbolic Constraints in Integer Linear Programming. *Lect Notes Comput Sc*. 2002;2461:75–87.

[7] Marriott K, Stuckey PJ. *Programming with Constraints: An Introduction*. Cambridge: MIT Press; 1998. xiv, 467 p. p.

[8] Van Hentenryck P, Michel L. *Constraint-Based Local Search*. Cambridge: MIT Press; 2005. xix, 422 p. p.

[9] Hooker J. *Logic-Based Methods for Optimization: Combining Optimization and Constraint Satisfaction*. New York: John Wiley & Sons; 2000. xvi, 495 p. p.

[10] Hooker J. *Integrated Methods for Optimization*. New York: Springer; 2007. xiv, 486 p. p.

[11] Bockmayr A, Pisaruk N. Detecting Infeasibility and Generating Cuts for Mixed Integer Programming Using Constraint Programming. *Computers & Operations Research*. 2006;33 (10):2777–2786.

[12] Jain V, Grossmann IE. Algorithms for Hybrid MILP/CP Models for a Class of Optimization Problems. *INFORMS Journal on Computing*. 2001;13(4):258–276.

[13] Hooker JN. Planning and Scheduling by Logic-Based Benders Decomposition. *Oper Res*. 2007;55(3):588–602.

[14] Sadykov R, Wolsey LA. Integer Programming and Constraint Programming in Solving a Multimachine Assignment Scheduling Problem with Deadlines and Release Dates. *INFORMS Journal on Computing*. 2006;18(2):209–217.

[15] Harjunkoski I, Grossmann IE. Decomposition Techniques for Multistage Scheduling Problems Using Mixed-Integer and Constraint Programming Methods. *Comput Chem Eng*. 2002;26(11):1533–1552.

[16] Roe B, Papageorgiou LG, Shah N. A Hybrid MILP/CLP Algorithm for Multipurpose Batch Process Scheduling. *Comput Chem Eng*. 2005;29(6):1277–1291.

[17] Balas E, Lancia G, Serafini P, Vazacopoulos A. Job Shop Scheduling with Deadlines. *J Comb Optim*. 1998;1(4):329–353.

[18] Balas E, Vazacopoulos A. Guided Local Search with Shifting Bottleneck for Job Shop Scheduling. *Manage Sci*. 1998;44(2):262–275.

[19] Castro PM, Aguirre AM, Zeballos LJ, Mendez CA. Hybrid Mathematical Programming Discrete-Event Simulation Approach for Large-Scale Scheduling Problems. *Ind Eng Chem Res*. 2011;50(18):10665–10680.

[20] Chu Y, Wassick JM, You F. Efficient Scheduling Method of Complex Batch Processes with General Network Structure via Agent-Based Modeling. *AIChE J*. 2013;59(8):2884–2906.

[21] Chu Y, You F, Wassick JM. Hybrid Method Integrating Agent-Based Modeling and Heuristic Tree Search for Scheduling of Complex Batch Processes. *Comput Chem Eng.* 2014;60:277–296.

[22] Castro PM, Harjunkoski I, Grossmann IE. Optimal Short-Term Scheduling of Large-Scale Multistage Batch Plants. *Ind Eng Chem Res.* 2009;48(24):11002–11016.

[23] Castro PM, Harjunkoski I, Grossmann IE. Greedy Algorithm for Scheduling Batch Plants with Sequence-Dependent Changeovers. *AlChE J.* 2011;57(2):373–387.

[24] Kopanos GM, Mendez CA, Puigjaner L. MIP-Based Decomposition Strategies for Large-Scale Scheduling Problems in Multiproduct Multistage Batch Plants: A Benchmark Scheduling Problem of the Pharmaceutical Industry. *Eur J Oper Res.* 2010;207(2):644–655.

13 Solution Methods: Network Environments

We discuss four solution methods for problems in the general network production environment. Specifically, after some background and motivation, in Section 13.1 we present (1) preprocessing and tightening methods, in Section 13.2; (2) reformulations, in Section 13.3; (3) an approach to formulate models that employ multiple discrete time grids, in Section 13.4; and (4) a three-stage algorithm that employs both a discrete and continuous time models, in Section 13.5. For simplicity, we do not consider shared utilities nor special processing features such as storage in processing units and multiple material transfers. The methods presented in Sections 13.2 and 13.3 are applicable to both discrete and continuous time models, but to keep the presentation short, we apply them to discrete time models, though we comment on their application to their continuous counterparts. The reader can study each section, after Section 13.1, independently, that is, Section 13.2 is not prerequisite for Section 13.3, and so on. Finally, we note that due to the large number of parameters needed to describe the methods in the present chapter, many Greek letters are reused to denote new entities.

13.1 Background and Motivation

For completeness, we reintroduce the problem statement, in Section 13.1.1, and the basic STN-based model, in Section 13.1.2. Then, in Section 13.1.3, we introduce a series of examples to illustrate some of the computational challenges in the solution of discrete time models and motivate the development of the solution methods in the subsequent sections.

13.1.1 Problem Statement

We consider scheduling in network environments, represented using the STN representation. We also consider a discrete time grid, with points $n \in \mathbf{N} = \{0, 1, 2, \ldots |\mathbf{N}|\}$, where period n (of length δ) starts at point $n-1$ and ends at point n. Without loss of generality, we assume that all time-related data are integers and that $\delta = 1$. Also, as already explained in Chapter 7, we use the term *material* instead of *state*. The processing facility is defined in terms of the following sets, subsets, and parameters (see Section 7.1.1 for details):

Indices/Sets

$i \in \mathbf{I}$	Tasks
$j \in \mathbf{J}$	Processing units
$k \in \mathbf{K}$	Materials

Subsets

$\mathbf{I}_k^+/\mathbf{I}_k^-$	Tasks producing/consuming material k
$\mathbf{I}_j$	Tasks that can be executed on unit j
$\mathbf{J}_i$	Processing units that can process task i
$\mathbf{K}_i^+/\mathbf{K}_i^-$	Materials produced/consumed by task i

Parameters

$\beta_j^{MIN}/\beta_j^{MAX}$	Minimum/maximum capacity of unit j
$\gamma_{ij}^F/\gamma_{ij}^V$	Fixed/variable cost for carrying out task i in unit j
ξ_{kn}	Net shipment of material k at time point n
π_k	Price of material k
ρ_{ik}	Conversion coefficient of material k produced (>0) or consumed (<0) by task i
τ_{ij}	Processing time of task i in unit j
χ_k^M	Capacity of storage vessel dedicated to material k

13.1.2 Basic STN-Based Model

We use the following variables:

- $X_{ijn} \in \{0,1\}$: denotes the start of the execution of a batch of task i on unit j at time point n.
- $B_{ijn} \in \mathbb{R}_+$: batchsize of a batch of task i that starts on unit j at time point n.
- $I_{kn} \in \mathbb{R}_+$: inventory level of material k during period n.
- $S_{kn} \in \mathbb{R}_+$: sales, in addition to fixed orders, of product $k \in \mathbf{K}^P \subset \mathbf{K}$ at time point n.

Only one batch can be executed at any time in a given unit:

$$\sum\nolimits_{i \in \mathbf{I}_j} \sum\nolimits_{n' \in \mathbf{N}_{in}^U} X_{in'} \leq 1, \quad j, n, \tag{13.1}$$

where $\mathbf{N}_{in}^U = \{n - \tau_i + 1, \ldots, n\}$. Batchsize variables are subject to unit capacity constraints:

$$\beta_j^{MIN} X_{ijn} \leq B_{ijn} \leq \beta_j^{MAX} X_{ijn}, \quad i, j, n. \tag{13.2}$$

The inventory level of material i during period n is defined and bounded as follows:

$$I_{k,n+1} = I_{kn} + \sum\nolimits_{i \in \mathbf{I}_k^+} \sum\nolimits_{j \in \mathbf{J}_i} \rho_{ik} B_{ij,n-\tau_{ij}} + \sum\nolimits_{i \in \mathbf{I}_k^-} \sum\nolimits_{j \in \mathbf{J}_i} \rho_{ik} B_{ijn} + \xi_{kn} - S_{kn} \leq \chi_k^M, \quad k, n. \tag{13.3}$$

In cost minimization problems, a set of orders are given, resulting in negative shipments ($\xi_{kn} < 0$), which have to be met at minimum cost. The objective function is

$$\min \sum_{i,j} \left\{ \gamma_{ij}^F \sum_n X_{ijn} + \gamma_{ij}^V \sum_n B_{ijn} \right\}. \tag{13.4}$$

In profit maximization, a set of orders may or may not be given. The objective function is

$$\max \sum_{k,n} \pi_k S_{kn} - \sum_{i,j} \left\{ \gamma_{ij}^F \sum_n X_{ijn} + \gamma_{ij}^V \sum_n B_{ijn} \right\}. \tag{13.5}$$

Note that the cost minimization objective function includes variables X_{ijn}, multiplied by positive γ_{ij}^F; and B_{ijn} variables, which enter variable lower bound constraints ($\beta_j^{MIN} X_{ijn} \le B_{ijn}$), also multiplied by positive γ_{ij}^V. Thus, conceptually, the objective function *pushes* X_{ijn} variables to zero because this would minimize, directly, term $\gamma_{ij}^F \sum_n X_{ijn}$; and, indirectly, term $\gamma_{ij}^V \sum_n B_{ijn}$.

Conversely, in profit maximization instances, the objective function *pushes* variables X_{ijn} to 1, because this would lead to larger (variable) upper bounds on variables B_{ijn} ($B_{ijn} \le \beta_j^{MAX} X_{ijn}$), which would in turn lead to larger production, and thus sales, S_{kn}.[1]

13.1.3 Motivating Examples

Example 13.1 illustrates how the variable lower bound constraint in (13.2) becomes ineffective in the LP-relaxation of the MIP model, leading to poor relaxations; and how logic inference can be used to calculate bounds on the total number of batch executions of a task. Example 13.2 illustrates how a large number of mathematical solutions correspond to essentially the same scheduling solution. Example 13.3 illustrates how vastly different processing times can lead to unnecessarily large MIP formulations.

Example 13.1 Tightening through Logic Inference We consider an instance using the facility shown, along with all associated data, in Figure 13.1. Products A and B can be produced by tasks TA and TB, respectively. The demand for A and B at the end of a sufficiently long horizon is 90 and 25 kg, respectively. Our objective is to minimize cost. The minimum cost obtained using the LP-relaxation of the model presented in the previous subsection is \$76.7, while the number of batches producing materials INT, A, and B, are 1.9, 1.8, and 0.5, respectively. In other words, if X_{ijn}^L are the

[1] The reasoning for profit maximization problems is based on the assumption that the revenue from sales exceeds the total cost of producing this material. In other words, π_k is larger than the cost of producing one unit of k. If this is not the case, then the product will not be produced if there is no demand for it.

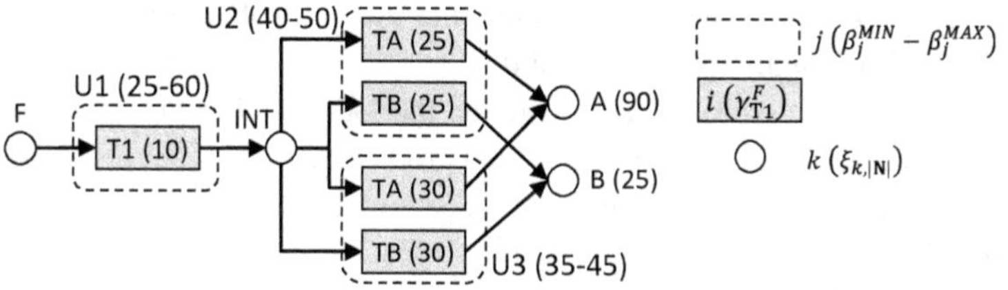

Figure 13.1 STN representation and associated data for Example 13.1; batchsize bounds (kg), fixed costs ($), and demands (kg) given next to corresponding units, tasks, and materials, respectively.

optimal values of X_{ijn} variables in the LP-relaxation, then $\sum_{j,n} X^L_{\mathrm{TA},j,n} = 1.9$, $\sum_n \left(X^L_{\mathrm{TA,U2},n} + X^L_{\mathrm{TA,U3},n}\right) = 1.8$, and $\sum_n \left(X^L_{\mathrm{TB,U2},n} + X^L_{\mathrm{TB,U3},n}\right) = 0.5$.[2]

However, if we take into account the capacities of processing units, we can infer the following:

- TA, in both units U2 and U3, must produce 90 kg of A to satisfy demand, and since the largest batchsize is 50 kg, this requires at least two batches, that is,

$$\sum_n (X_{\mathrm{TA,U2},n} + X_{\mathrm{TA,U3},n}) \geq 2. \tag{13.6}$$

- TB must produce at least 25 kg of B, and since the minimum batchsize is 35 kg, at least 35 kg of B will be produced, which also means that at least 35 kg of INT will be converted to B. We can also trivially infer

$$\sum_n (X_{\mathrm{TB,U2},n} + X_{\mathrm{TB,U3},n}) \geq 1. \tag{13.7}$$

- The *demand* for intermediate INT is 125 kg: 90 kg, to be converted to A; plus 35 kg, to be converted to B. Since the maximum batchsize of T1 is 60, this means that in any feasible solution at least three batches of T1 will be required:

$$\sum_n X_{\mathrm{T1,U1},n} \geq 3. \tag{13.8}$$

Using this *constraint propagation*[3] procedure, we calculated lower bounds on the number of batches and cumulative production for each task. If (13.6) through (13.8) are added to the LP-relaxation of the STN-based model, then the optimal objective function value of the LP-relaxation becomes $105, which, in this simple instance, is equal to the optimal objective function value of the MIP model. Also, (13.6) through (13.8) are

[2] Can you see why $\sum_n \left(X^L_{\mathrm{TA,U2},n} + X^L_{\mathrm{TA,U3},n}\right) = 1.8$? Which ratio is equal to 1.8? Based on this, can you tell which task, between T2A and T2B, is preferred in the solution of the LP-relaxation? Similarly, can you see why $\sum_n \left(X^L_{\mathrm{T2B},n} + X^L_{\mathrm{T3B},n}\right) = 0.5$? Which task, between T2B and T3B, is executed in the solution of the LP-relaxation? Also, can you say anything about the horizon length? It is sufficient to meet the demand, but would a shorter horizon also be sufficient? How would you expect the solution of the LP-relaxation to change if the horizon was shorter (but still sufficient)?

[3] We term the procedure *constraint propagation* because it starts from a demand constraint, which is propagated to generate other constraints. The procedure is similar to search algorithms used in constraint programming (see discussion in Chapter 12).

satisfied as equalities in the LP relaxation and the optimal solution of the MIP model satisfies (13.6) through (13.8) as equalities.[4]

Note that, in addition to lower bounds on the total number of executed batches, we also calculated (1) lower bounds on the total production of an intermediate (i.e., we need at least 135 kg of INT); and (2) the total processing amount ($\sum_n B_{in}$) of a task (i.e., any feasible schedule should satisfy $\sum_n B_{\mathrm{T1},n} \geq 135$).

Example 13.1 shows, using a trivial instance, that the addition of constraints based on bounds calculated using constrain propagation during a form of preprocessing can lead to significant tightening. Can similar bounds be calculated for general problems? If yes, will the resulting constraints be effective? We will address these two questions in Section 13.2.

Example 13.2 Multiple Scheduling Solutions We consider the simple network shown in Figure 13.2, along with processing costs, processing times, unit capacities (note that batchsizes are fixed, $\beta_j^{MIN} = \beta_j^{MAX} = \beta_j$) and price of final products. The objective is to maximize profit over a 120-hour horizon.

The upper bound at the root node of the search (i.e., LP-relaxation of the STN-based model) is \$689.7, while the optimal integer solution is \$686, that is, the integrality gap is only 0.5%. Nevertheless, this simple instance cannot be easily solved to optimality. Specifically, if the default options in GAMS 23.9.2/CPLEX 12.4[5] are used, after hours of computations and over 95 million nodes explored, the bound improves to only \$689.5. Thus, the gap closes by less than 6% ($\frac{689.7-689.5}{689.7-686} = \frac{0.2}{3.7} = 5.4\%$).

The reason behind the large size of the branch-and-bound tree is the *multiplicity* of solutions with identical objective function values. Specifically, a schedule consisting of a given number of batches with fixed batch sizes can be represented by multiple solutions to the MIP model. One way to generate these solutions is by shifting tasks around, for example, moving a batch in a noncritical unit with idle time forward or

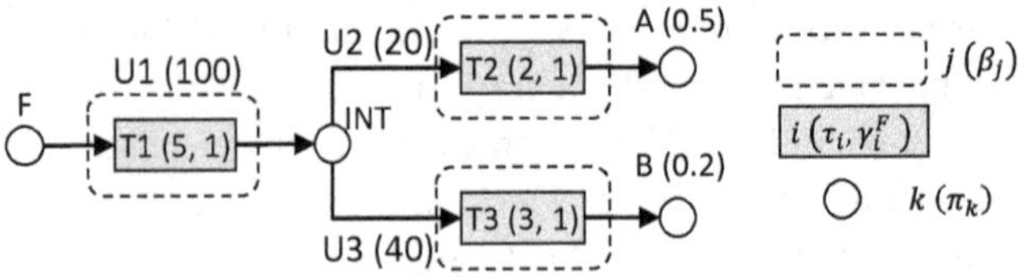

Figure 13.2 Network and associated data for Example 13.2; processing times in h, capacities in kg, costs in \$, and prices in \$/kg.

[4] The optimal objective function value of the MIP model is identical to the corresponding value of its LP-relaxation, and the summations of binary variables are integers. However, this does not mean that the LP-relaxation returns an integer solution. Can you see why?

[5] Note the versions of GAMS and CPLEX used here. If the reader tries to reproduce these results using more recent versions, they will likely obtain different numbers. This is true for all the computational statistics presented in the book.

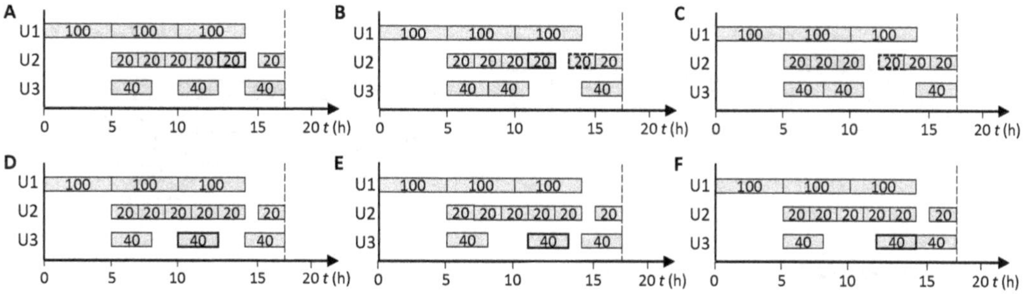

Figure 13.3 Illustration of a multiplicity of mathematical solutions with the same objective function value (batchsizes given inside blocks); the block representing a batch that is moved to generate the next solution is shown in a thick solid line; a block representing the same batch in the resulting schedule is shown in a dashed line.

backward. To illustrate, we consider profit maximization for the same network but over a 18 h horizon. Figure 13.3 shows six different solutions, in terms of values of X_{ijn} variables, which, however, correspond to essentially the same schedule. Removing index j from variables X_{ijn} and B_{ijn}, since each task can be carried out only in one unit, we observe that all solutions have the same number of batches ($\sum_n B_{T1,n} = 3$, $\sum_n B_{T2,n} = 6$, and $\sum_n B_{T3,n} = 3$) and lead to the same objective function value, \$42. (Can you calculate the same value?) Interestingly, there are more than 200 different *mathematical* solutions that correspond to the same *scheduling* solution with an objective function value of \$42.

In addition to having multiple equivalent[6] integer solutions, the relaxation of a discrete time STN-based model has many fractional equivalent solutions which, conceptually, can be obtained by moving *fractional* batches.[7] Thus, branching on binary variables does not lead to different *scheduling* solutions; it simply results in moving the beginning of a batch one time period earlier or later. Thus, no improvement to the upper bound (in profit maximization) is made, and the algorithm must search through a large number of nodes before improving the bound. In Section 13.3, we discuss how the introduction of new variables can facilitate effective branching.

Example 13.3 Processing at Different Time Scales Consider a facility consisting of three types of processes: upstream fermentation, followed by separation and purification, and finally packaging (see Figure 13.4A). Fermentation tasks, which lead

[6] We use the term *equivalent* to refer to solutions that, although obtained from different solution vectors, are very similar from a scheduling standpoint: they have the same objective function value, and the same batching, assignment, and sequencing decisions; they differ only in their timing decisions.

[7] These fractional nodes have the same objective function, which does not deteriorate when branching on a fractional binary variable because the fractional batch will be moved forward or backward in time, leading to very similar (equivalent) fractional schedules in the child nodes. This is why the optimality gap in Example 13.2 cannot be closed effectively.

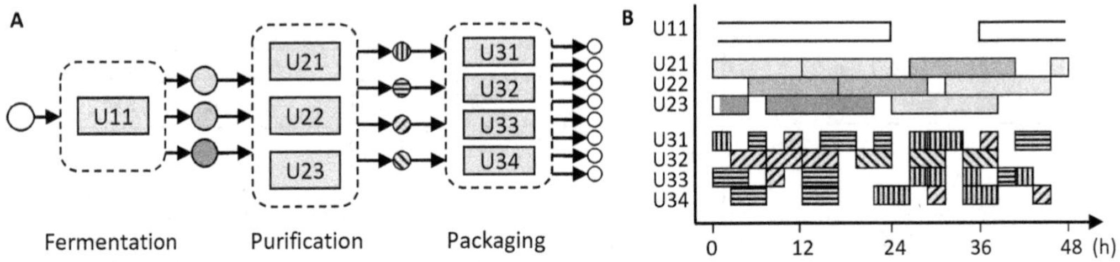

Figure 13.4 Production facility employing processes operating at different time scales. (A) Example facility. (B) Illustrative Gantt chart of a typical solution.

to the production of a limited number of types of broths, are carried out in large volumes and are slow – they last days. Separation and purification tasks, which lead to a larger number of intermediates, are shorter – they typically last 4–8 h. Finally, packaging tasks, which result in a large number of finished products, are fast – they last 1–2 h.

First, we note that while separation and packaging pose some constraints on fermentation, it appears that it is the scheduling of fermentation tasks that will mostly impact the quality of the overall solution and the scheduling of the two downstream stages. Second, while a fine time discretization, with say $\delta = 1$ h, is necessary for accurate scheduling of packaging tasks, such discretization does not appear to be necessary for the scheduling of fermentation tasks, for which a granularity of a grid with $\delta = 12$ h would suffice (see Figure 13.4B). Nevertheless, if the discretization is chosen so that all processing times are approximated with sufficient accuracy, a fine discretization will be used for a common time grid. Can an alternative discrete time model, employing multiple time grids, be developed? In Section 13.4, we describe systematic methods to develop such models.

13.2 Preprocessing and Tightening

The method described in this section is based on a constraint propagation algorithm for the calculation of parameters, which are then used to formulate valid inequalities. The algorithm depends on the structure of the process network. Networks fall into four categories (see Figure 13.5A):

(1) Networks with no loops (Figure 13.5B)
(2) Networks with loops but no recycle materials (Figure 13.5C)
(3) Networks with recycle materials in a single loop (Figure 13.5D)
(4) Networks with a recycle material and multiple nested loops (Figure 13.5E)

A loop is any closed path (moving only in the direction of the stream arrows) within the STN representation of the facility, and a recycle material is a material in a loop that can be produced by multiple tasks. Each block in Figure 13.5A has its own specific

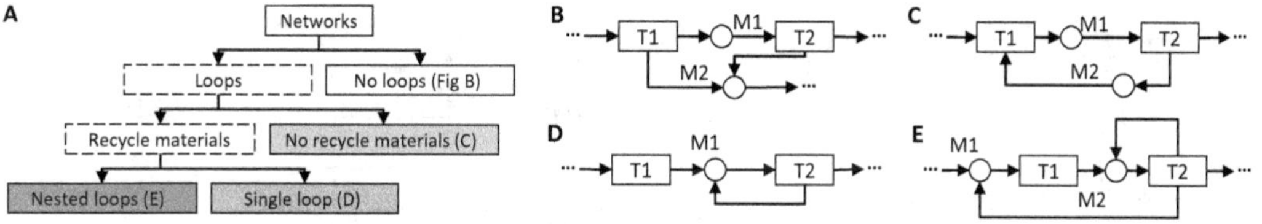

Figure 13.5 Classification of networks for preprocessing algorithm. (A) Classification based on the existence of loops, recycle materials, and nested loops. Example of a general network with no loops (B); with a loop and no recycle materials (C); with a recycle material (D); and with nested loops (E).

tightening procedures. Networks are first divided into networks with and without loops. In Section 13.2.1, we discuss methods that can be applied to all networks. Networks with loops are further divided into networks with or without recycle materials. The additional methods that are applicable to networks with loops that do not contain recycle materials are described in Section 13.2.2. Loops with a recycle material can either consist of a single loop or nested loops. The algorithm can be modified to exploit the structure containing loops with recycle materials, but this is beyond the scope of the book. The interested reader can find more information in Section 13.6.

After presenting the methods for networks with and without loops, the algorithm is described in Section 13.2.3, and the tightening constraints are discussed in Section 13.2.4. Finally, we close in Section 13.2.5 with some extensions.

13.2.1 General Networks

Backward Propagation. We introduce the following parameters:

- ω_k: lower bound on the amount of material k required to meet final demand.
- μ_i^0/μ_i^1: lower bounds on the production of task i required to meet final demand, $\mu_i^0 \le \mu_i^1$
- ν_{ik}: lower bound on the amount of material k that must be produced by task i

First, we calculate ω_k for all materials and μ_i^1 for all tasks sequentially, by backward propagating final product ($k \in \mathbf{K}^P$) demands. We calculate ω_k once μ_i^1 is known for all tasks consuming material k ($i \in \mathbf{I}_k^-$). Similarly, we calculate μ_i^1 after we have calculated ω_k for all materials produced by i ($k \in \mathbf{K}_i^+$). The sequence of calculations for a simple network is shown in Figure 13.6.

Parameter ω_k is calculated as follows:

$$\omega_k = \begin{cases} -\sum_n \xi_{kn}, & k \in \mathbf{K}^P \\ -\sum_{i \in \mathbf{I}_k^-} \rho_{ik}\mu_i^1, & k \notin \mathbf{K}^P \end{cases}, \tag{13.9}$$

where the top expression in the RHS is the total demand, if k is a final product; while the bottom expression is the amount of intermediate or raw material required by all tasks

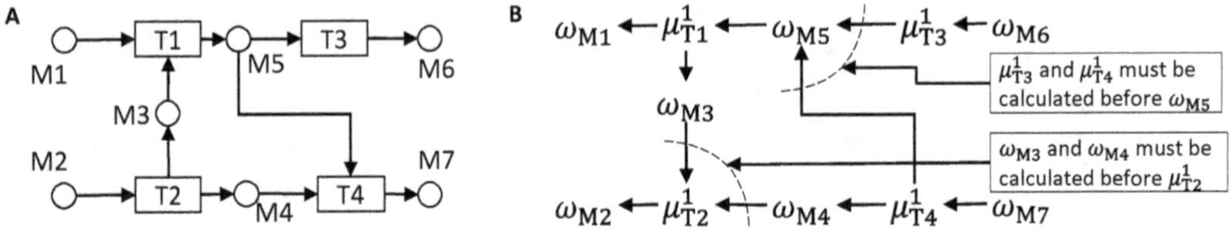

Figure 13.6 Illustration of backward propagation of demand. (A) STN representation of example network. (B) Sequence of calculations.

consuming it minus the initial inventory (we assume zero initial inventory for $k \in \mathbf{K}^P$). If the initial inventory exceeds the amount required, ω_k will be negative, and the process does not need to produce the material.

The minimum amount of material k produced by task i, ν_{ik}, is calculated as follows:

$$\nu_{ik} = \begin{cases} \max\{0, \omega_k\}, & k \in \mathbf{K}^{ST} \\ 0, k \in \mathbf{K}^{MP} \backslash \mathbf{K}^R \end{cases}, \quad i \in \mathbf{I}_k^+, \tag{13.10}$$

where $\mathbf{K}^{ST}$ is the set of materials produced by a single task, $\mathbf{K}^{MT}$ is the set of materials that can be produced by multiple tasks, and $\mathbf{K}^R$ is the set of recycle materials. Note that if multiple tasks can produce a material, then any single task can meet the demand and therefore $\nu_{ik} = 0$. An alternative approach for these materials is described later. Also, note that ν_{ik} is not calculated for recycle materials using (13.10).

Once ν_{ik} for all materials produced by a task are calculated, we calculate μ_i^0:

$$\mu_i^0 = \max_{k \in \mathbf{K}_i^+} \{\nu_{ik}/\rho_{ik}\}, \quad i, \tag{13.11}$$

where ν_{ik}/ρ_{ik} is the amount of material k that must be produced by task i. The maximum over all materials is used to ensure that demand for all materials is met. While (13.11) provides a lower bound on the total production amount of task i, it may be impossible to produce exactly μ_i^0, thus we can adjust (tighten) as follows (with $\Delta\mu_i \geq 0$):

$$\mu_i^1 = \mu_i^0 + \Delta\mu_i, \quad i. \tag{13.12}$$

The calculation of $\Delta\mu_i$ is described next.

Attainable Production Amounts. In general, the total production of a task, that is, the sum of batchsizes of the executed batches of the task, lies in a union of intervals (ranges). Accordingly, we say that the *attainable region* of the total production amount of a batch is a union of *ranges*. When only one unit can process a task $(|\mathbf{J}_i|) = 1$, it is straightforward to find the attainable production amounts for any number of batches and to check if the required production, μ_i^0, is in one of those ranges. If μ_i^0 falls in the attainable region, demand can be met exactly and $\Delta\mu_i = 0$; otherwise, $\Delta\mu_i$ is the difference between the start of the next attainable range and μ_i^0.

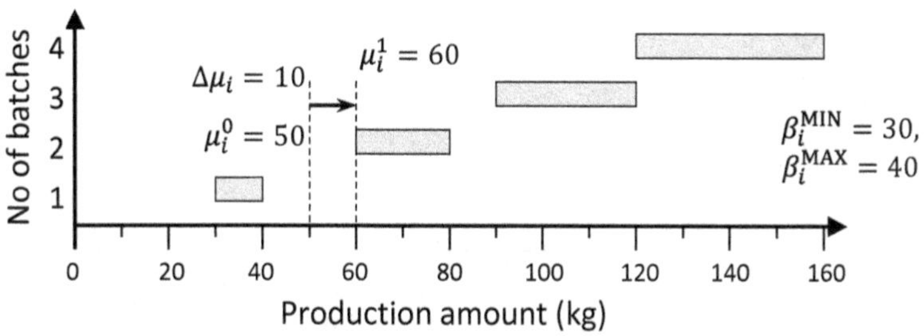

Figure 13.7 Attainable production region determination for unit with capacity between 30 and 40 kg; gray rectangles represent attainable ranges.

To illustrate, consider a task that can be carried out in only one unit with $\beta_j^{MIN} = 30$ and $\beta_j^{MAX} = 40$. The resulting attainable region, shown in Figure 13.7, is $[30, 40] \cup [60, 80] \cup [90, \infty)$. Thus, if $\mu_i^0 = 75$, then $\Delta\mu_i = 0$ and $\mu_i^1 = 75$; but if $\mu_i^0 = 50$, then $\Delta\mu_i = 10$ and $\mu_i^1 = 60$.

When multiple units can process a task ($|\mathbf{J}_i| > 1$), then the attainable ranges for every possible combination of batches in units should be identified and checked. To reduce the number of combinations, an upper bound on the number of batches in a particular unit can be found by dividing μ_i^0 by the largest possible batchsize unit j can process and rounding up:

$$v_{ij}^{MAX} = \left\lceil \mu_i^0 / \beta_j^{MAX} \right\rceil, \quad i, j \in \mathbf{J}_i. \tag{13.13}$$

When the number of batches for a particular unit is at its upper bound, we are guaranteed that the corresponding attainable range meets or exceeds demand with just one unit, and the number of batches in all other units is set to zero to eliminate more combinations. There are $v_i^{MAX} = \prod_{j \in \mathbf{J}_i} (v_{ij}^{MAX}) + |\mathbf{J}_i|$ total ranges to check, and for each range $r \in \mathbf{R}_i = \{0, 1, \ldots, v_i^{MAX} - 1\}$ there is a unique combination of number of batches v_{ij}^r in unit j. We check if μ_i^0 falls into an attainable range by looping over $r \in \mathbf{R}_i$,

$$\sum_{j \in \mathbf{J}_i} v_{ij}^r \beta_j^{MIN} \leq \mu_i^0 \leq \sum_{j \in \mathbf{J}_i} v_{ij}^r \beta_j^{MAX}, \quad r \in \mathbf{R}_i, \tag{13.14}$$

where the LHS/RHS gives the lower/upper bound of attainable range r. If (13.14) is satisfied for some r, the required amount μ_i^0 can be met exactly and $\Delta\mu_i = 0$. If (13.14) is not satisfied for any r, we identify the *closest* range that can meet μ_i^0 and calculate

$$\Delta\mu_i^r = \begin{cases} \sum_{j \in \mathbf{J}_i} v_{ij}^r \beta_j^{MIN} - \mu_i^0 & \text{if } \sum_{j \in \mathbf{J}_i} v_{ij}^r \beta_j^{MIN} \geq \mu_i^0 \\ \infty & \text{if } \sum_{j \in \mathbf{J}_i} v_{ij}^r \beta_j^{MIN} < \mu_i^0 \end{cases} \tag{13.15}$$

and then

$$\Delta\mu_i = \min_r \{\Delta\mu_i^r\}. \tag{13.16}$$

An example calculation, for a task that can be carried out in two units, U1 and U2, is shown in Figure 13.8, where it is assumed that (13.11) has yielded $\mu_i^0 = 55$. We

r	$v_{i,U1}^r$	$v_{i,U2}^r$
1	0	1
2	0	2
3	1	0
4	1	1
5	2	0
6	2	1
7	3	0

$\beta_{U1}^{MIN} = 20, \beta_{U2}^{MAX} = 25$
$\beta_{U2}^{MIN} = 45, \beta_{U2}^{MAX} = 50$

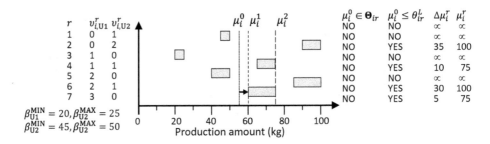

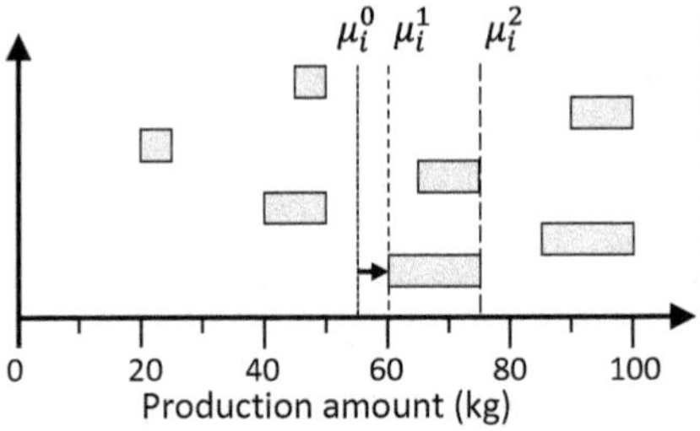

$\mu_i^0 \in \Theta_{ir}$	$\mu_i^0 \le \theta_{ir}^L$	$\Delta\mu_i^r$	μ_i^r
NO	NO	∞	∞
NO	NO	∞	∞
NO	YES	35	100
NO	NO	∞	∞
NO	YES	10	75
NO	NO	∞	∞
NO	YES	30	100
NO	YES	5	75

Figure 13.8 Example calculation of μ_i^1 and μ_i^2.

calculate $v_{i,U1}^{MAX} = 3$ and $v_{i,U2}^{MAX} = 2$ from (13.13) and $v_i^{MAX} = 8$. Figure 13.8 shows all v_{ij}^r on the left and the corresponding attainable ranges in the graph in the middle (gray blocks). For simplicity, we use $\theta_{ir}^L = \sum_{j \in \mathbf{J}_i} v_{ij}^r \beta_j^{MIN}$ and $\theta_{ir}^U = \sum_{j \in \mathbf{J}_i} v_{ij}^r \beta_j^{MAX}$ to define the bounds of each range r, $\boldsymbol{\Theta}_{ir} = \left[\theta_{ir}^L, \theta_{ir}^U\right]$, and express the check carried out in (13.13) as $\mu_i^0 \in \boldsymbol{\Theta}_{ir}$. Since $\mu_i^0 = 55$ does not fall into any range, (13.14) is never satisfied. The minimum excess production is achieved when three batches in U1 are used ($r = 7$, $\boldsymbol{\Theta}_{i,7} = [60, 75]$). We calculate $\Delta\mu_i = 5$ and $\mu_i^1 = 60$.

13.2.2 Networks with Loops

When there are loops, the backward propagation presented in the previous subsection will not work. (Can you see why?) To address this, we use an idea from process flowsheeting and simulation – we *tear* a stream to break a loop $l \in \mathbf{L}$. A tear stream consists of a task, referred to as the *tear task*, and a material produced by that task, referred to as the *tear material.* We use one tear stream in every loop. If a loop has a task that produces a material outside the loop, this task should be used as the tear task; otherwise, any task can be chosen. The set of tasks/materials in loop l is denoted by $\mathbf{I}_l/\mathbf{K}_l$, while the tear task and material of loop l are denoted by $i^{TR}(l)$ and $k^{TR}(l)$. The sets of all tear tasks and materials are $\mathbf{I}^{TR}$ and $\mathbf{K}^{TR}$, respectively.

Figure 13.9 shows an illustrative calculation. The STN representation of the network, along with the necessary process data and algorithmic sets, is given in Figure 13.9A, while the sequence of calculations is shown in Figure 13.9B, where the value of μ_i^0/μ_i^1 is given inside a task block, the value of ω_k inside the node representing material k, and the value of ν_{ik} next to the stream connecting task i to material k. For simplicity, we consider one unit per task with the capacity range, $[\beta_j^{MIN}, \beta_j^{MAX}]$, given above each task block. We chose T3 as the tear task, because it produces M4 outside the loop, and M5 as the tear material. We initialize $\nu_{T3,M5} = 0$ for the tear stream and we estimate μ_{T3}^1 as soon as ω_{M4} is known. We backward propagate demand as follows: $\omega_{M4} = 50 \rightarrow \mu_{T3}^1 = 100 \rightarrow \omega_{M3} = 100 \rightarrow \mu_{T2}^1 = 100 \rightarrow \omega_{M2} = 100 \rightarrow \mu_{T1}^1 = 110$ $(\mu_{T1}^0 = 100, \Delta\mu_{T1} = 10) \rightarrow \omega_{M1} = 110 - 45 = 65 \rightarrow \mu_{T4}^1 = 65 \rightarrow \omega_{M5} = 65 \cdot 0.8 = 52$. We stop after calculating $\omega_{M5} = 52$ for the tear material. Since we need a minimum of 52 kg of M5, but T3

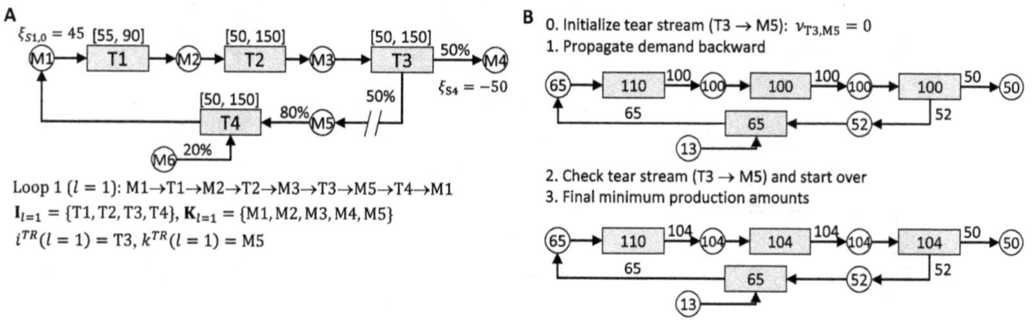

Figure 13.9 Example of back propagation in networks with loops. (A) STN representation of example network, associated process parameters, and loop information. (B) Sequence of calculations.

produces only 50 kg $(= 100 \cdot 0.5)$, we set $\nu_{T3,M5} = 52$, delete ω_k and μ_i^1 for all other materials and tasks, and repeat the propagation using the new value $\nu_{T3,M5} = 52$. We see that now T3 produces the required 52 kg of M5.

In general, we start by initializing $\nu_{ik} = 0$ for all tear streams. We backward propagate demand until ω_k is estimated for a tear material. We update ν_{ik} for the tear stream and compare it to the production of the tear task. If production can meet demand $(\nu_{ik} \leq \rho_{ik}\mu_i^1)$, we continue with the backward propagation. If demand is greater than production $(\nu_{ik} > \rho_{ik}\mu_i^1)$, we reset all other ω_k and μ_i^1, and restart the backward propagation using the new initial value for the tear stream. If, on the second pass, demand for a tear material still exceeds production, the problem is infeasible with the given initial inventories.

The algorithm can be extended to networks with recycle materials and networks with multiple nested loops, but this requires the introduction of additional algorithmic sets and parameters as well as a more complex algorithmic structure, so these extensions are not presented. The interested reader can find some discussion and references in Section 13.6. However, it is important to note that these additional extensions allow us to calculate tighter ω_k and μ_i^1 bounds, in the presence of recycle materials and nested loops, but are not necessary. In other words, the bounds calculated based on the procedures discussed so far in Section 13.2 are valid for all networks.

13.2.3 Preprocessing Algorithm

The complete algorithm combines the tightening procedures for the two network types we have discussed. The flowchart of the algorithm is shown in Figure 13.10, where the procedures for the two types of networks are color coded, following the shades of gray used in Figure 13.5. We define the following new sets:

- $\mathbf{L}_k$: loops containing material k
- $\mathbf{K}_k^{SL}$: materials in any loop containing material k, $\mathbf{K}_k^{SL} = \cup_{l \in \mathbf{L}_k} \mathbf{K}_l$
- $\mathbf{I}_k^{SL}$: tasks in any loop containing material k $\mathbf{I}_k^{SL} = \cup_{l \in \mathbf{L}_k} \mathbf{I}_l$

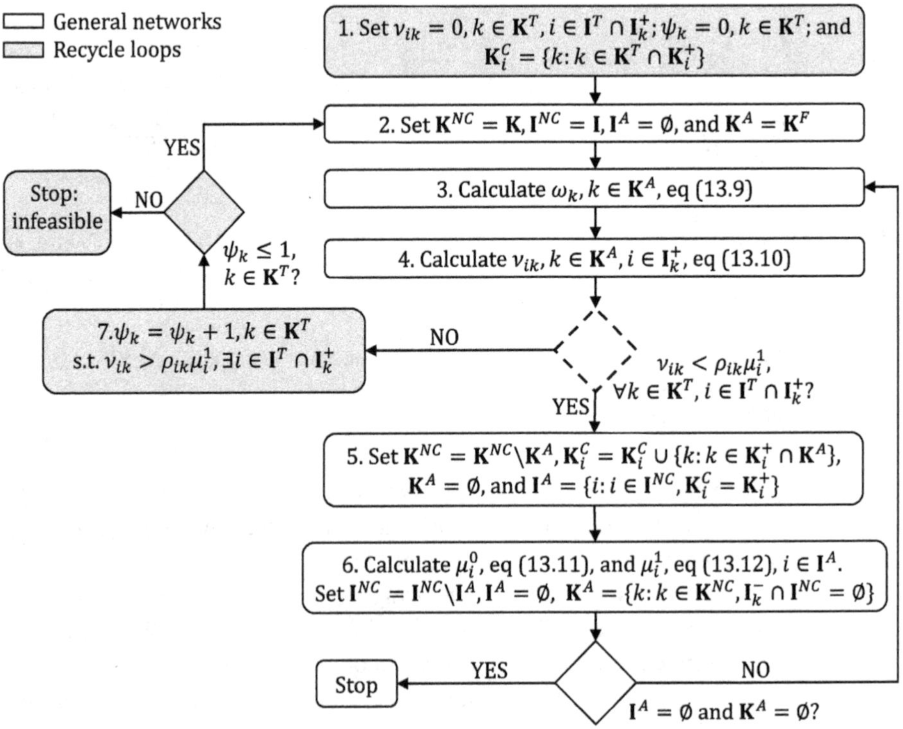

Figure 13.10 Flowchart of the preprocessing algorithm.

- $\mathbf{I}^{NC}$: tasks for which μ_i^0 is not known
- $\mathbf{I}^A$: tasks available for the calculation of μ_i^0 (i.e., ω_k has been calculated for all $k \in \mathbf{K}_i^+$).
- $\mathbf{K}^{NC}$: materials for which ω_k is not known
- $\mathbf{K}^A$: materials available for calculating ω_k (i.e., μ_i^0 has been calculated for all $i \in \mathbf{I}_k^-$).
- $\mathbf{K}_i^C$: materials for which v_{ik} has been calculated for task i

Also, we use parameter ψ_k to keep track of how many times tear material $k \in \mathbf{K}^T$ has been updated. If the tear material is updated more than once, the instance is infeasible.

The algorithm is applicable to all types of facilities, including facilities with multiple tasks producing or consuming the same material(s) and multiple units capable of performing the same task(s). The only inputs to the algorithm are customer demand (ξ_{kn}), conversion coefficients (ρ_{ik}), unit capacities ($\beta_j^{MIN}/\beta_j^{MAX}$), and the network structure, which means that the algorithm is applicable to problems with a wide range of characteristics and constraints, including processes with variable processing times, utility constraints, changeovers, and so forth. Also, note that the algorithm is independent of the type of MIP model used to address the instance at hand. The output of the algorithm is parameters μ_i^1 and ω_k for all tasks and materials, respectively. These parameters are used for the generation of tightening constraints, as discussed next.

13.2.4 Valid Inequalities

Strong valid inequalities can be written once μ_i^1 and ω_k are calculated. The inequalities are written using time-indexed assignment variables, so they are applicable to all (discrete or continuous time) models. For illustration, we will use the discrete time model introduced in Section 13.1.2.

First, we find the minimum number of required batches of a task, λ_i, to meet demand,[8]

$$\lambda_i = \left\lceil \mu_i^1 \Big/ \max_{j\in \mathbf{J}_i}\{\beta_j^{MAX}\} \right\rceil, \quad i, \tag{13.17}$$

and write the following constraint:

$$\sum_{j\in \mathbf{J}_i} \sum_n X_{ijn} \geq \lambda_i, \quad i. \tag{13.18}$$

When multiple tasks produce a material, (13.18) may not be effective. Thus, we calculate lower bounds, κ_k, on the number of batches of all tasks producing a material, by dividing the required amount by the largest possible amount of that material produced in a single batch and rounding up:

$$\kappa_k = \left\lceil \omega_k \Big/ \max_{i\in \mathbf{I}_k^+, j\in \mathbf{J}_i}\{\rho_{ik}\beta_j^{MAX}\} \right\rceil, \quad i. \tag{13.19}$$

The corresponding tightening constraint, involving all units, is

$$\sum_{i\in \mathbf{I}_k^+} \sum_{j\in \mathbf{J}_i} \sum_n X_{ijn} \geq \kappa_k, \quad k \tag{13.20}$$

When a task can be processed in units with very different capacities, (13.18) and (13.20) may not be effective (can you see why?), so the bounds on the production amounts, rather than the bounds on the number of batches, can be used instead to write the following tightening constraint,

$$\sum_{j\in \mathbf{J}_i} \sum_n \beta_j^{MAX} X_{ijn} \geq \mu_i^2, \quad i, \tag{13.21}$$

where $\mu_i^2 \geq \mu_i^1$ is the tightest bound that can be used in the RHS (see the procedure to calculate μ_i^2 in the next paragraph). If a material is produced by multiple tasks, then the counterpart of (13.21) is

$$\sum_{i\in \mathbf{I}_k^+} \sum_{j\in \mathbf{J}_i} \sum_n \rho_{ik}\beta_j^{MAX} X_{ijn} \geq \omega_k, \quad k. \tag{13.22}$$

Parameter μ_i^1 was calculated using β_j^{MIN} because the lower bound of a range was based on the β_j^{MIN} of units $j \in \mathbf{J}_i$ – see the top expression in the RHS of (13.15). However, μ_i^1 would not provide a tight RHS in (13.21) because its LHS includes β_j^{MAX}. To find a tighter bound, we calculate μ_i^r,

$$\mu_i^r = \begin{cases} \sum_{j\in \mathbf{J}_i} v_j^r \beta_j^{MAX} & \text{if } \sum_{j\in \mathbf{J}_i} v_j^r \beta_j^{MAX} \geq \mu_i^0 \\ \infty & \text{otherwise} \end{cases}, \tag{13.23}$$

[8] For continuous tasks, λ_i would be the minimum number of time periods the tasks must run.

which is the counterpart of $\mu_i^0 + \Delta\mu_i^r$ but based on β_j^{MAX} instead of β_j^{MIN} (note the similarities between (13.15) and (13.23)). In other words, μ_i^r corresponds to the upper, rather than the lower, bound of a range that can meet the demand. Once μ_i^r are available, we calculate

$$\mu_i^2 = \min_r \{\mu_i^r\}, \quad i. \tag{13.24}$$

Equation (13.21) using μ_i^2 is valid because the LHS can take only a discrete set of values when $X_{ijn} \in \{0,1\}$ and these values are equal to the upper bounds of the ranges defined for different values of $v_j^r, j \in \mathbf{J}_i$. In general, μ_i^2 is the smallest range upper bound that meets demand μ_i^1. Figure 13.8 illustrates how μ_i^r and then μ_i^2 are calculated for a task with $\mathbf{J}_i = \{\mathrm{U1},\mathrm{U2}\}$, $\beta_{\mathrm{U1}}^{MIN} = 20$, $\beta_{\mathrm{U1}}^{MAX} = 25$, $\beta_{\mathrm{U2}}^{MIN} = 45$, $\beta_{\mathrm{U2}}^{MAX} = 50$, and $\mu_i^0 = 55$. We observe that four ranges can satisfy production of 55 ($r \in \{2,4,6,7\}$), and among them, the smallest upper bound corresponds to ranges 4 and 7 ($\mu_i^4 = \mu_i^7 = 75$) and thus $\mu_i^2 = 75$.

The effect of tightening through (13.18) and (13.21), the latter using different RHS, is illustrated in Figure 13.11, using the task studied in Figure 13.8. We have calculated $\mu_i^0 = 55$, $\mu_i^1 = 60$, and $\mu_i^2 = 75$, and from (13.17) we also calculate $\lambda_i = 55/50 = 2$. For simplicity, we drop index i and define $N_{\mathrm{U1}} = \sum_n X_{\mathrm{U1},n}$ and $N_{\mathrm{U2}} = \sum_n X_{\mathrm{U2},n}$. The tightening constraints can then be rewritten as follows:

$$N_{\mathrm{U1}} + N_{\mathrm{U2}} \geq 2 \tag{13.25}$$

$$25N_{\mathrm{U1}} + 50N_{\mathrm{U2}} \geq \mu_i^m, \quad m, \tag{13.26}$$

where $m \in \{0,1,2\}$. Figure 13.11 shows the feasible space of the LP defined by $N_{\mathrm{U1}} \geq 0$, $N_{\mathrm{U2}} \geq 0$, (13.25), and (13.26). We observe that even the weakest version of (13.26), with $\mu_i^0 = 55$, cuts off a region not excluded by (13.25), but the strongest among the three constraints is the one employing $\mu_i^2 = 75$. We also note that in this simple example, the feasible space of the LP defined using μ_i^2 in (13.26) is integral.

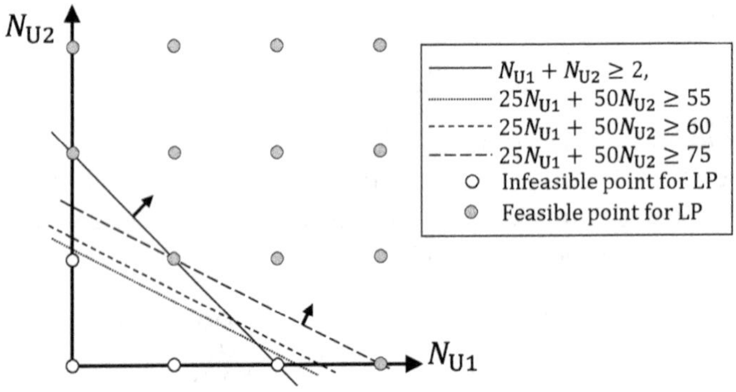

Figure 13.11 Effect and relative strength of valid inequalities.

13.2.5 Extensions

Materials Produced by Multiple Tasks. When multiple tasks can produce a material (i.e., $\left|\mathbf{I}_k^+\right| > 1$), (13.10) yields $\nu_{ik} = 0$ for each $i \in \mathbf{I}_k^+$, which may mean $\mu_i^0 = 0$ for upstream tasks that, collectively, should be producing a nonzero amount. To address this, we can solve a simple linear program, $(\mathbb{L}_i)$, to find ν_{ik}:

$$\left.\begin{aligned} &\min Q_i \\ &st \quad \xi_{k,0} + \sum\nolimits_{i' \in \mathbf{I}_k^+} \rho_{i'k} Q_{i'} \geq -\sum\nolimits_{i' \in \mathbf{I}_k^-} \rho_{i'k} Q_{i'}, k \in \mathbf{K}, \\ &\sum\nolimits_{i' \in \mathbf{I}_k^+} \rho_{i'k} Q_{i'} \geq \omega_k, \quad k \notin \mathbf{K}^{NC} \end{aligned}\right. \quad i \in \mathbf{I}\,(\mathbb{L}_i), \tag{13.27}$$

where $Q_i \in \mathbb{R}_+$ represents the total amount of material task i produces. The first constraint requires that, for each material, the amount produced plus any initial inventory is greater than the amount consumed. The second constraint, which is only written for materials for which ω_k is known, enforces that the amount produced of a material must exceed ω_k. In step 4 of the algorithm, we solve $(\mathbb{L}_i)$ for task i and then multiply the optimal objective value by ρ_{ik} to get ν_{ik}. When a task (other than a tear task) produces multiple materials, we only need to solve $(\mathbb{L}_i)$ once and multiply the optimal value by ρ_{ik} for all $k \in \mathbf{K}_i^+$. For tear tasks that produce multiple materials, we solve $(\mathbb{L}_i)$ twice, first to calculate ν_{ik} for any nontear materials produced by the task, and second to update ν_{ik} for the tear material. When we use this method, we solve $(\mathbb{L}_i)$ to find ν_{ik} for all tasks in the network. Demand is still backward propagated according to the algorithm in Figure 13.10, but we calculate ν_{ik} in step 4 of the algorithm with $(\mathbb{L}_i)$ instead of with (13.10).

Continuous Processes. The algorithm is also applicable to continuous processes. As discussed in Chapter 9, a continuous process can be modeled using a batch task with a duration of one time period and many batches of this task occurring sequentially. The minimum and maximum batchsizes (calculated from the corresponding rate bounds; see Chapter 9) can be used to calculate, essentially, the minimum total duration a continuous task should be executed.

Orders Due at Intermediate Times. We can use intermediate due times to generate additional tightening inequalities that, in some cases, may be stronger. The parameters needed to generate such inequalities are conceptually the same, but are also indexed by n: λ_{in}, κ_{kn}, μ_{in}^2, and ω_{kn}. They are nonozero only for time points, $n \in \mathbf{N}^{MS-}$, at which there are orders (negative material shipments) due. The algorithm to calculate them is similar to the one already presented but with two key differences: (1) the algorithm is executed separately for each time point an order is due (to calculate the corresponding parameters) and (2) the demand that is propagated corresponds to the cumulative demand up to that time point; i.e., if the algorithm is run for $n \in \mathbf{N}^{MS-}$, then the demand for final products that is used is $\sum_{n' \leq n} \xi_{kn'}$.

The generalization of (13.18), based on intermediate due times, is

$$\sum\nolimits_{j \in \mathbf{J}_i} \sum\nolimits_{n' \leq n - \tau_{ij}} X_{ijn'} \geq \lambda_{in}, \quad i, n \in \mathbf{N}_i^{\Lambda}, \tag{13.28}$$

Table 13.1 Parameter values calculated through preprocessing based on intermediate due times.

	ξ_{kn}		ω_{kn}	μ_{in}^0			λ_{in}		
n	M3	M4	M2	T1	T2	T3	T1	T2	T3
6	30	0	35	35	35	0	1	1	0
9	90	25	125	125	90	35	3	2	1

where $\mathbf{N}_k^A$ is the set of times points for which parameter λ_{in} is positive.[9] Note that the tasks accounted for, in the LHS, should be completed by n because the calculation of λ_{in} was based on demand that was due up until n[10]. One of (13.28) can be tighter than (13.18), for the same i, if, for example, all the orders of all products that require task i are due before the end of the horizon. (Can you see why?) Equations (13.20) through (13.22) can be generalized in a similar manner.

To illustrate, we consider the instance introduced in Example 13.1 with a horizon of 9 h and an intermediate order for 30 kg of M3 at $n = 6$. Table 13.1 gives the parameters calculated for both $n = 6$ and $n = \eta = 9$.

Forward Propagation. The presented valid inequalities lower bound combinations of assignment binary variables (X_{ijn}) and thus are expected to be effective in, for example, cost minimization problems where the objective function *pushes* (combinations of) X_{ijn} variables to obtain small values. The reader may be wondering, then, if these or similar methods can be effective for, say, profit maximization problems, where the objective function pushes X_{ijn} variables to obtain large values. The answer is that very similar ideas can be used to *forward propagate* other information, leading to the calculation of parameters that can be used to upper bound (combinations of) X_{ijn} variables. Specifically, such ideas can lead to parameters regarding the maximum production that can be calculated given the available initial inventories, and then used to upper bound combinations of X_{ijn} variables.

13.3 Reformulations

The motivation behind the development of the reformulations presented in this section is to facilitate efficient branching that overcomes the challenge of *equivalent* solutions, as discussed in Example 13.2. In Section 13.3.1, we show three such reformulations,

[9] While the algorithm is executed for all points at which there is demand, a task may not participate in the production of a product (i.e., it may not produce the product or any intermediate necessary for that product) whose demand is due at given n, and thus λ_{in} will be zero. The same may occur with κ_{kn}, μ_{in}^2, and ω_{kn}.

[10] A tighter constraint can be obtained if the *tail* of task i, with respect to the production of the product that is due at given n, is considered. For example, if λ_{in} is calculated because product A is due at n and it takes at least five periods to convert the intermediate produced by i to product A, then the domain of the summation over n' should be $n' \leq n - \tau_{ij} - 5$. (Can you see why?) However, calculating minimum task tails in network environments is nontrivial and thus omitted.

while in Section 13.3.2 we illustrate why these reformulations are effective and discuss some additional solution aspects.

13.3.1 New Variables and Branching Strategies

Our goal is to lead the branch-and-bound algorithm to branch on a solution characteristic, rather than binary variables, which is likely to lead to the generation of child nodes that correspond to different solutions, and therefore result in fast bound improvement. This can be accomplished through the introduction of new variables, which can then be used for branching. There are two approaches.

In the first, we introduce nonnegative integer variable N_{ij} to denote the total number of batches of task i executed on unit $j \in \mathbf{J}_i$:

$$N_{ij} = \sum\nolimits_n X_{ijn}, \quad i,j, \tag{13.29}$$

where a valid upper bound on N_{ij} is $\lfloor \eta / \tau_{ij} \rfloor$.[11]

In the second, we introduce $Z_{ijm} \in \{0, 1\}$, which is 1 if there are m batches of i executed on unit $j \in \mathbf{J}_i$:

$$\sum\nolimits_{m \in \mathbf{M}_{ij}} m Z_{ijm} = \sum\nolimits_n X_{ijn}, \quad i,j, \tag{13.30}$$

where $\mathbf{M}_{ij} = \{0, 1, \ldots, \lfloor \eta / \tau_{ij} \rfloor\}$. Clearly, only one Z_{ijm} can be active in any feasible solution:

$$\sum\nolimits_{m \in \mathbf{M}_{ij}} Z_{ijm} = 1, \quad i,j. \tag{13.31}$$

Equation (13.29) and variables N_{ij} or, alternatively, Equations (13.30) and (13.31) and variables Z_{ijn} can be readily added to an STN- or RTN-based model, regardless of the processing features and constraints the model accounts for. Also, the same variables and constraints can be added to continuous grid-based models. Finally, we note that the addition of these constraints does not lead to a tighter formulation but, as we will see in the next subsection, leads to more effective branching. In fact, it can lead, depending on the characteristics of the instance, to dramatic computational improvement with speed-ups up to three orders of magnitude (see discussion in Section 13.6). Nevertheless, for instances that remain computationally challenging even after the reformulation(s), one could also explore branching strategies through commercial MIP solvers. Some potential approaches are the following.

Branching Priorities. Priorities determine the order in which branching is performed – the solver can be *instructed* to branch first on variables with higher priority. Clearly, N_{ij} (or Z_{ijm}) variables should have higher priority than the *native* X_{ijn} variables, but this is (expected to be) automatically detected by MIP solvers. More interestingly, different priorities can be set among N_{ij} variables based on the *importance* of tasks and

[11] Can you think of a valid lower bound on N_{ij}? In what types of problems one would be able to calculate such lower bound?

units. For example, one would expect that the scheduling of a bottleneck unit, j^{BN}, would affect the quality of the schedule significantly; thus, setting higher priorities for all $N_{i,j=j^{BN}}$ may lead to computational enhancements.

Special Order Set 1 Branching. Variables Z_{ijm} are binary and satisfy (13.31), which means that only one of them will be equal to one in any feasible solution. Therefore, they are *special order set* of type 1 (SOS1) variables (see discussion in Section 2.2.1). Thus, if declared as SOS1 variables, the MIP solver will employ special branching schemes that have been shown to be effective in many problems.

Strong Branching. It is a branching approach where different variables are tested in terms of bound improvement by temporarily branching on all of them, estimating the improvement resulting from branching on each one and then selecting the most promising one. It requires more computations per node, but often results in shorter computational times due to fast bound improvement and thereby reduced tree size.

The preceding techniques can be combined, in various ways, with the two reformulations. For example, one could introduce Z_{ijm} variables as SOS1 variables and instruct the solver to use strong branching. To select the best combination, for a given network, testing should be carried out.

13.3.2 Remarks

We revisit Example 13.2 to illustrate why branching on variables N_{ij} can significantly reduce the size of the search tree.

Example 13.2 (continued) We introduce variables N_{ij} through (13.29) and then implement a basic branch-and-bound search, where we solve the LP-relaxation, select a fractional N_{ij} variable to branch on, add bounds, solve the LP-relaxation of the resulting child nodes, and repeat the process for the next node. We drop index j because each task can be carried out in only one unit. The results of this *manual* search are shown in Figure 13.12, where the node number is given inside the circle representing a node, z^L represents the optimal objective function value of a node, and

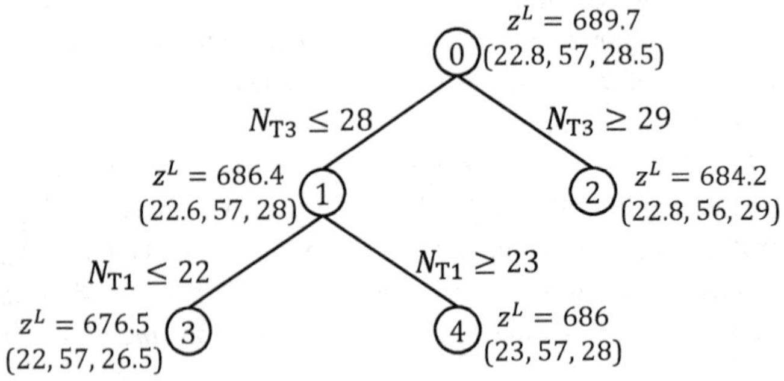

Figure 13.12 Branch-and-bound tree resulting from branching on N_i variables.

N_i^L represents the optimal value of variable N_i at a node. The values of z^L and vector $[N_{T1}^L, N_{T2}^L, N_{T3}^L]$ are given next to each node. Recall that optimal objective function value of the LP-relaxation is 689.7, the optimal objective function value of the MIP is 686, and that the optimality gap could not be closed after hours of computation and millions of explored nodes when using the standard STN-based model. After 95 million nodes, the upper bound is 689.5, that is, the gap closes by less than 6%.

We observe that after branching once, on N_{T3}, the upper bound improves to 686.4, that is, the gap closes by almost 90% $\left(\frac{689.7-686.4}{689.7-686}\right)$. In other words, the improvement based on solving three LPs, based on branching on N_{ij}, is significantly larger than the improvement after millions on LPs, based on branching on X_{ijn} variables. The $[N_{T1}^L, N_{T2}^L, N_{T3}^L]$ vector is fractional on both nodes 1 and 2. We choose node 1 to branch from. (Can you explain why?) After branching on N_{T1} the upper bound (best possible) decreases to 686, that is, the gap closes to 0% after solving five LPs![12] While such dramatic improvement is not expected in all instances, this simple example shows why branching on this new variable leads to significant improvement of the bound of the child nodes.

The idea of introducing new variables to represent a solution characteristic can be extended to other features. For example, in instances with setups or changeovers and long scheduling horizons, where a large number of setups is expected in the optimal solution, integer variables representing the total number of setups in a unit or the total number of setups for a task can be introduced (see some discussion in Section 13.6). Finally, similar ideas can be applied to problems in sequential environments, even when batching decisions are fixed. For example, integer variables can be introduced to represent the total number of batches assigned to a unit.

13.4 Models Based on Multiple Discrete Time Grids

In this section, we discuss models that employ multiple unit-, task-, and material-specific discrete grids. The motivation behind the development of these models is that different subsystems in a facility may operate under different timescales. Furthermore, there are cases where it may be beneficial to change the discretization of a single grid. For example, it may be necessary to consider a fine grid during the early portion of the scheduling horizon, when a detailed and accurate schedule is necessary, but a coarser grid may be sufficient for the late portion of the horizon that is subject to significant uncertainty.

13.4.1 Time Windows

The time grid of a unit or task may depend on unit availability. For example, time points need not be defined for units during (predetermined) maintenance periods. Also, the

[12] Note that an integer solution may not be available after solving these five LPs because the solution in node 4 can be fractional despite $[N_{T1}^L, N_{T2}^L, N_{T3}^L]$ being integral.

grid of a material, which determines when material balance and storage constraints are enforced, depends not only on the processing times of the tasks producing or consuming this material but also on the times that this material is shipped to customers. Finally, different levels of accuracy can be used during the scheduling horizon. For example, exact processing times can be used during the first few days of the horizon, while approximations can be employed during the latter part of the horizon, the scheduling during which will be determined by a different optimization problem solved at a future point. The approach discussed in this section accounts for all these features to generate unit, task, and material grids that employ fixed (i.e., discrete) but not necessarily equally spaced time points. If none of the three aforementioned features is present, then the approach results in grids with fixed points uniformly spaced. However, the resulting grids may still employ different subsets of time points, defined using different time steps (i.e., periods of different length).

First, the scheduling horizon is divided into user-supplied windows, $w \in \mathbf{W}$:

- $\mathbf{W}_j^U$: windows related to unit j
- $\mathbf{W}_i^T$: windows related to task i

with the corresponding start and end times (lower and upper bounds):

- $\omega_{jw}^L/\omega_{jw}^U$: start/end of window $w \in \mathbf{W}_j^U$
- $\omega_{iw}^L/\omega_{iw}^U$: start/end of window $w \in \mathbf{W}_i^T$

Note that *window 1* ($w = 1$) for unit U1 may be different from *window 1* for unit U2 or task T1. Also, each final product due time is represented as a material-related window, $\mathbf{W}_k^M$, consisting of a single time point:

- ω_{kw}: due time for material k in window $w \in \mathbf{W}_k^M$

We divide the scheduling horizon into windows $w \in \mathbf{W}^S$ during which different processing time approximations are made. We will refer to these windows as *accuracy* windows. We define the start/end of each such window as ω_w^L/ω_w^U. In general, accuracy windows can include times where a task or unit is unavailable.

Using unit and task availability data (ω_{jw}^L, ω_{jw}^U, ω_{iw}^L, and ω_{iw}^U), material due date data (ω_{kw}), and the parameters defining the windows during which different levels of accuracy are used (ω_w^L, ω_w^U), we generate grids for tasks, units, and materials, defined in terms if the following subsets of $\mathbf{N}$:

- $\mathbf{N}_{ij}^{TU}$: time points at which task i in unit j can start
- $\mathbf{N}_j^U$: time points at which some task can start on unit j
- $\mathbf{N}_k^M$: time points at which material k can be consumed or shipped

Point n occurs at time t_n. Points are unique ($t_n = t_{n'}$ only if $n = n'$) and not necessarily ordered (it is possible that $t_n < t_{n'}$ for $n > n'$), though they can be trivially ordered after the grids have been generated. Figure 13.13 illustrates the generation of grids for a simple instance with one unit, U1; two tasks, T1 and T2, in $\mathbf{J}_{\mathrm{U1}}$; and two *accuracy* windows.

In many instances, adopting multiple grids can lead to smaller MIP models without introducing any approximations, though, in general, multiple grids can be combined

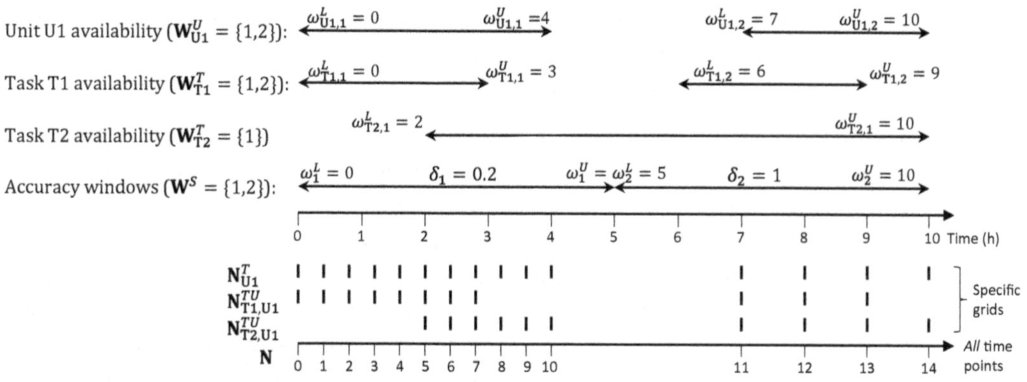

Figure 13.13 Illustration of unit-specific and task-specific grid generation; grid points are shown as vertical thick lines in the bottom panel.

with time-related data approximations, as in single-grid models, leading to even smaller formulations. In the subsequent sections, we discuss the details of the case where no approximations are introduced and outline the approach with approximations. For more details of the approach that combines multiple grids with approximations, the reader can find references in Section 13.6.

13.4.2 Exact Task and Unit Time Discretization

We discuss multigrid models that yield an optimal solution with an objective function value equal to the optimal objective function of a solution that would have been obtained using a single-grid model that introduces no approximations, though the two solutions may be different. The general principle that allows us to generate such grids is that when a task in the solution obtained by the single-grid model starts at a time point not in the multigrid model, an equivalent schedule can be generated by shifting the task earlier to an earlier grid point. Intermediates are only produced at grid points, so they can always be consumed at a grid point.

In general, if certain assumptions hold, multiple grids that will not introduce any approximations can be defined. First, since shifting tasks may change which materials are stored and for how long, storage constraints should be handled carefully. Second, we assume that the holding cost of intermediates does not affect the optimality of the solution. Third, we assume that we are always able to shift tasks earlier without overlapping a time when either the task or the unit is unavailable. If these three assumptions are met, then a multigrid model will yield a solution with the same optimal objective function value as the single-grid model, provided that we choose a discretization, δ^{TU}_{ijw}, for task i in unit j during window $w \in \mathbf{W}^S$ such that:

(1) The step size is the same for all tasks processed on the same unit.
(2) The step size is a factor of the processing times so that tasks finish at time points.

(3) The step size decreases by an integer factor moving downstream; that is, if i produces a material that is then consumed by i', the step size for i' must be an integer factor of the step size for i.

The first requirement ensures that when one batch finishes in a unit, another batch can immediately start in that unit. The second requirement ensures that solutions with no downtime between batches are feasible. The final requirement allows materials to be consumed as soon as they are produced. If the step sizes do not satisfy any of these requirements, there may be downtime between batches. For a single accuracy window (i.e., $|\mathbf{W}^S| = 1$), step sizes, δ_{ijw}^{TU}, meeting these requirements are calculated using Algorithm 1, where $\mathbf{J}_j^+/\mathbf{J}_j^-$ is the subset of units that consume/produce a material produced/consumed by unit j, $\mathbf{J}^C$ is the subset of units for which requirement (3) must be checked, δ_j^U is the step size of the grid of unit j, and $gcf(\cdot)$ returns the greatest common factor (GCF) of a set of numbers.[13]

Algorithm 1

set $\delta_j^U = gcf\{\tau_{ij} : i \in \mathbf{I}_j\}$ and $\mathbf{J}^C = \mathbf{J}$

while $\mathbf{J}^C \neq \emptyset$

for $j \in \mathbf{J}^C$

$\mathbf{J}^C = \mathbf{J}^C \backslash \{j\}$

for $j' \in \mathbf{J}_j^-$

if $\delta_{j'}^U < \delta_j^U$ or $\delta_{j'}^U \, mod \, \delta_j^U \neq \emptyset$

then $\delta_j^U = gcf\left\{\delta_{j'}^U, \delta_j^U\right\}$ and $\mathbf{J}^C = \mathbf{J}^C \cup \mathbf{J}_j^+$

$\delta_{ijw}^{TU} = \delta_j^U, \quad i,j \in \mathbf{J}_i, w \in \mathbf{W}^S.$

Initially, we set the step size for a unit equal to the GCF of the processing times of all tasks that can be processed in that unit to ensure requirements (1) and (2) are met. Then we loop over units to check if requirement (3) is met. We first remove j from $\mathbf{J}^C$ because we will check requirement (3) for j and its upstream units, $j' \in \mathbf{J}_j^-$; then we loop over $j' \in \mathbf{J}_j^-$ and check requirement (3). If it is not met, then we set $\delta_j^U = gcf\{\delta_{j'}^U, \delta_j^U\}$, which is less than and an integer factor of $\delta_{j'}^U$ so that requirement (3) is met. The new value of δ_j^U is a factor of the old value, so requirement (2) is still met. If we change δ_j^U we must repeat the check of requirement (3) for j and all units whose step size was calculated based on j ($j'' \in \mathbf{J}_j^+$), so we add these units back to $\mathbf{J}^C$. Finally, we define δ_{ijw}^{TU} so that the step sizes for all tasks in a unit are the same to satisfy requirement (1). Note that the value of δ_j^U can only decrease during the execution of the algorithm, and it may become smaller than the GCF of the processing times. Algorithm 1 works for networks

[13] Recall from Chapter 3, that we define the GCF of a set of numbers as the greatest number that divides all numbers in the set with zero remainder, so that the GCF is defined for all rational numbers and may be fractional

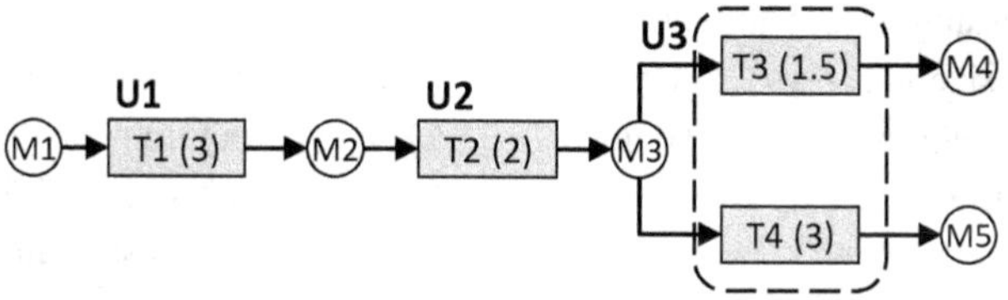

$\mathbf{J}^-_{U1} = \emptyset, \mathbf{J}^+_{U1} = \{U2\}$

$\mathbf{J}^-_{U2} = \{U1, U2\}, \mathbf{J}^+_{U2} = \{U3\}$

$\mathbf{J}^-_{U3} = \{U2\}, \mathbf{J}^+_{U3} = \{U2\}$

Figure 13.14 Instance used to illustrate the application of Algorithm 1.

A (1): $\mathbf{J}^C = \{U1, U2, U3\}$,
$\delta_{U1} = 3, \delta_{U2} = 2, \delta_{U3} = gcf(1.5,3) = 1.5$
(2): $\mathbf{J}^C \neq \emptyset$;
(3): $j = U1$;
(4): $\mathbf{J}^C = \{U2, U3\}$
(5): $\mathbf{J}^-_{U1} = \emptyset$

B (2): $\mathbf{J}^C \neq \emptyset$;
(3): $j = U2$;
(4): $\mathbf{J}^C = \{U3\}$
(5) $j' = U1$
(6): True
(7): $\delta_{U2} = gcf(3,2) = 1; \mathbf{J}^C = \{U3\}$

C (5) $j' = U3$
(6): True
(7): $\delta_{U2} = gcf(1,1.5) = 0.5; \mathbf{J}^C = \{U3\}$

D (2): $\mathbf{J}^C \neq \emptyset$;
(3): $j = U3$;
(4): $\mathbf{J}^C = \emptyset$
(5) $j' = U2$
(6): True
(7): $\delta_{U3} = gcf(0.5,1.5) = 0.5; \mathbf{J}^C = \{U2\}$

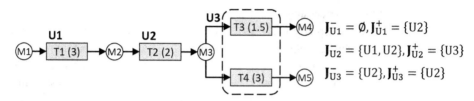

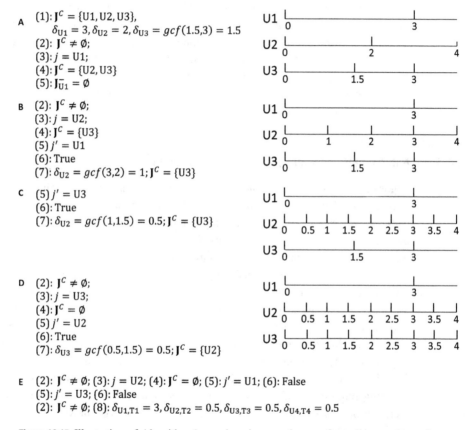

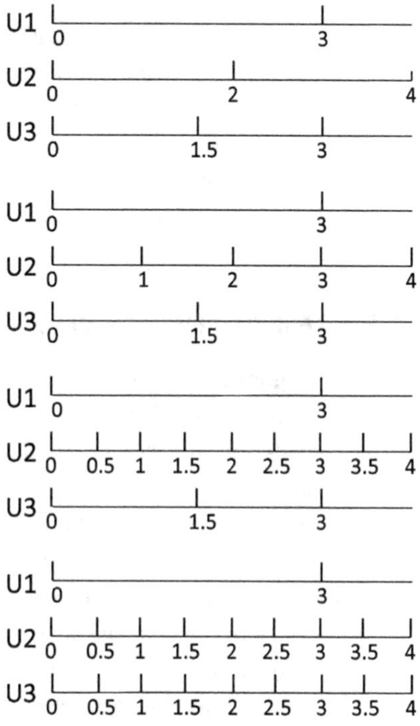

E (2): $\mathbf{J}^C \neq \emptyset$; (3): $j = U2$; (4): $\mathbf{J}^C = \emptyset$; (5): $j' = U1$; (6): False
(5): $j' = U3$; (6): False
(2): $\mathbf{J}^C \neq \emptyset$; (8): $\delta_{U1,T1} = 3, \delta_{U2,T2} = 0.5, \delta_{U3,T3} = 0.5, \delta_{U4,T4} = 0.5$

Figure 13.15 Illustration of Algorithm 1; numbers in parentheses refer to line numbers of Algorithm 1. (A) Lines 1–5 in the first iteration. (B) Lines 2–7: update U2 grid based on U1 grid. (C) Lines 5–7: update U2 grid based on U3 grid. (D) Lines 2–7: update U3 grid based on U2 grid. (E) Complete algorithm: no changes made. Grid points shown as vertical thick lines.

with recycle streams because the step sizes are rechecked in line 7. When there are recycle loops, δ^U_j will be the same for all units in the loop.

To illustrate, we apply the algorithm to the simple network shown in Figure 13.14. The calculations made in six iterations of Algorithm 1 are shown in Figure 13.15. Initially the step sizes are set to the GCF of the processing times for all tasks that take place in a unit. Units U1 and U2 can only process one task each, so their step size is equal to the processing time. Unit U3 processes two tasks with processing times of 1.5 and 3, so its step size is set to 1.5. Unit U1 is removed from $\mathbf{J}^C$ because its step size currently satisfies requirement (3) (Figure 13.15A). Next, requirement (3) has to be satisfied for U2 and all its upstream units. We set δ^U_{U2} to the GCF of δ^U_{U1} and δ^U_{U2} (Figure 13.15B). For U3 and U2, we set δ^U_{U2} to the GCF of the corresponding stepsizes, δ^U_{U3} and δ^U_{U2}, to get $\delta^U_{\mathrm{U2}} = 0.5$ (Figure 13.15C). We remove U2 from $\mathbf{J}^C$ because its step size currently satisfies requirement (3). Next, only requirement (3) for U3 and its upstream units (U2) has to be checked. We set δ^U_{U3} to the GCF of $\delta^U_{\mathrm{U2}} = 0.5$ and $\delta^U_{\mathrm{U3}} = 1.5$ to get $\delta^U_{\mathrm{U3}} = 0.5$ and remove U3 from $\mathbf{J}^C$ (Figure 13.15D). Since the value of δ^U_{U3} changed, we should ensure that U2 still satisfies requirement (3), so we add it to $\mathbf{J}^C$ and repeat the check for U2 to determine that it still satisfies the requirement. We calculate the final values of δ^{TU}_{ij} ensuring that all tasks carried out on the same unit have the same step, so requirement (1) is satisfied (Figure 13.15E). Note that $\delta_{\mathrm{U2,T2}} = \delta_{\mathrm{U3,T3}} = \delta_{\mathrm{U4,T4}} = 0.5$ because T2 and T3 are part of a recycle loop and T3 and T4 are executed on the same unit.

13.4.3 Approximate Task and Unit Time Discretization

Algorithm 1 will give the greatest reduction in the number of time points when using a step size equal to the processing time satisfies requirements (1) through (3). For some networks, the step sizes calculated using Algorithm 1 will not give a significant (or any) decrease in the number of time points. For example, tasks and units in a recycle loop must all have the same step size to satisfy requirement (3), so networks with many recycle loops may require a small step size. Also, when a unit can process tasks with different processing times, a small step size may be required to satisfy requirement (1). To obtain significant model reduction in such networks, we can choose step sizes that may eliminate the optimal solution, but can yield good solutions quickly by relaxing or eliminating the three requirements introduced in the previous subsection. Relaxing requirement (1) is useful when there are many tasks that can take place in a single unit. Relaxing requirement (2) is useful when the processing times have a very small GCF. Relaxing requirement (3) is useful when there are recycle streams; processing times increase moving downstream; or processing times decrease moving downstream, but not by an integer factor.

Instead of completely eliminating the three requirements, we relax them using parameters μ, ν, and α_{ij}; and require, instead:

(4) The grids for any two tasks in a unit must share a point at least every μ points.
(5) The step size for task i in unit j is an integer factor of $\alpha_{ij}\lceil \tau_{ij}/\alpha_{ij} \rceil$.
(6) The grid for a task producing a material shares a point with the grid for a task consuming that material at least every ν points of the finer grid.

Using the network shown in Figure 13.16, an illustration of the approximate grid generation is given in Figure 13.17. Figure 13.17A shows how enforcing requirement (4), using different μ values, leads to different grids, while Figure 13.17B shows how

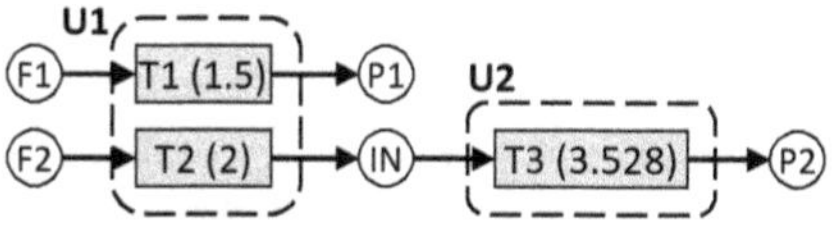

Figure 13.16 Network used for illustrating how user-defined parameters can be used to define different grids.

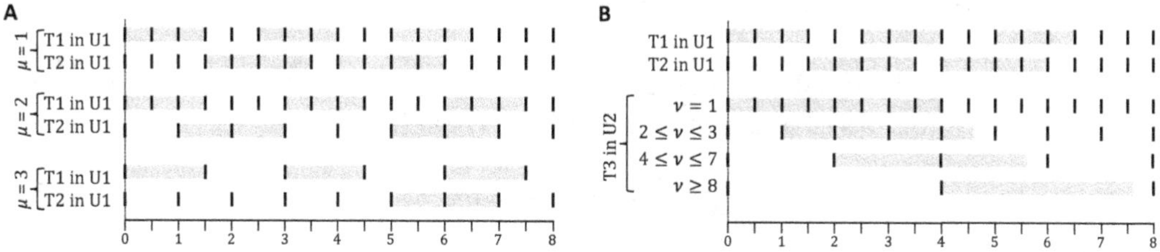

Figure 13.17 Illustration of approximate grids for the network given in Figure 13.16. (A) Grids for different values of μ. (B) Grids for different values of ν (grid points shown as vertical thick lines; shaded blocks, representing executed batches, are shown for illustration).

enforcing requirement (6) for task T3 on unit U2, using different ν values, leads to different grids for T3 in U2.

Parameter α_{ij} is used to round the processing time to the next multiple of α_{ij} before finding the step size, which is equivalent to selecting the step size, δ, for a single uniform grid. However, with multiple grids we can choose different values of α_{ij} for each task and unit. If α_{ij} is the same for all tasks and units, there is always a valid step size greater than or equal to α_{ij}, so this sets a minimum step size for all tasks and units. Setting $\mu = 1$, $\nu = 1$ and $\alpha_{ij} = \tau_{ij}$ is equivalent to enforcing requirements (1) through (3). Setting $\mu = \infty$, $\nu = \infty$ and $\alpha_{ij} = \infty$ removes requirements (1) and (2) so that $\delta_{ijw}^{TU} = \tau_{ij}$. Based on requirements (4) through (6), it is possible to develop an algorithm that yields grids that lead to significantly smaller models, albeit at the cost of yielding suboptimal solutions. Nevertheless, an advantage of expressing requirements (4) through (6) in terms of user-defined parameters μ, ν, and α_{ij} is that the user *controls* the extent of the approximations as well as the units for which better approximations are made. The interested reader can find more information in Section 13.6.

When the time horizon is very long, different time steps can be used; for example, a finer grid can be used for earlier times and a coarser grid for later times when there is more uncertainty. Generating such time-varying grids can be accomplished by dividing the horizon into windows $w \in \mathbf{W}^S$ and using different sets of μ, ν, and α_{ij} parameters in each window. The concept is illustrated in Figure 13.18 using the network introduced in Figure 13.16. Note that the resulting discretization is fixed (discrete) but nonuniform.

Subsets $\mathbf{N}_{ij}^{TU}$ are defined based on window information and the step size, δ_{ijw}^{TU}, calculated in Algorithm 1 or its counterpart when approximations are considered. A task can start at a time point if the task and the unit are available during the entire

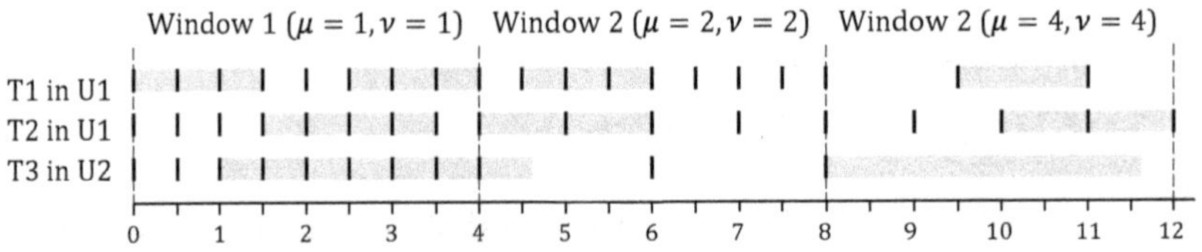

Figure 13.18 Use of different combinations of parameters to generate time-varying task grids (shaded blocks, representing executed batches shown for illustration).

execution of the task, and the spacing between time points has the desired step size. The set $\mathbf{N}_j^U$ contains all points where any task $i \in \mathbf{I}_j$ can start:

$$\mathbf{N}_j^U = \left\{ n \in \mathbf{N}_{ij}^{TU} : i \in \mathbf{I}_j \right\}, \quad j. \tag{13.32}$$

13.4.4 Material Grids

The grid of a raw material or intermediate consists of all points when the material can be consumed. When a material can be consumed by multiple tasks, potentially carried out in different units, its grid is generated by combining the grids for all of those tasks and units. For raw materials and intermediates, $\mathbf{N}_k^M$ includes all points when a task that consumes them can start:

$$\mathbf{N}_k^M = \left\{ n \in \mathbf{N}_{ij}^{TU} : i \in \mathbf{I}_k^-, j \in \mathbf{J}_i \right\}, \quad k. \tag{13.33}$$

The concept is illustrated in Figure 13.19 using an intermediate that is produced by one task (T1) and consumed by two downstream tasks (T2 and T3). The grid is generated based on the time points at which INT can be consumed by T2 and T3. The grid of a final product includes points where it can be produced or shipped to the customers, that is, it consists of all points when it can be produced by some task and due times.[14]

13.4.5 Model

The model based on multiple grids adopts exactly the same variables as the models based on a single uniform grid. The only difference is that the variables are defined for a subset of time points, as follows:

- $X_{ijn} \in \{0, 1\}$: = 1 if a batch of task i starts on unit j at time point n (time t_n); defined for $n \in \mathbf{N}_{ij}^{TU}$
- $B_{ijn} \in \mathbb{R}_+$: batchsize of batch of task i that starts on unit j at time point n; defined for $n \in \mathbf{N}_{ij}^{TU}$
- $I_{kn} \in \mathbb{R}_+$: inventory level of material k during period $n \in \mathbf{N}_k^M$

[14] Recall that single uniform-grid models require rounding the due times down to the nearest grid point, which may result in inaccurate holding or backlog cost calculation. With a multigrid model, due times are explicitly included in the grid of a final product, giving the most accurate cost calculation.

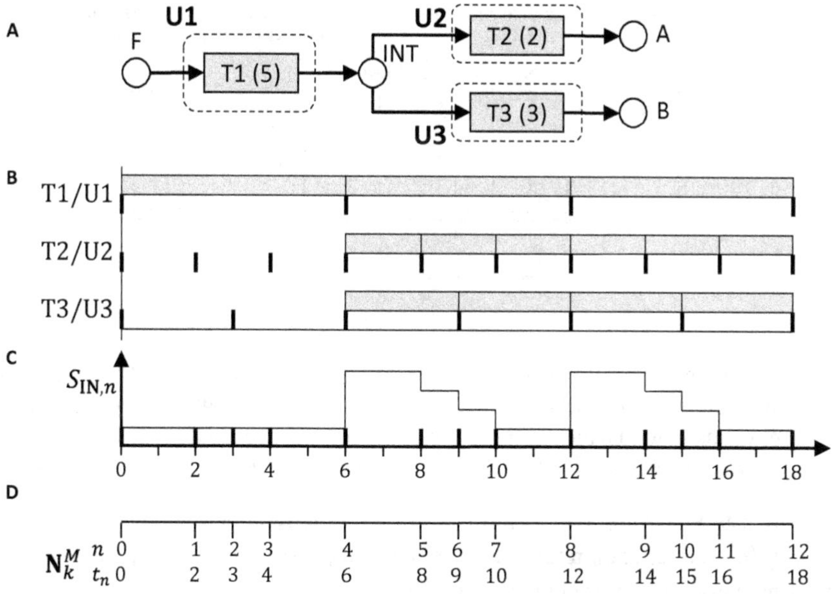

Figure 13.19 Material grid generation. (A) Network used for illustration (processing times are given in parentheses). (B) Gantt chart of an example solution embedded on task/unit grids. (C) Inventory profile of intermediate INT, and points in grid of INT. (D) Time axes: the top axis shows absolute time; the bottom axis shows time points in the grid of INT, along with corresponding timing of points.

Interestingly, the multigrid model consists of exactly the same types of constraints present in the STN-based model presented in Chapter 7, but these constraints are expressed for different subsets of points. A unit can process only one batch at a time:

$$\sum\nolimits_{i \in \mathbf{I}_j} \sum\nolimits_{n' \in \mathbf{N}_{ijn}^A} X_{ijn'} \leq 1, \quad j, n \in \mathbf{N}_j^U, \tag{13.34}$$

where $\mathbf{N}_{ijn}^A = \left\{ n' \in \mathbf{N}_{ij}^{TU} : t_n - \tau_{ij} < t_{n'} \leq t_n \right\}$ includes the points in $\mathbf{N}_{ij}^{TU}$ at which if a batch of task i starts, then it is not finished by point n; and thus, no other batch can be assigned to start on j at point n. Note that the number of variables included in each assignment constraint depends not only on the processing times of the tasks in $\mathbf{I}_j$, but also on the grid of each task. Thus, two tasks with the same processing time may have different number of binaries included in (13.34) when expressed for different time points. Also, the number of binaries for the same task may vary, as the step size between points in $\mathbf{N}_{ij}^{TU}$ changes. Figure 13.20 illustrates how different step sizes for different tasks as well as the time-varying changes in task grids impact the number of variables included.

Batchsizes are constrained as follows:

$$\beta_j^{MIN} X_{ijn} \leq B_{ijn} \leq \beta_j^{MAX} X_{ijn}, \quad i, j, n \in \mathbf{N}_{ij}^{TU}. \tag{13.35}$$

The logic behind the material balance is the same as in (13.3), but certain conditions should be checked. Specifically, the production term should include material produced

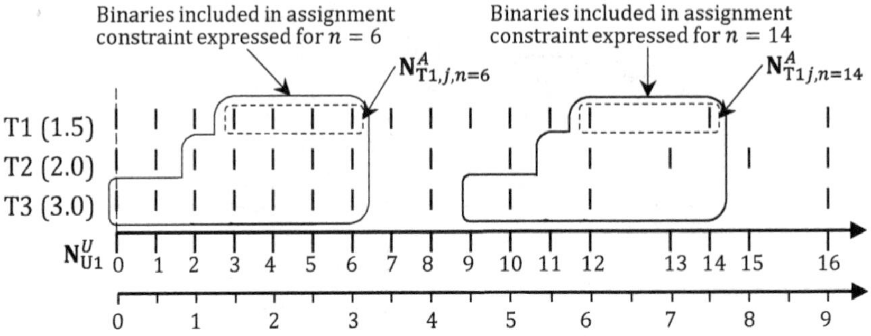

Figure 13.20 Inclusion of different number of binary variables in the LHS of the assignment constraint; variables included in the assignment constraint written at two different points shown inside shape in solid line; corresponding $\mathbf{N}^A_{ijn}$ sets also shown (tasks T1, T2, and T3 are carried out in the same unit; processing times given in parentheses next to each task).

by a task starting at or before $t_n - \tau_{ij}$ but strictly after $t_{n'} - \tau_{ij}$, where $n' = n_k^-(n)$ and $n_k^-(n)$ is a function returning the point immediately preceding point n in set $\mathbf{N}_k^M$, while the consumption or delivery occurs exactly at n:

$$I_{kn} = I_{kn'} + \sum_{\substack{i \in \mathbf{I}_k^+ \\ j \in \mathbf{J}_i}} \sum_{n'' \in \mathbf{N}_{ijn}^{MBP}} \rho_{ik} B_{ijn''} + \sum_{\substack{i \in \mathbf{I}_k^-, j \in \mathbf{J}_i: \\ n \in \mathbf{N}_{ij}^{TU}}} \rho_{ik} B_{ijn} + \xi_{kn}, \ k, n \in \mathbf{N}_k^M, \ n' = n_k^-(n), \tag{13.36}$$

where $\mathbf{N}_{ijn}^{MBP} = \{n'' \in \mathbf{N}_{ij}^{TU} : t_{n'} < t_{n''} + \tau_{ij} \leq t_n\}$ includes the points at which a batch of task i can start on unit j and produce material before point n. Alternatively, the point preceding n can be defined as n': $t_{n'} = t_n - \delta_{kn}^-$, where δ_{kn}^- is the length of the period between point $n \in \mathbf{N}_k^M$ and the immediately preceding point in $\mathbf{N}_k^M$ (see Figure 13.21). Note that tasks starting at different time points can finish between the same two points, so the second term on the RHS adds the production of all tasks that can finish between time points n' and n (i.e., during $(t_{n'}, t_n]$). Figure 13.21 shows that, if a task finishes between time points, then I_{kn} may underestimate the actual inventory (see inventory differences during $[2.5, 3.0]$).

In single uniform-grid models, processing times are rounded up, which means material is produced before it is included in the inventory variable and could violate material storage constraints. The multigrid model can enforce the inventory capacity constraint during all times by using unrounded processing times. This is accomplished by adding the amount produced strictly between time point n' and the next point to the inventory at point n':

$$I_{kn'} + \sum_{i \in \mathbf{I}_k^+, j \in \mathbf{J}_i} \sum_{n'' \in \mathbf{N}_{ijn}^{SC}} \rho_{ik} B_{ijn''} \leq \gamma_k^M, \quad k, n' \in \mathbf{N}_k^M, \tag{13.37}$$

where $\mathbf{N}_{ijn}^{SC} = \{n \in \mathbf{N}_{ij}^{TU} : t_{n'} < t_{n''} + \tau_{ij} < t_{n'} + \delta_{kn'}^+\}$ includes the points at which if a batch of task i starts, then it will end between n' and the next point in the grid of material k; and $\delta_{kn'}^+$ is the time between point $n' \in \mathbf{N}_k^M$ and the point immediately following n' in $\mathbf{N}_k^M$.

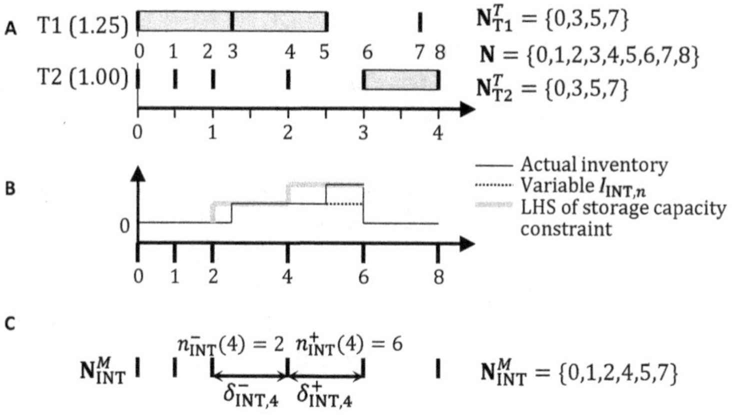

Figure 13.21 Inventory level calculations: intermediate INT is produced by T1 and consumed by T2. (A) Schedule with two batches of T1 and one batch of T2 (processing times given next to task names in the axis of the Gantt chart); a superset of grid points and task grids is also given. (B) Inventory-level profiles of INT calculated by different methods. (C) Points of material grid and illustration of associated parameters and functions.

Note that the LHS of (13.37) represents the maximum inventory level during interval $[t_{n'}, t_n)$, where $n = n_k^+(n')$ is the point immediately following n' in $\mathbf{N}_k^M$. The only difference compared to the first two terms in the RHS of (13.36) is that it does not include production at t_n because this production is added to the inventory at point n. The difference between the actual inventory, the value of variable I_{kn} calculated in (13.36), and the LHS of (13.37) is illustrated in Figure 13.21.

All the objective functions considered previously using models based on a common discrete grid can also be addressed using the model based on a multiple grids, after small modifications, similar to the ones necessary for (13.34) through (13.37), are made.

The model described in this subsection consists of exactly the same four types of constraints comprising models employing a common uniform discrete time grid: unit-task allocation (13.34), which is a clique constraint; unit capacity, (13.35), which is a variable lower/bound constraint; material balance (13.36), which is a flow-balance constraint; and storage capacity (13.37). This means that the proposed formulation is tight and that solution methods that have been proposed for traditional discrete time models can be readily applied. Most importantly, the synchronization between the different grids is accomplished through the expression of all constraints using carefully selected logic conditions based on the timing of time points (t_n), which are known prior to optimization. These exact same conditions are used to define sets $\mathbf{N}_{ijn}^A$, $\mathbf{N}_{ijn}^{MBP}$, and $\mathbf{N}_{ijn}^{SC}$ to express the assignment (A), production term in material balance (MBP), and storage capacity (SC) constraints, respectively.

13.4.6 Types of Time Grids

Time-grid-based models have been traditionally classified as *discrete* or *continuous time* models. While nonuniform discrete grids have been used in specific scheduling applications, the general discrete time models employ a uniform and common grid for all tasks, units, and materials. On the other hand, various continuous time models have been proposed employing a wide range of types of grids, including unit-specific grids, as well as a range of assumptions regarding the relative positioning of events with respect to the corresponding grids (see discussion in Sections 3.6.3 and 7.3.3). The approach in this section leads to the adoption of multiple unit-specific discrete grids, as well, as task- and material-specific discrete grids. Furthermore, it was shown that a grid with fixed time points does not have to be necessarily uniform. It may include unequally spaced points or, equivalently, periods of nonuniform duration. Thus, to be accurate, the term *discrete time* should be used to describe the broad class of models employing one or more grids with fixed points, and reserve the term *uniform* for the subset of these models employing one or more grids with fixed and equally spaced points. More generally, discrete time models should be further classified into single (common) and multiple grid models, and then into uniform and nonuniform grid models (see Figure 13.22). Note that (1) a single grid can be nonuniform and (2) all grids of a multigrid model can be uniform. Finally, we note that continuous time models always lead to solutions with periods of unequal length, so it is more accurate to use the term *nonuniform* to describe discrete time models only.

13.5 Discrete-Continuous Algorithm

As already discussed, one potential disadvantage of discrete time models is that the obtained solution may be inaccurate. In the previous chapter, we discussed a two-step method that employs a coarse discrete time grid to obtain an approximate solution, and then, with assignment and sequencing decisions fixed, employs an LP to obtain an accurate solution. The method is quite powerful, so the natural question is: Can a similar method be developed for problems in network environments? This is the topic of the present section. In Section 13.5.1, we present some preliminary concepts and the outline of the method. In Section 13.5.2, we present an algorithm for the mapping of a discrete

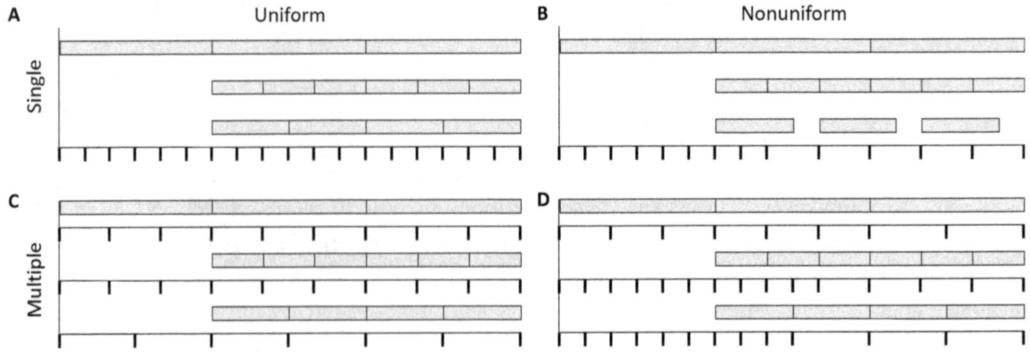

Figure 13.22. Modeling of time using discrete time grids. (A) Single uniform grid. (B) Single nonuniform grid. (C) Multiple uniform grids. (D) Multiple nonuniform grids.

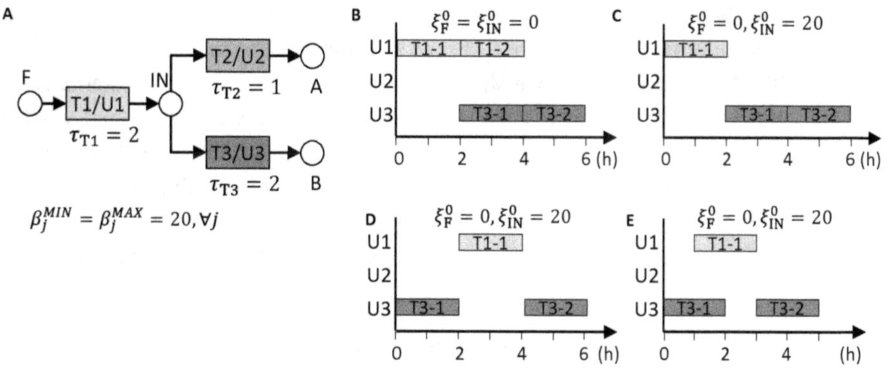

Figure 13.23 Challenge in identifying precedence constraints between batches in network environments. (A) Example network used for illustration. (B) Feasible schedules starting from no initial inventory for F and INT. (C) through (E). Alternative feasible schedules of instance with 20 units of initial inventory of IN.

time solution onto multiple grids, which is a new component with respect to the method we discussed in the previous chapter. In Section 13.5.3, we present the LP model, and in Section 13.5.4 we discuss several extensions.

13.5.1 Preliminaries and Outline

What makes the development of a similar algorithm more challenging in the case of network environments is that there are no direct precedence constraints between executed batches. The three reasons for that are that (1) the number of executed batches are unknown prior to optimization, (2) different tasks may share one or more materials as their inputs or outputs, and (3) multiple batches are executed for each task. These reasons make it impossible to predefine precedence constraints that have to be satisfied in any feasible solution. Furthermore, even if we were given a set of batches, unit-batch assignment and batch–batch sequencing decisions, generating a solution, using continuous time, that satisfies all material balances is not trivial.[15]

Figure 13.23 illustrates that precedence constraints are difficult to establish even in the case where batchsizes are fixed, tasks produce and consume a single material, and the first-stage solution includes batches of only two tasks. Figure 13.23A shows the example network used for illustration; note that batchsizes are fixed and same across tasks. A feasible schedule starting from no initial inventory for F and INT is shown in Figure 13.23B, where batch T1–1 should be followed by T3–1, and T1–2 by T3–2; that is, precedences T1–1 → T3–1 and T1–2 → T3–2 should be enforced. A feasible schedule of an instance with 20 units of initial inventory of IN is shown in Figure 13.23C; batch T3–1 follows T1–1, although this is not necessary (i.e., T1–1

[15] In problems in sequential environments with no batching decisions, which is the problem we studied in the previous chapter, material balances are not taken into account; they are automatically satisfied if the precedence constraints are satisfied.

→ T3–1 should not be enforced); and batch T3–2 should follow T1–1. A different schedule for the case with $\xi_{IN}^{0} = 20$ is shown in Figure 13.23D, where T3–1 precedes T1–1, but this precedence should not be enforced. Finally, an alternative schedule for the case with $\xi_{IN}^{0} = 20$ is shown in Figure 13.23E, where T1–1 and T3–1 are executed in parallel. Note that establishing precedence relationships becomes harder when batchsizes can vary (can you think why?) and, obviously, when tasks consume and produce multiple materials.

To address these challenges, we develop a solution method that consists of three stages:[16]

(1) Obtain a solution using a discrete time MIP model.
(2) Generate continuous time material and unit grids and map key attributes of the first-stage solution onto these grids using a mapping algorithm.
(3) Obtain an accurate solution, potentially with different timing as well as batching decisions, using an LP.

The second-stage algorithm maps attributes of the discrete time solution onto the two types of continuous time grids: material- and unit-specific grids. The former are introduced to ensure that material balances are satisfied in the final solution; batches of tasks that produce or consume a material are mapped onto the corresponding grid. Note that a batch may be mapped onto multiple material-specific grids because a task may produce and consume multiple materials. The sequencing of batches consuming and producing each material is identified by the algorithm. Unit-specific grids are introduced to ensure that a unit carries at most one batch at a time; the batches assigned to each unit in the first-stage solution are mapped onto the unit grid. The algorithm identifies the assignment of batches to units and their relative sequence within each unit.

The mapping of a discrete time solution onto grids is illustrated in Figure 13.24. The solution used for illustration, based on the network introduced in Figure 13.23A, is shown in Figure 13.24A. The generation of the continuous time grid for material INT (produced by T1 and consumed by both T2 and T3) is shown in Figure 13.24B; note that six time points are introduced to map the production and consumption of INT and that if production and consumption occur at the same time in the discrete time solution, then, in the continuous time grid, the production point precedes consumption. Figure 13.24C shows the generation of the grid for unit U2, which includes three time points because three batches (of T2) are executed in the first-stage solution.

The batches mapped onto grids are assigned specific material or unit grid points. The assignment and sequencing decisions are stored in algorithmic sets and parameters and later used to ensure feasible material balance and unit utilization in the third-stage model. The former is achieved by enforcing batches producing a material to finish before the start of the next batch that consumes the material. The latter is achieved by

[16] In the previous chapter, we used the term *steps* to describe the two components of the algorithm. We did this to avoid confusion with the term *stage* used to describe a feature of the multistage environment. In the present chapter, since network environments are not defined in terms of stages, we use the more appropriate term *stage* to describe the components of the method.

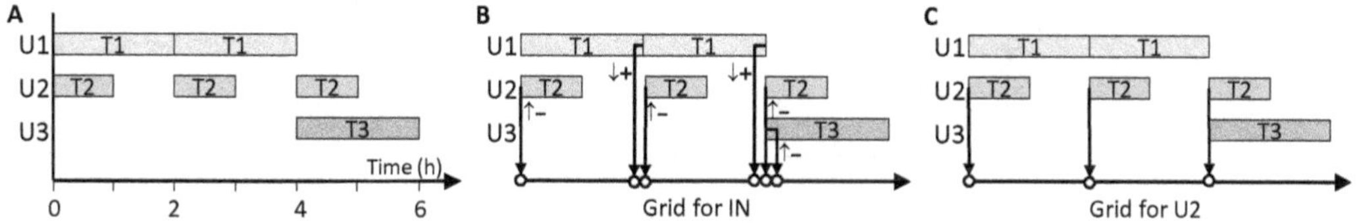

Figure 13.24 Material- and unit-specific grid generation. (A) Gantt chart of the discrete time (first-stage) solution used for illustration. (B) Material grid generated for IN: (↓+)/(↑−) denotes material production/consumption by batch mapped onto the grid. (C) Unit grid generated for U2.

ensuring that the batches assigned to a unit are executed in the sequence they were in the first-stage solution and do not overlap. Any discrete time model can be used to obtain a first-stage solution, so we will not discuss this stage.

13.5.2 Mapping

We introduce $m \in \mathbf{M}$ to denote the set of grid points on the unit- and material-specific grids. Index pair (j, m) is used to denote the mth grid point in the grid of unit j. The start of batches executed on j are mapped onto a point (j, m). Index pair (k, m) is used to denote the mth grid point in the grid of material k. We map the end of batches producing material k and the start of batches consuming material k. Subsets $\mathbf{M}_k^P/\mathbf{M}_k^C$ are introduced to denote the subset of points where material k is produced/consumed. In addition, we introduce the following algorithmic parameters:

- M_j^U: number of batches performed in unit j
- M_k^M: number of batches producing or consuming material k
- φ_{ikm}^1: = 1 if a batch of task i is mapped to material grid point (k, m)
- φ_{jkm}^2: = 1 if a batch mapped to material grid point (k, m) is performed in unit j
- $\varphi_{kmm'}^3$: = 1 if point m, where material k is produced is before point m' where it is consumed
- $\varphi_{ijkmm'}^4$: = 1 if batch of task i is mapped to unit grid point (j, m) and material grid point (k, m)

The assignment of batches to unit and material grids is stored in parameters φ_{ikm}^1, φ_{jkm}^2, and $\varphi_{ijkmm'}^4$, while the relative sequence of batches assigned to a grid is stored using index m. Given a batch of task $i \in \mathbf{I}_k^-$, the start of which is mapped on the grid of material k, we define its *preceding batches* as the batches that produce material k before the given batch starts. The batches of tasks $i \in \mathbf{I}_k^-$, whose start, mapped onto the grid of material k, is after the end of a batch of $i \in \mathbf{I}_k^+$, also mapped onto the same material grid, are termed the *succeeding* batches. Information regarding the sequencing of preceding and succeeding batches is stored in parameter $\varphi_{kmm'}^3$.

Figure 13.25 illustrates the generation of two unit and three material grids for a partial solution to an instance defined using the network in Figure 13.23A. The values of the algorithmic parameters (φ_{ikm}^1, φ_{jkm}^2, $\varphi_{kmm'}^3$, $\varphi_{ijkmm'}^4$) and subsets ($\mathbf{M}_k^P$, $\mathbf{M}_k^C$) used for mapping are also given. The calculation of the aforementioned parameters can be

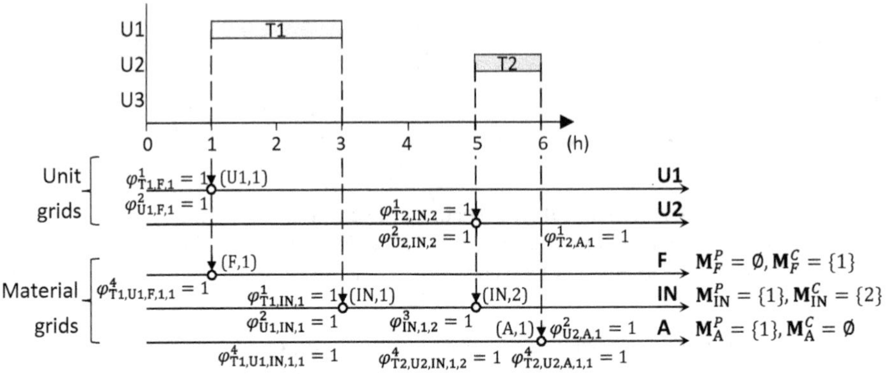

Figure 13.25 Illustration of a mapping algorithm: a Gantt chart of first-stage solution is shown in the top panel; corresponding unit and material grids are shown in the bottom panel (parameters used for mapping and subsets of material grid points are also given).

carried out, algorithmically, in a fraction of a second once a first-stage solution is available. The reader interested in the details of such an algorithm can find more information in Section 13.6.

13.5.3 Third-Stage Linear Programming Model

The model used in the third stage has a fixed number of batches with predetermined batch-unit assignments and sequencing of batches in each unit. However, the timing of batch execution and the batchsizes are optimization variables, that is, they can be different from the corresponding values in the first-stage optimal solution, which in turn means that the inventory profiles can also be different.

The following variables are used in the third-stage LP model:

- $B^C_{km} \in \mathbb{R}_+$: batchsize of batch mapped to material grid point (k, m)
- $I^C_{km} \in \mathbb{R}_+$: inventory level of material k at grid point (k, m)
- $T^U_{jm} \in \mathbb{R}_+$: timing of unit grid point (j, m)
- $T^M_{km} \in \mathbb{R}_+$: timing of material grid point (k, m)

First, the model consists of a set of timing constraints enforcing precedence relationships that are consistent with the first-stage solution. Given a material grid, the finish time of a preceding batch should be less than or equal to the start time of its successors:

$$T^M_{km} \leq T^M_{km'}, \quad k, m, m' : \varphi^3_{kmm'} = 1. \tag{13.38}$$

The sequence of batches processed on the same unit is enforced by

$$T^U_{jm} + \left(\tau^F_{ij} + \tau^V_{ij} B^C_{km'}\right) \leq T^U_{j,m+1}, \quad i, j, k, m, m' : \varphi^4_{ijkmm'} = 1, \tag{13.39}$$

which accounts for variable processing times, if necessary. The following two equations enforce that the start of a batch is the same across all grids the batch is assigned to:

$$T^{U}_{jm} = T^{M}_{km'}, \quad i,j,k,m,m' \in \mathbf{M}^{C}_{k} : \varphi^{4}_{ijkmm'} = 1 \tag{13.40}$$

$$T^{U}_{jm} = T^{M}_{km'} + \left(\tau^{F}_{ij} + \tau^{V}_{ij} B^{C}_{km'}\right), \quad i,j,k,m,m' \in \mathbf{M}^{P}_{k} : \varphi^{4}_{ijkmm'} = 1, \tag{13.41}$$

with (13.40)/(13.41) written for batches consuming/producing material k.

Second, batchsize variables are constrained to be within the operating range of the corresponding unit,

$$\beta^{MIN}_{j} \leq B^{C}_{km} \leq \beta^{MAX}_{j}, \quad j,k,m : \varphi^{2}_{jkm} = 1, \tag{13.42}$$

and should also be equal across the grids the same batch is assigned to:

$$B^{C}_{km'} = B^{C}_{k'm''}, \quad i,j,k,k',m,m',m'': \varphi^{4}_{ijkmm'} = \varphi^{4}_{ijk'm''} = 1, \quad k < k'. \tag{13.43}$$

Third, the inventory balance equation calculates the inventory at a point as the sum of the initial inventory plus the net production up to that point:

$$I^{C}_{km'} = \xi^{0}_{k} + \sum_{\mathbf{M}^{P}_{k} \ni m' < m} \sum_{i \in \mathbf{I}^{+}_{k} : \varphi^{1}_{ikm'} = 1} \rho_{ik} B^{C}_{km'} + \sum_{\mathbf{M}^{C}_{k} \ni m' < m} \sum_{i \in \mathbf{I}^{-}_{k} : \varphi^{1}_{ikm'} = 1} \rho_{ik} B^{C}_{km'}, \quad k, m \in \mathbf{M}^{C}_{k}. \tag{13.44}$$

Note that (13.44) is written only for grid points at which consumption occurs (i.e., $m \in \mathbf{M}^{C}_{k}$) to avoid negative inventories. (Can you see why?) If the total demand, ξ^{TOT}_{k}, for a final product is due at the end of the horizon, then demand satisfaction is enforced by

$$I^{C}_{k,m=M^{M}_{k}} \geq \xi^{TOT}_{k}, \quad k. \tag{13.45}$$

Finally, the third-stage LP model can accommodate various objective functions. For makespan minimization, we have

$$\min MS \tag{13.46}$$

subject to,

$$MS \geq T^{M}_{km}, \quad k, m \in \mathbf{M}^{P}_{k}. \tag{13.47}$$

For profit maximization, we have

$$\max \sum_{k} \pi_{k} I^{C}_{k,m=M^{M}_{k}} - \sum_{i,j,k} \sum_{m \in \mathbf{M}^{P}_{k} \cap \mathbf{M}^{12}_{ijk}} \frac{1}{\left|\mathbf{K}^{+}_{i}\right|} \left(\gamma^{F}_{ij} + \gamma^{V}_{ij} B^{C}_{km}\right), \tag{13.48}$$

where $\mathbf{M}^{12}_{ijk} = \{m : \varphi^{1}_{ikm} = \varphi^{2}_{jkm} = 1\}$. The division by the cardinality of $\mathbf{K}^{+}_{i}$ in the second summation (cost term) is necessary to negate the effect of double counting the cost of batches mapped to multiple material grids. For cost minimization problems, subject to demand constraints, (13.48) can be modified by removing the first summation (revenue term).

13.5.4 Extensions

Intermediate Shipments. If intermediate deliveries and orders are present, then they can be readily considered in the first-stage model. In terms of material grids, intermediate deliveries and orders can be simply treated as batches of dummy tasks producing and consuming, respectively, the corresponding materials. The only differences are that these batches (1) are not assigned to any units; (2) have zero processing times; and (3) are fixed, that is, their batchsizes and timing are not subject to optimization. Otherwise, they are treated as other batches of production and consumption tasks, and thus the precedence constraints (identified by the algorithm) are enforced in the third-stage LP model.

Limited Shared Utilities. We have seen that shared utilities can be readily addressed by discrete time models used in the first stage. If we limit the discussion to the case where utilities are consumed at the beginning of a task and produced at its end, then utilities can, in principle, be treated as materials, except that the type and amount of utility *consumed* and *produced* (using the RTN terminology in Chapter 7) by a given task are the same. Thus, adopting the ideas behind the generation of material-specific grids, we introduce utility-specific grids onto which batches of tasks that consume and produce utilities are mapped. The mapping algorithm can be modified to identify the relative sequence of utility consumption and production on each utility grid, which is then used in the third-stage LP model to ensure feasible utility balance. Furthermore, time-varying utility availability can be handled similarly to intermediate material shipments: increase/decrease in resource availability are modeled as utility production/consumption by batches of dummy tasks that only produce/consume the corresponding utilities.

First-Stage Model Infeasibility. Given a feasible instance (i.e., an instance for which a feasible solution exists if the accurate processing times are used), the use of approximate processing times in the first stage leads to the formulation of a feasible MIP model for most problem classes. In profit maximization problems with no orders or orders that can be met easily (so excess capacity is used for additional sales), the first-stage problem is always feasible. In makespan minimization problems, which typically have no deadlines, the first-stage problem is also always feasible provided that a sufficiently long horizon is considered. The first-stage problem can become infeasible when a set of orders with tight deadlines have to be met at, say, minimum cost. One way to address this is by replacing hard deadlines with due dates and introducing backlogs, something which is in fact often necessary in practice regardless of the model employed. An alternative approach, especially useful when tight deadlines have to be considered, is to delay the deadlines and use a longer scheduling horizon, thereby allowing the identification of first-stage solutions that, although they violate the original due dates, will lead to feasible solutions in the third stage, when the actual (shorter) processing times are used. In summary, all problem classes and instances can be addressed through either backlog introduction or horizon and deadline relaxation.

Model Parameters. The method offers flexibility in the selection of model parameters, namely, the discretization step length (δ) and the horizon relaxation (η). The choice of δ impacts the size of the first-stage model, which is the most computationally expensive stage. A large δ leads to faster solution time, but it might result in a less accurate solution; while a small δ leads to more accurate solutions, but results in models that are more difficult to solve. The choice of η is more relevant in profit maximization and cost minimization problems. When maximizing profit, the quality of the first-stage solution determines the quality of the third-stage solution, since no batches are added or removed in the third stage. High-quality first-stage solutions can potentially be missed because of the overestimated processing times. By relaxing the horizon, discretization errors are offset, allowing high-quality solutions (with more batches) to be found. In the case of cost minimization, appropriately chosen η allows the first-stage model to find the solution with the batch-unit assignments with the lowest cost, which may be infeasible if the original horizon is used. Nevertheless, finding the right horizon relaxation is critical, and nontrivial, as excessive relaxation leads to infeasible solutions. Systematic methods for the selection of these two model parameters have been proposed (see discussion in Section 13.6).

First-Stage Solution Pool. The proposed method has been shown to lead to dramatic computational improvements (three to four orders of magnitude), especially for hard large-scale instances. However, it has one limitation: it is not guaranteed to return the best[17] solution. This is because if the *optimal* number of batches and assignment and sequencing decisions are not identified in the first stage, then the third-stage LP cannot return the best solution. One approach to mitigate this limitation is to generate multiple first-stage solutions (solution pool). The obtained solutions can then be used to generate multiple third-stage solutions, in seconds, and the best among all third-stage solutions can be chosen. Using solution pools significantly improves the possibility of finding a (near)-best solution. This method takes advantage of the fact that the second and third stage of the discrete-continuous algorithm (DCA) are computationally inexpensive, and that solution pools can be efficiently generated using commercial MIP solvers.

13.6 Notes and Further Reading

(1) The preprocessing and tightening methods discussed in Section 13.2 are based on the methods proposed by Velez et al. [1]. The original preprocessing algorithm includes extensions to yield tighter bounds in networks with recycle materials and nested loops. Speedups of three orders of magnitude can be achieved in certain problem classes.

(2) The approach of Velez et al. [1] was extended and applied to a number of continuous time models by Merchan et al. [2]. The authors also showed that alternative constraints can be developed depending on the variables used (recall that in continuous time models, both starting and ending binary variables are typically employed). The methods were shown to be effective, leading again to order-of-magnitude computational improvements.

[17] We use the term *best* to describe the best solution to the original instance, as opposed to the *optimal* solutions found by the first (MIP) and third (LP) stage models.

(3) The approach of Velez et al. [1] was extended to address problems in facilities that consist of network and sequential subsystems in [3].

(4) A preprocessing algorithm based on forward propagation, the idea of which is discussed in Section 13.2.5, was proposed by Merchan and Maravelias [4]. In addition to initial inventories, the length of the horizon is also taken into account to calculate tighter bounds. Interestingly, the tightening methods described in the previous chapter can be viewed as a special case of the methods described in [4]. Recall that the tightening constraints discussed, for example, in Sections 12.2.1 and 12.2.2, enforce upper bounds on linear combinations of assignment binaries based on, essentially, the available processing time of a unit.

(5) Other tightening methods have been developed by Burkard and Hatzl [5] and Janak and Floudas [6]. The former are rather fast but do not consider the minimum unit capacity (β_j^{MIN}), so they lead to weaker constraints. The latter result in tight bounds but require the solution of a series of MIPs that are computationally expensive; they were shown to be effective for only one continuous time model.

(6) In problems with constant consumption coefficients, ρ_{ik}, material demand ω_k can be readily calculated by the bound on task production μ_i^1 via (13.9). If the conversion coefficients are variable, bounded within a range $\left[\rho_{ik}^{MIN}, \rho_{ik}^{MAX}\right]$, then preprocessing and tightening can still be done, based on a conservative bound of the range [7].

(7) In blending problems, there are typically no lower bounds on the consumption of a given stream toward a given product; for example, a stream may have zero contribution toward satisfying demand for a product at time point n, but a positive contribution at a later point $n' > n$. Thus, no bounds can be generated based on backward propagation of demand amounts. However, property specifications coupled with demand can be used to generate bounds on the demand of specific streams [8].

(8) The idea of decomposing a scheduling problem into subproblems with, at a first level, fixed number of batches, was discussed first in Ferris et al. [9], in the context of problems with batching decisions in sequential environments. In summary, the original problem was partitioned into subproblems with different (constraints on) assignment decisions; then, at a second level, each such subproblem could be further partitioned into subproblems with different assignment decisions; and finally, at a third level, hard second-level subproblems could be further decomposed into subproblems with constraints on a subset of sequencing decisions. The decomposition algorithm was implemented on a grid of computers.

(9) Motivated by the work of Ferris et al. [9], Velez and Maravelias developed a similar decomposition algorithm for models developed to address problems in network environments [10]. The three main ideas were that (1) MIP subproblems will be generated by branching on a constraint that lower/upper bounds $\sum_i X_{ijn}$; (2) the subproblems will be solved at different cores of a multicore computer; and (3) subproblems are used in an outer branch-and-bound algorithm, that is,

nonpromising subproblems are pruned, whereas promising subproblems can be further decomposed.

(10) The presentation of the reformulation methods in Section 13.3 is based on the methods first developed by Velez and Maravelias [11]. The underlying idea was identical to the one in [10] (branch on the number of batches of a task carried out in a unit rather than X_{ijn} variables), but performed automatically by the MIP solver through the introduction of N_{ij} (or Z_{ijm}) variables. Additional branching strategies were also discussed, and computational enhancements of two to three orders of magnitude were reported.

(11) The reformulations discussed in Section 13.3 can be readily implemented in continuous time models, as shown by Merchan and Maravelias [12]. However, since the number of different mathematical solutions in terms of X_{ijt} variables is not as large as in discrete time models (why?), the computational enhancements are smaller.

(12) Reformulations based on other variables (e.g., number of setups carried out in a unit) are discussed in Velez et al. [8].

(13) The discussion in Section 13.4 is based on the work of Velez and Maravelias [13], where the interested reader can find the actual algorithms for generating (1) exact unit-specific grids; (2) approximate unit-specific grids, based on user-defined parameters, μ, ν, and α_{ij}; and (3) material-specific grids, given all unit grids.

(14) Velez and Maravelias proposed a general framework for the formulation of models based on multiple grids, addressing problems with (1) general shared resources, potentially with time-varying cost and availability; (2) no or limited intermediate storage; and (3) changeovers [14]. A series of theoretical results regarding the feasibility and optimality of the solutions obtained by the multigrid models was also given.

(15) The idea behind the discrete-continuous algorithm presented in Section 13.5 is the same as the idea used to develop the method described in the previous chapter [15].

(16) The discussion in Section 13.5 is based on the methods proposed in [16] where the interested reader can find the algorithms for calculating the parameters (φ^1_{ikm}, φ^2_{jkm}, $\varphi^3_{kmm'}$, $\varphi^4_{ijkmm'}$) and subsets ($\mathbf{M}^P_k$, $\mathbf{M}^C_k$) used for mapping. The paper also describes extensions, in all three stages, necessary to address problems with (1) intermediate deliveries and orders and (2) limited shared resources with time-varying availability.

(17) Systematic methods for the selection of model parameters that would result in the best performance of DCA, both in terms of solution quality and computational performance, have been proposed [17]. The method for selecting δ is based on the concept of the error evaluation function, which essentially measures the amount of discretization error introduced by the selected δ. The selection of η is based on an estimation of how much additional time is needed to offset the cumulative discretization errors in each unit.

(18) A generalized discrete-continuous algorithm for problems with a wide range of processing features and constraints often found in practice (setups, changeovers, different storage policies, storage of materials in processing units, multiple material transfers, etc.) was presented in Lee and Maravelias [18].

(19) Methods to generate multiple alternative solutions to the same instance were proposed by Lee et al. [19]. These methods can be employed in the first stage of DCA to yield solutions that can then be refined in stages two and three, and the best solution be chosen.

13.7 Exercises

(1) The goal of this exercise is to help you understand how the preprocessing algorithm described in Section 13.2 works. If developing code for the algorithm is challenging, you can skip part (a) and calculate the parameters manually.

- (a) Develop code for the algorithm described in Section 13.2.3, but without accounting for loops.
- (b) Find all the necessary data for the instances based on network N1 from Velez et al. [1].
- (c) Calculate, manually or using the code you developed in part (a), the parameters necessary to generate all valid inequalities described in Section 13.2.4 for a cost minimization instance based on network N1 over 40 h (problem class a); generate the corresponding constraints.
- (d) Solve the preceding instance using a discrete time STN-based model with (i) no tightening constraints; (ii) constraints (13.18) and (13.20), (iii) constraints (13.21) and (13.22); and (iv) all tightening constraints.
- (e) Repeat for the instances described in Velez et al. [1] as problem class c. In what ways are the calculated parameters different? Can you predict the impact of the different constraints before you run the corresponding models?

(2) Based on the data for N1 from Velez et al. [1], repeat the tasks described in Exercise 1 for the following:

- (a) Cost minimization instances over a horizon of 80 and 120 h.
- (b) Makespan minimization instances over a horizon of 40, 80, and 120 h.

(3) Based on the data for N4 from Velez et al. [1], repeat the tasks described in Exercise 1 for the following:

- (a) Cost minimization instances over a horizon of 40, 80, and 120 h.
- (b) Makespan minimization instances over a horizon of 40, 80, and 120 h.

(4) Write code for the complete algorithm in Section 13.2.3 (i.e., accounting for loops). Is it applicable to the instances from Velez et al. [1] based on networks N2 and N3? If yes, run it to calculate the necessary parameters for the cost minimization instance over a horizon of 120 h. Solve the four different STN-based models described in Exercise 1(d).

(5) Solve all instances in Problems (1) through (4) using your favorite continuous time STN-based model with (i) no tightening constraints; (ii) constraints (13.18) and (13.20); (iii) constraints (13.21) and (13.22); and (iv) all tightening constraints.

(6) Retrieve the discrete time STN-based model you developed to solve problems in Chapter 7. Using it as a starting point, develop three different model variants: (i) the original model (referred to as base model); (ii) the base model augmented by (13.29); and (iii) the base model augmented by (13.30) and (13.31). Using the preceding models (three in total), solve the following instances:
   (a) Instances defined in Problem (1) of Chapter 7.
   (b) Instances defined in Problem (2) of Chapter 7.
   (c) Instances defined in Problem (3) of Chapter 7.

(7) Repeat Exercise 6 using different branching priorities as described in Section 13.3. Additional information can be found in Velez and Maravelias [11].

(8) Find the data for the instances used by Velez and Maravelias [11] to perform computational testing of the reformulations and branching strategies. Solve the following instances (for horizon equal to 240 h):
   (a) Makespan minimization for Networks 2, 5, and 6.
   (b) Cost minimization for Networks 2, 5, and 6.
   (c) Profit maximization for Networks 2, 5, and 6

(9) Repeat Exercises 6 and 7 starting from an RTN-based discrete time model.

(10) Repeat Exercises 6 and 7 starting from your favorite STN-based continuous time model.

(11) Develop code for Algorithm 1 described in Section 13.4.2. Use it to define the grids for the motivating example in Section 13.4.2 (see Figures 13.14 and 13.15). Develop the MIP model described in Section 13.4.5. for a profit maximization instance over 120 h using arbitrary material prices.

(12) Reproduce the results for the first instance in section 7 of Velez and Maravelias [13].

(13) Reproduce the results in section 8.1 of Velez and Maravelias [13].

References

[1] Velez S, Sundaramoorthy A, Maravelias CT. Valid Inequalities Based on Demand Propagation for Chemical Production Scheduling MIP Models. *AIChE J.* 2013;59 (3):872–887.

[2] Merchan AF, Velez S, Maravelias CT. Tightening Methods for Continuous-Time Mixed-Integer Programming Models for Chemical Production Scheduling. *AIChE J.* 2013;59 (12):4461–4467.

[3] Velez S, Maravelias CT. Mixed-Integer Programming Model and Tightening Methods for Scheduling in General Chemical Production Environments. *Ind Eng Chem Res.* 2013;52 (9):3407–3423.

[4] Merchan AF, Maravelias CT. Preprocessing and Tightening Methods for Time-Indexed MIP Chemical Production Scheduling Models. *Comput Chem Eng*. 2016;84:516–535.
[5] Burkard RE, Hatzl J. Review, Extensions and Computational Comparison of MILP Formulations for Scheduling of Batch Processes. *Comput Chem Eng*. 2005;29 (8):1752–1769.
[6] Janak SL, Floudas CA. Improving Unit-Specific Event Based Continuous-Time Approaches for Batch Processes: Integrality Gap and Task Splitting. *Comput Chem Eng*. 2008;32(4–5): 913–955.
[7] Velez S, Merchan AF, Maravelias CT. On the Solution of Large-Scale Mixed Integer Programming Scheduling Models. *Chem Eng Sci.* 2015;136:139–157.
[8] Chen Y, Maravelias CT. Preprocessing Algorithm and Tightening Constraints for Multiperiod Blend Scheduling: Cost Minimization. *Journal of Global Optimization*. 2020; 77, 603–625.
[9] Ferris MC, Maravelias CT, Sundaramoorthy A. Simultaneous Batching and Scheduling Using Dynamic Decomposition on a Grid. *INFORMS Journal on Computing*. 2009;21 (3):398–410.
[10] Velez S, Maravelias CT. A Branch-and-Bound Algorithm for the Solution of Chemical Production Scheduling MIP Models Using Parallel Computing. *Comput Chem Eng*. 2013;55 (0):28–39.
[11] Velez S, Maravelias CT. Reformulations and Branching Methods for Mixed-Integer Programming Chemical Production Scheduling Models. *Ind Eng Chem Res*. 2013;52 (10):3832–3841.
[12] Merchan AF, Maravelias CT. Reformulations of Mixed-Integer Programming Continuous-Time Models for Chemical Production Scheduling. *Ind Eng Chem Res*. 2014;53 (24):10155–10165.
[13] Velez S, Maravelias CT. Multiple and Nonuniform Time Grids in Discrete-Time MIP Models for Chemical Production Scheduling. *Comput Chem Eng*. 2013;53:70–85.
[14] Velez S, Maravelias CT. Theoretical Framework for Formulating MIP Scheduling Models with Multiple and Non-Uniform Discrete-Time Grids. *Comput Chem Eng*. 2015;72:233–254.
[15] Merchan AF, Lee H, Maravelias CT. Discrete-Time Mixed-Integer Programming Models and Solution Methods for Production Scheduling in Multistage Facilities. *Comput Chem Eng*. 2016;94:387–410.
[16] Lee H, Maravelias CT. Combining the Advantages of Discrete- and Continuous-Time Scheduling Models: Part 1. Framework and Mathematical Formulations. *Comput Chem Eng*. 2018;116:176–190.
[17] Lee H, Maravelias CT. Combining the Advantages of Discrete- and Continuous-Time Scheduling Models: Part 2. Systematic Methods for Determining Model Parameters. *Comput Chem Eng*. 2019; 128: 557–573.
[18] Lee H, Maravelias CT. Combining the Advantages of Discrete- and Continuous-time Scheduling Models. Part 3: General Algorithm. *Comput Chem Eng*. 2020;139:106848.
[19] Lee H, Gupta D, Maravelias CT. Systematic Generation of Alternative Production Schedules. *AIChE J*. 2020: e16926.

14 Real-Time Scheduling

The focus of the book so far has been on the development of models and solution methods to obtain a (near) optimal *predicted* schedule, that is, obtain a schedule that will be implemented in the future. While these two elements, modeling and solution, are critical and necessary toward the implementation of optimization-based methods for chemical production scheduling, they are not sufficient by themselves. Specifically, in an industrial setting the optimization model has to be solved repeatedly, in *real time,*[1] taking into account new information and disturbances. The goal of the present chapter is to provide some high-level understanding on how the optimization model should be modified and then used in real time, along with solution methods, to obtain a *real-time scheduling algorithm* that yields high-quality implemented schedules.

In Section 14.1, we explain why repeated optimization is necessary, introduce necessary notation, and present the overall framework we will use in the present chapter. In Section 14.2, we present a state-space scheduling formulation that offers a natural way to formulate the optimization model that is updated and solved in real time. In Section 14.3, we present the basic considerations and a general simulation-based framework for designing real-time scheduling algorithms, and close, in Section 14.4, with a discussion on how integration with other functions can offer *early* feedback leading to faster recourse and thereby better implemented schedules.

We use models and examples based on network environments, but all the ideas and methods are directly applicable to problems in sequential environments. Also, to keep the presentation simple, we do not consider utilities and special processing features such as storage in processing units, multiple and/or resource-constrained material transfers, and so on. We use models based on a single common discrete time grid with $\delta = 1$ h, unless otherwise specified.

[1] The term *real time* is also used for a different, closely related, problem, namely, real-time optimization (RTO). While both RTO and real-time scheduling are solved in real time, the former is, typically, a single-period (i.e., steady-state) optimization problem, whereas the latter is a multiperiod problem. In that respect, real-time scheduling is, conceptually, more similar to the optimization problem solved in model predictive control.

14.1 Motivation and Background

We first present two examples, in Sections 14.1.1 and 14.1.2, to illustrate why repeated real-time scheduling is necessary.[2] Next, we present general notation, in Section 14.1.3; and a general classification of approaches to real-time scheduling, in Section 14.1.4.

14.1.1 Uncertainty versus New Information

We consider the network shown in Figure 14.1 along with the associated process parameters. We first solve, at $n = t = 0$, a cost minimization problem with horizon $\eta = 16$ h to meet orders of 12 tons for A and B each due every 6 h starting at $n = t = 12$.[3] The resulting schedule is shown in Figure 14.2A. As this schedule is executed, it becomes known to the scheduler, at $t = 5$, that the amount of A due at $t = 12$ has changed from 12 to 15 tons. This triggers a reoptimization, assumed to be executed instantaneously, to obtain a new schedule with a horizon length $\eta = 11$ between $t = 5$ and $t = 16$, to react to the order change. The new solution, shown in Figure 14.2B, results in changes in the batchsizes of batches of T1 (starting at $t = 6$) and T2 (starting at $t = 10$), but no other changes.

If instead of the change in the order due at $t = 12$, the orders were to remain at their original values, but instead a reoptimization is carried out at $t = 5$ considering a 16 h horizon (from $t = 5$ to $t = 21$), the schedule shown in Figure 14.2C would be obtained. We see that considering a longer horizon (and thus accounting for a new order, due at $t = 18$), leads to a schedule between $t = 5$ and $t = 12$, which is more different from the original schedule in Figure 14.2A than the schedule obtained when reacting to uncertainty (order size) shown in Figure 14.2B. This simple example shows that accounting for new information as soon as it becomes available can be even more (or at least

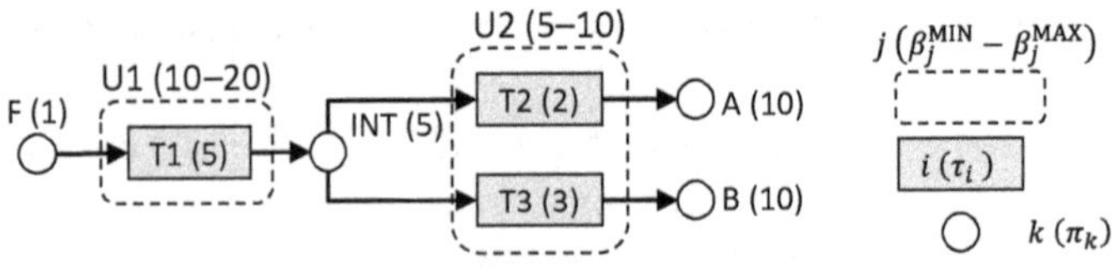

Figure 14.1 STN representation and parameters for motivating example.

[2] The term *repeated* is used here to refer to the process of performing optimization repeatedly, often in a periodic fashion, that is, every certain amount of elapsed time. While the process is periodic, for now, we use the term *repeated*, and not *periodic*, to avoid confusion with the term *periodic scheduling* used in Chapter 10.

[3] Since we use $\delta = 1$, the time point and the actual time are the same if we consider a single optimization. However, in the present chapter, we will consider repeated optimizations, carried out at different times, so the time points, at a given optimization model, will not coincide with the actual (absolute) time. We will continue to use n for the time points of the optimization *model*, solved at each iteration, and t to refer to the absolute time, which is used as common reference across optimizations.

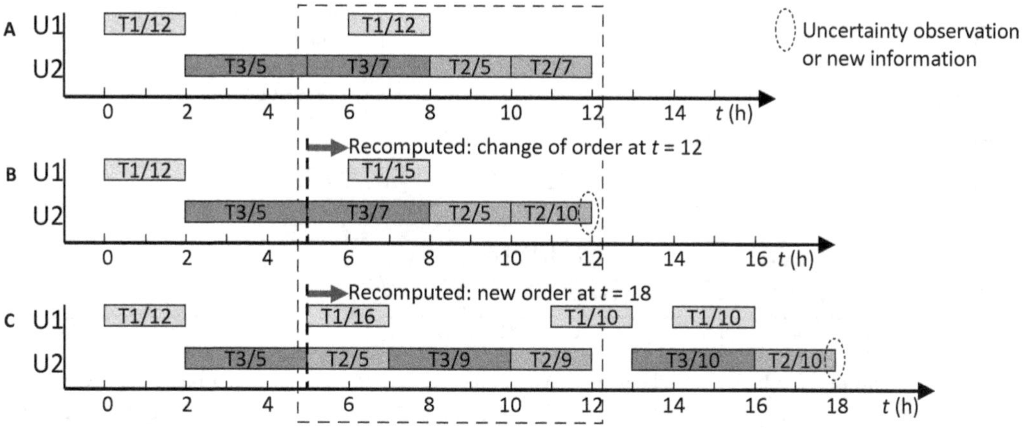

Figure 14.2 Computed schedules for motivating example; dashed box indicates time window where comparison across schedules is made. (A) Schedule obtained at $t = 0$. (B) Schedule obtained at $t = 5$, using horizon of 11 h, to react to change in the order of A at $t = 12$ (from 12 to 15 tons); only batchsize changes are observed. (C) Schedule obtained at $t = 5$ as a result of a reoptimization using a horizon of 16 h; both timing and batchsize changes are observed.

equally) important than accounting for uncertainty within the scheduling horizon, when revising scheduling decisions. This further suggests that repeated (periodically) reoptimization, even in the absence of major events that would render the previously obtained schedule infeasible or suboptimal, can lead to better implemented solutions. In the remainder of the chapter, we will refer to the process of reoptimizing in a periodic fashion as periodic reoptimization, periodic rescheduling, or periodic (real-time) scheduling.[4]

14.1.2 Event Triggered versus Periodic Rescheduling

In the previous example, we showed that rescheduling should be performed to react to the arrival of new information. However, since this new information was in the form of a new order, it can be argued that this was a disturbance or, more generally, an event that should trigger rescheduling. We now explore if rescheduling is necessary even if there are no trigger events. We consider the same network as in the previous example (Figure 14.1), and schedule for orders of 12 tons of A and B due at $t = 16$ with a horizon of 16 h, and no other orders due till at least $t = 40$.

Note that the order is within the considered horizon and hence the optimization (at $t = 0$) has the opportunity to take decisions in a timely manner to meet the order. For example, a feasible solution is to carry out one batch of T1 (of size 12 tons) and two batches of T2 and T3 each (each of size 6 tons), which require a total of 12 h. We

[4] In other words, in the present chapter, we will use the term periodic to refer to the periodic execution of an optimization, rather than the identification of a solution that can be executed periodically (i.e., the problem discussed in Chapter 10).

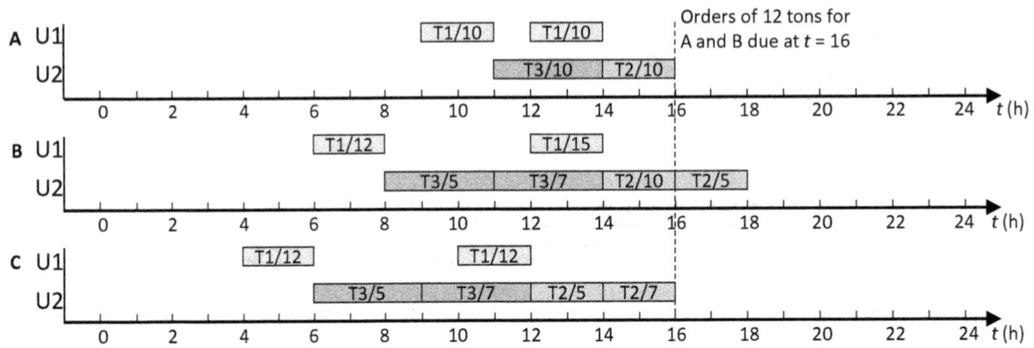

Figure 14.3 Implemented schedules to meet orders at $t = 16$. (A) Schedule obtained at $t = 0$, followed by no reoptimization (cost = $34,004). (B) Schedule obtained from reoptimization every 6 h (cost = $11,256). (C) Schedule obtained from reoptimization every 3 h (cost = $9,056).

assume there is no uncertainty, and since there are no new orders (after $t = 16$), there are no possible events to trigger rescheduling, thus we do not reschedule. Surprisingly, the resulting schedule, shown in Figure 14.3A, results in only partial order fulfillment, and hence has considerable backlog costs, if evaluated over a 24 h horizon. The cost of the resulting implemented 24 h schedule is $34,004.

Hoping to correct this puzzling situation, we decide to reschedule four times a day, that is, every 6 h, irrespective of any *trigger* events, with a horizon of 16 h for every computation. The resulting implemented schedule, based on three optimizations, is shown in Figure 14.3B. We observe that the quality of the schedule improves; the order is met, but part of it late. In addition, there is excess inventory of 3 tons of materials INT and A each, which is carried till the end of the day. The cost of the schedule is $11,256. Since rescheduling helped reduce the cost, we decide to reschedule faster, now eight times a day (every 3 h). The resulting implemented schedule is shown in Figure 14.3C. We see that with this faster rescheduling, the full order is met on time with an implemented schedule cost of $9,056. (Given this explanation, can you see why the schedule in Figure 14.3B leads to the production of three additional units of B?)

To explain the preceding observations, we study the solutions that we would obtain from rescheduling every hour. The Gantt charts for these solutions are shown in Figure 14.4. We see that in the first solution (iteration 0), the optimization assesses that the holding cost of inventory produced from early batches of T2 and T3, which would be necessary to meet the orders, outweighs the backlog costs of not meeting the orders fully. This is because the penalty for not meeting an order is due to accumulated backlog cost over time, but the horizon ends at $t = 16$ itself, so there is backlog cost only for one time period (due to terminal backlog costs in the model). Thus, the orders are not met fully. When we reoptimize at $t = 1$ (iteration 1), the horizon *advances* by 1 h, that is, we consider decisions between $t = 1$ and $t = 17$, which means that there is an additional time period beyond the due time. Thus, the cumulative backlog cost could have increased, leading to new batches executed to meet a larger fraction of the demand for product B. When reoptimization occurs at $t = 2$ (iteration 2), there are enough

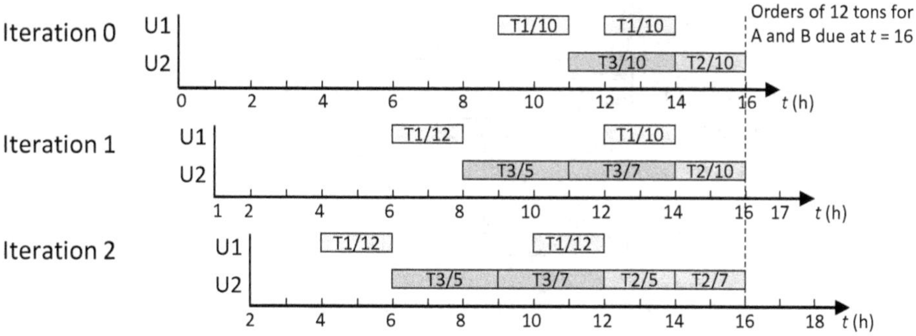

Figure 14.4 Solutions obtained from reoptimizations at $t = 0$, 1, and 2 (iterations 0, 1, and 2, respectively).

periods beyond the due date making the accruing of backlog costs disadvantageous, and thus the solution includes additional early batches to meet orders fully and on time. This schedule does not change further in any subsequent iterations; and the implemented schedule cost is $9,056.

Hence, we observed that purely event-triggered rescheduling is insufficient. Importantly, the consideration of new time periods even with no new orders can also require rescheduling. This is in spite of having a horizon length that is long enough to have the first event (the order) within its scope and having enough time to produce in accordance with this event. We emphasize that the discussion here is more general than what might appear to be a trivial inventory holding cost versus backlog cost issue. For example, if the network facility was such that byproducts were also made, then the interplay of inventory holding costs between different materials by themselves, even without backlogs, can also lead to revised decisions with the advancement of the horizon. Also, in profit maximization problems, the produced product mix may need revisions. Obviously, periodic rescheduling is even more critical in complex facilities where there are various interconnections and several ancillary decisions are considered simultaneously with the major scheduling decisions.

14.1.3 Notation

In general, decisions obtained from a finite horizon planning model are implemented in a setting where the system operates indefinitely and under uncertainty, hence, what the model returns as *optimal predicted* solution may not be optimal for the long-term operation of the actual system. Scheduling is an ongoing process, but a finite horizon must be used in any (optimization or other) model because it is computationally expensive to solve large long-term scheduling models and, additionally, information only for finite future is known with reasonable (un)certainty. As time passes by, more information becomes available, and this information should be accounted for as soon as possible to determine new decisions. Thus, real-time scheduling is a generalization of rescheduling, since it is based on a recomputation that is carried out not only upon the

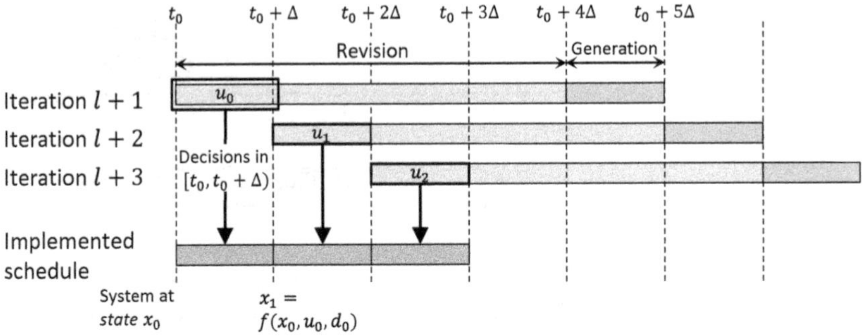

Figure 14.5 Real-time scheduling using a moving (rolling) horizon.

realization of trigger events but also periodically to consider new information (Figure 14.5). The nature of real-time scheduling requires the recomputation step to be carried out periodically, though nonperiodic but regular recomputation may also be acceptable. We note that the fact that a recomputation is carried out regularly does not necessarily imply that all the decisions are also changed at this frequency. In that respect, there are parallels between real-time production scheduling and process control. When real-time scheduling employs an optimization model solved in real time, it becomes analogous to model predictive control (MPC).

Following the nomenclature for MPC, at each iteration an *open-loop* (optimization) problem is solved. The resulting schedule from solving several open-loop problems, to account for disturbances (feedback), results in the implemented schedule which is also called the *closed-loop* schedule (or closed-loop solution). As opposed to MPC, wherein the prediction horizon can be longer than the control horizon, in real-time scheduling the two are typically identical and referred to as the scheduling horizon or just the horizon (η).[5] The time between two real-time scheduling iterations is analogous to the time-step size in MPC, hence, we also refer to it as the real-time scheduling time-step or just the time-step (Δ), which should not be confused with δ, which denotes the granularity of the time grid employed by the discrete time scheduling model. The length of time for which the closed-loop schedule is computed can be referred to as the *simulation time*. Analogous to the control input move penalties in MPC, there can be penalties (nervousness[6] costs) imposed in the objective function to minimize differences between schedules computed in consecutive iterations. The scheduling model solved in real time can be nominal, robust, stochastic, parametric, and so on. It is worth noting that real-time scheduling can also be carried out using dispatch rules, and in abstract terms is analogous to using nonoptimization-based approaches for control, such as,

[5] In most control applications, a constant input can be used beyond the control horizon (e.g., zero input, in terms of deviation variables or nominal period solution). Real-time scheduling is naturally applied in systems where short-term scheduling is necessary, and thus no periodic solution is available to be implemented beyond the control horizon.

[6] Plant *nervousness* is used to describe the operators' dislike of frequent changes to the deployed schedule.

proportional integral derivative (PID) control. For example, a dispatch rule such as "shortest-task-first" can be used when information rapidly becomes obsolete.

14.1.4 Approach Classification

The various approaches to real-time scheduling can be classified based on six distinguishing characteristics. The first one is whether uncertainty is explicitly considered at each computation, and if yes, how. Methods that employ a nominal (deterministic) model do not consider uncertainty though account for it through feedback. Methods that account for uncertainty, can be subdivided into stochastic programming, fuzzy programming, and so on. The second characteristic is the subset of uncertainties/disturbances a method handles, such as demand, processing times, yields, and so on. The third characteristic is what triggers a new computation. The trigger can be (1) the detection of an event, (2) a (periodic) predefined or online calculated time-step, or (3) a combination of the previous two, referred to as *hybrid* triggering. The time-step for recomputation is typically based on clock time, but can also be based on total equipment use time (variable time intervals); that is, recomputation occurs more often when units are heavily used as opposed to when they are relatively idle. The fourth characteristic is how each computation is performed, and specifically the technology that is employed. Deterministic methods can employ mathematical programming as well as heuristic and metaheuristic methods, while methods that account for uncertainty typically rely on mathematical programming. The fifth characteristic concerns the changes that are allowed in each iteration or, alternatively, the constraints enforced (often, but not always, to reduce nervousness). Methods vary from full recomputation, to partial recomputation, to local fixing (e.g., time shifting). Naturally, computational methods and allowed changes are coupled. Finally, methods can be classified in terms of problem classes they are applicable to (see classification in Chapter 1).

The first and fourth characteristics define essentially the type of model solved at each iteration, that is, the computational engine of the algorithm. The third and fifth characteristics define how often the model is solved and what revisions are allowed and/or favored. Finally, the second and sixth characteristics define, essentially, the system that is being studied. Figure 14.6 shows how the first four characteristics define the real-time scheduling algorithm, which can be further customized based on the characteristics of the specific production system.

14.2 State-Space Scheduling Model

The scheduling models we have presented so far were not necessarily designed with an emphasis on being natively ready for implementation in a real-time setting. Thus, they typically require many ad hoc (heuristic) adjustments to be able to represent and resolve a disturbance. To address this limitation, in this section we introduce a state-space scheduling formulation, which alleviates many of these issues. In Section 14.2.1, we present the general form of state-space models, often used in process control. In Section

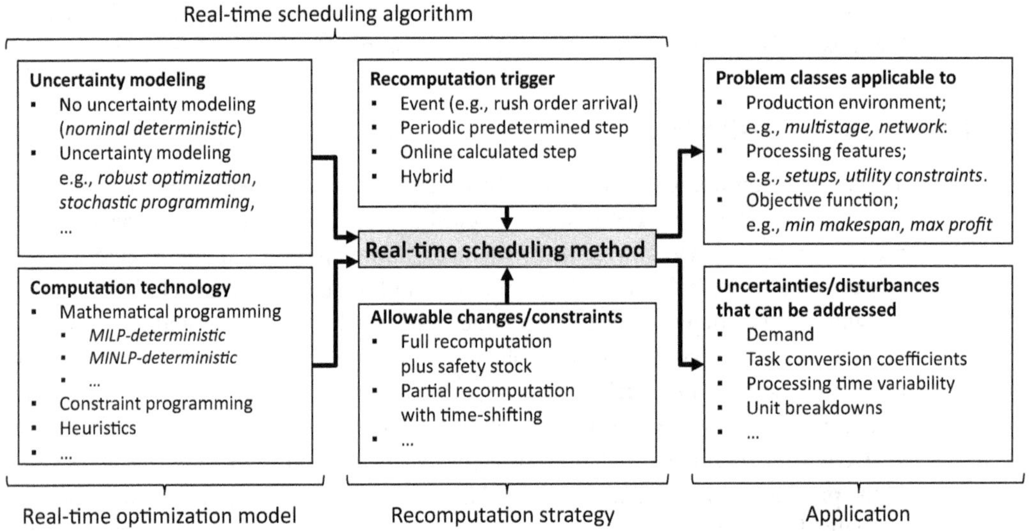

Figure 14.6 Framework for classification of real-time scheduling methods.

14.2.2, we reintroduce the basic STN-based model we studied in Chapter 7 and present its state-space counterpart. In Section 14.2.3, we model a range of common disturbances, and in Section 14.2.4, we discuss some extensions. In this subsection, we assume that the open-loop model is solved every δ h, that is, $\Delta = \delta$.

14.2.1 Preliminaries

State-space model formulations have been used in process control, and as optimization-based control and economic MPC are becoming the new standard, state-space models have become ubiquitous. In the most general form, a state-space based model can be written as $\frac{dx}{dt} = f(x, u, d)$, where x are the states, u are the manipulated inputs, and d are the disturbances.[7] Function $f(\cdot)$ is not theoretically restricted to the class of linear functions but is typically approximated as linear due to computational tractability considerations. The linear difference equation form for $f(\cdot)$ yields

$$x(n+1) = Ax(n) + Bu(n) + B_d d(n), \quad n, \tag{14.1}$$

where n is the index for time (discrete periods); and A, B, and B_d are state-space matrices.[8] The states x need not be associated with a physically identifiable entity in the plant; some can have a direct physical meaning, while others can be artificial

[7] Note that the term *state* is used here to denote something different from the state in the state-task network formulation. Here, it means the condition/status of the system – it is a profile over time – whereas in the STN representation, it refers to a material. The use of the term in the current chapter was the reason we chose to use the term *material* instead of *state* in this book.

[8] To stay consistent with most literature on state-space models, in the present subsection (only), we use index n in parentheses, and not as a variable subscript.

constructs to enable the modeling exercise. The output (measurements y) is related to the states and inputs as $y = h(x, u)$, where $h(\cdot)$ can be nonlinear, but is typically linear and can be written as a difference equation, $y(n) = Cx(n) + Du(n)$, where C and D are coefficient matrices.

The control optimization model has to satisfy plant physical constraints as well as other constraints that may be added to follow an operational strategy or enable better closed-loop properties. When linear, these constraints take the general form

$$E_x x(n) + E_u u(n) + E_d d(n) \leq 0, \quad n, \tag{14.2}$$

where E_x, E_u, and E_d are coefficient matrices. If there are any equality constraints, these can also be represented as two opposite inequality constraints. For example, the following constraints are equivalent:

$$(E_x x(t) + E_u u(t) + E_d d(t) = 0) \Leftrightarrow \left(\begin{bmatrix} E_x \\ -E_x \end{bmatrix} x(t) + \begin{bmatrix} E_u \\ -E_u \end{bmatrix} u(t) + \begin{bmatrix} E_d \\ -E_d \end{bmatrix} d(t) \leq 0 \right). \tag{14.3}$$

Finally, the objective function is written as follows:

$$z = \min_{u(0),\, u(1),\, \ldots,\, u(N-1)} g(x(0),\, u(0),\, x(1),\, u(1), \ldots,\, x(N-1),\, u(N-1)), \tag{14.4}$$

where N is the number of discrete time points/periods in the real-time optimization horizon.

14.2.2 Basic Model

We present a basic state-space model based on the STN representation introduced in Chapter 7. We use a common, uniform, discrete time grid, with points $n \in \mathbf{N} = \{0, 1, \ldots, N\}$, and periods of length δ. For brevity, we present the formulation for constant batchsizes ($\beta_{ij}^{MIN} = \beta_{ij}^{MAX} = \beta_{ij}$). There are two distinct features of this model. First, the complete *status* (state) of the plant can be interpreted solely from the variables (states) at that moment in time. This is made possible by *lifting* past actions/inputs (i.e., task start variables, $X_{ijn} \in \{0, 1\}$), which have a lagged effect on the *state* of the plant. Second, observed uncertainties are treated as disturbances and represented as parameters in the model. These two features, together, allow for the model to be kept identical in each real-time iteration without any ad hoc adjustments (due to observation of uncertainty). Thus, the model is in *real-time ready* form. In addition, due to the use of the state-space formulation, this model also happens to be a very suitable candidate for integration of scheduling and control.

To enable lifting of inputs, new task states (variables) X_{ijn}^{m} are introduced, where index m denotes the state of a batch of task i that started at time point n. Although this increases the number of variables, it is matched by an equal increase in the number of equations (the lifting equations), so no new degrees of freedom are introduced. When a

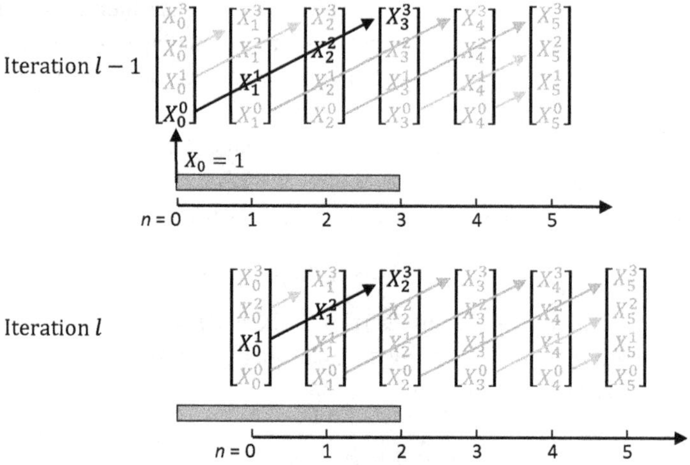

Figure 14.7 Illustration of lifting using a task with processing time equal to 3. Task states are shown for two real-time iterations, $l-1$ and l. Each iteration uses its own local time grid that is reset to start from 0. Lifting of past inputs enables determination of the complete status of the plant by looking at the states (variables) only at the current moment in time. Arrows show which variables are equal due to the lifting equations (in the absence of delays and unit breakdowns). Variables in black (gray) have a value of 1 (0).

batch starts, $m = 0$, and when it finishes, $m = \tau_{ij}$. The normal lifting and update of the state of a batch are written as follows:

$$X_{ijn}^{m} = \begin{cases} X_{ijn}, m = 0 \\ X_{ij,n-1}^{m}, m > 1 \end{cases} \quad i, j, n. \tag{14.5}$$

The lifted variables are defined only for $m \in \{0, 1, \ldots, \tau_{ij}\}$ because a *look-back* beyond τ_{ij} periods is not needed. The concept is illustrated in Figure 14.7, where we use index l to denote iterations.[9]

Using the new lifted variables, the assignment constraint ((7.4) in Chapter 7) can be rewritten as follows:

$$\sum\nolimits_{i \in \mathbf{I}_j} \left\{ X_{ijn} + \sum\nolimits_{m \in \mathbf{M}_i^*} X_{ijn}^{m} \right\} \leq 1, \quad j, n \tag{14.6}$$

or equivalently, given (14.5), as

$$\sum\nolimits_{i \in \mathbf{I}_j} \sum\nolimits_{m \in \mathbf{M}_i} X_{ijn}^{m} \leq 1, \quad j, n, \tag{14.7}$$

[9] In previous chapters, we used index l to denote batches of the same order (e.g., in Chapter 4) or utilities (e.g., in Chapter 7). To continue using letters for indices from $\{i,j,k,l,m,n,t\}$, in the present chapter only, we use index l to denote iterations. Recall, that, in the present chapter, we use STN representation, and thus batches, in the manner defined in sequential environments, are not used, and utilities are not considered in the present chapter.

where $\mathbf{M}_i = \{0, 1, \ldots, \tau_{ij} - 1\}$ and $\mathbf{M}_i^* = \mathbf{M}_i \backslash \{0\}$. The variable lower/upper bounding constraint of the batchsize, B_{ijn}, is replaced by the following constraint, because of the fixed capacity:

$$B_{ijn} = \beta_{ij} X_{ijn}, \quad i, j, n, \tag{14.8}$$

which can be used to project B_{ijn} variables out from the material balance ((7.6) in Chapter 7), which is now written as follows:

$$I_{k,n+1} = I_{kn} + \sum\nolimits_{i \in \mathbf{I}_k^+} \sum\nolimits_{j \in \mathbf{J}_i} \rho_{ik} \beta_{ij} X_{ijn}^{\tau_{ij}} + \sum\nolimits_{i \in \mathbf{I}_k^-} \sum\nolimits_{j \in \mathbf{J}_i} \rho_{ik} \beta_{ij} X_{ijn} + \xi_{kn} - S_{kn} \le \chi_k^M, \quad k, n, \tag{14.9}$$

where I_{kn} is the inventory level of material k during period n, and S_{kn} the corresponding extra sales.

Thus, the new STN-based state-space model consists of (14.5), (14.6) or (14.7), (14.9), and the objective function of choice. Note that all constraints are either in the form of (14.1) (i.e., the RHS contains variables indexed by time/period n only) or in the form of (14.3), (i.e., they contain variables indexed by time/period n only).

If there are no disturbances to impact the normal batch progression, and we assume that the real-time optimization is carried out every δ time units (i.e., $\Delta = \delta$), then the update of the status of the system follows:

$$_lX_{ij,n=0}^m = {}_{l-1}X_{ij,n=1}^m, \quad i, j, m \in \{1, 2, \ldots, \tau_{ij}\} \tag{14.10}$$

$$_lI_{k,n=0} = {}_{l-1}I_{k,n=1}, \quad k, \tag{14.11}$$

where variables $_lX_{ij,n=0}^m$ and $_lI_{k,n=0}$, which represent the initial state $(n = 0)$ of the system at iteration l, are calculated (offline, between optimizations) from the calculated (predicted) state of the system at previous iteration $(l - 1)$. The concept is illustrated in Figure 14.7.

14.2.3 Modeling of Disturbances

To express task delay and unit breakdown disturbances, new parameters are introduced:[10]

- $\dot{Y}_{ij}^m \in \{0, 1\}$: = 1 if a batch (of task i in unit j) currently at progress status m is delayed by δ.
- $\dot{Z}_{ij}^m \in \{0, 1\}$: = 1 if unit j, carrying out a batch (of task i) currently at progress status m breaks down.

These two parameters are used to update the status of currently executed batches (see Figure 14.8):

$$_lX_{ij,n=0}^m = {}_{l-1}X_{ij,n=0}^{m-1} - \dot{Y}_{ij}^{m-1} + \dot{Y}_{ij}^m - \dot{Z}_{ij}^{m-1}, \quad i, j, m \in \{1, 2, \ldots, \tau_{ij}\}. \tag{14.12}$$

[10] In the previous chapters, we used, exclusively, lowercase Greek letters for parameters. While disturbances are parameters (as opposed to variables) for the optimization, we use uppercase letters to differentiate them from the (nominal) parameters used to define an instance and the nominal model.

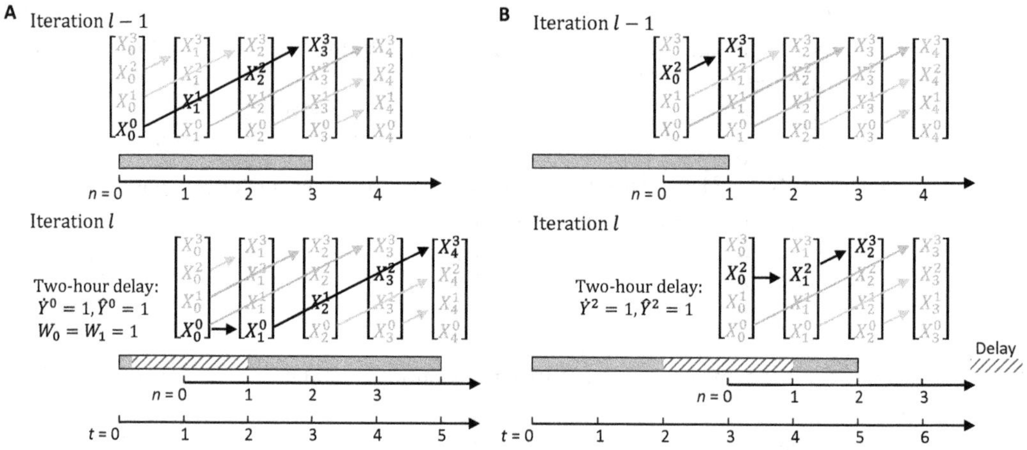

Figure 14.8 Modeling of delays; a two-period delay is observed on a batch with a three-period processing time; variables and parameters in black have a value of 1, rest have value 0; all indices except n and m are omitted for simplicity. (A) Delay is observed immediately after the batch starts. (B) Delay is observed two periods after the start of the batch. Absolute time also shown to illustrate use of index n and parameter t.

In addition, we introduce a new binary parameter ${}_lW_{ij,n=0}$, which, when 1, captures the information about delays in a task with progress status $m = 0$; that is, a batch gets delayed right after it starts.

$$ {}_lW_{ij,n=0} = \dot{Y}_{ij}^0, \quad i,j. \tag{14.13}$$

We also introduce parameter $\hat{Z}_{jn}$, which is 1 when the unit is unavailable during $[n\delta, (n+1)\delta)$. For a multiperiod delay, lasting φ periods, in a batch of task i carried out on unit j, in addition to $\dot{Y}_{ij}^m$ parameters, we employ $\hat{Y}_{ijn}^m$ parameters, which are set to 1 for $n \in \{0, 1, \ldots, \varphi - 2\}$. For unit breakdowns with downtime duration of φ periods, in addition to $\dot{Z}_{ij}^m$ parameters, we also set $\hat{Z}_{jn} = 1$ for $n \in \{0, 1, \ldots, \varphi - 1\}$. Thus, single-period delays do not result in activation of any $\hat{Y}_{ijn}^m$ parameters, but single-period breakdowns require activation of $\hat{Z}_{j,n=0}$.

The lifting equations of the model with disturbances are

$$W_{ij,n+1} = \hat{Y}_{ijn}^0, \quad i,j,n \tag{14.14}$$

$$X_{ijn}^0 = X_{ijn} + W_{ijn}, \quad i,j,n \tag{14.15}$$

$$X_{ij,n+1}^m = X_{ijn}^{m-1} - \hat{Y}_{ijn}^{m-1} + \hat{Y}_{ijn}^m, \quad i,j,n,m \in \mathbf{M}_i^*. \tag{14.16}$$

When there is a multiperiod delay in a batch with progress $m = 0$, the update step determines $W_{ij,n=0} = 1$ and $\hat{Y}_{ijn}^0 = 1$ for $n \in \{0, 1, \ldots, \varphi - 2\}$, ensuring that X_{ijn}^0 stays activated for the next $\varphi - 1$ periods, but with $X_{ijn} = 0$; that is, the batch is not erroneously interpreted as a new batch. If there are no delays , then, through (14.12) and (14.13), $W_{ijn} = 0$ and any new batch that starts ($X_{ijn} = 1$) results in $X^0_{ijn} = 1$ through (14.15). Note that (14.15) can be viewed as a constraint on the inputs (X_{ijn}) given the states (W_{ijn} and X^0_{ijn}), and if needed can be converted to inequality form.

The modeling of delays is illustrated in Figure 14.8 using a batch of a task with $\tau_{ij} = 3$. All indices except n and m are omitted for simplicity. Figure 14.8A shows the batch status evolution if the delay occurs immediately after the batch starts, while Figure 14.8B shows the same when the delay is observed two periods after the start of the batch. We see that through the lifting equations, X^m_{ijn} evolve over the black arrows leading to the batch correctly finishing two periods late in iteration l. In Figure 14.8A, parameters $\dot{Y}^0$ and $\hat{Y}^0$, and X_0 are all equal to 1 from the update; and hence, $W_1 = 1$ and $X^0_0 = X^0_1 = 1$. The batch in iteration l is expected to finish at $t = 4\delta = 4$ instead of $t = 2$. In Figure 14.8B, the update step ensures that the true status of the batch is reflected in the state variables, that is, $m = 2$. The batch in iteration l is expected to finish at $t = 5$.

The modeling of breakdowns is illustrated in Figure 14.9, where indices except n and m are again omitted. A breakdown that lasts two periods during the execution of a batch of a task with $\tau_{ij} = 3$ is used for illustration. Figure 14.9A shows the evolution of the task states when the breakdown occurs just after the batch starts, and Figure 14.9B shows the evolution when the breakdown occurs two periods after the start of the batch. In both panels, the currently executed batch is suspended at $n = 0$ of iteration l, and a new batch starts, in iteration l, once the unit downtime, assumed to be (rounded up to) 3 h, is over at $n = 2$. As we will see in the next paragraph, through the modified assignment constraint, (14.17), no batch can be assigned to the unit at $n = 0$ and $n = 1$ in iteration l. In Figure 14.9A, parameter $\dot{Z}^0$, through (14.12), prevents the

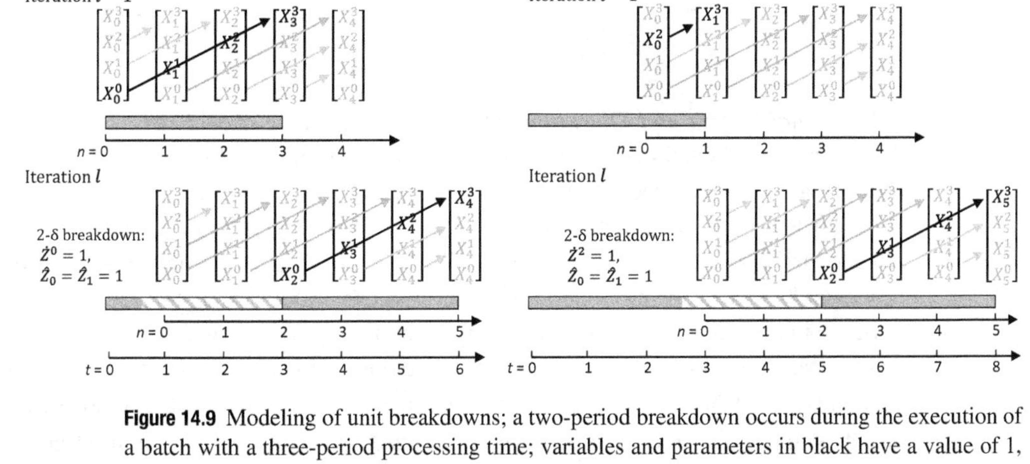

Figure 14.9 Modeling of unit breakdowns; a two-period breakdown occurs during the execution of a batch with a three-period processing time; variables and parameters in black have a value of 1, the rest have value 0; all indices except n and m are omitted. (A) Breakdown occurs immediately after the batch starts (at $t \in [0, 1]$). (B) Breakdown occurs two periods after the start of the batch (at $t \in [2, 3]$).

state from evolving from ${}_{l-1}X_0^0$ to ${}_lX_0^1$. Similarly, in Figure 14.9B, parameter $\dot{Z}^2$ prevents the task state to evolve from ${}_{l-1}X_0^2$ to ${}_lX_0^3$.

The assignment constraint is modified to include parameter $\hat{Z}_{jn}$ to account for unit downtime, and variable X_{ijn}^0 on the LHS, and not variable X_{ijn}, to correctly account for the unit being busy, specifically when a delay in a task with progress $m = 0$ is observed:

$$\sum\nolimits_{i \in \mathbf{I}_j} \left\{ X_{ijn}^0 + \sum\nolimits_{m \in \mathbf{M}_i^*} X_{ijn}^m \right\} \leq 1 - \hat{Z}_{jn}, \quad j, n. \tag{14.17}$$

The inventory balance, (14.9), does not require any corrective delay or breakdown terms because the *start* (X_{ijn}) and *end* $(X_{ijn}^{\tau_{ij}})$ states are active at most once, even if delays or breakdowns are observed.

The complete optimization model, solved in each iteration, consists of lifting equations (14.14) through (14.16); assignment constraint, (14.17); inventory balance, (14.9); and an objective function of choice.

14.2.4 Extensions

The update equations and the state-space model can be extended to handle additional optimization decisions and disturbances. Next, we outline some of these extensions. The sources of the complete formulations are discussed in Section 14.5.

Conversion Disturbances. If the consumption of input materials deviates from the nominal values, $\rho_{ik}\beta_{ij}$, then a *consumption* disturbance, $\hat{\beta}_{ijkn}^C$, is introduced that can be used to update the inventory balance. Similarly, deviations from the nominal values in the production of output materials are modeled using a *production* disturbance, $\hat{\beta}_{ijkn}^P$. The two disturbances are then used to *correct* the inventory balance, which is now written as follows:

$$I_{k,n+1} = I_{kn} + \sum_{\substack{i \in \mathbf{I}_k^+ \\ j \in \mathbf{J}_i}} \left(\rho_{ik}\beta_{ij} X_{ijn}^{\tau_{ij}} + \hat{\beta}_{ijkn}^P \right) + \sum_{\substack{i \in \mathbf{I}_k^- \\ j \in \mathbf{J}_i}} \left(\rho_{ik}\beta_{ij} X_{ijn} + \hat{\beta}_{ijkn}^C \right) + \xi_{kn} - S_{kn}, \quad k, n. \tag{14.18}$$

Fractional Delays and Downtimes. In Section 14.2.3, we considered cases where delays and unit downtimes are integer multiples of the time-grid spacing δ, and unit breakdowns were assumed to take place (almost) at a time point. However, if δ is not small, then these assumptions may not be good. Here, we present a simple update scheme that, given any fractional delays (φ^D), downtimes (φ^{DT}), or time a breakdown occurs (φ'^{BD})[11], ensures a rounding of these to integer values, so as to keep the task-finish and unit-availability times in sync with the time grid. A single batch may have many separate delays, so we introduce index r for delay occurrence during a batch with delay time φ_r^D. A breakdown can occur only once, at φ'^{BD}, during a batch, followed by unit downtime, φ^{DT}. For the first delay, a rounded up value, $\lceil \varphi_r^D / \delta \rceil$, is used in the

[11] All parameters denoted by φ represent duration. We use an accent for φ'^{BD} because it denotes timing of an event.

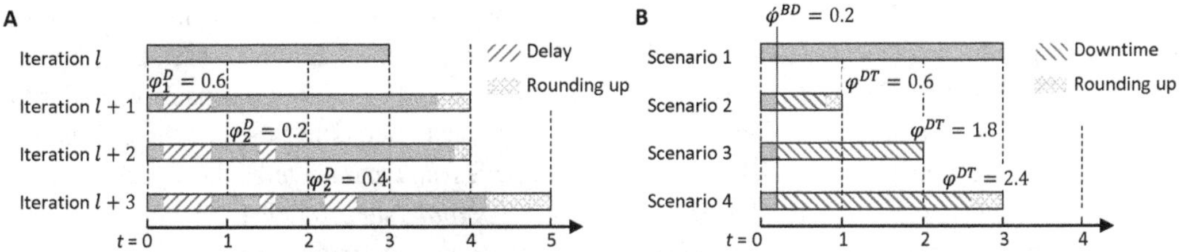

Figure 14.10 Modeling of fractional delays and downtimes; batch with nominal duration of three periods used for illustration; the time grid is not reset at every iteration; $\delta = 1$ h. (A) Delay of 0.3 h is observed between iteration l and $l+1$. (B) Breakdown occurs at 0.2 h, followed by downtime.

update step. For the subsequent delays, the integer, φ_r^D, that should be used in the update is determined as $\varphi_r^D = \lceil (\sum_{r' \le r} \varphi_{r'}^D)/\delta \rceil - \lceil (\sum_{r' < r} \varphi_{r'}^D)/\delta \rceil$.

Figure 14.10A shows how the occurrence of many fractional delays is modeled using a batch with processing time equal to 3. A delay of 0.6 h is observed between iteration l and $l+1$, and since $\lceil 0.6/1 \rceil = 1$, a delay of 1 h is used at the update before iteration $l+1$, leading to a modeled expected finish time at $t = 4$, although the actual finish time, if there is no other delay, would be 3.6. Next, a second delay with $\varphi_2^D = 0.2$ is observed between iterations $l+1$ and $l+2$. Because the total delay time so far is $0.6 + 0.2 < 1$ h, no new delay is introduced prior to iteration $l+2$. Finally, a third delay is observed between iterations $l+2$ and $l+3$ with $\varphi_3^D = 0.4$, bringing the total delay time up to 1.2, and since $\lceil 1.2/1 \rceil = 2$, a new 1 h delay is introduced, leading to a modeled finish time at $t = 5$.

When a unit breaks down, parameters $\dot{Z}_{ij}^m$ are always activated to suspend the running batch. The key challenge is to identify for how many periods the unit will be unavailable. This dictates if and how many of the $\hat{Z}_{jn}$ parameters are activated. On breakdown, after iteration $l-1$ (and before l), the unit becomes unavailable from φ'^{BD} to $\varphi'^{BD} + \varphi^{DT}$ (note that in the reset time grid for iteration l, we have $\varphi'^{BD} < 0$). Hence, all $\hat{Z}_{jn}$ that correspond to time points during the breakdown, that is, $n \in (\varphi'^{BD}, \varphi'^{BD} + \varphi^{DT})$, are activated. This also means that if $(\varphi'^{BD} - \lfloor \varphi'^{BD}/\delta \rfloor \delta + \varphi^{DT}) < \delta$, no $\hat{Z}_{jn}$ are activated, that is, the unit breaks down and becomes available again before the next time point.

The modeling of fractional downtimes is illustrated in Figure 14.10B using four scenarios. In scenario 1, there are no breakdowns or delays, so the batch is finished at $t = 3$. In all other scenarios, a breakdown is observed after 0.2 h ($\varphi'^{BD} = 0.2$) and, thus, $\dot{Z}^0 = 1$ for the update step. If $\varphi^{DT} = 0.6$ (scenario 2), then the unit will be available at the next iteration, so no $\hat{Z}_n$ are activated. Note that no time points are in $(0.2, 0.8)$. If $\varphi^{DT} = 1.8$, then the unit becomes available at $\varphi'^{BD} + \varphi^{DT} = 0.2 + 1.8 = 2$. Interval $(0.2, 2)$ contains 1, which means that we set $\hat{Z}_1 = 1$. Finally, if $\varphi^{DT} = 2.4$, then the unit becomes available at 2.6, which means that $(0.2, 2.6)$ contains 1 and 2, and therefore $\hat{Z}_1 = \hat{Z}_2 = 1$.

Variable Batchsizes. Variable batchsizes can be handled through the introduction of: (1) the standard batchsize variables, B_{ijn}, subject to the variable lower/upper bound constraints; (2) lifted batchsize variables, similar to X^m_{ijn}; (3) lifted batchsize variables to represent batchsizes of delayed batches; and (4) parameters to facilitate the expression of the update steps. Using these new variables and parameters, the update step is augmented to include the calculation of the *initial* (state of) batchsizes of ongoing batches; and the model is augmented with equations describing the evolution of the new state variables.

14.3 Design of Real-Time Scheduling Algorithm

We discuss the basic principles underpinning the design of a system for real-time scheduling, henceforth referred to as the *real-time scheduling algorithm.*[12] Due to space limitations, we do not present the details of the methods used to obtain some of the presented closed-loop results. In Section 14.3.1, we introduce some preliminary ideas and discuss the parameters that define the algorithm, whereas in Section 14.3.2, we discuss the system characteristics upon which the algorithmic parameters depend. In Section 14.3.3, we outline a simulation-based procedure for algorithm design for the deterministic case. In Section 14.3.4, we discuss how, in addition to the tuning of the algorithmic parameters, model modifications can be employed to improve closed-loop performance. In Section 14.3.5, we outline the counterpart of the simulation-based design procedure for the stochastic case. We close, in Section 14.3.6, with the presentation of the integrated framework for the design of real-time scheduling algorithms.

14.3.1 Algorithmic Parameters

Traditionally, rescheduling has been treated as an event-triggered activity where an event could be the arrival of a new order, a unit breakdown, or a processing delay. However, as we have reasoned, optimization should also be performed periodically. There is, however, an implicit cost associated with rescheduling due to the *plant nervousness* that it induces. Hence, it is important to quantify how the frequency of the optimization affects the quality of the implemented closed-loop schedule, and then decide how often to optimize.

In addition to rescheduling frequency, a key characteristic of the open-loop problem is the scheduling horizon length, because it dictates if orders further into the future are accounted for while computing the current schedule. A myopic horizon can lead to bad early decisions, necessitating costly revised schedules in the future. In general, one could argue that accounting for as long a horizon as possible is preferred. However, longer horizons increase problem size and might render finding good solutions in each

[12] The term *algorithm* here is not used to describe a method for the solution of an optimization model, as it has been used thus far. The term is used to describe the process of iteratively solving a scheduling (MIP) model, taking into account feedback, to obtain a closed-loop solution.

iteration challenging in real time. Thus, a natural question is what an appropriate horizon length is, and on what instance features does it depend.

Solving large-scale scheduling problems can be computationally expensive. Due to the limited time available for real-time computations, it might not be possible to solve the models to optimality. In general, while it is reasonable to expect that the closed-loop solution will deteriorate due to accumulation of moves determined by suboptimal solutions, it can also be argued that suboptimal moves will be corrected due to feedback, so the closed-loop schedule will remain minimally affected. Furthermore, due to the finite horizon of the open-loop problem, it cannot be directly deduced that the open-loop optimization should even be attempted to be optimally solved, especially since suboptimal solutions can be obtained faster, allowing us to reschedule more frequently.

Naturally, one question the reader might have is if there is coupling among the three aforementioned attributes (reoptimization frequency, scheduling horizon, and optimality gap). For example, in the deterministic case, a longer horizon would have already accounted for forthcoming orders, and hence might not require frequent rescheduling. A shorter moving horizon, on the other hand, does not account for impending orders outside its extent, and hence requires frequent rescheduling. In addition, if rescheduling is performed in real time with limited time available for computing, then an interesting question is whether a larger problem (long horizon) should be solved, potentially suboptimally, rather than a smaller problem (short horizon) solved to optimality. Also, would optimizing frequently compensate for suboptimal computations through frequent corrective feedback?

Finally, in addition to rescheduling frequency, moving horizon length, and suboptimal solution of the open-loop problem, there are additional approaches, such as objective function modifications and constraint addition, that can be adopted to improve the quality of the closed-loop solution. In summary, the design of the real-time scheduling algorithm seeks to address the following questions:

(1) How frequently should schedules be revised?
(2) Does the frequency depend on instance characteristics (e.g., frequency at which orders are due)?
(3) What is an appropriate scheduling horizon and on what instance characteristics does it depend?
(4) How does suboptimalilty in each open-loop optimization affect the quality of the closed-loop solution?
(5) Could the open-loop problem be modified in terms of its objective function and constraints to give better closed-loop schedules?

Before we outline a simulation-based approach to address these questions, we introduce some important system characteristics in the next subsection.

14.3.2 System Characteristics

Load. While unit capacities are known, it is not obvious what the overall capacity of a facility is in terms of rate of converting raw materials to final products when multiple batches of different tasks can be executed. For some networks, there can be a fixed bottleneck task; however, for most networks, the bottleneck can shift from one task to another, depending on the magnitude and composition (relative product mix) of the demand. In fact, even if the bottleneck unit is known and fixed, it could be a combination of tasks, rather than an individual task, executed on that unit that is bottlenecked. The details of the method that can be used to identify the overall capacity of a facility, given a ratio of demands, are beyond the scope of this book (the interested reader can find the source of the method in Section 14.5). However, it suffices to say that the load $\Lambda \in [0, 1]$ can be calculated and essentially represents the ratio of the demand over the capacity of the facility.

Order Variability. In short-term scheduling, order sizes can vary from one to another, that is, even when there is no uncertainty, there is variability. Assuming that order sizes are bounded, we define the max-mean relative difference (Σ), representing the difference between the maximum and mean order size relative to the mean order size.[13] For example, orders from a distribution with discrete values 4, 6, and 8 tons would have a mean order size of 6 tons and a $\Sigma = \frac{8-6}{6} = 0.33$. While other metrics of variability can be used, Σ has been shown to be appropriate and can be easily calculated. As we will discuss, higher Σ, in general, necessitates the use of a longer horizon to achieve good closed-loop performance. Finally, without loss of generality, for the numerical illustrations, we assume that order sizes follow the symmetric triangular probability distribution.

Time Constants. Time-related data include task processing times and the time between orders (Ω). We denote the scheduling horizon by η and the time between two iterations, the reoptimization time-step, by Δ. Note that Δ is different from δ, which denotes the time-grid spacing employed in discrete time scheduling model and dictates the resolution of the time-related data and provides a lower bound on the reoptimization time-step Δ. The processing times, coupled with the structure of the facility and the time between orders, together, determine how *fast* the system is, or more accurately, how fast changes occur and new information becomes available, and, consequently, how fast reoptimization should be performed. For example, if processing times are large and the orders are infrequent, the system is *slow*.

We further define the demand uncertainty observation time, H. It is the time ahead, with respect to time zero of each real-time iteration, at which the demand uncertainty is observed. It can be thought of as the length of the horizon, within which there is no uncertainty in any of the demand-related parameters. This implies that true order sizes are known within between $n = 0$ and $n = H$, while the orders due at $t > H$ are not deterministically known and modeled (estimated) to be at their mean. When the horizon moves forward, for the next iteration, the order sizes are updated as per the new observations (see Figure 14.11). When the demand uncertainty observation time is greater than or equal to the horizon length ($H \geq \eta$), the problem is deterministic, since the true size of all the orders within the horizon, irrespective of any variability, is known.

[13] Note that we use uppercase Greek letters to represent systems characteristics (Λ and Σ) and algorithmic parameters (to be further discussed in subsequent sections).

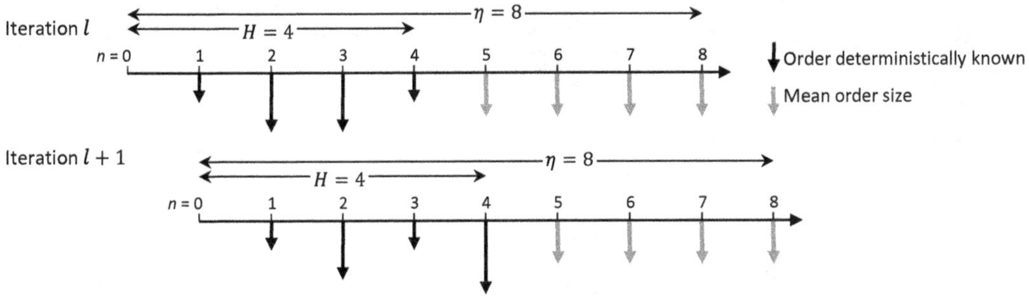

Figure 14.11 Illustration of concept of uncertainty observation time. As the horizon moves forward, the observation horizon also moves along, resulting in the observation of a new order (at $n = 4$ in iteration $l + 1$).

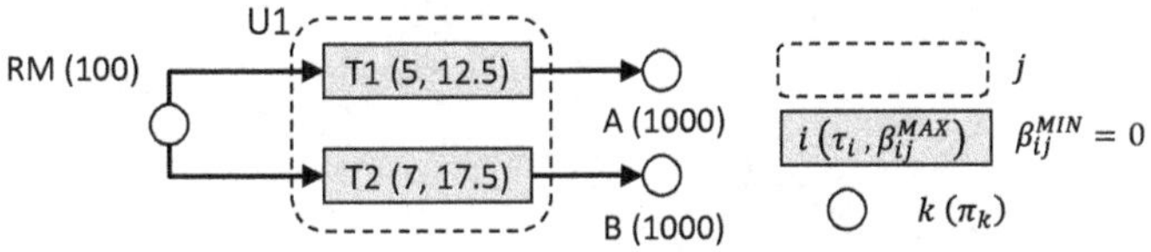

Figure 14.12 Network used for closed-loop simulations.

14.3.3 Design through Simulation: Deterministic Case

Our first goal is to find appropriate values for the horizon, η, and reoptimization time-step, Δ, given the characteristics of the system at hand. To illustrate qualitatively the interdependence between η and Δ, we show results from simulations using the network shown in Figure 14.12.

We compute closed-loop schedules for one week using $\eta \in \{1, 2, 3, \ldots, 48\}$ and $\Delta \in \{1, 2, 3, \ldots, 6\}$, with orders due every hour. The MIP scheduling model employs $\delta = 1$ and is solved to optimality. We choose order sizes such that $\Lambda = 0.125, 0.25, 0.5, 0.625, 0.75, 0.85$ and 1. In each simulation, the starting inventory is sufficient to meet orders for the first 12 h. To avoid any initial schedule ramp-up from impacting our conclusions, the closed-loop cost is evaluated from the start of day 3 (49 h) to the end of day 7 (168 h). The closed-loop cost of the simulations is shown in Figure 14.13 as a heat map with η on the abscissa and Δ on the ordinate. The maps for $\Lambda = 0.125, 0.25, 0.5, 0.625$ are like the one for $\Lambda = 0.75$, so we omit them. There are several interesting observations.

First, keeping Δ constant, a longer η can sometimes lead to worse closed-loop cost – see gray and black entries between white entries (e.g., in the plot for $\Lambda = 0.75$, the entries for $\eta = 18, 19, 20$ and $\Delta = 6$ are black, and are between the white entries for $\eta = 17$ and $\eta = 22$). This is in spite of the fact that orders are due at every time point, which means that it is not due to an asynchronization between the reoptimization step and the time between orders. However, with a sufficiently long horizon the minimum

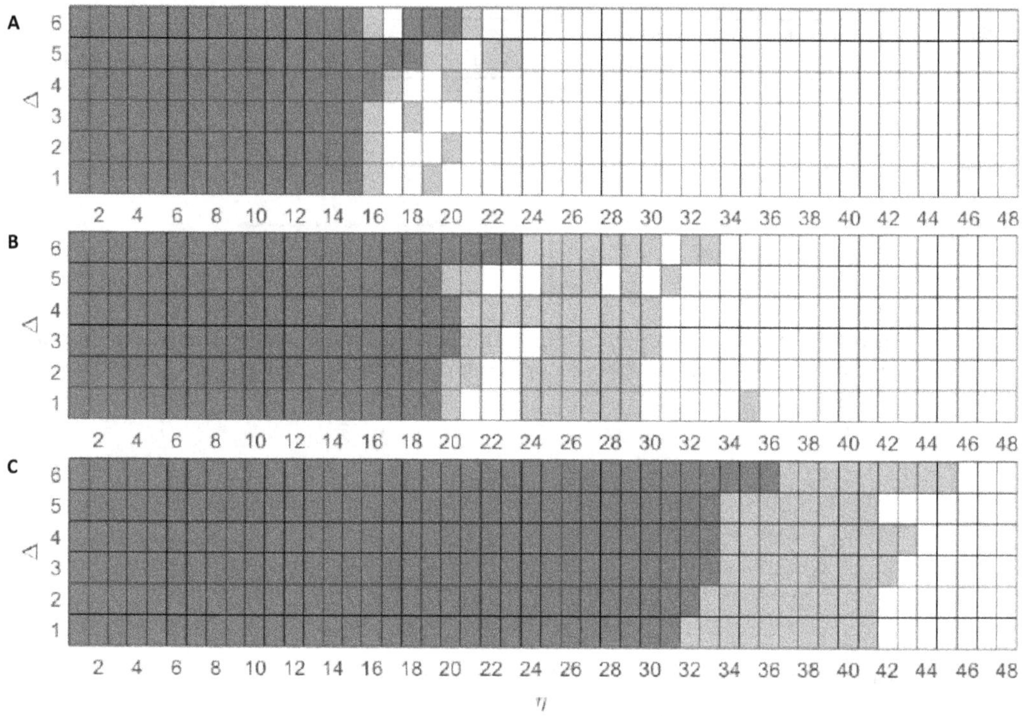

Figure 14.13 Closed-loop costs for different values of η and Δ: results for $\Lambda = 0.750$ (A), $\Lambda = 0.875$ (B), and $\Lambda = 1.000$ (C). The color for a simulation (for a pair of η and Δ) is white, gray, and black, if the cost is 0–5%, 5–20%, and > 20%, respectively, more than the least closed-loop cost for the corresponding load.

possible cost is still achieved (e.g., in the plot for $\Lambda = 0.75$, all entries are white for $\eta \geq 22$). From analyzing the open-loop solutions, it can be inferred that the reason for this closed-loop cost behavior is the trade-off between starting fewer but bigger batches versus more but smaller batches.

Second, we observe that when operating close to capacity (i.e., $\Lambda \to 1$), a much longer horizon is needed. This is because in this case there is almost no allowance for idle time and batches need to be *packed* optimally. Hence, a longer horizon, leading to good solutions, is necessary because a lost production opportunity cannot be recovered later through reoptimization.

Third, for any Δ, there always exists an η with which the best closed-loop performance can be achieved. Therefore, to determine the algorithmic parameters, one should first decide the desired Δ, and then experiment with different η to find the one which results in good closed-loop performance. The selection of Δ depends on practical considerations (e.g., how fast are operators willing to revise the schedules), as well as the computational capabilities (e.g., how fast can a good new schedule be computed). In general, for a larger Δ, a longer η is needed.

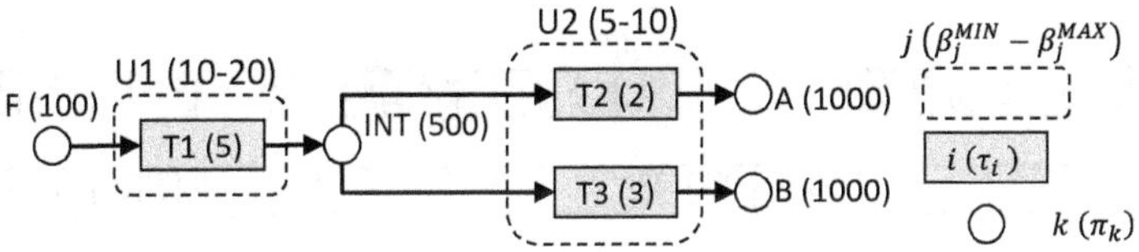

Figure 14.14 Motivating example used for illustration of model modifications.

14.3.4 Model Modifications

In this subsection, we explore how objective function modifications and constraint addition can be used to improve closed-loop quality when we are limited to use shorter η or larger Δ. We consider the network shown in Figure 14.14, where there are orders of size 3 due every 3 h ($\Omega = 3$), and excess sales are also allowed at order due times. The starting inventory of F is 3 and the objective function is profit maximization. The contractual/minimum demand of A and B over 1 week (168 h) is 168 tons, which can be satisfied by 17 batches of T2 and T3, respectively.

Since T3 takes 3 h to produce 10 tons of C and T2 takes only 2 h to produce 10 tons of A and they have the same price ($\pi_A = \pi_C = \$1,000/\text{ton}$), more value is created per hour for every batch of T2. Hence, in an optimal schedule, we expect to have the minimum number of batches to exactly meet the demand for B, and the excess time available on unit U2 to be used toward the production of (more valuable) A. Since the demand for B over a week can be satisfied by 17 batches, if, in the implemented schedule, there are more than 17 batches of T3, then the closed-loop solution can be improved, because time that could have been used to produce A is used to produce B. Further, the time required for 17 batches of T3 is 51 h, which means that, out of 168 h, 117 h are available to execute 58 batches of T2. The best[14] 168 h closed-loop schedule, with this number of batches, can be obtained if we use an open-loop problem with $\eta = 24$, resolved every hour ($\Delta = \delta = 1$). However, we next explore how the same good performance can be achieved if, due to practical reasons, we can only solve models with $\eta = 12$.

We first obtain the closed-loop schedule (with $\eta = 12$) without any model modifications. The solution includes 55 batches of T2 and 18 batches of T3 with a total unit utilization of $55{\cdot}2 + 18{\cdot}3 = 164$ h, that is, with idle time of 4 h (see Figure 14.15A). This is not the best solution because there are gaps (idle times) in the schedule. To reduce this idle time and thus increase profit, we modify the objective function to favor early sales of both final products.[15] This modification eliminates the idle times, but the number of batches of T2 decreases to 46 while the number of batches of T3 increases to 25.

[14] We have used the term *optimal* to refer to the solution of a single MIP model. We use the term *best* to describe the best possible closed-loop solution, obtained from the iterative solution (not necessarily to optimality) of multiple MIP models.

[15] This is accomplished by using time-dependent material price, π_{kn}, which is equal to $2\pi_k$ at $n = 0$ and linearly decreases to π_k at the end of the horizon.

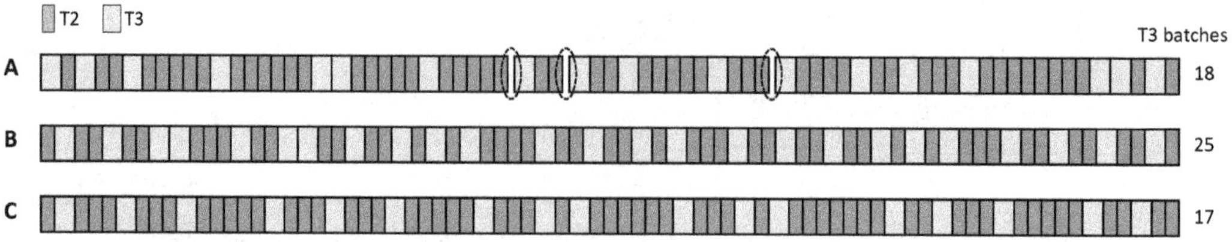

Figure 14.15 One-week closed-loop schedules using $\eta = 12$ h and $\Delta = 1$ h for U2 of the network in Figure 14.14; idle times are marked with ellipses. (A) With no objective function modifications: 55 batches of T2 and 18 batches of T3. (B) With early sales objective function modification only: 46 batches of T2 and 25 batches of T3. (C) With objective function modifications and constraints: the best schedule is obtained with 58 batches of T2 and 17 batches of T3.

Analyzing the closed-loop solution (Figure 14.15B) and the open-loop iterations (not shown), we infer that this is because excess inventory of B is sold at the earliest possible opportunity. This means that when a new order for B is observed, as the horizon rolls forward, a new batch must be executed to meet demand. However, had some inventory of B been maintained rather than immediately sold, it could be used to meet new orders for B, thereby negating the need to execute more than 17 batches of T3. To address this, we introduce a new constraint to disallow sales of B in the first half of the horizon. With this added constraint along with the early sales objective function modification, we obtain a closed-loop schedule with 58 batches of T2, 17 batches of T3, and no idle time (see Figure 14.15C). In other words, we were able to obtain the best possible closed-loop solution over 168 h using a reduced horizon coupled with an objective function modification and constraint addition. Finally, we note that adding only the no-early-sales constraint, without the objective function modification, results in an inferior closed-loop solution with idle times as shown in Figure 14.15A.

As demonstrated, model modifications can be used to improve closed-loop quality when the use of long η or short Δ is prohibitive. Examples of objective function modifications, other than incentivizing early sales, are (1) time-varying material pricing and (2) time- or unit-dependent production costs. In addition to disallowing sales in part of the horizon, other candidate constraints are (1) safety stocks that account for multi-product interactions; (2) time-dependent bounds on inventories or backlogs; and (3) disallowing specific tasks to be executed in succession or in parallel. The objective function modifications and constraint additions necessary to achieve good solution quality are instance specific and should be identified through simulations.

14.3.5 Design through Simulation: Stochastic Case

The contrasting observation we make when studying problems under uncertainty is that using a long horizon cannot necessarily overcome infrequent reoptimization. We demonstrate this through closed-loop simulations for the network shown in Figure 14.16. We study how the one-week closed-loop costs vary with η and Δ in Figure 14.17. We see that in order to achieve good closed-loop performance (white

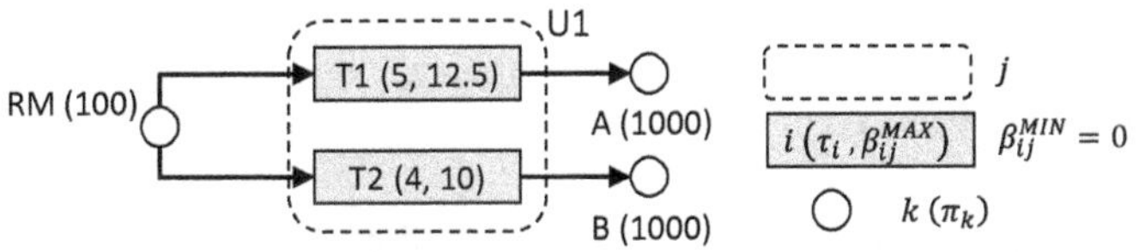

Figure 14.16 Process network and parameters used to obtain closed-loop results in Section 14.3.5.

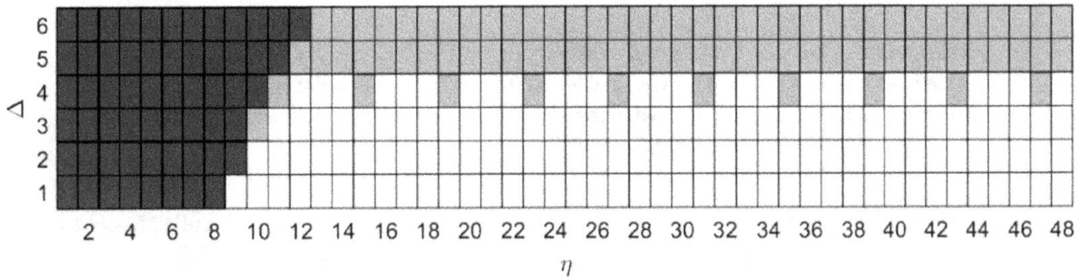

Figure 14.17 Closed-loop costs for different values of η and Δ. System characteristics: $\Omega = 1$ h, $\Lambda = 0, 875$, $H = 6$ h, and $\Sigma = 0.6$. Starting inventory of M1 and M2 is 13.125 tons each. The color coding for the heat map is as follows: white, gray, and dark gray, if the cost is 0–5%, 5–20%, and > 20%, respectively, more than the least closed-loop cost. Each closed-loop simulation is replicated 10 times with different demand samples.

region), the reoptimization time-step has to be smaller than a threshold value Δ^{THR} (in this case, $\Delta \leq 4$). If the reoptimization time-step is larger than Δ^{THR}, then no matter how long the horizon is, good closed-loop performance cannot be achieved. This is expected because, when uncertainty is present, it is important to react to it by reoptimizing promptly. To reduce the number of simulations required to find the shortest η and largest Δ that can result in good closed-loop performance, we recommend first choosing a long η (e.g., $\eta = 48$ here), and then finding Δ^{THR} for this horizon. Following this, simulations should be carried out keeping Δ fixed at the threshold so as to find the shortest η, which still results in good closed-loop performance.

Next, we study how demand uncertainty impacts closed-loop costs and draw further insights for selecting η and Δ to mitigate these effects. We model demand uncertainty as composed of two elements: (1) the demand uncertainty observation time (H) and (2) the relative maximum deviation of an order from its mean size (Σ). We carry out simulations using a network producing three products (not shown), with $\Omega = 6$, $\Lambda = 0.5, 0.75$, and 0.9. We perform multiple simulations (replications) to obtain different schedules (one-week length) for different samples of demand and discern statistical significance (p-value = 0.05) by carrying out two-way analysis of variance (ANOVA). The results are shown in Figures 14.18 and 14.19, where in each panel the average cost, over all replications, is scaled with respect to a single chosen mean value (described in the caption) within each panel. The one standard deviation intervals are not shown for legibility.

If the information about true demand comes almost as a surprise (here $H = 8$), then, interestingly, facilities with low loads are impacted more than facilities with heavier

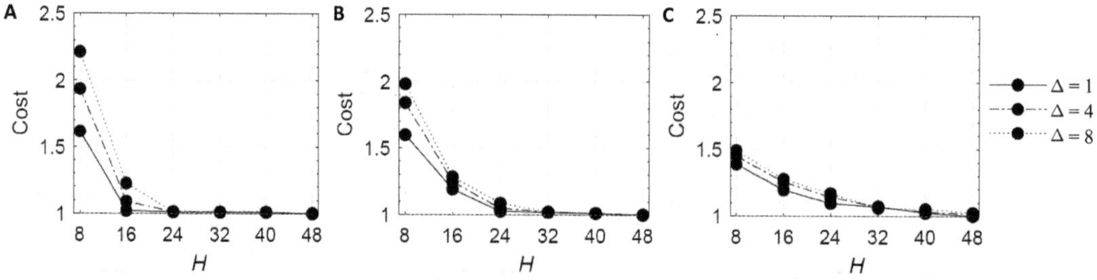

Figure 14.18 Effect of H and Δ on closed-loop cost for fixed $\eta = 48$ and $\Sigma = 30\%$. Results for loads $\Lambda = 0.5$ (A), $\Lambda = 0.75$ (B), and $\Lambda = 0.9$ (C). Each data point is an average cost of closed-loop schedules for 10 demand samples and is scaled by the mean closed-loop cost corresponding to $\eta = 48$ and $\Delta = 1$ within each panel.

loads. This counterintuitive trend can be seen in Figure 14.18, specifically, for $H = 8$. It is further magnified by larger Δ. This observation is explainable: a larger load requires us to operate at near production capacity, thus any order much larger than the mean size will lead to a backlog, irrespective of whether we know about it early or not. There is not enough slack in the schedule to adjust production in a reactive fashion. Although a large Σ implies that many orders can also be much smaller than the mean (symmetric triangular probability distribution), since backlog here is 10 times as expensive as holding inventory, backlog costs influence the closed-loop quality more than inventory costs do. In contrast, for smaller loads, idle time can be utilized to meet an order with size that has a larger deviation from its mean.

Next, we study how the reoptimization time-step can be selected to mitigate the effect of demand uncertainty. From Figure 14.18, we infer that a smaller reoptimization time-step (i.e., more frequent reoptimization) leads to better closed-loop solutions for any given demand uncertainty. Since the horizon is very long ($\eta = 48$), the role of reoptimization then is to tackle disturbances due to uncertainty and not new information. Given that uncertainty is less important for heavier loads, we see correspondingly the cost advantage from shorter Δ, under heavier loads, is less pronounced.

Finally, we focus on the choice of η. Figure 14.19 shows how closed-loop solution quality is affected by η, Σ, and H. We see that the importance of employing any longer η than a threshold value gradually diminishes. This threshold is smaller for lighter loads. This merely confirms yet again that the choice of η depends on Λ. Importantly, looking closer, we see that the isolines in Figure 14.19 do not intersect with each other, which leads to an interesting corollary: a longer η is not able to compensate for higher Σ when H is fixed and smaller than η (i.e., we have uncertainty). This makes sense, as a longer horizon is unlikely to help us with a surprise that we get at the early portion of the horizon.

14.3.6 Integrated Framework

We present an overall framework outlining the necessary steps for designing an efficient real-time scheduling algorithm. With this framework in mind, scheduling application

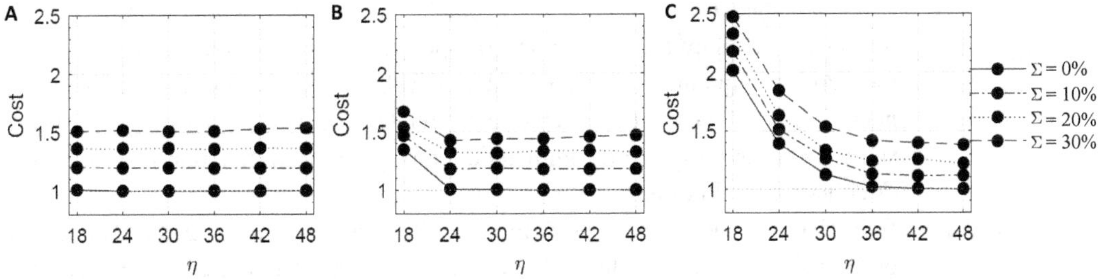

Figure 14.19 Effect of η and Σ on closed-loop cost for fixed $H = 8$ and $\Delta = 1$. Results for loads $\Lambda = 0.5$ (A), $\Lambda = 0.75$ (B), and $\Lambda = 0.9$ (C). Each data point is average closed-loop cost for 10 demand samples and is scaled by the mean closed-loop cost corresponding to $\eta = 48$ and $\Sigma = 0$ within each panel.

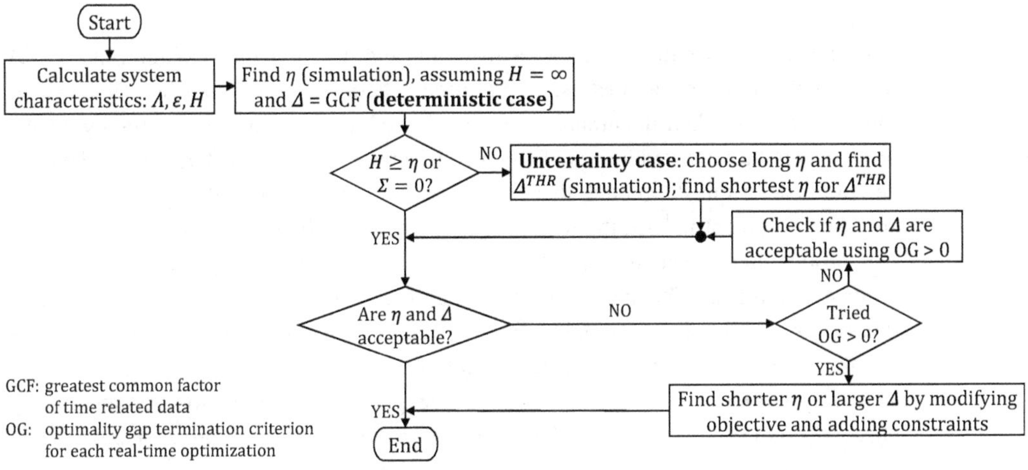

Figure 14.20 Framework for designing real-time scheduling algorithm.

developers can systematically design their algorithm. The flowchart of the proposed design process is shown in Figure 14.20.

(1) We calculate the system characteristics, that is, the load (Λ), order size max-mean relative difference (Σ), and demand uncertainty observation time (H). It is important that $\Lambda \leq 1$ so that there is an η for which backlogs can be avoided. A $\Lambda > 1$ implies that the demand exceeds production capacity, hence backlogs are unavoidable.

(2) We set Δ equal to the GCF of time-related data and, assuming $H = \infty$ (i.e., deterministic case), we find the necessary η through closed-loop simulations (as discussed in Section 14.3.3). Next, we check if the instance is indeed deterministic, that is, if $H \geq \eta$ or $\Sigma = 0$. If the problem is under uncertainty, then we find threshold frequency Δ^{THR} (as discussed in Section 14.3.5). Using this Δ^{THR}, we find the shortest η that does not lead to any closed-loop solution quality decrease.

(3) If the obtained η and Δ are not acceptable due to, for example, computational reasons, we explore the use of nonzero optimality gap. We determine the closed-loop solution quality deterioration due to suboptimal optimization. If it is not negligible, or there are other deployment-related reasons necessitating the use of shorter η or larger Δ, we recommend exploring objective function modifications or constraints addition.

(4) If the demand pattern changes significantly (and especially if Λ, Σ, or Ω increases or H decreases), new values of η and Δ should be determined. If H increases beyond the currently used η, then a longer η could result in improvements by accounting for additional deterministic information.

14.4 Feedback through Integration with Other Functions

In Section 14.2.3, we incorporated feedback through the update calculation in (14.12) based on disturbance parameters. If real-time scheduling is viewed as a standalone planning function, then disturbances can be viewed as the result of uncertainty observation: when the observed values of the uncertain parameters are different from the ones used in the deterministic model, then a disturbance is *injected* into the system. Through this scheduling-centric lens, the scheduler is presented with disturbances after they have been observed. But can the scheduler *predict* some of these disturbances and take recourse action fast? The answer is yes, if the system boundary is expanded to include additional modeling/decision-making layers. Accordingly, in Section 14.4.1, we discuss how incorporating information from the automation system can lead to better implemented schedules; and in Section 14.4.2, we present a formulation for integrating scheduling and process control. The former is applicable to batch processes, while the latter is applicable, primarily, to continuous processes.

14.4.1 Integration with Automation Logic

One source of new information that may lead to disturbances, from a scheduling standpoint, is the automation system that makes the low-level discrete decisions that drive the operation of the plant and therefore leads to a mismatch between the model of the plant used in the scheduling model and the actual behavior of the plant. For example, a batch reaction may be modeled as a single *task* with a fixed duration in the scheduling model, while the automation system includes a sequence of five or more distinct *steps*, with the transitions between them dependent upon the state of other pieces of equipment in the plant. Thus, a schedule computed using the coarser model may become infeasible when executed, with one or more of the batches taking longer in reality than in the (predicted) open-loop schedule. We next outline an approach to address the challenges stemming from the mismatch between the scheduling model and the actual process.

To monitor the execution of a schedule and detect disturbances, a finite transition system that models the plant's automation logic can be constructed and augmented with

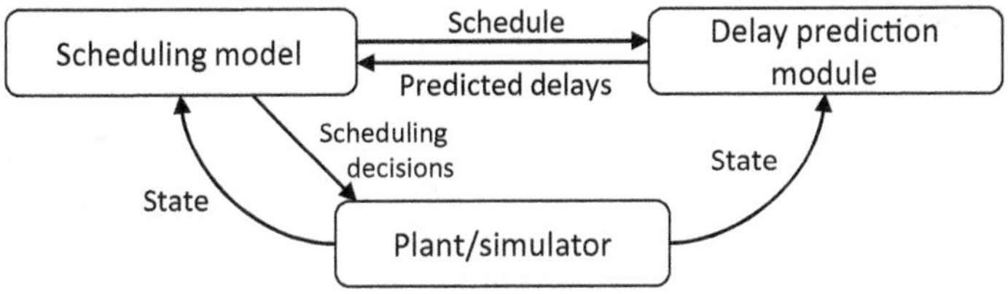

Figure 14.21 Interaction across different components of integrated scheduling-automation system.

historic data on the amount of time the system spends in each step. With this detailed model of plant's dynamics, the current schedule can be treated as a specification of the desired system behavior. If the specification is violated (i.e., the current schedule is infeasible), the delay caused by the infeasibility is estimated and used to guide the next optimization. The information can be fed to the generalized discrete time state-space model presented in Section 14.2 or any other state-space scheduling model (see Section 14.5 for more information).

To determine transitions between steps comprising a task, a delay prediction module (DPM) that analyzes automation logic should be developed (see Section 14.5). Assuming that such tool is available, delays detected by DPM are transmitted to the scheduling layer. Other disturbances such as yield changes or disruptions from nominal/predicted rates can also be communicated to the scheduling model from process measurements. The architecture of the integrated system is shown in Figure 14.21. We note that, to evaluate this methodology, the process can be replaced with a simulator mimicking the process dynamics and automation logic. To show that feedback based on the automation system's dynamics (specifically, when informed of an upcoming delay) can improve the quality of the closed-loop solution, we present an example.

Example 14.1 Feedback from Automation Logic We consider a system, with the STN representation shown in Figure 14.22A, converting two raw materials (FA and FB) into two final products (A and B). The process consists of a reactor, which can produce two different materials (I1A, I1B), a solvent recovery unit, and a check tank. For each product, the reaction task has three steps: React, Decant, and Sample. Depending on the result of the Sample step, the Decant and Sample steps may need to be repeated. The time required to complete some of the steps is product dependent. The solvent recovery unit processes each batch produced by the reactor. Over time, the filter in the solvent recovery unit becomes fouled, requiring routine cleaning that is performed by an operator. Processing I1A fouls the filter faster than does processing I1B. The check tank receives materials from the solvent recovery unit so that quality checks can be performed. These checks are more detailed than the sampling performed in the reactor and require the operator. The steps considered by the automation system, for all tasks, are shown in Figure 14.22B.

We consider the maximization of profit realized before a deadline at $t = 32$ h. The scheduling formulation does not account for the individual steps in the reactor, and instead treats reaction, for each product, as a single task with a processing time equal to

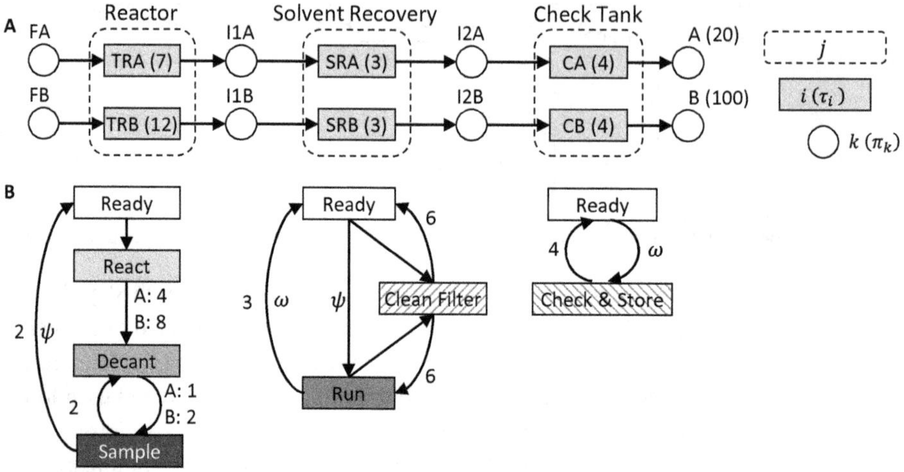

Figure 14.22 Process used in Example 14.1. (A) STN representation (partial) of the system; nominal processing times given in parentheses inside task blocks; the filter is a nonrenewable discrete resource (not shown), consumed by SRA (two units) and SRB (one unit); a Clean Filter task, which requires the operator, restores its availability at three units. (B) Sequence of steps that the automation logic implements appearing below the corresponding tasks; time required for transitions between steps is given along corresponding transition arcs; transitions that occur simultaneously (enforced by the automation system) are labeled with the same Greek letter; steps requiring the operator (unary resource) are denoted using patterned fill.

the total time required to complete the three steps, assuming that the Decant and Sample steps will not have to be repeated (which is the most common case).

Figure 14.23A shows the nominal optimal schedule, in which two batches of TRB are completed before the deadline. The nominal schedule is feasible if neither batch requires a second decant-and-sample cycle. If such a disturbance does occur, then the processing time for that batch will be extended by 4 h. If the disturbance occurs in the first batch, simply pushing the schedule back will result in the second batch not being completed before the deadline, as shown in Figure 14.23B. Figure 14.23C demonstrates that, because the filter in the solvent recovery unit does not have enough capacity to process one TRA and one TRB batch without maintenance, replacing the second TRB batch with a TRA batch at $t = 16$ will not allow two batches to be completed before the deadline.

Clearly, waiting long enough to observe the delay caused by the second decant-and-sample cycle that starts at $t = 12$ (i.e., waiting until at least time 14) results in a poor solution. How can delay monitoring, based on automation logic, help? If the delay of 4 h is detected at $t = 12$, when the automation system receives the signal to return to the Decant step, then it is already known, at $t = 12$, that the current schedule is no longer feasible, and rescheduling at that point, with simply updating the end time of the first batch to be 16 (through the activation of the corresponding delay parameters; see discussion in Section 14.2.3), leads to the schedule shown in Figure 14.23D. The filter is cleaned proactively, starting at $t = 12$, to overlap with the second cycle of

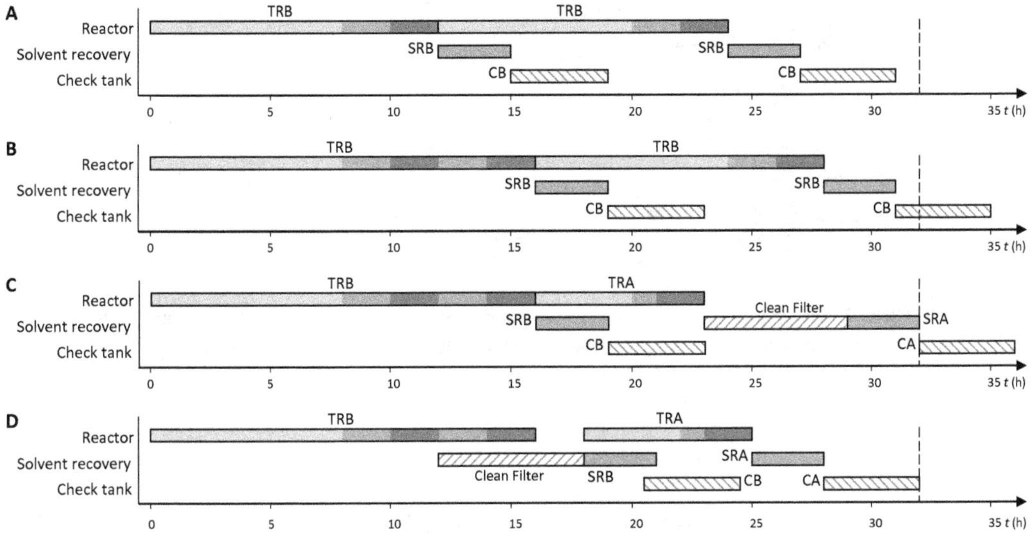

Figure 14.23 Illustration of the impact of a disturbance (second decant-and-sample cycle for the first batch of TRB); steps are shaded differently to match the shading in Figure 14.22. (A) Open-loop schedule obtained at $t = 0$. (B) Schedule obtained by simply pushing all batches back; the deadline is missed. (C) Alternative response to replace the second batch of TRB with a batch of TRA, necessitating filter cleaning; cleaning cannot start earlier because the operator is occupied with task CB, leading to a delay in SRA and CB. (D) Schedule obtained by accounting for information coming from the automation system, at $t = 12$; filter cleaning starts earlier, and while the batch of TRA starts later, it is finished at the deadline.

decanting and sampling in the reactor, which allows the batch of TRA to be completed before the deadline. Note that the makespan of the revised schedule is 32 h, with no time to spare. Obtaining this schedule was only possible because the delay was predicted at $t = 12$. Attempting to reschedule after $t = 12$ would be too late, as the second batch could not be completed before $t = 33$, thereby missing the deadline.

14.4.2 Integration with Process Control

Another way to introduce early feedback to scheduling is through the integration of scheduling and process control, that is, the *inclusion* of some (typically viewed as) control variables into the scheduling layer. This integration informs the scheduling layer of possible deviations and thus enables faster recourse. To carry out such integration, a unified state-space formulation for integrated scheduling and control is necessary, which is the topic of this subsection.

Representation. Without loss of generality, we consider a continuous parallel-unit (single-stage) system with units $j \in \mathbf{J}$ producing products $i \in \mathbf{I}$ (see Chapter 3 and its continuous processing counterpart in Chapter 9). We say that a unit can produce product i when it is in mode i, and $\mathbf{I}_j$ is the set of compatible products/modes for unit j.

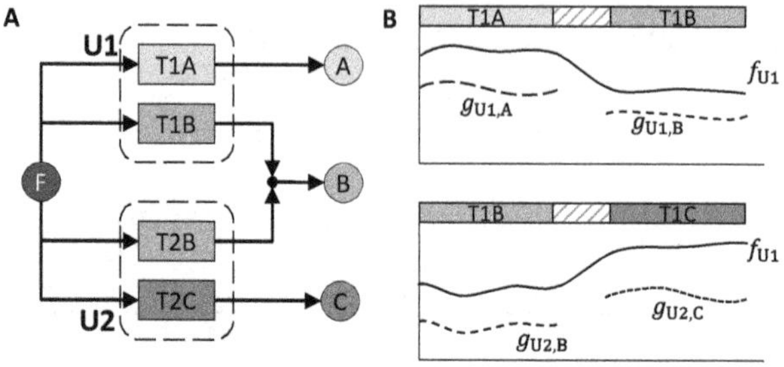

Figure 14.24 System representation for integrated scheduling-control problem. Each unit evolves dynamically while producing different products. Evolution of unit state variables (f_j) and production levels of individual products (g_{ij}) shown schematically.

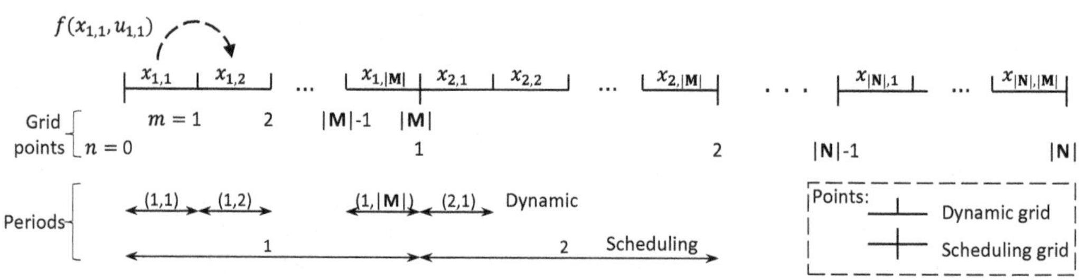

Figure 14.25 Illustration of coarse (scheduling) and fine (dynamic) time grids, and dynamic evolution of state vector. Indices i and j are omitted for simplicity.

We further assume that all products require the same raw material, which is unlimited, so we do not account for material consumption. Unit production is based on the dynamic evolution of the unit. The STN representation of an example instance and the dynamic production are illustrated in Figure 14.24A.

Time Grids. To allow for potential differences in timescales between the scheduling decisions and the dynamic control decisions, we define two different time resolutions. For the scheduling decisions, we use a *coarse* grid with points $n \in \mathbf{N} = \{0, 1, \ldots, |\mathbf{N}|\}$. For the dynamic decisions, each scheduling period $n \in \mathbf{N}^* = \{1, \ldots, |\mathbf{N}|\}$ is divided into subperiods (*fine* time grid) indexed by $m \in \mathbf{M} = \{1, 2, \ldots, |\mathbf{M}|\}$, where we assume that all time periods (both n and m) are constant in duration and number. A generalization, compatible with the presented model, is to allow each unit to have its own fine discretization, that is, have $m \in \mathbf{M}_j$. Figure 14.25 illustrates the two grids.

Dynamic Layer. The underlying dynamics of each unit j are defined by a (time-invariant) state-space model $f_j(x, u)$ that evolves on the fine timescale. Thus, there is a state vector x_{jnm} for each unit and time point (n, m):

$$x_{jn,m+1} = f_j\left(x_{jnm}, u_{jnm}\right), \quad j, n, m, \tag{14.19}$$

where states and inputs must satisfy constraints $e_j(x_{jnm}, u_{jnm}) \leq 0$. We further assume that the resolution of the fine grid is such that intrasample dynamic effects can be neglected. The dynamic evolution is illustrated in Figure 14.24B.

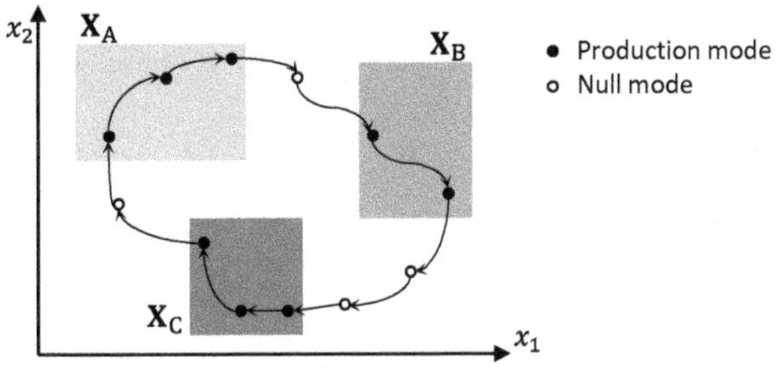

Figure 14.26 Operating regions of unit that can produce three products (A, B, and C). Sample dynamic evolution is shown, with solid disks representing state values within a production region and hollow disks representing states in the null mode.

Whether unit j is operated in a particular mode $i \in \mathbf{I}_j$ is determined by whether or not the dynamic state of the unit is in a particular region, that is, whether $x_{jnm} \in \mathbf{X}_{ij}$. In the case of steady-state production, these sets could be singletons $\mathbf{X}_{ij} = \{x_{ij}^{SS}\}$, but, more generally, they could be large operating regions. The goal of the control system is to keep the dynamic state inside $\mathbf{X}_{ij}$ during production of product i, and to move the system from $\mathbf{X}_{ij}$ to $\mathbf{X}_{i'j}$ during an $i \rightarrow i'$ transition. The concept is illustrated in Figure 14.26.

The operating mode of the unit is determined by the scheduling layer using binary variables X_{ijn}: if $X_{ijn} = 1$, then $x_{jnm} \in \mathbf{X}_{ij}$ should be satisfied.[16] That is, if a unit is in a given mode during a period n, then dynamic constraints must be satisfied at each point m within that period. Note that the converse is not true, and the scheduling layer may decide to not activate a given mode despite being in the suitable operating region. A *null* mode is used to model the state when the unit is not in a *production* mode.[17] Assuming that sets $\mathbf{X}_{ij}$ can be described by linear constraints, that is, $\mathbf{X}_{ij} = \{x: A_j^i x \leq b_j^i\}$ for suitably chosen A_j^i and b_j^i, the aforementioned implication can be described as follows, using the convex hull reformulation (see Chapter 2):

$$x_{jnm} = \sum_i x_{jnm}^i, \quad j, n, m \tag{14.20}$$

$$A_j^i x_{jnm}^i \leq b_j^i X_{ijn}, \quad j, i \in \mathbf{I}_j, n, m \tag{14.21}$$

$$x_{jnm}^i \leq M_j^i X_{ijn}, \quad j, i \in \mathbf{I}_j, n, m, \tag{14.22}$$

16 Please note that in the current section we use letter X (x) in multiple ways: (1) X_{ijn}, assignment (scheduling) binary variables; (2) x_{jnm}, (dynamic) states, and their disaggregated counterparts, x_{jnm}^i, defined in (14.20) through (14.22); and (3) $\mathbf{X}_{ij}$, operating regions.

17 While the idea is similar, the *null* mode is different from the *neutral* mode we have used earlier in the book to facilitate the modeling of setups (and that is why a different term is used). The former represents a *real* region in which the state of the system can be in, while the latter was only a modeling convention – if a setup is not executed, then a unit is always set up for a task (i.e., is in a production mode).

where M_j^i is a sufficiently large parameter, and binary variables X_{ijn} satisfy the scheduling constraints. The inputs u_{jnm} can be disaggregated, in a similar way, into mode-specific ones, u_{jnm}^i, including inputs when in the null mode.

The (dynamic) product inventory balance then becomes (see Figure 14.24B):

$$I_{in,m+1} = I_{inm} + \sum\nolimits_{j \in \mathbf{J}_i} g_{ij}\left(x_{jnm}^i, u_{jnm}^i\right) + \xi_{inm} - S_{inm}, \quad i, n, m, \tag{14.23}$$

where $\mathbf{J}_i$ is the set of units where product i can be produced; shipments, $\xi_{inm}(< 0)$, and additional sales, S_{inm}, are allowed to be positive only at the points of the coarse grid ($m = |\mathbf{M}|$); and functions $g_{ij}(x_{jnm}^i, u_{jnm}^i)$ describe production.[18] If necessary, raw material consumption can be modeled in a very similar way. Also, using the ideas of RTN, general resource balances can be expressed over the fine grid, where resources are, in general, dynamically consumed and produced by units. Equation (14.23) ensures that dynamic production is accurately accounted for and is also the main source of coupling between units.

Scheduling Layer. In principle, any discrete time scheduling model that employs assignment variables $X_{ijn} \in \{0, 1\}$ can be integrated with the dynamic model. Variables X_{ijn}, which in this setting can also be viewed as mode selection variables, should satisfy all unit assignment constraints. In addition, (1) continuous batchsize-type variables (e.g., B_{ijn}) are replaced by dynamic inputs because we now have more degrees of freedom and production can be controlled at the fine time grid and (2) batchsize constraints are replaced by (14.20) through (14.22) and their counterparts for inputs. The material balance of the scheduling layer is now replaced by (14.23).[19] Finally, any objective function can be used, appropriately augmented to account for costs occurring at the dynamic layer (e.g., resource costs based on dynamic profiles).

14.5 Notes and Further Reading

(1) The discussion in Section 14.1 is modified from Gupta et al. [1]. The interested reader can also find some background material and preliminary concepts in Gupta and Maravelias [2].

(2) The first state-space formulation for scheduling problems was proposed by Subramanian et al. [3]. The state-space model presented in Section 14.2 is based on the model presented in Gupta and Maravelias [4]. Specifically, the discussion

[18] We have already used a simple production model $g_{ij}(\cdot)$ in this book. Treating task rates (batchsizes), B_{ijn}, as inputs, the production model of a continuous task is $\rho_{ij}B_{ijn}$, where $\rho_{ij} > 0$ is the production coefficient of task i when executed in unit j. In Chapter 9, we in fact utilized this *model* to express the product material balance. Here, we apply the same idea at the finer time grid, and also consider more general, potentially nonlinear, production models.

[19] Note that the material balance at the coarse level can be obtained from (14.23) for $m = |\mathbf{M}|$. Thus, the equations in (14.23) could be partitioned into the dynamic component, expressed for $m < |\mathbf{M}|$, where the RHS does not contain demand and sales; and a scheduling component, expressed for $m = |\mathbf{M}|$, containing all terms.

in Section 14.2 employs the same approach for the update equations, but does not cover the modeling of the wide range of disturbances discussed in [4].

(3) The concepts for the design of the real-time scheduling algorithm in Section 14.3 are primarily from Gupta and Maravelias [5].

(4) The model presented in Section 14.4.1 is a simplification of the model proposed by Rawlings et al. [6], where implementation details for the use of delay monitoring, based on process automation, to provide early feedback to scheduling are also presented.

(5) The model introduced in Section 14.4.2 is a simplification of the general formulation for integrated scheduling-control problems presented in Risbeck et al. [7]. The same paper also presents solution methods for the resulting large-scale mixed-integer model; theoretical properties regarding the quality of the closed-loop schedule; and insights into the characteristics of the implemented schedules.

(6) If events that render the current schedule infeasible or suboptimal are viewed strictly as the result of uncertainty observation, then one natural approach to mitigate the impact of uncertainty is to account for it when generating the original schedule, that is, consider an *optimization problem under uncertainty* (or *stochastic optimization*). A wide range of methods adopting this approach have been studied, including robust optimization [8–13]; adjustable robust optimization [14, 15]; stochastic programming [16–22]; parametric programming [23–28]; and fuzzy programming [29, 30].

(7) It is important to note that, while stochastic optimization methods can potentially reduce the negative impact of uncertainty, they do not remove the need for iterative recomputation. One of the reasons for this, often not appreciated, is that the stochastic optimization approach considers an approximate problem. For example, a scenario-based stochastic programming model will employ a limited number of discrete scenarios; thus, the observed parameter values will not in general be the same as the values used for the scenarios, which means that there is always an advantage in reoptimizing upon uncertainty observation. Another reason is the arrival of new information, as discussed in Section 14.1.1.

(8) Real-time scheduling is the generalization of *reactive scheduling* (or *rescheduling*), which, in general, has been thought of as a recomputation, upon realization of a disruptive event that would render the previously obtained schedule infeasible of significantly suboptimal. Many methods have been proposed for reactive scheduling, including heuristics [31–35], time-automata [36], knowledge-based system [37–39], and reoptimization (i.e., solution of a new/modified MIP model) [40–45].

(9) The discussion in Section 14.3 focuses on systems where the only source of uncertainty/disturbances are the product orders/demand. While this is the most studied class of problems, there are many other types of disturbances such processing time delays, yield changes, and unit breakdowns. These disturbances are often more critical because they are observed during the course of *currently executed* batches (i.e., between real-time iterations) as opposed to being injected at a future time point. Interestingly, the observation of these disturbances depends

on the decisions of the scheduler; for example, a batch will be delayed only if it has been scheduled to be executed in the first place. Thus, if the disturbances that are injected to the currently executed batches are viewed as uncertain parameters, then the real-time scheduling problem under consideration becomes an optimization problem *under endogenous observation of uncertainty*, in which the decision maker determines if and when uncertainty will be observed. Methods to address this class of problems have been explored for other application areas, such as the exploration oil fields [46–48] and the scheduling of research and development activities [49–51].

(10) A framework to address endogenous observation of uncertainty in the context of chemical production scheduling is discussed in Gupta and Maravelias [52]. The paper discusses how endogenous observation of uncertainty can result in infeasibilities in the incumbent schedule and presents a model for systematic schedule adjustment to restore feasibility. The paper also presents a formal procedure for carrying out closed-loop simulations and evaluating closed-loop performance under endogenous observation of uncertainty.

(11) In Section 14.3.4, it was discussed how modeling modifications can be used to obtain better closed-loop solutions. Interestingly, such modifications have been routinely used in the MPC literature to achieve the same and, importantly, show various theoretical properties regarding system stability and solution quality [53]. Since no traditional reference *steady state* exists in scheduling problems, the use of a reference solution (trajectory) as a terminal constraint is of particular interest.

(12) If periodic scheduling is considered, then the periodic solution[20] can be viewed as the reference trajectory, at which the system should terminate, and real-time scheduling ensures that disturbances are rejected so the system stays (as close as possible) to the reference solution.

(13) If real-time scheduling is viewed as an MPC problem (see discussion in Section 14.1.3), then it is a problem with both discrete and continuous inputs (actuators). Rawlings and Risbeck showed that, under a set of assumptions and provided that a high-quality reference trajectory is available to be used as terminal constraint, many properties that have been established for MPC with continuous actuators hold also for MPC with both discrete and continuous actuators [54].

(14) The integration of scheduling and control decisions is an area that has attracted significant attention, though it is important to make a distinction between two different problems: (1) simultaneous scheduling and dynamic optimization and (2) integrated scheduling control. The former focuses on the formulation of the open-loop problem (without accounting for feedback), while the latter employs a formulation that includes scheduling and dynamic optimization in an iterative manner while accounting for feedback.

(15) The first papers on combined scheduling and dynamic optimization appeared in the 2000s [55–57], while the first papers on integrated scheduling-control appeared in the early 2010s [58–61]. Many papers addressing scheduling and

[20] Here we use the term *periodic solution* to describe the schedule obtained from the one-time solution of a model (Chapter 10).

dynamic optimization were based on (extensions of) the state-space formulation of Subramanian et al. [3]; see, for example [62, 63]. For a broader treatment of the integrated scheduling-control problem, the reader is pointed to a review by Engell and Harjunkoski [64] and Risbeck et al. [7]. Finally, we note that when the present chapter was being written, the integration of scheduling with process control was continuing to be a highly investigated problem, with many scientific papers appearing almost every month.

(16) In addition to providing early feedback, as discussed in Section 14.4.1, automation logic has been used directly for chemical production systems [36, 65] and chemical process systems verification [66]. Delay monitoring has also been used for performance measurements and real-time monitoring [67, 68].

(17) The impact of process dynamics [69] as well as process-level faults [70] on production schedules has also been explored recently.

(18) From an implementation standpoint, it is important to note that the scheduling solution should be integrated with the process control system to enable real-time fully closed-loop operation. Therefore, the obtained schedule, in each iteration, should be imported to the control system for implementation without human intervention. Some discussion of industrial implementation of real-time scheduling solutions can be found in Harjunkoski et al. [71] and Gupta et al. [1].

(19) The aspect of nervousness has long been discussed in the context of discrete manufacturing [72–75]. More recently, nervousness in chemical production facilities has been discussed [39, 76], including systematic methods to reduce it [77–79].

14.6 Exercises

(1) Develop the state-space model proposed by Subramanian et al. [3] (note that it is different from the one presented in Section 14.2). Use it to solve the instances defined in Exercises 1 through 3 of Chapter 7, that is, obtain the first open-loop solution (iteration 0). You should find the same solutions. What do you observe in terms of computational requirements?

(2) Use the model you developed in Exercise (1) to reproduce the results in section 5 of Subramanian et al. [3]. (Hints: you will have to develop automated methods to (1) move the horizon forward and (2) to *inject* the described disturbances.)

(3) Use the model you developed in Exercise (1) to reproduce the results in subsection 6.4 of Subramanian et al. [3]. (Hint: you will have to develop a method that ensures that the state at the last time point is one of the states in the identified periodic solution – see figure 14 of the paper.)

(4) Develop the state-space model described in Section 14.2. What differences do you see when compared to the models proposed by Subramanian et al. [3]? Do you see what the role of update equations is? Use the model to obtain the results for the motivating example in Sections 14.1.1 and 14.1.2.

(5) Use the model you developed for Exercise 4 to reproduce all the results used for motivation in Gupta et al. [1] and Gupta and Maravelias [2].

(6) Study the model proposed by Gupta and Maravelias [4]. Reproduce their case study (section 5).

References

[1] Gupta D, Maravelias CT, Wassick JM. From Rescheduling to Online Scheduling. *Chem Eng Res Des.* 2016;116:83–97.

[2] Gupta D, Maravelias CT. On Deterministic Online Scheduling: Major Considerations, Paradoxes and Remedies. *Comput Chem Eng.* 2016;94:312–330.

[3] Subramanian K, Maravelias CT, Rawlings JB. A State-Space Model for Chemical Production Scheduling. *Comput Chem Eng.* 2012;47:97–110.

[4] Gupta D, Maravelias CT. A General State-Space Formulation for Online Scheduling. *Processes.* 2017;5(4):69.

[5] Gupta D, Maravelias CT. On the Design of Online Production Scheduling Algorithms. *Comput Chem Eng.* 2019:106517.

[6] Rawlings BC, Avadiappan V, Lafortune S, Maravelias CT, Wassick JM. Incorporating Automation Logic in Online Chemical Production Scheduling. *Comput Chem Eng.* 2019;128:201–215.

[7] Risbeck MJ, Maravelias CT, Rawlings JB. Unification of Closed-Loop Scheduling and Control: State-Space Formulations, Terminal Constraints, and Nominal Theoretical Properties. *Comput Chem Eng.* 2019; 129: 106496,

[8] Mignon DJ, Honkomp SJ, Reklaitis GV. A Framework for Investigating Schedule Robustness under Uncertainty. *Comput Chem Eng.* 1995;19:615–620.

[9] Sanmartí E, Espuña A, Puigjaner L. Effects of Equipment Failure Uncertainty in Batch Production Scheduling. *Comput Chem Eng.* 1995;19:565–570.

[10] Vin JP, Ierapetritou MG. Robust Short-Term Scheduling of Multiproduct Batch Plants under Demand Uncertainty. *Ind Eng Chem Res.* 2001;40(21):4543–4554.

[11] Lin X, Janak SL, Floudas CA. A New Robust Optimization Approach for Scheduling under Uncertainty: I. Bounded Uncertainty. *Comput Chem Eng.* 2004;28(6):1069–1085.

[12] Janak SL, Lin X, Floudas CA. A New Robust Optimization Approach for Scheduling under Uncertainty: II. Uncertainty with Known Probability Distribution. *Comput Chem Eng.* 2007;31(3):171–195.

[13] Bonfill A, Espuna A, Puigjaner L. Addressing Robustness in Scheduling Batch Processes with Uncertain Operation Times. *Ind Eng Chem Res.* 2005;44(5):1524–1534.

[14] Shi H, You F. A Computational Framework and Solution Algorithms for Two-Stage Adaptive Robust Scheduling of Batch Manufacturing Processes under Uncertainty. *AlChE J.* 2016;62(3):687–703.

[15] Lappas NH, Gounaris CE. Multi-stage Adjustable Robust Optimization for Process Scheduling under Uncertainty. *AlChE J.* 2016;62(5):1646–1667.

[16] Orçun S, Kuban Altinel İ, Hortaçsu Ö. Scheduling of Batch Processes with Operational Uncertainties. *Comput Chem Eng.* 1996;20:S1191–S1196.

[17] Petkov SB, Maranas CD. Multiperiod Planning and Scheduling of Multiproduct Batch Plants under Demand Uncertainty. *Ind Eng Chem Res.* 1997;36(11):4864–4881.

[18] Balasubramanian J, Grossmann IE. A Novel Branch and Bound Algorithm for Scheduling Flowshop Plants with Uncertain Processing Times. *Comput Chem Eng*. 2002;26(1):41–57.
[19] Balasubramanian J, Grossmann IE. Approximation to Multistage Stochastic Optimization in Multiperiod Batch Plant Scheduling under Demand Uncertainty. *Ind Eng Chem Res*. 2004;43(14):3695–3713.
[20] Bonfill A, Bagajewicz M, Espuña A, Puigjaner L. Risk Management in the Scheduling of Batch Plants under Uncertain Market Demand. *Ind Eng Chem Res*. 2004;43 (3):741–750.
[21] Bonfill A, Espuña A, Puigjaner L. Proactive Approach to Address the Uncertainty in Short-Term Scheduling. *Comput Chem Eng*. 2008;32(8):1689–1706.
[22] Sand G, Engell S. Modeling and Solving Real-Time Scheduling Problems by Stochastic Integer Programming. *Comput Chem Eng*. 2004;28(6):1087–1103.
[23] Ryu J-H, Pistikopoulos EN. A Novel Approach to Scheduling of Zero-Wait Batch Processes under Processing Time Variations. *Comput Chem Eng*. 2007;31(3):101–106.
[24] Ryu J-h, Dua V, Pistikopoulos EN. Proactive Scheduling under Uncertainty: A Parametric Optimization Approach. *Ind Eng Chem Res*. 2007;46(24):8044–8049.
[25] Li Z, Ierapetritou MG. Process Scheduling under Uncertainty Using Multiparametric Programming. *AlChE J*. 2007;53(12):3183–3203.
[26] Li Z, Ierapetritou MG. Reactive Scheduling Using Parametric Programming. *AlChE J*. 2008;54(10):2610–2623.
[27] Kopanos GM, Pistikopoulos EN. Reactive Scheduling by a Multiparametric Programming Rolling Horizon Framework: A Case of a Network of Combined Heat and Power Units. *Ind Eng Chem Res*. 2014;53(11):4366–4386.
[28] Li ZK, Ierapetritou MG. Reactive Scheduling Using Parametric Programming. *AlChE J*. 2008;54(10):2610–2623.
[29] Balasubramanian J, Grossmann IE. Scheduling Optimization under Uncertainty – an Alternative Approach. *Comput Chem Eng*. 2003;27(4):469–490.
[30] Petrovic D, Duenas A. A Fuzzy Logic Based Production Scheduling/Rescheduling in the Presence of Uncertain Disruptions. *Fuzzy Sets and Systems*. 2006;157(16):2273–2285.
[31] Cott BJ, Macchietto S. Minimizing the Effects of Batch Process Variability Using Online Schedule Modification. *Comput Chem Eng*. 1989;13(1):105–113.
[32] Kanakamedala KB, Reklaitis GV, Venkatasubramanian V. Reactive Schedule Modification in Multipurpose Batch Chemical Plants. *Ind Eng Chem Res*. 1994;33(1):77-90.
[33] Huercio A, Espuña A, Puigjaner L. Incorporating On-Line Scheduling Strategies in Integrated Batch Production Control. *Comput Chem Eng*. 1995;19:609–614.
[34] Sanmartí E, Huercio A, Espuña A, Puigjaner L. A Combined Scheduling/Reactive Scheduling Strategy to Minimize the Effect of Process Operations Uncertainty in Batch Plants. *Comput Chem Eng*. 1996;20:S1263–S1268.
[35] Ko D, Moon I. Rescheduling Algorithms in Case of Unit Failure for Batch Process Management. *Comput Chem Eng*. 1997;21:S1067–S1072.
[36] Panek S, Engell S, Subbiah S, Stursberg O. Scheduling of Multi-product Batch Plants Based upon Timed Automata Models. *Comput Chem Eng*. 2008;32(1):275–291.
[37] Henning GP, Cerdá J. Knowledge-Based Predictive and Reactive Scheduling in Industrial Environments. *Comput Chem Eng*. 2000;24(9):2315–2338.
[38] Palombarini J, Martínez E. SmartGantt – an Interactive System for Generating and Updating Rescheduling Knowledge Using Relational Abstractions. *Comput Chem Eng*. 2012;47:202–216.

[39] Novas JM, Henning GP. Reactive Scheduling Framework Based on Domain Knowledge and Constraint Programming. *Comput Chem Eng*. 2010;34(12):2129–2148.

[40] Elkamel ALI, Mohindra A. A Rolling Horizon Heuristic for Reactive Scheduling of Batch Process Operations. *Engineering Optimization*. 1999;31(6):763–792.

[41] Vin JP, Ierapetritou MG. A New Approach for Efficient Rescheduling of Multiproduct Batch Plants. *Ind Eng Chem Res*. 2000;39(11):4228–4238.

[42] Sand G, Engell S, Märkert A, Schultz R, Schulz C. Approximation of an Ideal Online Scheduler for A Multiproduct Batch Plant. *Comput Chem Eng*. 2000;24(2):361–367.

[43] Méndez CA, Cerdá J. Dynamic Scheduling in Multiproduct Batch Plants. *Comput Chem Eng*. 2003;27(8):1247–1259.

[44] Munawar SA, Gudi RD. A Multilevel, Control-Theoretic Framework for Integration of Planning, Scheduling, and Rescheduling. *Ind Eng Chem Res*. 2005;44(11):4001–4021.

[45] Janak SL, Floudas CA, Kallrath J, Vormbrock N. Production Scheduling of a Large-Scale Industrial Batch Plant. II. Reactive Scheduling. *Ind Eng Chem Res*. 2006;45(25):8253–8269.

[46] Goel V, Grossmann IE. A stochastic Programming Approach to Planning of Offshore Gas Field Developments under Uncertainty in Reserves. *Comput Chem Eng*. 2004;28(8):1409–1429.

[47] Goel V, Grossmann IE, El-Bakry AS, Mulkay EL. A Novel Branch and Bound Algorithm for Optimal Development of Gas Fields under Uncertainty in Reserves. *Comput Chem Eng*. 2006;30(6-7):1076–1092.

[48] Tarhan B, Grossmann IE, Goel V. Stochastic Programming Approach for the Planning of Offshore Oil or Gas Field Infrastructure under Decision-Dependent Uncertainty. *Ind Eng Chem Res*. 2009;48(6):3078–3097.

[49] Colvin M, Maravelias CT. A Stochastic Programming Approach for Clinical Trial Planning in New Drug Development. *Comput Chem Eng*. 2008;32(11):2626–2642.

[50] Colvin M, Maravelias CT. Scheduling of Testing Tasks and Resource Planning in New Product Development Using Stochastic Programming. *Comput Chem Eng*. 2009;33(5):964–976.

[51] Colvin M, Maravelias CT. Modeling Methods and a Branch and Cut Algorithm for Pharmaceutical Clinical Trial Planning Using Stochastic Programming. *Eur J Oper Res*. 2010;203(1):205–215.

[52] Gupta D, Maravelias CT. Framework for Studying Online Production Scheduling under Endogenous Uncertainty. *Comput Chem Eng*. 2019:135, 106670.

[53] Rawlings JB, Mayne DQ. *Model Predictive Control: Theory and Design*. Madison: Nob Hill Pub., 2009.

[54] Rawlings JB, Risbeck MJ. Model Predictive Control with Discrete Actuators: Theory and Application. *Automatica*. 2017;78:258–265.

[55] Nystrom RH, Franke R, Harjunkoski I, Kroll A. Production Campaign Planning Including Grade Transition Sequencing and Dynamic Optimization. *Comput Chem Eng*. 2005;29(10):2163–2179.

[56] Flores-Tlacuahuac A, Grossmann IE. Simultaneous Cyclic Scheduling and Control of a Multiproduct CSTR. *Ind Eng Chem Res*. 2006;45(20):6698–6712.

[57] Terrazas-Moreno S, Flores-Tlacuahuac A, Grossmann IE. Simultaneous Cyclic Scheduling and Optimal Control of Polymerization Reactors. *AIChE J*. 2007;53(9):2301–2315.

[58] Zhuge J, Ierapetritou MG. Integration of Scheduling and Control with Closed Loop Implementation. *Ind Eng Chem Res*. 2012;51(25):8550–8565.

[59] Chu Y, You F. Integration of Scheduling and Control with Online Closed-Loop Implementation: Fast Computational Strategy and Large-Scale Global Optimization Algorithm. *Comput Chem Eng*. 2012;47:248–268.
[60] Gutiérrez-Limón MA, Flores-Tlacuahuac A, Grossmann IE. MINLP Formulation for Simultaneous Planning, Scheduling, and Control of Short-Period Single-Unit Processing Systems. *Ind Eng Chem Res*. 2014;53(38):14679–14694.
[61] Du J, Park J, Harjunkoski I, Baldea M. A Time Scale-Bridging Approach for Integrating Production Scheduling and Process Control. *Comput Chem Eng*. 2015;79:59–69.
[62] Nie Y, Biegler LT, Wassick JM, Villa CM. Extended Discrete-Time Resource Task Network Formulation for the Reactive Scheduling of a Mixed Batch/Continuous Process. *Ind Eng Chem Res*. 2014; 53(44):17112–17123.
[63] Nie Y, Biegler LT, Villa CM, Wassick JM. Discrete Time Formulation for the Integration of Scheduling and Dynamic Optimization. *Ind Eng Chem Res*. 2015;54 (16):4303–4315.
[64] Engell S, Harjunkoski I. Optimal Operation: Scheduling, Advanced Control and Their Integration. *Comput Chem Eng*. 2012;47:121–133.
[65] Kim J, Kim J, Moon I. Error-Free Scheduling for Batch Processes Using Symbolic Model Verifier. *Journal of Loss Prevention in the Process Industries*. 2009;22(4):367–372.
[66] Rawlings BC, Wassick JM, Ydstie BE. Application of Formal Verification and Falsification to Large-Scale Chemical Plant Automation Systems. *Comput Chem Eng*. 2018;114:211–220.
[67] Suresh P, Wassick JM, Ferrio J, editors. Real Time Performance Measurement for Batch Chemical Plants. *Wint Simul C Proc*; 2011;12325–2335.
[68] Faggian A, Facco P, Doplicher F, Bezzo F, Barolo M. Multivariate Statistical Real-Time Monitoring of an Industrial Fed-Batch Process for the Production of Specialty Chemicals. *Chemical Engineering Research and Design*. 2009;87(3):325–334.
[69] Pattison RC, Touretzky CR, Harjunkoski I, Baldea M. Moving Horizon Closed-Loop Production Scheduling Using Dynamic Process Models. *AIChE J*. 2017;63 (2):639–651.
[70] Touretzky CR, Harjunkoski I, Baldea M. Dynamic Models and Fault Diagnosis-Based Triggers for Closed-Loop Scheduling. *AIChE J*. 2017;63(6):1959–1973.
[71] Harjunkoski I, Maravelias CT, Bongers P, Castro PM, Engell S, Grossmann IE, et al. Scope for Industrial Applications of Production Scheduling Models and Solution Methods. *Comput Chem Eng*. 2014;62(0):161–193.
[72] Blackburn JD, Kropp DH, Millen RA. A Comparison of Strategies to Dampen Nervousness in MRP Systems. *Manage Sci*. 1986;32(4):413–429.
[73] Sridharan V, Berry WL. Master Production Scheduling Make-to-Stock Products: A Framework for Analysis. *International Journal of Production Research*. 1990;28 (3):541–558.
[74] Wu SD, Storer RH, Pei-Chann C. One-Machine Rescheduling Heuristics with Efficiency and Stability as Criteria. *Computers & Operations Research*. 1993;20(1):1–14.
[75] Kazan O, Nagi R, Rump CM. New Lot-Sizing Formulations for Less Nervous Production Schedules. *Computers & Operations Research*. 2000;27(13):1325–1345.
[76] Kopanos GM, Capon-Garcia E, Espuna A, Puigjaner L. Costs for Rescheduling Actions: A Critical Issue for Reducing the Gap between Scheduling Theory and Practice. *Ind Eng Chem Res*. 2008;47(22):8785–8795.

[77] McAllister RD, Rawlings JB, Maravelias CT. Rescheduling Penalties for Economic Model Predictive Control and Closed-Loop Scheduling. *Ind Eng Chem Res*. 2019; 59 (6):2214–2228.

[78] Lee H, Gupta D, Maravelias CT. Systematic Generation of Alternative Production Schedules. *AIChE J*. 2020; 66(5):e16926.

[79] Mathur P, Swartz CLE, Zyngier D, Welt F. Uncertainty Management via Online Scheduling for Optimal Short-Term Operation of Cascaded Hydropower Systems. *Comput Chem Eng*. 2020;134:106677.

15 Integration of Production Planning and Scheduling

In Chapter 1, we introduced the supply chain (SC) planning matrix and its different planning functions, discussed how scheduling fits within this matrix, and mentioned that integration across planning functions can lead to better solutions. Chemical production scheduling interacts directly with two other functions: (1) production planning and (2) process automation and control, though the latter are not typically defined as a function of the SC matrix. The integration with automation and control were discussed in the previous chapter. In the present chapter, we will discuss the integration of production planning and scheduling.

We start, in Section 15.1, with some preliminary concepts and motivation for the need to integrate planning with scheduling. In Section 15.2, we present a formulation for a simplified, introductory planning-scheduling problem. We continue, in Section 15.3, with an approach for more complex problems, both in single- and multiunit environments. Finally, in Section 15.4, we overview a more general but also algorithmically more advanced approach that is applicable to any production environment. For simplicity, in Sections 15.2 and 15.3, we do not consider special processing features, such as complex storage policies and utility constraints. The method in Section 15.4 can in principle be applied to any facility with any processing feature.

15.1 Preliminaries

A formal definition of production planning is given in Section 15.1.1. An example is introduced in Section 15.1.2 to motivate why it is often necessary to integrate production planning and scheduling. Then, in Section 15.1.3, we reintroduce lot sizing, a basic MIP problem already discussed in Chapter 2, which is often used as basis for addressing production planning problems.

15.1.1 Production Planning

The objective in production planning is to fulfill customer demand at minimum total (i.e., production plus inventory) cost. Formally, we are given the following:

(1) A planning horizon divided into a set **T** of planning (time) periods.[1]
(2) A set **I** of products (items)[2]; product $i \in \mathbf{I}$ has unit inventory holding cost θ_i [\$/(kg·period)], and total customer demand $\xi_{it}(< 0)$ [kg], due at the end of period $t \in \mathbf{T}$.
(3) Resource capacities.
(4) Production costs.

The optimization decisions include the following:

(1) Production amount (target) $P_{it} \in \mathbb{R}_+$ of item i in period t.
(2) Inventory level $I_{it} \in \mathbb{R}_+$ of item i at the end of period t.

Note that only final products are considered, although multiple raw materials and intermediates may be necessary for their production.

If we assume that demand can always be satisfied, then a general formulation for production planning consists of (15.1) through (15.5), where vector $\boldsymbol{P}_t = \left[P_{1,t}, P_{2,t},.., P_{|\mathbf{I}|,t}\right]^T$ represents all production targets at the end of period t:

$$\min \sum\nolimits_t \{C_t^{PRD} + C_t^{INV}\} \tag{15.1}$$

$$f(\boldsymbol{P}_t) \le 0, \quad t \tag{15.2}$$

$$C_t^{PRD} = g(\boldsymbol{P}_t), \quad t \tag{15.3}$$

$$I_{it} = I_{i,t-1} + P_{it} + \xi_{it} \le \chi_i^M, \quad i,t \tag{15.4}$$

$$C_t^{INV} = \sum\nolimits_i \theta_i I_{it}, \quad t. \tag{15.5}$$

The region of feasible targets is described, generically, by (15.2); while the corresponding production cost during period t, C_t^{PRD}, given a set of production targets, is calculated in (15.3). Equation (15.4) is a material balance for item i across periods, where we assume that all the production during period t becomes available at the end of the period; and (15.5) calculates inventory cost in period t. The objective function includes production and inventory costs,

Functions $f(\boldsymbol{P}_t)$ and $g(\boldsymbol{P}_t)$ depend on the characteristics of the process network and often involve a large number of constraints as well as additional variables. In general, the complexity of these functions increases with the complexity of the production facility and the desired accuracy of the feasibility and cost information provided by $f(\boldsymbol{P}_t)$ and $g(\boldsymbol{P}_t)$. Interestingly, the differences across the various methods for the

[1] Recall that in previous chapters we used index t (n) to denote points/periods in grid-based continuous (discrete) time models. In the present chapter, we use index t to denote points and planning periods because we do not discuss grid-based continuous time scheduling models, whereas we briefly discuss, in Section 15.4, discrete time models. Thus, in the present chapter index t is reserved for planning periods/points that are assumed to be given, that is, discrete.

[2] If a facility is expressed using the STN representation, then final products (items) are a subset of materials. However, to stay consistent with the notation in production planning literature, and since STN-based scheduling models are not discussed extensively, we use the terms *products* and *items*.

integrated planning-scheduling problem lies exactly in the form of functions $f(\boldsymbol{P}_t)$ and $g(\boldsymbol{P}_t)$.

If the demand cannot be satisfied in every period, then two variants can be considered (see also discussion in Section 7.2.2). In the first one, unsatisfied demand is backlogged and a backlog penalty is paid until the backlogged demand is satisfied. In this case, (15.4) is replaced by

$$I_{it} = I_{i,t-1} + P_{it} - V_{it} \leq \chi_i^M, \quad i, t, \tag{15.6}$$

where demand $(-\xi_{it})$ is replaced by actual shipments, V_{it}; backlogs, I_{it}^B, are calculated as follows:

$$I_{it}^B = I_{i,t-1}^B - (\xi_{it} + V_{it}), \quad i, t. \tag{15.7}$$

The total backlog cost in period t is calculated by (15.8), where π_i^{BCK} is the unit backlog cost [\$/(kg·period)] ,

$$C_t^{BCK} = \sum\nolimits_i \pi_i^{BCK} I_{it}^B, \quad t, \tag{15.8}$$

and the objective function becomes

$$\min \sum\nolimits_t \left\{ C_t^{PRD} + C_t^{INV} + C_t^{BCK} \right\}. \tag{15.9}$$

Note that if (15.7) is used to calculate $V_{it} = I_{i,t-1}^B - I_{it}^B - \xi_{it}$, then we can replace (15.6) and (15.7) with a single overall material balance:

$$I_{it} - I_{it}^B = I_{i,t-1} - I_{i,t-1}^B + P_{it} + \xi_{it} \leq \chi_i^M, \quad i, t. \tag{15.10}$$

In the second one, unsatisfied demand is discarded at a lost-sales penalty. Compared to the basic model consisting of (15.1) through (15.5), material balance is replaced by (15.6); lost sales, I_{it}^{LS}, are calculated by

$$I_{it}^{LS} = -\xi_{it} - V_{it}, \quad i, n, \tag{15.11}$$

the lost sales cost is calculated as

$$C_t^{LS} = \sum\nolimits_i \pi_i^{LS} I_{it}^{LS}, \quad t, \tag{15.12}$$

and the objective function becomes

$$\min \sum\nolimits_t \left\{ C_t^{PRD} + C_t^{INV} + C_t^{LS} \right\}. \tag{15.13}$$

Finally, we note that production planning can be represented as a network problem, with a node for each item and time period, and arcs for the production, demand, and inventory variables (see Figure 15.1A). Specifically, the material balance for an item at a time point, (15.6), can be viewed as flow balance around a node. The network representation can be extended to include backlog arcs (Figure 15.1B). In this case, the flow balance represents the overall material balance for an item at time point, as

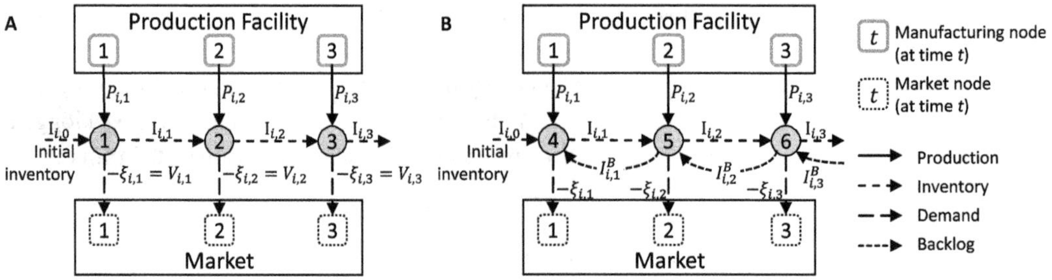

Figure 15.1 Network representation of production planning for a single item. (A) Representation when demand can be met. (B) Representation when demand cannot be met, so backlogs occur.

given in (15.10). To illustrate this, we rewrite (15.10) for $t = 2$ so as all outgoing/incoming flows are in the LHS/RHS (recall that $\xi_{it} < 0$):

$$I_{i,2} + I^B_{i,1} + (-\xi_{i,2}) = I_{i,1} + I^B_{i,2} + P_{i,2}, \quad i. \tag{15.14}$$

Finally, we note that manufacturing nodes are subject to resource constraints bounding production flows (P_{it}) and market nodes generate demand flows $(-\xi_{it})$.

15.1.2 Motivation

To illustrate why integration between production planning and scheduling is often necessary, we study a simple motivating example.

Example 15.1 Why Integrate Planning with Scheduling? We consider the process network shown in Figure 15.2. Raw material RM is converted by continuous task T1, carried out in unit U1, into unstable intermediate INT (no storage available), which should be immediately converted by continuous task T2, carried out in unit U2, into final product A or by continuous task T3, in U3, into final product B. Since all tasks have dedicated units, we denote each task as $j - i$ (e.g., U1-T1). The maximum rate of unit U1 is 300 kg/day, while the minimum rate is 200 kg/day. Similarly, unit U2 is operated between 150 and 200 kg/day or else turned off. If on, then unit U3 is operated at a fixed rate of 150 kg/day. Note that U1 is able to achieve an *average* rate lower than 200 kg/day by being turned off for some time. If we use the techniques presented in Chapter 9, then all tasks can be modeled as batch tasks having duration of one scheduling period (δ). However, here we are interested in understanding what combinations of production of products A and B can be produced in an one-week planning period.

Given the structure of the network and unit processing rates, task U1-T1 appears to be the bottleneck. Thus, we can use the maximum rate of unit U1 to generate a weekly aggregate capacity constraint which should be satisfied by all feasible $(P_{A,t}, P_{B,t})$:

$$P_{A,t} + P_{B,t} \le 2,100. \tag{15.15}$$

The feasible region for $(P_{A,t}, P_{B,t})$ defined by (15.15) is shown in Figure 15.3A.

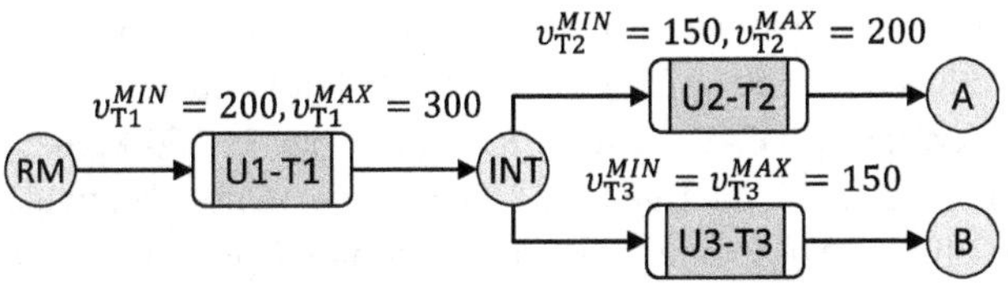

Figure 15.2 STN representation and associated data for motivating example; processing rate bounds (kg/day) given above corresponding units-tasks.

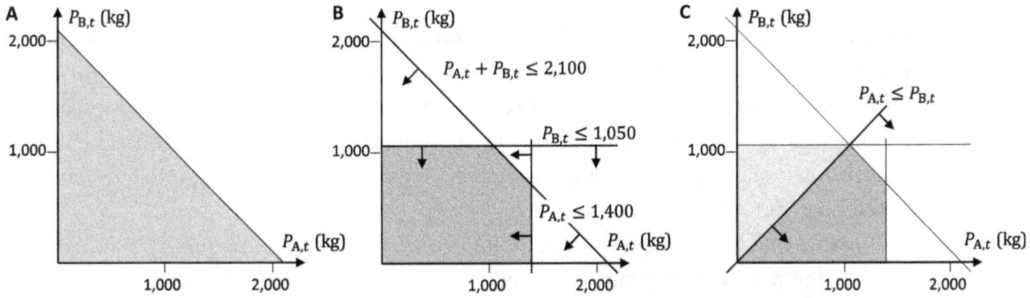

Figure 15.3 Approximations of the feasible region of (weekly) production amounts for the motivating example. (A) Approximation based on capacity of U1-T1. (B) Approximation based on capacities of all units. (C) Approximation based on constraints in B plus the impact of the no storage constraint for INT.

Further, the production of products A and B are bounded by the rates of U2-T2 and U3-T3, respectively, so we can generate two additional valid constraints, which lead to the feasible region shown in Figure 15.3B:

$$P_{A,t} \leq 1,400 \tag{15.16}$$

$$P_{B,t} \leq 1,050 \tag{15.17}$$

At this point, we have considered the processing rates of all tasks, so one would think that we have obtained the true feasible region of production amounts, henceforth referred to as process attainable region (PAR) and denoted by $\mathbb{P}$. However, it turns out that the region shown in Figure 15.3B is not a good approximation of $\mathbb{P}$; for example, if a detailed scheduling model is used to check if a given point $(P'_{A,t}, P'_{B,t})$ is feasible, then we can readily verify that no point above the 45^0 line is feasible. In other words, all feasible points should also satisfy

$$P_{A,t} - P_{B,t} \leq 0, \tag{15.18}$$

which results in the updated approximation of $\mathbb{P}$ shown in Figure 15.3C. (Can you see why (15.18) is valid?)

The reason that the seemingly counterintuitive (15.18) is valid is the lack of storage for intermediate INT, which causes A and B to be produced at equal amounts whenever B is produced. This is because U1-T1 cannot provide exactly

150 kg/day (since its minimum rate is 200 kg/day) to U3-T3, which means that whenever B is produced, at 150 kg/day, then U1-T1 should operate at 300 kg/day to also provide 150 kg/day to U2-T2, whose minimum rate is 150 kg/day. But is the region shown in Figure 15.3C the true PAR? The answer is no. Can you see why, for example, point (1,400, 700), which is the intersection of $(P_{A,t} + P_{B,t} = 2{,}100)$ and $(P_{A,t} = 1{,}400)$, would be infeasible?

Example 15.1 illustrates that even for simple networks it is often challenging to develop constraints that will yield a good approximation of $\mathbb{P}$, that is, develop constraints $f(\boldsymbol{P}_t) \leq 0$ that can be used directly in (15.2) to solve the planning problem. Similarly, it is often difficult to generate accurate constraints $g(\boldsymbol{P}_t)$, in terms of $\boldsymbol{P}_t$ variables only. To address these challenges, the production planning formulation should be augmented with a submodel that will provide scheduling (feasibility and cost) information, thus leading to the integrated planning-scheduling model.

We note that from an operational point of view, production targets P_{it} obtained by solving the production planning-scheduling problem are used as inputs to scheduling, thus leading to a hierarchical flow of information from planning to scheduling. However, to obtain feasible and (near-) optimal production targets, P_{it}, process capacity and cost information have to be communicated from the scheduling to the production planning formulation via the integration of some form of (an approximation of) a scheduling model. In this context, the integration of planning and scheduling is necessary primarily in order to obtain a feasible and (near-) optimal production planning solution, not to obtain a scheduling solution that will be readily implemented.

15.1.3 Lot Sizing

The first basic formulation that can be used to address planning problems is the multi-item capacitated lot-sizing (LS) formulation with setup times and costs, where all items are assumed to be produced by a single resource.[3] In addition to unit holding costs, θ_i, and demands, ξ_{it}, the resource constraints (item 3 in the problem statement in Section 15.1.1) are described in terms of production rates υ_i [kg/h] for all items, and availability η_t [h] of the resource during period t; and the production costs (item 4 in the statement in Section 15.1.1) include item-specific unit processing, γ_i^{PR} [\$/kg], and setup, γ_i^{SET} [\$], costs.

[3] Note that here we use the more general term *resource* rather than *unit*. This is because the LS model can be used to address problems in more *complex* systems. For example, a production line, which can consist of multiple units, often has product-specific but unique (i.e., independent of the production mix) processing rates, which means that it can be treated as a unit. Similarly, a facility may have a bottleneck unit that determines the overall rate; thus, an LS model applied to the bottleneck unit may be sufficient to provide a planning solution for the entire facility.

The objective function includes processing[4] and inventory costs,

$$\min \sum\nolimits_{i,t} \left(\gamma_i^{PR} P_{it} + \theta_i I_{it} \right). \tag{15.19}$$

The material balance for item i is written as in (15.4), and the resource capacity constraint is

$$\sum\nolimits_i (1/v_i) P_{it} \leq \eta_t, \quad t, \tag{15.20}$$

where $(1/v_j)P_{it}$ represents the time allocated to production of item i during period t, and the LHS represents the total processing time.

If a setup time σ_i is required if item i is produced in a period and/or this setup incurs a cost γ_i^{SET}, then $X_i \in \{0, 1\}$ is one if item i is produced during period t. The objective function becomes

$$\min \sum\nolimits_{i,t} \left(\gamma_i^{PR} P_{it} + \gamma_i^{SET} X_{it} + \theta_i I_{it} \right), \tag{15.21}$$

the capacity constraint includes setup time,

$$\sum\nolimits_i \left[\left(\frac{1}{v_i} \right) P_{it} + \sigma_i X_{it} \right] \leq \eta_t, \quad t, \tag{15.22}$$

and the following constraint is used to activate X_{it},

$$P_{it} \leq \left(\frac{\eta_t - \sigma_i}{v_i} \right) X_{it}, \quad i, t, \tag{15.23}$$

where the ratio in the RHS represents the maximum production of i during period t.

Unmet demand can be readily handled using the extensions presented in Section 15.1.1.

15.2 Generalized Capacitated Lot Sizing

Lot sizing is the backbone of many production planning systems and has been studied extensively in the literature. It has been extended to account for a series of features, such as overtime, product substitutes, and capacity utilization. In this section, we present a model that employs ideas from LS but overcomes some of the limitations of the basic model presented in the previous section, thereby illustrating how flexible LS-based models can be in addressing a range of problems. Specifically, we consider the so-called generalized capacitated lot- sizing (GCLS) problem and propose a MIP model for its solution.

In Section 15.2.1, we introduce the concepts of setup *cross over* and setup *carry over*, and explain why they should be taken into account. In Section 15.2.2, we present some

[4] Note that we use the term *production* (PRD) cost to include all production-related costs (e.g., setup, changeover, processing); and the term processing (PR) to represent the costs that are proportional to the produced amount, P_{it}.

basic modeling concepts, and, in Section 15.2.3, we present properties that all optimal solutions have. In Sections 15.2.4 and 15.2.5, we utilize the concepts introduced in Section 15.2.2 and exploit the properties presented in Section 15.2.3 to formulate the model for the GCLS problem.

15.2.1 Motivation

Some of the key features and assumptions of the basic LS model are the following:

(1) The sequence in which items are processed during a period is undetermined.
(2) There is no *memory* across periods; for example, if only one item is produced during two consecutive periods, the corresponding setup time will be included in (15.22) for both periods, although only one setup will be necessary.
(3) Setup times should start and finish within a period.
(4) All setup times are shorter than the length of all planning periods (i.e., $\sigma_i < \eta_t, \forall i, t$).

The first feature is common across most LS-based models, and does not typically impact the quality and accuracy of planning solutions. The remaining ones, however, can lead to suboptimal solutions or even models that are infeasible, although the underlying actual planning problem has a feasible solution. We illustrate using a series of small instances based on a two-item, three-period example with the setup, processing, and backlog costs, as well as processing rates given in Figure 15.4A, and zero unit holding costs.

First, we consider an instance with $\sigma_A = \sigma_B = 1$ and the demands given in the top panel of Figure 15.4B. Note that item A has demand in periods 1 and 2 only, and item B has demand in period 3 only. The fact that B is not produced in periods 1 and 2 means that the setup for A has to occur only once, leading to the solution shown in the Gantt chart in the middle panel of Figure 15.4B, labeled as "Best," with a total production cost equal to 22 (two setups at 5, plus variable cost of 12). However, according to (15.23) in the LS formulation, if production of an item occurs during a period, then the corresponding setup should occur, which means that the optimal solution that can be obtained

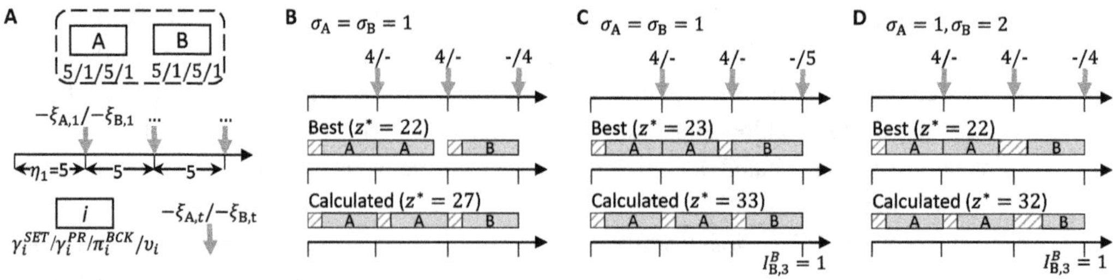

Figure 15.4 Example illustrating limitations of basic lot-sizing formulation; figures (B) through (D) show setup times and demands (top), best feasible solution (middle), and optimal solution of the LS model (bottom). (A) Data for two products (A and B), three period instances, and demand representation. (B) Instance 1. (C) Instance 2. (D) Instance 3.

using the LS model is the one shown in the Gantt chart in the bottom panel of Figure 15.4.B, labeled as "Calculated," with a total cost of 27 (three setups plus variable processing costs of 12).

Second, we consider an instance with $\sigma_A = \sigma_B = 1$ and the demands given in the top panel of Figure 15.4C. The only difference compared to previous instance is that the demand for item B in period 3 is 5 instead of 4. The increased demand can be met if the setup for item B occurs at the end of period 2, so that the entire period 3 is devoted to producing B. This solution is shown in the Gantt chart in the middle panel of Figure 15.4C, and has a total cost of 23. However, this feasible, in reality, solution is infeasible for the LS model, because if an item is produced in a period, then its setup should occur in the same period. Thus, the optimal solution obtained using the LS model is the one shown in the bottom panel of Figure 15.4C, with one unit of backlog for item B in period 3, and a total cost of 33 (three setups at 5 each plus variable costs of 13, plus backlog cost of 5).

Third, we consider an instance with $\sigma_A = 1$, $\sigma_B = 2$, and the demands given in the top panel of Figure 15.4D. The only difference compared to the first instance is that the setup for item B is 2 instead of 1. The solution shown in the Gantt chart in the middle panel of Figure 15.4D, where the setup for item B starts at the end of period 2 and ends within period 3, satisfies demand and has a total cost of 22. However, this solution is infeasible for the LS model, the optimal solution of which is shown in the Gantt chart in the bottom panel of Figure 15.4D, with a total cost of 32.

These three instances illustrate two major limitations of the basic LS model:

- If an item is produced last in period $t - 1$ and first in period t, then there are two setups; that is, setup *carry over* across the boundary[5] between the two periods is not allowed.
- A setup for the production of an item in period t cannot start in period $t - 1$ (and end at or after the beginning of period t); that is, setup *cross over* is not allowed.

These two limitations can lead to suboptimal solutions, especially if setup times are relatively large compared to the length of a time period, as is the case, for example, in biotech and fine chemical manufacturing. In the next subsection, we present the basic ideas underpinning the development of a model that overcomes these limitations.

15.2.2 Basic Concepts

The following assumptions are made:

(1) Production within each time period instantaneously becomes available at the end of that time period, and demand fulfillment occurs only at time period boundaries. Thus, inventory changes occur only at period boundaries.

[5] The *boundary* between two periods is essentially what we have termed *time point* in the previous chapters. We use this term because the model presented in this section is based on a representation where this *dividing* point is moved to facilitate modeling and accounting of the various costs.

(2) Production of each product is continuous, that is, it is not restricted to discrete increments (e.g., batches). This assumption is reasonable for batch processes as long as batchsizes are relatively small and processing times are relatively short compared to the time period length.

(3) Setup involves product-dependent but sequence-independent time and cost. Setup cost is accrued at the beginning of the setup.

(4) Processing rates are constant, but production may be interrupted by idle time. Also, no setup is required to begin or end idle time, but this assumption can be relaxed.

The modeling of setup cross over is challenging because setups can be partially completed at time period boundaries, which means that we must deduct, in correct proportion, from the capacity of every time period that a setup is located in. To address this limitation, a "modified" time period representation is introduced.

Modified (Planning) Time Period Representation. The main idea is to move time period boundaries such that every setup takes place entirely within a modified period, while the production levels and setup costs of the modified representation match exactly the corresponding levels and costs of the "unmodified" one. In other words, we map any feasible solution of the original problem onto an equivalent solution/representation that allows us to effectively handle setup cross over and carry over. Figure 15.5 shows how the boundary of a period is moved so that every setup (barely) fits into the period it begins in. Note that setups begin (therefore setup costs accrue) in the same time period in both representations, and production amounts are also the same. Therefore, the two representations differ only in where boundaries are located. Figure 15.6 illustrates why a boundary that is crossed over by a setup should be moved exactly to the end of the setup.

Modes and Setups. A unit is in "mode i" if it is producing or is ready to produce product i. The unit is in no mode during setups because no product may be produced. While in mode i, the unit may either produce item i or be idle. Idle time does not affect setup and may occur in any mode. Since boundaries are moved such that they are not crossed over by setups, modified time periods must start in some mode and end in some mode. Furthermore, the ending mode of a modified period is necessarily the starting mode of its succeeding modified period. Relative to period t, we classify setups as "short" if $\sigma_i \leq \eta_t$ and "long" if $\sigma_i > \eta_t$. Note that some setups may sometimes be short and sometimes long, depending on the time period of interest.

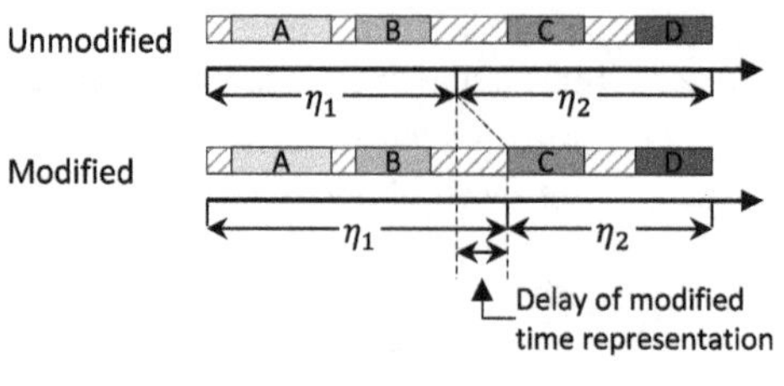

Figure 15.5 Illustration of modified time representation.

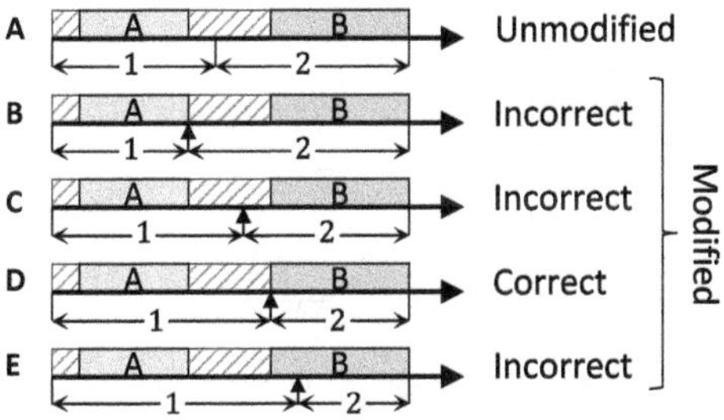

Figure 15.6 Illustration of correct boundary placement. (A) Gantt chart of solution and unmodified representation. (B) Boundary is moved to (or before) the beginning of the setup; setup cost is incorrectly assigned to period 2. (C) Boundary is moved somewhere between the beginning and finish of setup; setup cross over is not removed. (D) Boundary moved to the end of setup; the two representations are consistent. (E) Boundary moved past the end of setup; some production of B is incorrectly assigned to period 1.

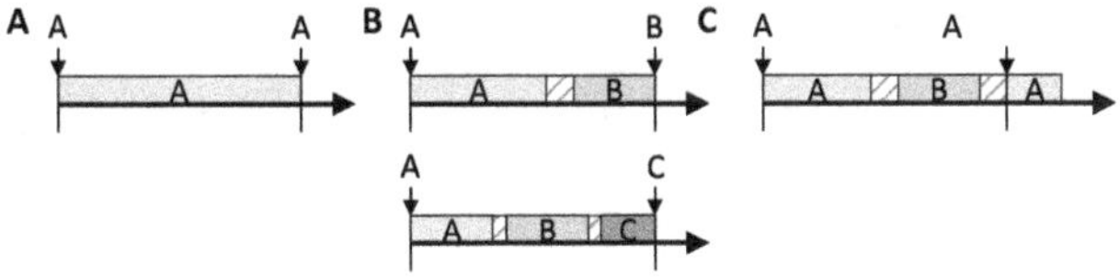

Figure 15.7 Examples of modified planning periods operated under SP (A), MP (B), and RP (C) campaigns.

Operating Campaigns.[6] If we use the modified time representation, then every feasible solution falls in exactly one of the following three types of production campaigns (see Figure 15.7):

- *Single product* (*SP*): The starting and ending modes are the same, and the unit visits no other mode.
- *Multiple product* (*MP*): The starting and ending modes are different.
- *Return product* (*RP*): The starting and ending modes are the same, but at least one other mode is visited in between.

To account for setups whose lengths are longer than a time period, we introduce the setup in progress (SIP) campaign, which is a special case of the SP campaign. In SIP, the modified period has zero length, so no production or idle time is allowed. Figure 15.8 shows how the SIP campaign is triggered when the same setup crosses over two consecutive boundaries. Note that a short setup ($\sigma_i \leq \eta_t$) cannot completely span a period and therefore cannot trigger an SIP campaign.

[6] The term *campaign* is used somewhat liberally here, to describe a relationship between the starting and ending mode, rather than the complete sequence of modes a unit visits during a (modified) planning period. Thus, its use differs from its use in the context of periodic scheduling (Chapter 10).

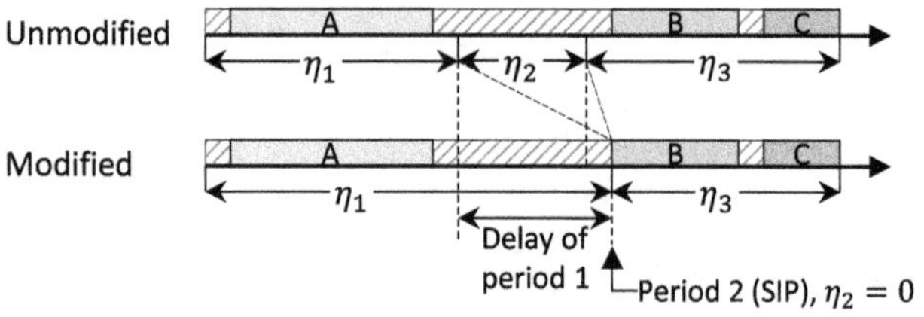

Figure 15.8 Illustration of setup in progress (SIP) campaign. Modified period 2 has zero length; setup for B starts in period 1, in both representations, and production of B occurs only in period 3, in both representations.

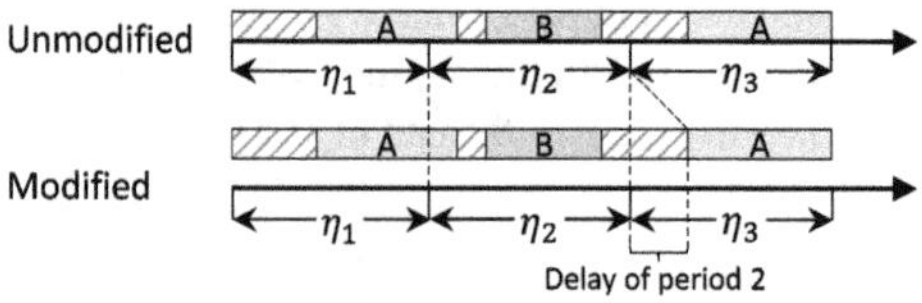

Figure 15.9 Illustration of RP campaign.

15.2.3 Solution Properties

Based on the assumptions and concepts presented in the previous subsection, we present a number of properties that are exploited by the formulation for the GCLS, presented in Section 15.2.4.

Sequencing. The sequencing of modes that are visited after the first and before the last modes of a period (henceforth referred to as *middle* modes) does not have to be considered because setups are sequence independent and production is assumed to become available at the end of the period. Thus, if four or more modes are visited, the sequencing of the middle modes does not affect the objective function. As will be discussed in the next subsection, the GCLS model exploits this by tracking only the starting and ending state of each modified time bucket.

Return Product (RP) Campaign. The RP mode is a special case in which a product can optimally be visited twice during the same time period. This can occur only if (1) production of an item i in period t is continued from the previous period $(t - 1)$, (2) at least one other product i' is visited in t, and (3) product i is setup in preparation for production in $t + 1$. Thus, product i is visited twice (starting and ending mode), but only setup once (ending mode). Note that an equivalent solution can be obtained by moving the production from the ending mode to the starting mode. The modified representation under RP campaign is illustrated in Figure 15.9. It is assumed that production occurs only in the starting mode, so that a matching unmodified representation (top panel) exists. Note that production levels and the beginning of setups stay the same in the modified representation.

Number of Setups. As we have discussed, the ending mode of a modified period is necessarily the starting mode of its succeeding modified period. This implies that the starting mode of a period (except the first one) never requires a setup. Furthermore, if a product is produced during two or more distinct intervals in a period,[7] then an equivalent or better solution can be found by moving all production to one interval and then removing redundant setups. Therefore, in an optimal solution each product is set up at most once in each modified period. Clearly, in an SP campaign the number of modes visited is one and the number of setups is zero. In an MP campaign, the number of setups required is one less than the number of products visited because every product is visited once and there is no setup for the first product. In an RP campaign, the number of setups required is equal to the number of products visited because the starting and ending mode involve the same product and the starting mode does not require a setup.

Delayed Time Boundaries. The ending boundary of a period can be delayed in the modified representation only if a setup crosses over it. Since such a delay is reserved for setups, to forbid production during delay, the ending boundary is moved exactly to the end of the setup (see Figure 15.6). Based on the campaign under which a modified period is operated, we can infer the following about its ending boundary:

- *SP but not an SIP campaign*. The ending boundary is not delayed because there is no setup during SP campaign.
- *MP or RP campaign*. The ending boundary may be delayed by a continuing setup beginning in *that* time period.
- *SIP campaign*. The ending boundary is necessarily delayed by a continuing setup beginning in a *previous* time period.

Note that a modified period is lengthened by delay of its ending boundary but shortened by delay of its starting boundary.

15.2.4 Basic Model

We first present the model assuming that all setup times are shorter than all planning periods, and then extend it to handle unrestricted setup times, that is, allow SIP campaigns. We introduce the following variables:

- $X_{it} \in \{0,1\}$: $=$ 1 if the unit is in mode i during the modified time period t.
- $X^E_{it} \in \{0,1\}$: $=$ 1 if mode i is the ending mode during modified period t.
- $\hat{X}^E_{it} \in \{0,1\}$: $=$ 1 if the setup for item i crosses over the ending boundary of unmodified period t.
- $Y_{it} \in \{0,1\}$: $=$ 1 if modified period t starts in mode i, visits another mode, and ends in mode i.
- $Z_{it} \in \{0,1\}$: number of setups for item i during period t.

The following two inequalities couple the modes at the boundaries of a period with X_{it}:

$$X^E_{i,t-1} \leq X_{it}, \quad i,t \tag{15.24}$$

$$X^E_{it} \leq X_{it}, \quad i,t, \tag{15.25}$$

[7] We use the term intervals here to refer to the separate occurrences of production within the same period, that is, intervals are *portions* of a period.

while the following should also hold true,

$$\sum_i X^E_{it} = 1, \quad t, \tag{15.26}$$

where we assume that no initial setup is available and thus $X^E_{i,t=0} = 0, \forall i$, though this assumption can be trivially relaxed.

The following four equations are used to define the RP campaign, by activating variable Y_{it} only when the three conditions in its definition hold:

$$Y_{it} \leq X^E_{i,t-1}, \quad i,t \tag{15.27}$$

$$Y_{it} \leq X^E_{it}, \quad i,t \tag{15.28}$$

$$Y_{it} \leq \sum_{i' \neq i} X_{i't}, \quad i,t \tag{15.29}$$

$$Y_{it} \geq X^E_{i,t-1} + X^E_{it} + X_{i't} - X_{it} - 1, \quad i,i' \neq i,t. \tag{15.30}$$

Furthermore, if a setup for item i crosses over the unmodified period, then the ending mode should be i:

$$\hat{X}^E_{it} \leq X^E_{it}, \quad i,t. \tag{15.31}$$

Equation (15.32) computes the number of setups, Z_{it}, for item i in period t:

$$Z_{it} = X_{it} - X^E_{i,t-1} + Y_{it}, \quad i,t. \tag{15.32}$$

Note that (1) if a product is produced but not in the initial mode ($X_{it} = 1, X^E_{i,t-1} = 0$), it requires one setup and (2) if we have an RP campaign ($X_{it} = 1, X^E_{i,t-1} = 1$), then an additional setup is needed so Y_{it} is added as correction. Also, note that (15.24) and (15.27) prevent the RHS of (15.32) from being –1 or 2, respectively, so Z_{it} is forced to take on binary values, even if defined as continuous in [0,1].

A short setup should start during period t to cross over the ending boundary of t:

$$\hat{X}^E_{it} \leq Z_{it}, \quad i,t, \tag{15.33}$$

and, if the unit is under RP campaign during period t, then the setup of the ending mode must cross over into period $t+1$:

$$Y_{it} \leq \hat{X}^E_{it}, \quad i,t. \tag{15.34}$$

Next, we define $T^{LATE}_t \in \mathbb{R}_+$ to denote the delay of the ending boundary of period t in the modified representation. It is positive if a setup crosses over the boundary of the unmodified period:

$$T^{LATE}_t \leq \sum_i (\sigma_i - \epsilon)\hat{X}^E_{it}, \quad i,t, \tag{15.35}$$

where ϵ is a small number.

The production of item i during period t, P_{it}, is bounded by

$$\frac{P_{it}}{v_i} \leq \eta_i \left(X_{it} - \hat{X}^E_{it} + Y_{it} \right), \quad i,t, \tag{15.36}$$

in which the RHS is tighter than the more intuitive $\eta_i X_{it}$. For items not in the ending mode ($\hat{X}^E_{it} = 0$, $Y_{it} = 0$), (15.36) enforces that a setup must occur ($X_{it} = 1$) to have positive production. For SP and MP campaigns, it forbids production in the ending mode ($X_{it} = 1$, $Y_{it} = 0$), if there is setup cross over ($\hat{X}^E_{it} = 1$). For RP campaigns ($X_{it} = 1$, $Y_{it} = 1$, $\hat{X}^E_{it} = 1$), production can be nonzero because item i is produced during the starting mode.

The overall resource constraint, which is essentially the counterpart of (15.22) but based on the modified period, is expressed as follows:

$$\sum_i \left(\frac{P_{it}}{\upsilon_i} + \sigma_i Z_{it} \right) + T^{IDL}_t = \eta_t - T^{LATE}_{t-1} + T^{LATE}_t, \quad t, \tag{15.37}$$

where $T^{IDL}_t \in \mathbb{R}_+$ is the total idle time during modified period t.

The only difference in the objective function is that setup cost now is equal to $\gamma^{SET}_i Z_{it}$:

$$\min \sum_{i,t} \left(\gamma^{PR}_i P_{it} + \gamma^{SET}_i Z_{it} + \theta_i I_{it} \right). \tag{15.38}$$

15.2.5 Model for Short and Long Setups

While most LS-based methods assume that setups are shorter than the length of the planning period, there are cases where this is not true because (1) small or nonuniform planning periods need to be considered or (2) setups are infrequent but rather long. The former arises when, for example, orders are due at high frequency and there is an advantage in considering the exact due dates (as planning period boundaries). The latter appears in, for example, sectors where extensive cleaning operations are necessary between different items (e.g., biotech facilities). Also, nonuniform planning periods and long setups result when multiple parallel units are considered simultaneously, a topic we will discuss in the next Section.

To account for long setups, we add the equations in the present subsection, while all variables and constraints introduced in Section 15.2.4 remain in the formulation. By allowing long setups, it becomes possible for continuing setups to trigger the SIP campaign whenever the remaining length of a setup beginning in a previous period $\left(T^{LATE}_{t-1}\right)$ is so large that it delays the ending boundary of the current period by $T^{LATE}_t > 0$. We introduce the following two constraints to enforce $W_t \in \{0, 1\}$ to be equal to 1 when period t is under the SIP campaign:

$$W_t \geq \frac{T^{LATE}_{t-1} - \eta_t}{\max_i \{\sigma_i\} - \eta_t}, \quad t: \max_i \{\sigma_i\} > \eta_t \tag{15.39}$$

$$W_t \leq \frac{T^{LATE}_{t-1}}{\eta_t}, \quad t: \max_i \{\sigma_i\} > \eta_t. \tag{15.40}$$

Note that (15.39) and (15.40) are written only for periods that can be in the SIP campaign, that is, for periods whose length is shorter than at least one setup

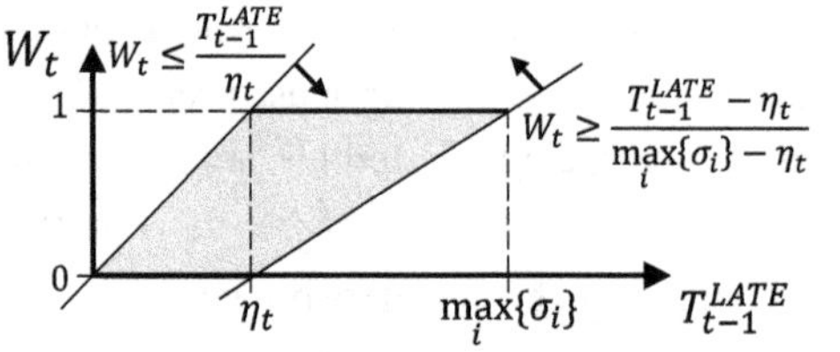

Figure 15.10 Feasible region defined by (15.39) and (15.40) and integrality of W_t shown as thick solid lines; note that when $T_{t-1}^{LATE} > \eta_t$, then $W_t = 1$.

(t: $\max_i \{\sigma_i\} > \eta_t$). Figure 15.10 shows how (15.39) and (15.40) force $W_{it} = 1$ when $T_{t-1}^{LATE} > \eta_t$ and $W_{it} = 0$ when $T_{t-1}^{LATE} < \eta_t$.

If in the SIP campaign ($W_{it} = 1$), then there is no available time for production, which means that the RHS of (15.37), the overall capacity constraint, should be zero:

$$\eta_t - T_{t-1}^{LATE} + T_t^{LATE} \leq \left(\max_i \{\sigma_i\} + \eta_t \right)(1 - W_t), \quad t\text{: } \max_i \{\sigma_i\} > \eta_t. \tag{15.41}$$

Furthermore, both boundaries are delayed ($T_{t-1}^{LATE} > 0$ and $T_t^{LATE} > 0$) by the same item, so $\hat{X}_{i,t-1}^E$ and $\hat{X}_{it}^E$ should be active for the same item. This is indirectly enforced by forbidding any new setups when in the SIP campaign

$$Z_{it} + W_t \leq 1, \quad t\text{: } \max_i \{\sigma_i\} > \eta_t. \tag{15.42}$$

Note that (15.35) couples positive T_t^{LATE} to some ending mode, and such setups are prevented from starting (but not from continuing) during the SIP campaign because the modified period length is zero.

Finally, a long setup should delay the ending boundary of any period it begins in:

$$Z_{it} \leq \hat{X}_{it}^E, \quad i, t\text{: } \sigma_i > \eta_t \tag{15.43}$$

and a long setup cannot delay the ending boundary of a period unless it begins in that period or SIP campaign is in effect:

$$\hat{X}_{it}^E \leq Z_{it} + W_t, \quad i, t\text{: } \sigma_i > \eta_t. \tag{15.44}$$

The proposed model for the GCLSP with long setups consists of (15.4) and (15.24) through (15.44).

Finally, we note that the models we presented in Section 15.2.4 and the present subsection can be extended to account for multiple parallel units as well as product families, but these extensions will be discussed in the next section.

15.3 Multiple Units Production Planning-Scheduling

The GCLS formulation discussed in Section 15.2 addresses some of the limitations of LS models. Specifically, it accounts for setup carry over, setup cross over, and long setups, thereby providing better and more accurate solutions. However, it does not provide *exact* scheduling information for all items,[8] and thus cannot yield accurate total production costs in systems that, for example, have sequence-dependent changeover costs. In this section, we present a model that does account for item sequencing through, essentially, embedding a detailed scheduling model onto the planning model. We consider continuous processing in the multiunit environment, as opposed to the single-unit environment addressed in in the previous section. Some necessary preliminary concepts and the problem statement are given in Section 15.3.1, the basic model is presented in Section 15.3.2, and extensions are discussed in Section 15.3.3.

15.3.1 Preliminaries

We study the general problem with product families, short planning periods that may lead to idle units for entire periods and changeovers spanning multiple periods, and maintenance activities. We next present the formal problem statement and an outline of the overall approach.

Problem Statement. We are given the following:

(1) A known planning horizon divided into periods, $t \in \mathbf{T}$.

(2) Parallel processing units, $j \in \mathbf{J}$, with available production time in period t equal to η_{jt}.

(3) Product families or simply families, $f \in \mathbf{F}$; $\mathbf{F}_j$ is the subset of families that can be assigned to unit j and $\mathbf{J}_f$ is the subset of units that can process family f.

(4) Products $i \in \mathbf{I}$ with demand $\xi_{it}(< 0)$ at the end of period t; backlog, π_i^{BCK}, inventory, θ_i, and processing, γ_{ij}^{PR}, costs; minimum/maximum production rates, $\upsilon_{ij}^{MIN}/\upsilon_{ij}^{MAX}$; and minimum run times, τ_{ij}^{MIN}.

(5) The subset of products in family f is $\mathbf{I}_f$; $\mathbf{I}_j$ is the subset of products that can be produced in unit j, and $\mathbf{J}_i$ is the subset of units that can produce product i. If $f(i)$ is the family product i belongs to, then $\mathbf{I}_j = \{i: f(i) \in \mathbf{F}_j\}$ and $\mathbf{J}_i = \mathbf{J}_{f(i)}$.

(6) A changeover is required between the processing of products in different families; the changeover time and cost are $\sigma_{jff'}^{CH}$ and $\gamma_{jff'}^{CH}$ respectively.

(7) A sequence-independent setup is required whenever product i is assigned to a unit; the setup time is σ_{ij}^{SET} and the setup cost is γ_{ij}^{SET}.

Our goal is to satisfy customer demand at the minimum total cost, including processing, changeover, and setup costs, as well as inventory and backlog costs. We have to determine the following for every unit and period: (1) assignment of product families to units; (2) sequencing between families assigned to the same unit; (3) assignment of products to units; and (4) production level, inventory, and backlog profiles for all products.

[8] The GCLS model discussed in Section 15.2 provides some detailed scheduling information – specifically, for the ending mode – because this is necessary to correctly account for setup carry over and cross over. The sequence and timing of production of items scheduled in the *middle* of the period, though, is not modeled.

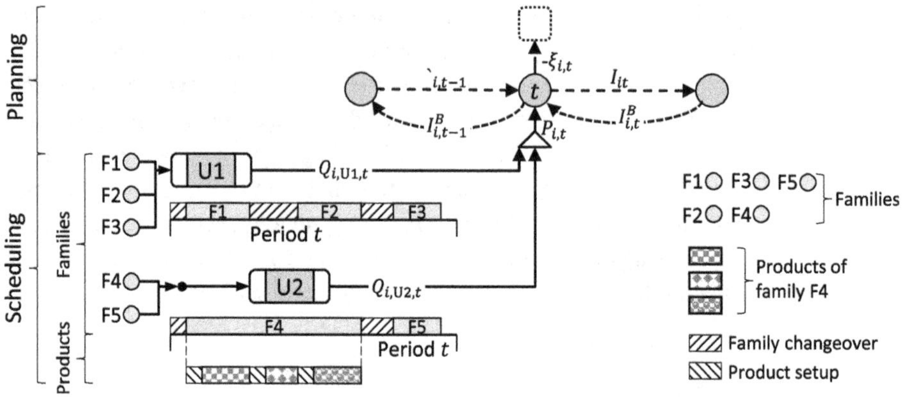

Figure 15.11 Illustration of three-level approach for the integrated planning-scheduling model.

Proposed Approach. The model combines three basic concepts. A discrete, potentially nonuniform, grid is used for the planning level (inventory and backlog cost calculation); a continuous-time sequence-based approach is used for the scheduling of families; and a lot-sizing-type approach is used for enforcing capacity constraints for products (see Figure 15.11). As we have already seen, material balances are expressed at the end of each planning period in terms of total production levels, P_{it}; inventory levels, I_{it}; and backlog levels, I^B_{it}. The communication between the planning and scheduling levels is accomplished through variables $Q_{ijt} \in \mathbb{R}_+$, which denote the production of item i in unit j during period t. At the planning level, variables Q_{ijt} are used to calculate P_{it}, while at the scheduling level they are subject to detailed sequencing and capacity constraints. The scheduling subproblem has two levels. At the first level, product families are scheduled on units using an immediate sequencing approach: $\bar{X}_{fjt} \in \{0,1\}$ is used to denote assignment of family to unit j during period t; and sequencing variable $Y_{jff't} \in \{0,1\}$ denotes immediate precedence $f \rightarrow f'$ in unit j during period t or across periods $t-1$ and t. At the second level, a lot-sizing-based approach is adopted to express capacity constraints: individual product setups and capacity constraints are modeled using $X_{ijt} \in \{0,1\}$, which is 1 if i is produced in unit j during period t. Note that when the same decision is made for both families and products (e.g., assignment to a unit), we use overbar for the decision regarding families.

15.3.2 Basic Model

In addition to the general variables for production planning (P_{it}, I_{it}, I^B_{it}) and the ones defined in the previous subsection ($Q_{ijt}, \bar{X}_{fjt}, X_{ijt}$, and $Y_{jf'ft}$), we define the following variables:

- $Y^F_{fjt} \in \{0,1\}: = 1$ if family f is the first to be processed in unit j during period t.[9]

[9] Note that, following the convention introduced in Chapter 3, superscript "F" in Y^F_{fjt} stands for "first" and not "family."

- $Y^L_{fjt} \in \{0,1\}: = 1$ if family f is the last to be processed in unit j during period t.
- $\hat{Y}_{jff't} \in \{0,1\}: = 1$ if there is an $f \rightarrow f'$ changeover cross over at the beginning of period t.
- $\bar{T}_{fjt} \in \mathbb{R}_+$: time allocated to production of items of family f in unit j during period t.
- $T_{ijt} \in \mathbb{R}_+$: time allocated to production of item i in unit j during period t.
- $C_{fjt} \in \mathbb{R}_+$: completion time of processing of family f in unit j in period t.
- $U^1_{jt} \in \mathbb{R}_+$: time allocated to a changeover cross over in unit j at the beginning of period t.
- $U^2_{jt} \in \mathbb{R}_+$: time allocated to a changeover cross over in unit j at the end of period t.

The total amount of item i produced in period t, P_{it}, comes from multiple units:

$$P_{it} = \sum\nolimits_{j \in \mathbf{J}_i} Q_{ijt}, \quad i,t \tag{15.45}$$

and its inventory balance, at the planning level, is (same as (15.10))

$$I_{it} - I^B_{it} = I_{i,t-1} - I^B_{i,t-1} + P_{it} + \xi_{it} \le \chi^M_i, \quad i,t. \tag{15.46}$$

A family f is assigned to a unit if at least one product belonging to this family is produced in the unit,

$$X_{ijt} \le \bar{X}_{fjt}, \quad f, j \in \mathbf{J}_f, i \in \mathbf{I}_f, t, \tag{15.47}$$

while it should not be assigned if no products in family f are produced in a unit:[10]

$$\bar{X}_{fjt} \le \sum\nolimits_{i \in \mathbf{I}_f} X_{ijt}, \quad f, j \in \mathbf{J}_f, t. \tag{15.48}$$

If family f is assigned to unit j ($\bar{X}_{fjt} = 1$), then it should have both a predecessor and a successor, unless it is the first/last family in the unit during period t:

$$\bar{X}_{fjt} = Y^F_{fjt} + \sum\nolimits_{f' \neq f, f' \in \mathbf{F}_j} Y_{jf'ft}, \quad f, j \in \mathbf{J}_f, t \tag{15.49}$$

$$\bar{X}_{fjt} = Y^L_{fjt} + \sum\nolimits_{f' \neq f, f' \in \mathbf{F}_j} Y_{jff't}, \quad f, j \in \mathbf{J}_f, t. \tag{15.50}$$

The correct number of precedence variables is activated through

$$\sum\nolimits_{f \in \mathbf{F}_j} \sum\nolimits_{f' \in \mathbf{F}_j} Y_{jff't} = \sum\nolimits_{f \in \mathbf{F}_j} \bar{X}_{fjt} - 1, \quad j,t. \tag{15.51}$$

To avoid family subcycles within a period, we use the following constraint:

$$C_{f'jt} \ge C_{fjt} + \sigma^{CH}_{ff'j} Y_{jff't} + \bar{T}_{f'jt} - \eta_{jt}\left(1 - Y_{jff't}\right), \quad f, j \in \mathbf{J}_f, t, \tag{15.52}$$

where η_{jt} is used as a big-M parameter, though tighter values can be calculated.

To model family changeovers across periods, we first activate variables $\hat{Y}_{jff't} \in \{0,1\}$, denoting an $f \rightarrow f'$ changeover cross over at the beginning of period

[10] A solution where $\bar{X}_{fjt} = 1$ while $\sum_{i \in \mathbf{I}_f} X_{ijt} = 0$ would normally not occur because X^F_{fjt} is associated with a changeover time (thus reduction in capacity) and cost. However, if the $f \rightarrow f'$ cost/time is larger than the sum of $f \rightarrow f''$ and $f'' \rightarrow f'$ costs/times, then this solution may be found if not excluded through (15.48).

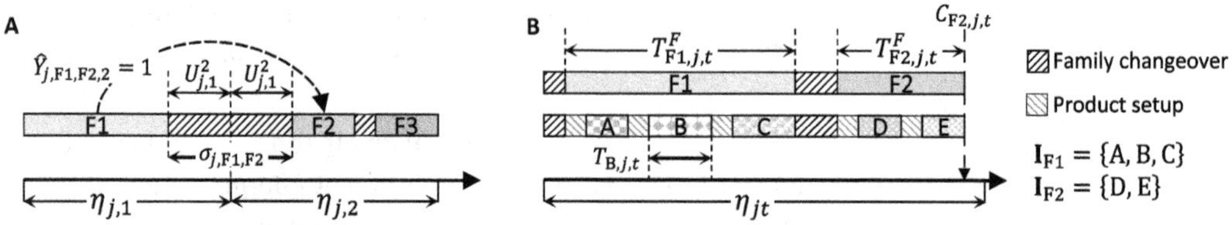

Figure 15.12 Illustration of model variables. (A) Modeling of changeover (F1 → F2) cross over. (B) Modeling of time allocated to families and products, as well as family completion time.

t, if family f was the last to be processed in period $t-1$ and family f' is the first to be processed in period t:

$$Y^F_{f'jt} = \sum_{f \in \mathbf{F}_j} \hat{Y}_{jff't}, \quad f', j \in \mathbf{J}_f, t > 1 \tag{15.53}$$

$$Y^L_{fj,t-1} = \sum_{f' \in \mathbf{F}_j} \hat{Y}_{jff't}, \quad f, j \in \mathbf{J}_f, t > 1. \tag{15.54}$$

If there is a changeover cross over at the beginning of period t, then a fraction of the changeover occurs at the end of period $t-1$ and the remaining at the beginning of period t. Let $U^2_{j,t-1}$ denote the former and U^1_{jt} the latter (see Figure 15.12A for an illustration). The following holds true:

$$U^2_{j,t-1} + U^1_{jt} = \sum_{f' \in \mathbf{F}_j} \sum_{f \in \mathbf{F}_j} \sigma^{CH}_{jff'} \hat{Y}_{jff't}, \quad j, t > 1. \tag{15.55}$$

The available time of a unit can be allocated to (1) families, (2) changeovers occurring within a period, and (3) changeover cross overs at the two boundaries. Thus, the overall capacity constraint can be written as follows:

$$\sum_{f \in \mathbf{F}_j} \bar{T}_{fjt} + \sum_{f \in \mathbf{F}_j} \sum_{f' \in \mathbf{F}_j} \sigma^{CH}_{jff'} Y_{jff't} + \left(U^1_{jt} + U^2_{jt}\right) \leq \eta_{jt}, \quad j, t. \tag{15.56}$$

The production of a product is constrained as follows:

$$\upsilon^{MIN}_{ij} T_{ijt} \leq Q_{ijt} \leq \upsilon^{MAX}_{ij} T_{ijt}, \quad i, j, t, \tag{15.57}$$

where the time, T_{ijt}, allocated to the production of item i, is subject to

$$\tau^{MIN}_{ij} Y_{ijt} \leq T_{ijt} \leq \min\left\{\eta_{jt}, -\sum_t \xi_{it}/\upsilon^{MIN}_{ij}\right\} Y_{ijt}, \quad i, j, t. \tag{15.58}$$

(Can you see why the min function is used?)

Since product setups are sequence independent, sequencing and timing decisions regarding products of the same family can be made postoptimization without affecting solution quality. Thus, the only capacity constraint coupling product production times is

$$\bar{T}_{fjt} = \sum_{i \in \mathbf{I}_f} \left(\sigma^{SET}_{ij} X_{ijt} + T_{ijt}\right), \quad f, j \in \mathbf{J}_f, t. \tag{15.59}$$

Note that no idle times are allowed within the time allocated to a family. However, since idle time can be allocated between families (note that (15.56) is an inequality), any feasible solution can be readily obtained postoptimization by keeping all T_{ijt} fixed but "sliding" product blocks as needed. The modeling using time-related variables $\bar{T}_{ffjt}$, T_{ijt}, and C_{fjt} is illustrated in Figure 15.12B.

The objective function is

$$\min \sum\nolimits_{i,t}\left(\theta_i I_{it} + \pi_i^{BCK} I_{it}^B\right) + \sum\nolimits_{i,j,t}\left(\gamma_{ij}^{SET} X_{ijt} + \gamma_{ij}^{PR} Q_{ijt}\right) + \sum\nolimits_{j,f,f',t}\gamma_{jff'}^{CH}\left(Y_{jff't} + \hat{Y}_{jff't}\right). \tag{15.60}$$

Finally, we note that the following domains can be used for the continuous nonnegative variables:

$$\bar{T}_{fjt} \in [0, \eta_{jt}]; T_{ijt} \in [0, \eta_{jt}]; U_{jt}^1 \in \left[0, \max_{f,f'}\left\{\sigma_{jff'}^{CH}\right\}\right]; U_{jt}^2 \in \left[0, \max_{f,f'}\left\{\sigma_{jff'}^{CH}\right\}\right]. \tag{15.61}$$

15.3.3 Extensions

Idle Units. The model in Section 15.3.2 is based on the assumption that units do not remain completely idle during any period. In other words, units produce at least one product in each period, except maintenance periods. Generally speaking, this assumption is valid for medium to long planning periods (e.g., a week). However, if short periods are used (e.g., a day) to accurately model frequent intermediate due dates, idle periods may be present in an optimal solution. To model unit idle periods, we define a dummy product for each family, i_f^{ID}, with zero setup time and cost. The processing times of dummy products are then constrained by (15.58) between 0 and η_{jt}. If a unit produces only a dummy product in a period, then the unit actually remains idle during that period. Note that defining *idle* products and the corresponding T_{ijt} variables means that (15.56) can be expressed as equality where $T_{i_f^{ID},jt}$ acts as slack variable. Figure 15.13A shows two equivalent solutions for an illustrative single-unit production plan over three periods, in which the unit produces products in family F1 in the first period, remains idle during the second period, and produces products in family F2 in the third period.

Maintenance Activities. When the planning horizon is long, then it is reasonable to expect that maintenance will be required. Maintenance activities can be readily

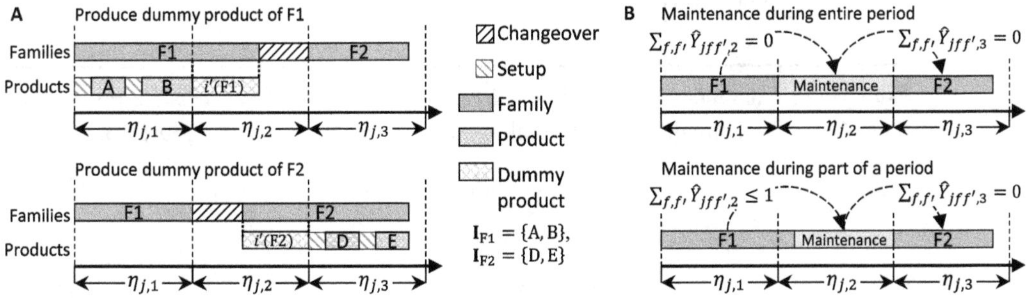

Figure 15.13 Illustration of modeling of extensions. (A) Unit remains idle during the entire period 2. (B) Modeling of maintenance activities with duration equal to (top panel) and shorter than (bottom panel) the length of a period.

addressed by fixing the corresponding changeover cross over variables to zero and modifying the available production time η_{jt} accordingly. Without loss of generality we can assume that the maintenance activity is carried out, and ends, at the end of a period. (Why?) The model can address cases where the duration of a maintenance activity is equal to the available production time and cases where its duration is smaller than the length of the period. Note that if maintenance is performed between the production of two different families, there is no need for a changeover operation. Figure 15.13B illustrates the approach. Finally, note that the model can also address maintenance with duration greater than a planning period. (Can you think how?)

15.4 Projection-Based Surrogate Methods

The methods presented in the previous two sections are based on the assumption that the region of feasible production targets can be approximated by the constraints describing the capacity of multiple parallel units. While this is a reasonable assumption for many systems, there are systems where this is not true, and thus an alternative method is necessary. In this section, we describe a method that is applicable to all production systems, regardless of structure, and processing features and constraints. In Section 15.4.1, we present the basic idea of projecting information from a detailed scheduling model to the space of production variables, P_{it}. In Section 15.4.2, we outline the algorithm for obtaining such projection, and we close in Section 15.4.3 with some remarks and extensions.

15.4.1 Feasible Region Projection

Our goal is the development of a systematic method for obtaining, using offline calculations based on a detailed scheduling model, production feasibility and cost information in a form that is amenable to integration with the production planning problem. Specifically, we aim to develop approximating functions $f(\boldsymbol{P}_t)$ and $g(\boldsymbol{P}_t)$ using linear inequalities that involve only planning variables P_{it}. To illustrate the basic idea, we revisit Example 15.1.

Example 15.1 (continued) We have already established that any feasible production point $(P_{A,t}, P_{B,t})$ should satisfy the following four inequalities (see Figure 15.14A):

$$P_{A,t} + P_{B,t} \leq 2{,}100 \tag{15.62}$$

$$P_{A,t} \leq 1{,}400 \tag{15.63}$$

$$P_{B,t} \leq 1{,}050 \tag{15.64}$$

$$P_{A,t} - P_{B,t} \leq 0 \tag{15.65}$$

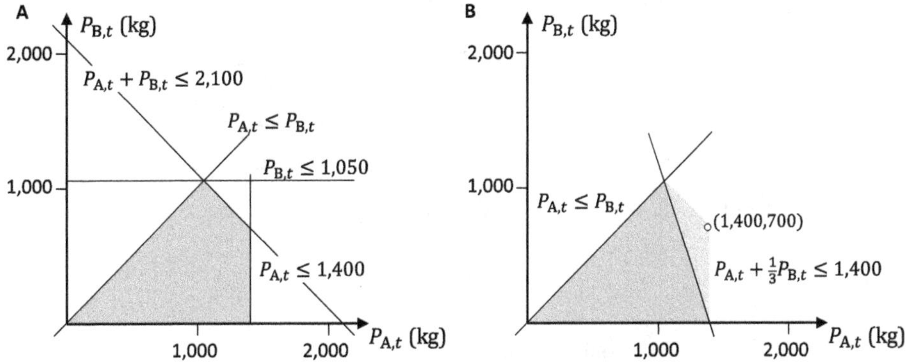

Figure 15.14 Approximations of PAR of network introduced in Example 15.1. (A) Region based on capacity considerations and the impact of the no storage restriction for intermediate INT. (B) True feasible region of production targets.

with the first three coming from capacity-based considerations, whereas the last is derived by recognizing that if product B is produced (at 150 kg/day), then product A should also be produced (at 150 kg/day) because of the no storage restriction for intermediate INT. But is the region described by these four constraints the true feasible region $\mathbb{P}$? It turns out that it is not. For a reason not immediately clear, all feasible production targets should also satisfy the following:

$$P_{A,t} + \frac{1}{3}P_{B,t} \leq 1,400. \tag{15.66}$$

Thus, a tighter overapproximation of the true feasible region $\mathbb{P}$ is described by (15.65), (15.66), and the nonnegativity of $P_{B,t}$ (see Figure 15.14B). Interestingly, none of these constraints is derived from individual unit capacity constraints.[11] One way to confirm that point (1,400, 700), for example, is infeasible is to solve a detailed scheduling model over a horizon of a week with demands, at the end of the week $-\xi_A = 1,400$, $-\xi_B = 1,050$. We will find that the given demand cannot be met, that is, the (1,400, 700) production target is infeasible.

Example 15.1 showed that even for a simple production environment, unit capacity considerations might not be sufficient to generate a good approximation of $\mathbb{P}$. However, it also suggested a way to do so: solve carefully selected scheduling problems to *interrogate* what production targets are feasible and then exploit this information to generate constraints of the general form given in (15.67), to be embedded in the integrated planning-scheduling formulation

[11] While (15.66) was not derived directly from a unit capacity restriction, unit capacities clearly play a role. Can you see why? (Hint: what is the intersection of (15.66) with (15.63)? What about the intersection of (15.66) and (15.64)?) What does the RHS of (15.66) depend on? What about the $\frac{1}{3}$ coefficient of $P_{B,t}$ in (15.66)?

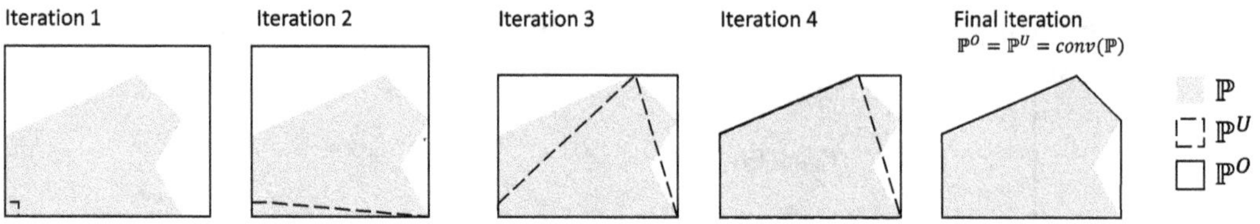

Figure 15.15 Illustration of iterative improvement of under- and overapproximation of PAR.

$$\sum_i \omega_i^l P_{it} \leq \omega^l, \quad l \in \mathbf{L}^F, \tag{15.67}$$

where $\mathbf{L}^F$ is the index set of $|\mathbf{L}^F|$ *feasibility* constraints.[12] While the derivation of the constraints in (15.67) may require the solution of a multiple instances of an expensive scheduling model, this can be done once as long as the underlying network remains the same over the planning horizon under consideration.

15.4.2 Method Outline

The goal of the method is to generate constraints, in the form of (15.67), that yield a good approximation of $\mathbb{P}$. The obtained region will be convex because (15.67) are linear inequalities, although $\mathbb{P}$ can be nonconvex.[13] The algorithm is based on the convergence of two convex polytopes: an underestimation, $\mathbb{P}^U$, and an overestimation, $\mathbb{P}^O$, of the convex hull of $\mathbb{P}$, $conv(\mathbb{P})$. The algorithm iteratively improves the two approximations, until $\mathbb{P}^O = \mathbb{P}^U = conv(\mathbb{P})$, or a resource/iteration criterion is met (see Figure 15.15 for an illustration). In this section, for simplicity, we drop index t from P_{it}.

The algorithm involves the iterative offline solution of a detailed scheduling model, henceforth referred to as (M1), with a scheduling horizon equal to the length of the (for now uniform) planning period. While any scheduling model, suitable for the production system under study, can be used, here, to illustrate, we use the STN model based on a common discrete time grid, where the final products $(\mathbf{K}^P \subset \mathbf{K})$ correspond to the items $i \in \mathbf{I}$ considered in the production planning problem.[14] The objective function of (M1) in iteration m is

$$\max \sum_{i \in \mathbf{I} = \mathbf{K}^P} \omega_i^m P_i, \tag{15.68}$$

where $P_i = S_{i,|\mathbf{N}|}$ is the inventory at the end of the scheduling horizon (assuming that there is unlimited storage for $i \in \mathbf{I} = \mathbf{K}^P$) and $\boldsymbol{\omega}^m = [\omega_i^m, i \in \mathbf{I}]$ is the *search direction* at iteration m.[15] The solution of (M1) at iteration m, even if suboptimal, provides a

[12] Note that, after rearranging (15.65), all the constraints of the example can be written in the form of (15.67).

[13] Region $\mathbb{P}$ is the projection of a higher-dimensional region (the feasible region of a scheduling model) onto the space of production target variables, P_i. The projection of a nonconvex region can be nonconvex.

[14] The STN-based model presented in Chapter 7 employs index $i \in \mathbf{I}$ to denote tasks. We choose to introduce this small inconsistency to keep the notation in the present chapter uniform, that is, we use $i \in \mathbf{I}$ to denote items throughout.

[15] The equality $P_i = S_{i,|\mathbf{N}|}$ is valid because by maximizing the inventory at the end of the scheduling horizon, no intermediate sales will occur if there is unlimited storage capacity. Thus, maximizing the weighted final inventories $\omega_i S_{i,|\mathbf{N}|}$ is equivalent to maximizing the weighted production targets $\omega_i P_i$.

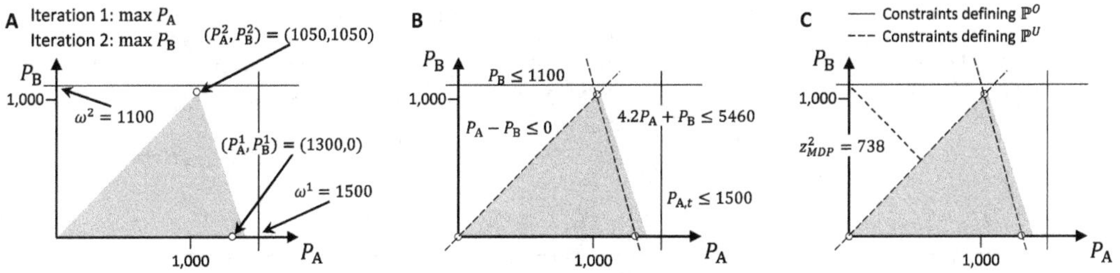

Figure 15.16 Information obtained by solving scheduling model (M1) and direction model (M2). (A) Results of (M1) over two iterations. (B) Generation of constraints based on (1) best bounds on (M1) solutions (for overapproximation) and (2) best solutions of (M1) (for underapproximation). (C) Regions $\mathbb{P}^O$ and $\mathbb{P}^U$ after two iterations and maximum distance between them.

feasible point of $\mathbb{P}^U$, $\boldsymbol{P}^m = \left[P_i^m, i \in \mathbf{I}\right]$, where P_i^m are the values of P_i in the best solution of (M1) in iteration m. The solution of (M1) also provides a valid inequality for $\mathbb{P}^O$:

$$\sum_i \omega_i^m P_i \le \omega^m, \tag{15.69}$$

where ω^m is the best upper bound on the optimal solution of (M1) at iteration m.

To illustrate, we consider the region of the network introduced in Example 15.1 (see Figure 15.16). It is assumed that (M1) is not solved to optimality. The first iteration ($m = 1$), with $(\omega_A^1, \omega_B^1) = (1, 0)$, returns $(P_A^1, P_B^1) = (1,300, 0)$ with an objective function value of 1,300 and a bound of $\omega^1 = 1,500$; thus, it is used to generate constraint $P_A \le 1,500$ for $\mathbb{P}^O$. The second iteration ($m = 2$), with $(\omega_A^2, \omega_B^2) = (0, 1)$, returns $(P_A^2, P_B^2) = (1,050, 1,050)$ with an objective function value of 1050 and a bound of $\omega^2 = 1,100$; thus, it is used to generate constraint $P_B \le 1,100$ for $\mathbb{P}^O$. For the description of $\mathbb{P}^U$, points $(0,0)$ and $(1,050, 1,050)$ are used to generate constraint $P_A - P_B \le 0$, and points $(1300, 0)$ and $(1,050, 1,050)$ are used to generate $4.2P_A + P_B \le 5,460$.

The initial $|\mathbf{I}|$ search directions (i.e., weights ω_i^m) are chosen to maximize individual production of each product to obtain $|\mathbf{I}|$ inequalities that set up the initial $\mathbb{P}^O$ and $|\mathbf{I}|$ (not necessarily unique) feasible points which along with the origin define the initial $\mathbb{P}^U$ (see Figure 15.16A,B). A different optimization problem (M2) is then solved to calculate the maximum perpendicular distance (MPD) between $\mathbb{P}^O$ and $\mathbb{P}^U$, z_{MDP}^m, and identify the next search direction for (M1). In the next iteration, (M1) is solved using the new search direction to obtain a new inequality for $\mathbb{P}^O$ and a new feasible point for $\mathbb{P}^U$. First, $\mathbb{P}^O$ is updated by adding the new inequality to the set of previously found inequalities. Second, $\mathbb{P}^U$ is updated by discarding all previously found inequalities and generating a new set of inequalities using all currently available feasible solutions.[16] Thus, all the

[16] Convex polytopes can be expressed by their vertices (V-representation) or by their facets (linear inequalities, H-representation).

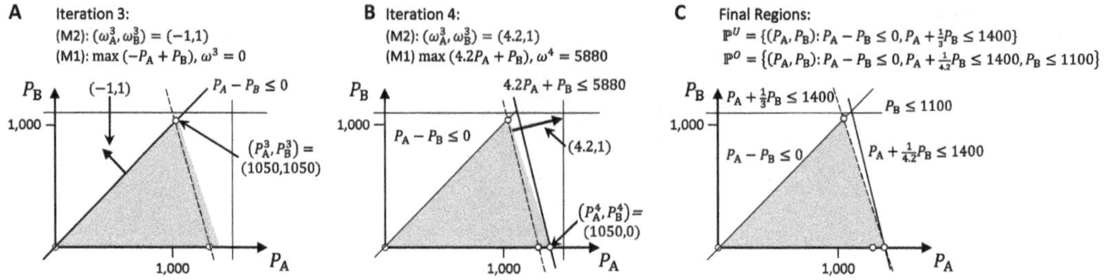

Figure 15.17 Solution of model (M2) and constraint generation from best bounds and solutions. (A) Iteration 3: search direction, optimal solution and objective function value, and new constraint. (B) Iteration 4: search direction, optimal solution and objective function value, and new constraint. (C) Final constraints describing $\mathbb{P}^U$ and $\mathbb{P}^O$.

feasible points, that is, best solutions of (M1), through the current iteration can be used to generate the inequalities defining $\mathbb{P}^U$. This can be accomplished using the Quickhull (QHull) algorithm (see discussion in Section 15.4.3). The process is repeated until a tolerance, in terms of convergence between $\mathbb{P}^O$ and $\mathbb{P}^U$, is met or the maximum number of iterations is met. The inequalities to be embedded in the planning problem can be either the ones describing $\mathbb{P}^O$ or the ones describing $\mathbb{P}^U$, depending on the preference of the user. If the former are to be used, then the last step would be to remove redundant constraints so the minimal number of constraints are added to the planning formulation.

The calculations performed after the initial $|\mathbf{I}|$ iterations, using the regions shown in Figure 15.16C as starting point, are illustrated in Figure 15.17. In iteration 3 (see Figure 15.17A), the solution of (M2) identifies $(\omega_A^3, \omega_B^3) = (-1, 1)$ as the next search direction; and the solution of (M1), which is now solved to optimality, returns $(P_A^3, P_B^3) = (1,050, 1,050)$ and $\omega^3 = 0$. The new constraint for $\mathbb{P}^O$ is thus $P_A - P_B \leq 0$. Since no new feasible point is found, the constraints describing $\mathbb{P}^U$ remain the same. In iteration 4 (see Figure 15.17B), the solution of (M2) identifies $(\omega_A^4, \omega_B^4) = (4.2, 1)$ as the next search direction; and the solution of (M1), which is again solved to optimality, returns $(P_A^4, P_B^4) = (1,400, 0)$ and $\omega^4 = 5,880$. The new constraint for $\mathbb{P}^O$ is $4.2P_A + P_B \leq 5,880$. Also, since a new feasible point is found, the QHull algorithm is executed again and returns new constaint $3P_A + P_B \leq 4,200$. In iteration 5, it is found that the maximum distance between $\mathbb{P}^O$ and $\mathbb{P}^U$ satisfies the desired tolerance, so the algorithm terminates. (The two regions are not the same. Can you calculate the maximum distance?) The final constraints defining $\mathbb{P}^O$, after the redundant constraints are removed, and $\mathbb{P}^U$, resulting from running QHull with all four found feasible points, rewritten so that $\omega_A^m = 1$, are shown in Figure 15.17C. Note that $P_A \leq 1,500$ is redundant and is removed, whereas $P_B \leq 1,100$ is kept, although the top vertex of $\mathbb{P}$ is $(1,050, 1,050)$. (Can you see why?) Also, in this case, we have $\mathbb{P}^U = \mathbb{P}$ whereas $\mathbb{P}^O$ strictly overestimates $\mathbb{P}$.

The pseudocode for generating the constraints defining $\mathbb{P}^O$ and $\mathbb{P}^U$ is as follows:

Algorithm 1

select tolerance, ε, for convergence between $\mathbb{P}^O$ and $\mathbb{P}^U$; and maximum iteration number m^{MAX}

choose $|\mathbf{I}|$ initial search directions $\omega_i^m = \begin{cases} 1, & \text{if } m = i \\ 0, & \text{if } m \neq i \end{cases}, \quad m \in \{1, \ldots, |\mathbf{I}|$

solve (M1) $|\mathbf{I}|$ times to obtain feasible points $\left[P_i^m, i \in \mathbf{I}\right]$ and bounds ω^m

using directions ω_i^m and bounds ω^m generate constraints describing $\mathbb{P}^O$

using points $\left[P_i^m, i \in \mathbf{I}\right]$ and Quickhull algorithm generate constraints describing $\mathbb{P}^U$

repeat until termination criteria are met

  $m = m + 1$

  solve (M2) to obtain MPD, z_{MDP}^m; and next search direction $\boldsymbol{\omega}^m = \left[\omega_i^m, i \in \mathbf{I}\right]$

  if $z_{MDP}^m > \varepsilon$ and $m < m^{MAX}$ then

    solve (M1) to obtain feasible point $\left[P_i^m, i \in \mathbf{I}\right]$ and bound ω^m

    generate a new constraint, (15.69), for $\mathbb{P}^O$ using $\boldsymbol{\omega}^m$ and ω^m

    generate new description of $\mathbb{P}^U$ by generating the convex hull of all feasible points

  else

    exit

  endif

remove redundant constraints

15.4.3 Remarks and Extensions

We first discuss three basic elements of the method (QHull, model (M2), and termination criteria) and then discuss three extensions (surrogate production cost model, periods of nonuniform length, and nonconvex regions).

Quickhull Algorithm. The method in Section 15.4.2 makes extensive use of the QHull algorithm (www.qhull.org) to convert convex polytopes from their V-representation to their H-representation, and vice versa; to calculate polytope volume; and to filter inequalities. At the heart of the QHull algorithm, an n-dimensional convex hull is constructed using $(n-1)$-dimensional facets that are each defined by n points. For $n = 2$, the convex hull is constructed from lines that are each defined by two points. For $n = 3$, the convex hull is constructed from planes that are each defined by three points. Every facet can be written as a linear inequality in the form, essentially, of (15.69), where P_i are coordinates, $\boldsymbol{\omega}^m = \left[\omega_i^m, i \in \mathbf{I}\right]$ is the outward-pointing (unit) normal vector of the mth facet, and ω^m is a scalar. For each facet, $\boldsymbol{\omega}^m$ and ω^m can be computed from the coordinates of the n points defining that facet. Different sets of points may lead to the same inequality. Note that the QHull algorithm automatically discards interior points. One method for filtering redundant inequalities from the set of $\mathbb{P}^O$ inequalities is to convert that set into a set of vertices and then back into inequalities.

Polytope Difference. New search directions should point toward *unexplored* regions of $\mathbb{P}$ where $\mathbb{P}^O$ and $\mathbb{P}^U$ are very different. To achieve this, we seek a point $\boldsymbol{P}^m$ that is within $\mathbb{P}^O$ but outside $\mathbb{P}^U$. To be outside $\mathbb{P}^U$, at least one $\mathbb{P}^U$ inequality must be violated. On the basis of the normal vector $\boldsymbol{\omega}^m$ and the RHS ω^m of the violated inequality, we can calculate the violation as $\max\{0, (\boldsymbol{\omega}^m)^T\boldsymbol{P}^m - \omega^m\}$. If $\mathbb{P}^O$ and $\mathbb{P}^U$ have not converged, then there is at least one point in $\mathbb{P}^O$ that violates at least one $\mathbb{P}^U$ inequality. Model (M2) chooses the next search direction, from among the outward-pointing normal vectors of current $\mathbb{P}^O$ inequalities, as the one that is violated the most. Selecting a search direction from among current $\mathbb{P}^U$ inequalities has several advantages: (1) $\mathbb{P}^U$ inequalities are generated from feasible points previously found by the algorithm, (2) $\mathbb{P}^U$ inequalities in the investigated region are likely to be replaced via the QHull algorithm by new tighter $\mathbb{P}^U$ inequalities, and (3) $\mathbb{P}^U$ inequalities away from the investigated region are unchanged.

Termination. The algorithm terminates if an iteration limit is reached or the MPD stops improving. MPD is attractive as a possible measure of convergence because it is already calculated in every iteration, and it is positive if $\mathbb{P}^O$ and $\mathbb{P}^U$ do not match (and zero if they do). However, the MPD does not necessarily decrease monotonically. An alternative convergence criterion that improves monotonically is the ratio, $Vol(\mathbb{P}^U)/Vol(\mathbb{P}^O)$, of the volumes of the two polytopes. In each iteration, solving (M2) results in one of the following cases:

(1) Model (M2) is solved to optimality.
(2) Model (M2) is not solved to optimality, yet $\mathbb{P}^O$, $\mathbb{P}^U$, or both are improved.
(3) Model (M2) is not solved to optimality, and neither $\mathbb{P}^O$ nor $\mathbb{P}^U$ is improved.

In cases (1) and (2), the new search direction is either new (which is desirable) or else a repeated one. The latter is possible if (M2) was previously solved suboptimally (case 2) using the same search direction. In case (3), neither polytope is improved, so solving (M1) yields the same active $\mathbb{P}^U$ inequality as in the previous iteration. The algorithm ceases to improve if the new search direction is a direction that has already been searched. Note that the algorithm can be modified to find a different search direction, or model (M2) can be solved using a higher computational resource limit when solving for the second time using a previously considered direction.

Production Cost. The method can be used, with almost no modifications, to generate linear inequalities of the following form to lower bound production cost,

$$C_t^{PRD} \geq \sum_i \omega_i^l P_{it} + \psi^l, \quad l \in \mathbf{L}^C \tag{15.70}$$

where ψ^l is a contstant and $\mathbf{L}^C$ is the set of *cost* inequalities. Equation (15.70) can be added to the planning model in lieu of $C_t^{PR} = g(\boldsymbol{P}_t)$. Since the planning problem seeks to minimize cost, for any production mix, at least one of the inequalities in (15.70) will be satisfied as equality in the optimal solution.

If $\omega^l = -\psi^l$ and $\omega_C^l = -1$, then (15.70) can be rewritten as

$$\sum_i \omega_i^l P_{it} + \omega_C^l C_t^{PR} \leq \omega^l, \quad l \in \mathbf{L}^C, \tag{15.71}$$

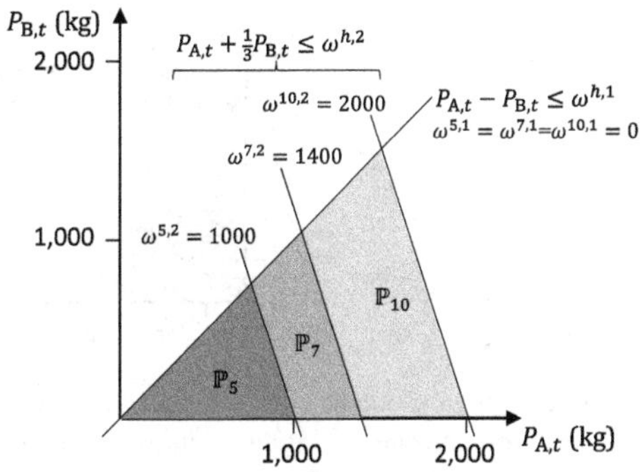

Figure 15.18 PARs of planning periods with length equal to five, seven, and ten days for the network introduced in Example 15.1. All three PARs are defined by two constraints and the nonegativity of P_B.

which has the same form as (15.67) and (15.69) but has one more dimension (cost). Thus, the algorithm can be adjusted to simultaneously generate the feasibility and cost inequalities.

Nonuniform Periods. As we have seen, it is often required to employ planning periods of different duration. One approach is to generate length-specific PARs. Specifically, if $\mathbf{H}$ is the set of period lengths (e.g., $\mathbf{H} = \{1, 2, 4\}$ in weeks), then $|\mathbf{H}|$ PARs are generated, having the following form,

$$\sum_i \omega_i^{h,l} P_{it} \le \omega^{h,l}, \quad h \in \mathbf{H}, l \in \mathbf{L}_h^F, t \in \mathbf{T}_h \tag{15.72}$$

where $\mathbf{L}_h^F$ is the set of constraints describing the PAR for a period with length h; $\mathbf{T}_h$ is the subset of periods with length h; and $\omega_i^{h,l}$ and $\omega^{h,l}$ are the LHS and RHS, respectively, coefficients of inequality l describing the PAR of periods with length h.

An alternative approach is to generate constraints that differ only in the RHS, that is, keep the same number of constraints for all periods, and only modify the RHS constant:

$$\sum_i \omega_i^l P_{it} \le \omega^{h,l}, \quad l \in \mathbf{L}^F, h \in \mathbf{H}, t \in \mathbf{T}_h. \tag{15.73}$$

The rationale for (15.73) is that the number of inequalities describing the PAR of a network, regardless of the duration of the period, will not change because these inequalities represent either a special feature or (combined) capacity constraints. Modifying the duration of the period will not change these features or the (combination of) bottleneck units, so the number of constraints should be the same. Furthermore, if the constraint-defining feature is qualitative (*intensive*), then the constraint will remain unchanged; whereas, if the feature is capacity-based, then the constraint will be translated, but not rotated. The concept is illustrated in Figure 15.18, using the PARs of the network introduced in Example 15.1 for periods with three different lengths (in days), $\mathbf{H} = \{5, 7, 10\}$, where $\mathbb{P}_h$ represents the PAR of a period with length h. Note the following:

(1) The RHS of the first constraint ($l = 1$) remains the same. (Can you think why?)
(2) The RHS of the second one ($l = 2$) changes with h. (Can you think why? Also, what is the rate of change with respect to h? Can you explain why?)

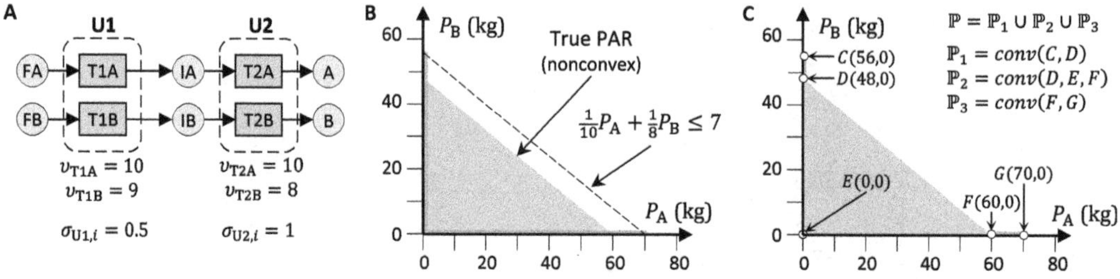

Figure 15.19 Nonconvex production attainable region due to long setups. (A) STN representation of system used for illustration and associated data (rates in kg/day, setups in days). (B) True nonconvex PAR and its tightest convex approximation. (C) Representation of nonconvex PAR as a union of convex polytopes.

Nonconvex Regions. The inequalities in (15.67) seek to provide a convex approximation of $\mathbb{P}$, or, more precisely, of $conv(\mathbb{P})$. However, in facilities whose capacity depends heavily on the loading of units or when large changeover and/or setup times are present, this approximation may be poor because $\mathbb{P}$ is nonconvex. However, a nonconvex region can be modeled as a union of convex regions using disjunctive programming. For example, if a nonconvex region can be represented as a union of r convex polytopes, each one with vertices $\{\hat{\boldsymbol{P}}_v^r = (\hat{P}_{v,1}^r, \ldots, \hat{P}_{v,|\mathbf{I}|}^r), v \in \mathbf{V}_r\}$, then a surrogate model of the following form can be developed:

$$\sum_r Z_{rt} = 1, \quad t \tag{15.74}$$

$$\sum_{v \in \mathbf{V}_r} W_{rvt} = Z_{rt}, \quad t, r \tag{15.75}$$

$$P_{it} = \sum_r \sum_{v \in \mathbf{V}_r} \hat{P}_{vi}^r W_{rvt}, \quad i, t, \tag{15.76}$$

where $Z_{rt} \in \{0, 1\} = 1$ if the production mix during period t lies within region r, and $W_{rvt} \in [0, 1]$ are the weights of the vertices. In general, there are many ways to partition a nonconvex region into convex regions and many ways to formulate a nonconvex region using discrete variables, but regardless of the specifics of the representation, the surrogate model is expected to be substantially simpler than the MIP scheduling model that would have to be otherwise used to provide feasibility information.

The idea is illustrated in Figure 15.19 using a network where units U1 and U2 can be used together to produce product A at a rate of 10 kg/day or product B at a rate of 8 kg/day (STN representation and data given in Figure 15.19A). A setup requires half a day in unit U1 and one day in unit U2. For a scheduling horizon of one week, region $\mathbb{P}$ is the gray, nonconvex region in Figure 15.19B and the tightest convex approximation of $\mathbb{P}$ is defined by the nonnegativity of production targets and the constraint shown by the dotted line in Figure 15.19B[17]. Note that points $(0, 56)$ and $(70, 0)$, which satisfy the

[17] Does this constraint remind you of any general constraint you have already seen?

constraint defining the tightest convex hull of the true PAR, are included, but the total production is reduced when both products are produced because a setup is required. The nonconvex PAR can be represented as the union of three convex regions, $\mathbb{P}_1$, $\mathbb{P}_2$, and $\mathbb{P}_3$ (see Figure 15.19C), which are defined by vertices (C, D), (D, E, F), and (F, G), respectively. In other words, if $conv(\cdot)$ returns the convex hull of the set(s) of points in the argument, then $\mathbb{P}_1 = conv(C, D)$, $\mathbb{P}_2 = conv(D, E, F)$ and $\mathbb{P}_3 = conv(F, G)$. The formulation for the nonconvex PAR becomes (where we drop index t)

$$Z_1 + Z_2 + Z_3 = 1 \tag{15.77}$$

$$W_{1,\mathrm{C}} + W_{1,\mathrm{D}} = Z_1; \quad W_{2,\mathrm{D}} + W_{2,\mathrm{E}} + W_{2,\mathrm{F}} = Z_2; \quad W_{3,\mathrm{F}} + W_{3,\mathrm{G}} = Z_3 \tag{15.78}$$

$$\begin{aligned} \begin{bmatrix} P_A \\ P_B \end{bmatrix} = \begin{bmatrix} 0 \\ 56 \end{bmatrix} W_{1,\mathrm{C}} + \begin{bmatrix} 0 \\ 48 \end{bmatrix} W_{1,\mathrm{D}} + \begin{bmatrix} 0 \\ 48 \end{bmatrix} W_{2,\mathrm{D}} + \begin{bmatrix} 0 \\ 0 \end{bmatrix} W_{2,E} \\ + \begin{bmatrix} 60 \\ 0 \end{bmatrix} W_{2,\mathrm{F}} + \begin{bmatrix} 60 \\ 0 \end{bmatrix} W_{3,\mathrm{F}} + \begin{bmatrix} 70 \\ 0 \end{bmatrix} W_{3,\mathrm{G}}. \end{aligned} \tag{15.79}$$

(Can you simplify the above formulation?)

15.5 Notes and Further Reading

(1) The introduction of the preliminary concepts in Section 15.1 is based in the review paper by Maravelias and Sung [1]. The interested reader can find more information about production planning and lot-sizing formulations in Miller et al. [2], and the book of Pochet and Wolsey [3], where a classification of problems is presented.

(2) The formulation of the generalized capacitated lot-sizing problem in Section 15.2 is adapted from Sung and Maravelias [4], where extensions for product families and parallel units are also presented. Earlier, less general variants of the GCLSP were studied by Suerie and Stadtler [5] and Suerie [6].

(3) The integrated planning-scheduling model in Section 15.3 is adapted from Kopanos et al. [7], where the reader can also find an application to a large-scale industrial case study with eight processing units, 22 product families, and 162 individual products. Similar planning-scheduling models have been proposed by Grossmann and coworkers [8, 9] and Papageorgiou, Pinto, and coworkers [10, 11].

(4) The presentation in Section 15.4 is adapted from [12], where the interested reader can find the details of a formulation for model (M2), for the identification of the next search direction; the details for generating a cost surrogate model, (15.70); the application of the method to identify bottleneck units and thus aid retrofit design; and a rolling horizon algorithm for the solution of the integrated planning-scheduling problem over long horizons (e.g., one year).

(5) An algorithm for the generation of an approximation of nonconvex feasible regions as a union of convex polytopes is presented in Sung and Maravelias [13].

(6) The idea of describing the feasible production amounts is similar to the attainable region approach for describing the space of chemical species concentrations that can be reached by a set of chemical reactions [14, 15]. However, in the case of process networks, attainable production depends not only on (intensive) stoichiometry, thermodynamics, and rates, but also on (extensive) equipment capacity, inventory capacity, utility constraints, changeover times, and other characteristics of the network.

(7) Conceptually, the idea of describing the attainable space is also similar to Benders decomposition [16], in that both the proposed method and Benders decomposition seek to replace the subproblem with linear constraints. However, the implementation of what and how linear inequalities are added is different. Benders decomposition is well defined for linear subproblems but remains an open topic for mixed-integer subproblems.

(8) The first contributions on chemical production planning-scheduling appeared in the 1990s [17–19].

(9) Ierapetritou and coworkers proposed methods based on Lagrangian decomposition–type approaches [20, 21], as well as a data-driven approach [22].

(10) Brunaud et al. proposed a method for communication between the planning and scheduling functions using inventory policies [23].

15.6 Exercises

(1) Formulate the model presented in Section 15.2. Reproduce the following results from Sung and Maravelias [4]: example 1 (subsection 6.1 of the paper); example 2; example 3; and example 4.

(2) Formulate the model presented in Section 15.3. Reproduce the results in subsection 5.1 of Kopanos et al. [7].

(3) Using the model you developed for Exercise 2 and, if necessary, augmenting it with new constraints, reproduce the results in subsection 5.2 of Kopanos et al. [7].

(4) Reproduce the results for the motivating example in [12]:
   (a) Find a tool you can use to run the Quickhull algorithm.
   (b) Using the discrete time STN-based model you have developed, use the six directions in table 2 of [12] to obtain six points and constraints.
   (c) Run the Quickhull algorithm to generate the projection of the production feasible region.

(5) Reproduce the results for the two-product example in [12]:
   (a) Formulate the model used to identify the next search direction (in the paper referred to as model (M1)).
   (b) Generate your own algorithm based on a discrete time STN-based model, the Quickhull algorithm, and model (M2).

(c) Run your algorithm for the network given on figure 14 in [12] to generate the region for a planning period with a length of 120 h and storage capacity for INT2 equal to 500.

(d) Repeat (c) for all the instances shown in figure 15 and 16 of the paper.

References

[1] Maravelias CT, Sung C. Integration of Production Planning and Scheduling: Overview, Challenges and Opportunities. *Comput Chem Eng*. 2009;33(12):1919–1930.

[2] Miller AJ, Nemhauser GL, Savelsbergh MWP. A Multi-item Production Planning Model with Setup Times: Algorithms, Reformulations, and Polyhedral Characterizations for a Special Case. *Math Program*. 2003;95(1):71–90.

[3] Pochet Y, Wolsey LA. *Production Planning by Mixed Integer Programming*. New York; Berlin: Springer; 2006. xxiii, 499 p. p.

[4] Sung C, Maravelias CT. A Mixed-Integer Programming Formulation for the General Capacitated Lot-Sizing Problem. *Comput Chem Eng*. 2008;32(1–2):244–259.

[5] Suerie C, Stadtler H. The Capacitated Lot-Sizing Problem with Linked Lot Sizes. *Manage Sci*. 2003;49(8):1039–1054.

[6] Suerie C. Modeling of Period Overlapping Setup Times. *Eur J Oper Res*. 2006;174 (2):874–886.

[7] Kopanos GM, Puigjaner L, Maravelias CT. Production Planning and Scheduling of Parallel Continuous Processes with Product Families. *Ind Eng Chem Res*. 2011;50(3):1369–1378.

[8] Erdirik-Dogan M, Grossmann IE. A Decomposition Method for the Simultaneous Planning and Scheduling of Single-Stage Continuous Multiproduct Plants. *Ind Eng Chem Res*. 2006;45(1):299–315.

[9] Castro PM, Erdirik-Dogan M, Grossmann IE. Simultaneous Batching and Scheduling of Single Stage Batch Plants with Parallel Units. *AIChE J*. 2008;54(1):183–193.

[10] Chen P, Papageorgiou LG, Pinto JM. Medium-Term Planning of Single-Stage Single-Unit Multiproduct Plants Using a Hybrid Discrete/Continuous-Time MILP Model. *Ind Eng Chem Res*. 2008;47(6):1925–1934.

[11] Liu SS, Pinto JM, Papageorgiou LG. A TSP-Based MILP Model for Medium-Term Planning of Single-Stage Continuous Multiproduct Plants. *Ind Eng Chem Res*. 2008;47 (20):7733–7743.

[12] Sung C, Maravelias CT. An Attainable Region Approach for Production Planning of Multiproduct Processes. *AIChE J*. 2007;53(5):1298–1315.

[13] Sung C, Maravelias CT. A Projection-Based Method for Production Planning of Multiproduct Facilities. *AIChE J*. 2009;55(10):2614–2630.

[14] Glasser D, Crowe C, Hildebrandt D. A Geometric Approach to Steady Flow Reactors: The Attainable Region and Optimization in Concentration Space. *Ind Eng Chem Res*. 1987;26 (9):1803–1810.

[15] Hildebrandt D, Glasser D. The Attainable Region and Optimal Reactor Structures. *Chem Eng Sci*. 1990;45(8):2161–2168.

[16] Benders JF. Partitioning Procedures for Solving Mixed-Variables Programming Problems. *Numerische Mathematik*. 1962;4(1):238–252.

[17] Sahinidis NV, Grossmann IE. Reformulation of Multiperiod MILP Models for Planning and Scheduling of Chemical Processes. *Comput Chem Eng*. 1991;15(4):255–272.
[18] Karimi IA, McDonald CM. Planning and Scheduling of Parallel Semicontinuous Processes. 2. Short-Term Scheduling. *Ind Eng Chem Res*. 1997;36(7):2701–2714.
[19] McDonald CM, Karimi IA. Planning and Scheduling of Parallel Semicontinuous Processes. 1. Production Planning. *Ind Eng Chem Res*. 1997;36(7):2691–2700.
[20] Shah NK, Ierapetritou MG. Integrated Production Planning and Scheduling Optimization of Multisite, Multiproduct Process Industry. *Comput Chem Eng*. 2012;37:214–226.
[21] Li ZK, Ierapetritou MG. Integrated Production Planning and Scheduling Using a Decomposition Framework. *Chem Eng Sci*. 2009;64(16):3585–3597.
[22] Dias LS, Ierapetritou MG. Data-Driven Feasibility Analysis for the Integration of Planning and Scheduling Problems. *Optimization and Engineering*. 2019;20(4):1029–1066.
[23] Brunaud B, Amaran S, Bury S, Wassick J, Grossmann IE. Novel Approaches for the Integration of Planning and Scheduling. *Ind Eng Chem Res*. 2019;58(43):19973–19984.

Index

CPSIA information can be obtained
at www.ICGtesting.com
Printed in the USA
LVHW051925030821
694430LV00004B/337